Mic
Mat

Microstructural Characterization of Materials

David Brandon and Wayne D. Kaplan
Technion, Israel Institute of Technology, Israel

JOHN WILEY & SONS
Chichester · New York · Weinheim · Brisbane · Singapore · Toronto

National 01243 779777
International (+44) 1243 779777
e-mail (for orders and customer service enquiries): cs-books@wiley.co.uk
Visit our Home Page on http://www.wiley.co.uk
or http://www.wiley.com

Other Wiley Editorial Offices

John Wiley & Sons, Inc., 605 Third Avenue,
New York, NY 10158-0012, USA

WILEY-VCH Verlag GmbH, Pappelallee 3,
D-69469 Weinheim, Germany

Jacaranda Wiley Ltd, 33 Park Road Milton,
Queensland 4064, Australia

John Wiley & Sons (Asia) Pte Ltd, Clementi Loop #02-01,
Jim Xing Distripark, Singapore 129809

John Wiley & Sons (Canada) Ltd, 22 Worcester Road,
Rexdale, Ontario M9W 1L1, Canada

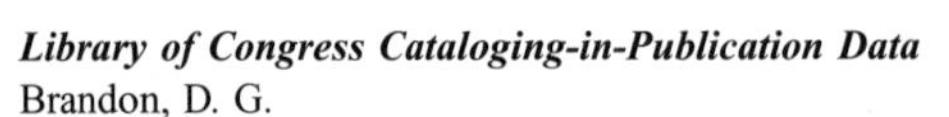

Library of Congress Cataloging-in-Publication Data
Brandon, D. G.
Microstructural characterization of materials / David Brandon and Wayne D. Kaplan.
p. cm.
Includes bibliographical references and index.
ISBN 0-471-98501-5 (cloth : alk. paper).—ISBN 0-471-98502-3 (paper : alk. paper)
1. Materials—Microscopy. 2. Microstructure. I. Kaplan, Wayne D. II. Title.
TA417.23.B73 1999
620.1′1299—dc 21 98-46589
CIP

British Library Cataloguing in Publication Data
A catalogue record for this book is available from the British Library
ISBN 0 471 98501 5 (cloth)
ISBN 0 471 98502 3 (paper)

Typeset in Times by Techset Composition Ltd, Salisbury, Wiltshire
Printed and bound in Great Britain by Biddles Ltd, Guildford, Surrey
This book is printed on acid-free paper responsibly manufactured from sustainable forestry, in which at least two trees are planted for each one used for paper production.

Contents

Preface

Most logical decisions rely on providing acceptable answers to precise questions, e.g. *what, why* and *how?* In the realm of scientific and technical investigation, the first question is typically *what is the problem* or *what is the objective*? This is then followed by a *why* question which attempts to pinpoint priorities, i.e. the urgency and importance of finding an acceptable answer. The third type of question, *how* is usually concerned with identifying means and methods, and the answers constitute an assessment of the available resources for resolving a problem or achieving an objective. The spectrum of problems arising in materials science and technology very often depends critically on providing adequate answers to these last two questions. The answers may take many forms, but when materials expertise is involved, they frequently include a need to characterize the *internal microstructure* of an engineering material.

This book is an introduction to the expertise involved in assessing the microstructure of engineering materials and to the experimental methods which are available for this purpose. For this text to be meaningful, the reader should understand why the investigation of the internal structure of an engineering material is of interest and appreciate why the microstructural features of the material are so often of critical engineering importance. This text is intended to provide a basic grasp of both the methodology and the means available for deriving qualitative and quantitative microstructural information from a suitable sample.

There are two ways of approaching *materials characterization*. The first of these is in terms of the engineering properties of materials, and reflects the need to know the physical, chemical and mechanical properties of the material before we can design an engineering system or manufacture its components. The second form of characterization is that which concerns us in this book, namely the *microstructural characterisation* of the material. In specifying the internal microstructure of an engineering material we include the chemistry, the crystallography and the structural morphology, with the term materials characterization being commonly taken to mean just this specification.

Characterization in terms of the chemistry involves an identification of the chemical constituents of the material and an analysis of their relative abundance, that is a determination of the chemical composition and the distribution of the chemical elements within the material. In this present text, we consider methods which are available for investigating the chemistry on the microscopic scale, both within the bulk of the material and at the surface.

Crystallography is the study of atomic order in the crystal structure. A crystallographic analysis serves to identify the phases which are present in the structure, and to describe the atomic packing of the various chemical elements within these phases. Most phases are highly ordered, so that they are *crystalline* phases in which the atoms are packed together in a well-ordered, regularly repeated array. Many solid phases possess no such long-range order, and their structure is said to be *amorphous* or *glassy*. Several *quasicrystalline* phases have also been discovered in which classical long-range order is absent, but the material nevertheless possesses well-defined rotational symmetry.

The *microstructure* of the material also includes those *morphological* features which are revealed by a microscopic examination of a suitably prepared specimen sample. A study of the microstructure may take place on many levels, and will be affected by various parameters associated with specimen preparation and the operation of the microscope, as well as by the methods of data reduction used to interpret results. Nevertheless, *all* microstructural studies have some features in common. They provide an image of the internal structure of the material in which the image contrast depends upon the interaction between the specimen and some incident radiation used to probe the sample morphology. The image is usually magnified, so that the region of the specimen being studied is small compared to the size of the specimen. Care must be exercised in interpreting results as being 'typical' of the bulk material. While the specimen is a three-dimensional object, the image is (with few exceptions) a two-dimensional projection. Even a qualitative interpretation of the image requires an understanding of the spatial relationship between the two-dimensional imaged features and the three-dimensional morphology of the bulk specimen.

Throughout this book we are concerned with the interpretation of the interaction between the probe and a sample prepared from a given material, and we limit the text to probes of X-rays, visible light or energetic electrons. In all cases, we include three stages of investigation, namely specimen preparation, image observation and recording, and the analysis and interpretation of recorded data. We will see that these three aspects of materials characterization interact: the microstructural morphology defines the phase boundaries, and the shape and dimensions of the grains or particles, the crystallography determines the phases present and the nature of the atomic packing within these phases, while the microchemistry correlates with both the crystallography of the phases and the microstructural morphology.

This text is intended to demonstrate the versatility and the limitations of the more important laboratory tools available for microstructural characterization. It is *not* a compendium of all of the possible methods, but rather a teaching outline of the most useful methods commonly found in student laboratories, university research departments and industrial development divisions.

Most of the data in this book are taken from work conducted in collaboration with our colleagues and students at the Technion. We wish to thank the following for their contributions: Moshe Eizenberg, Arnon Siegmann, Menachem Bamberger, Christina Scheu, Gerhard Dehm, Ming Wei, Ludmilla Shepelev, Michal Avinun, George Levi, Mike Lieberthal, and Oren Aharon.

D. B.
W. D. K.

Chapter 1

The Concept of Microstructure

This text is intended to provide a basic introduction to the most commonly used methods of microstructural characterization. It is intended for students of science and engineering whose course requirements, or perhaps their curiosity, lead them to explore beyond the causal connection between the properties of engineering materials and their microstructural features, and prompt them to ask how the microstructures of diverse materials are characterized in the laboratory.

Most introductory textbooks in materials science and engineering go to some lengths in order to emphasize that the processing route used to manufacture a component (shaping processes, thermal treatment, mechanical working, etc.) effectively determines the microstructural features (Fig. 1.1). They also explore the interrelation between the microstructure and the chemical, physical, and mechanical properties of materials, developing expressions for the dependence of these properties on such microstructural concepts as *grain size*, *preferred orientation* or *precipitate volume fraction*. What they *do not* usually do is to give more than a cursory description of either the methods used to determine the microstructural features, or the analysis required to convert a microstructural observation into a parameter with some useful engineering significance.

This book covers three aspects of microstructure (Table 1.1). First, the identification of the crystallographically distinct phases which are present in engineering materials, secondly, the morphology of these phases (their size, shape and distribution in space), and thirdly the chemical composition of these phases, again determined on the microstructural scale.

In all three cases, we will explore the characterization of the microstructure at both the qualitative and the quantitative level. Thus, in terms of crystallography, we will be concerned not only with qualitative phase identification, but also with elementary aspects of applied crystallography, used to determine crystal structure, as well as with the quantitative determination of the volume fraction of a second phase. As for the microstructure, we will introduce the *stereological relationships* which are needed to convert a qualitative observation of a morphological feature, such as the individual grain cross-sections, into a clearly defined microstructural parameter, i.e.

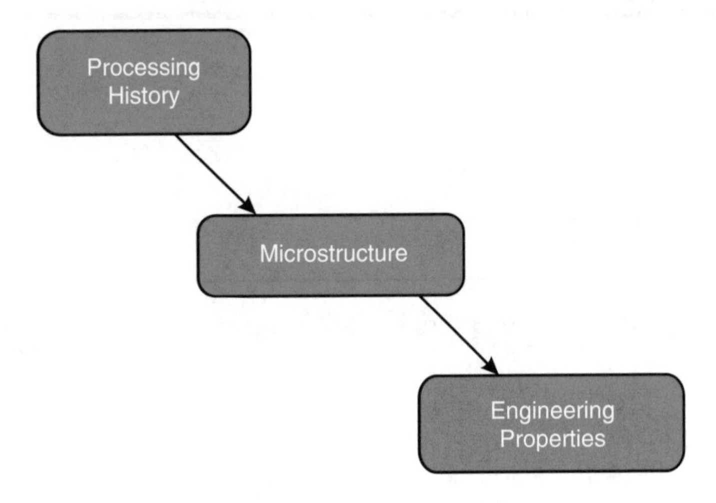

Figure 1.1 The microstructure of an engineering material is a result of its chemical composition and processing history. The microstructure determines the chemical, physical and mechanical properties of the material, and hence limits its engineering performance

the grain size. Similarly, we shall not just be satisfied with the microanalytical identification of the chemical elements present in a specific microstructural feature, but rather we shall seek to determine the local chemical composition through microanalysis.

In general terms (Fig. 1.2), microstructural characterization is achieved by allowing some form of probe to interact with a carefully prepared specimen sample. The most commonly used probes are visible light, X-ray radiation and high-energy electron beams. These three types of probe form the basis for optical microscopy, X-ray diffraction and electron microscopy, respectively. Once the probe has interacted with the sample, the scattered signal is collected and processed into a form where it can be interpreted, either qualitatively or quantitatively. Thus, in microscopy, a two-dimensional image of the specimen is obtained, while in microanalysis a spectrum is collected in which the signal intensity is displayed as a

Table 1.1 On the qualitative level, microstructural characterization is concerned with the identification of the phases present, their morphology (size and shape), and the identification of the chemical constituents in each phase. At the quantitative level, it is possible to determine the atomic arrangements (applied crystallography), the spatial relationships between microstructural features (stereology), and the microchemical composition (microanalysis)

Qualitative Analysis	Phase identification	Microstructural morphology	Microchemical identification
Quantitative Analysis	Applied crystallography	Stereology	Microchemical analysis

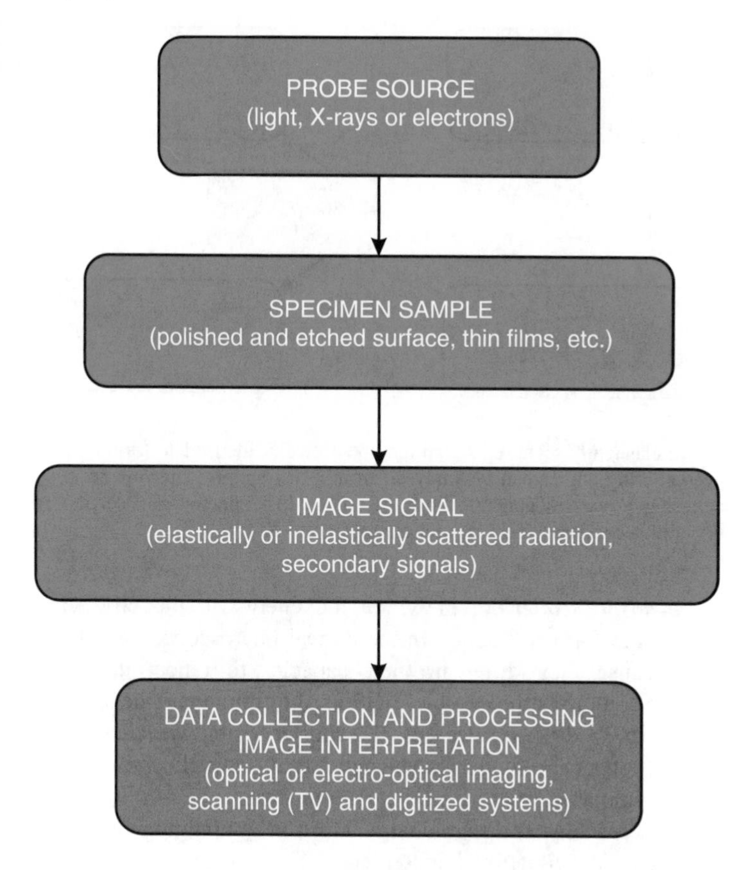

Figure 1.2 Microstructural characterization relies on the interaction of a material sample with a probe, where the latter is usually visible light, X-rays or a beam of high-energy electrons. The resultant signal must be collected and interpreted. If the signal is elastically scattered, an image can be formed by an optical system while if the signal is inelastically scattered, or generated by secondary emission, the image is formed by a scanning raster (as in a television monitor)

function of either its energy or wavelength. In diffraction, the signal may be displayed as either a diffraction pattern or a diffraction spectrum.

In all of the methods of characterization which we shall be discussing here, two forms of interaction between the probe and the specimen need to be considered (Fig. 1.3).

First, we shall discuss *elastic scattering processes*. These lead to intensity peaks in X-ray diffraction patterns which are characteristic of the phases present and their orientation in the sample. They also lead to diffraction contrast in transmission electron microscopy, which is directly related to the nature of the defects present in crystals (grain boundaries, dislocations and other microstructural features).

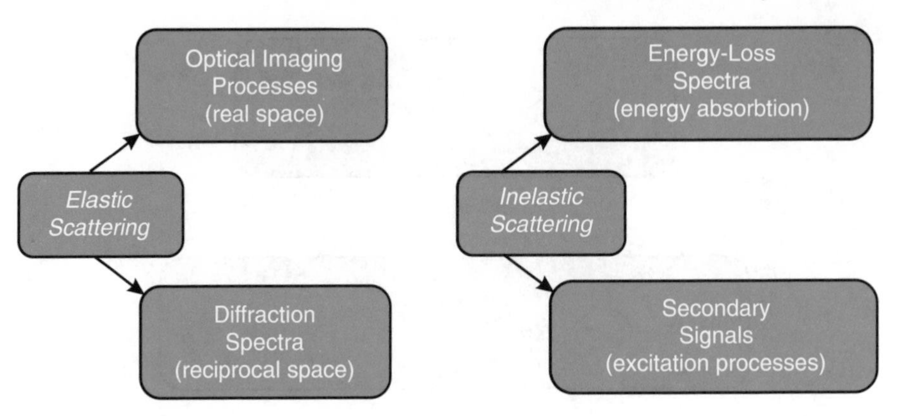

Figure 1.3 An elastically scattered signal may be optically focused, to form an image in real space, or the scattering angles can be analysed from a diffraction pattern in reciprocal space. Inelastic scattering processes generate both an energy-loss spectra, and secondary, excited signals, especially secondary electrons and characteristic X-rays

Secondly, *inelastic scattering*, in which the energy in the probe is partially converted into other forms of signal. In the optical microscope, some features may be revealed because they differ in their capacity to reflect the spectrum of wavelengths that comprises the 'visible' light used to illuminate the specimen. Gold and copper absorb, to differing degrees, the shorter visible wavelengths (blue and green light) and reflect the longer wavelengths (red and yellow). Reflection is an *elastic* process while absorption is an *inelastic* process.

In electron microscopy, the high-energy electrons impinging on the specimen often lose energy in well-defined increments. These inelastic energy losses are characteristic of the electron energy levels of the atoms which comprise the sample, and the energy-loss spectra can be analysed in order to identify the chemical composition of the sample beneath the electron beam (the 'probe'). Certain energy losses are accompanied by the emission of characteristic X-rays. These X-rays can also by analysed, by either energy dispersive or wavelength dispersive spectroscopy, to yield highly accurate information on the distribution of the chemical elements in the sample.

Elastic scattering processes are characteristic of optical or electro-optical systems which form an image in *real space* (the three dimensions in which we live), but elastic scattering is also a characteristic of diffraction phenomena, which are commonly analysed in *reciprocal space*. In real space, we are primarily concerned with the size and shape of the features observed and the distances between them, but in reciprocal space it is the angles through which the signal is scattered by the sample which are significant. These angles are inversely related to the size or separation of the features responsible for the characteristic intensity peaks observed in the diffraction pattern or diffraction spectrum. The elastically scattered signals which are generated by optical imaging and diffraction are compared in Fig. 1.4.

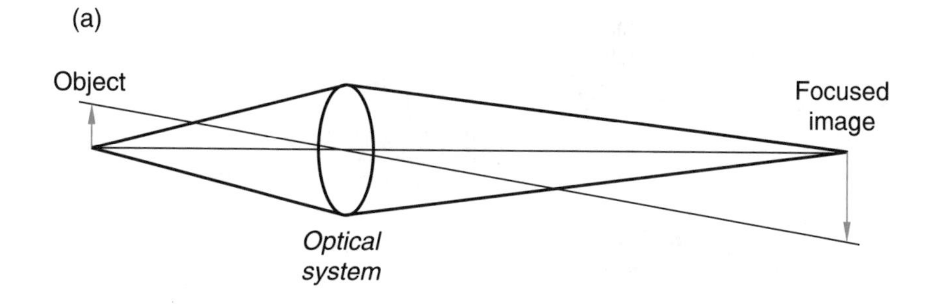

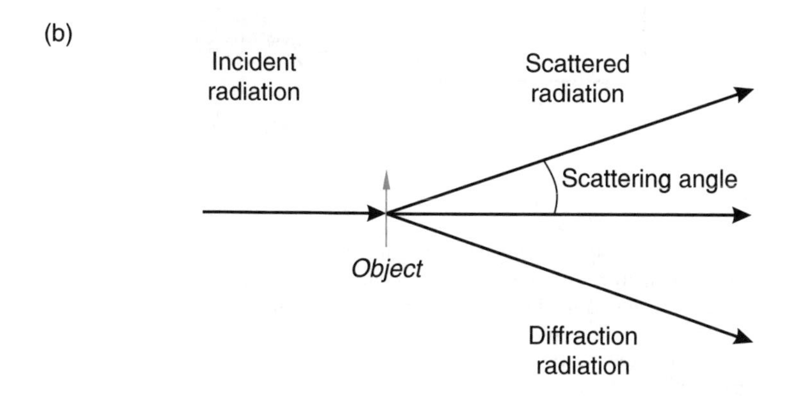

Figure 1.4 Schematic representations of an optical image (a) and a diffraction pattern (b). In the former, distances in the image are directly proportional to distances in the object, and the constant of proportionality is equal to the magnification. In the latter, the scattering angle for the diffracted radiation is inversely proportional to the scale of the features in the object, so that distances in a diffraction pattern are inversely proportional to the separation of features in the object

Inelastic scattering processes dominate the contrast in *scanning* electron imaging systems (as in the *scanning electron microscope* (SEM), see Fig. 1.5). It is possible to detect either the loss spectra (the energy distribution in the original probe after it has interacted with the sample), or the secondary signal (the excited particles or photons generated by the probe as a result of this inelastic interaction).

Large numbers of secondary electrons are also emitted when an energetic electron beam strikes a solid sample, and it is the detection of this secondary electron signal that forms the basis of the aesthetically striking high-resolution images of rough surfaces which are a common feature of scanning electron microscopy.

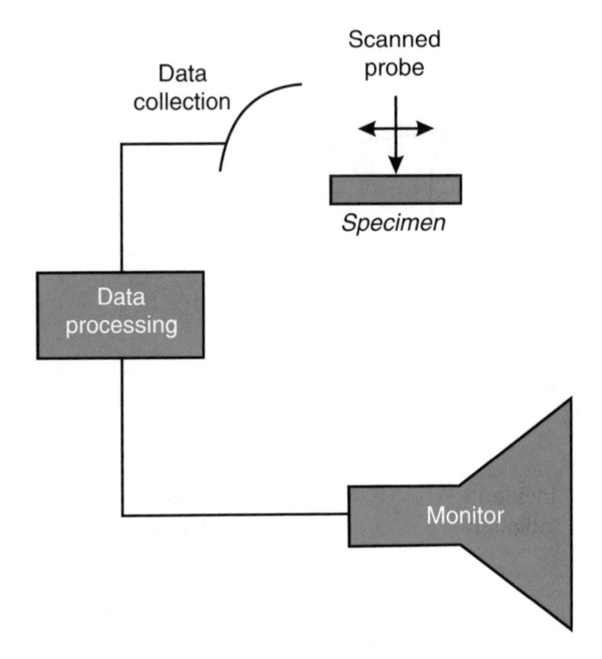

Figure 1.5 A scanning image is formed by scanning a focused probe over the specimen and collecting a data signal from the sample. The signal is then processed and displayed on a cathode ray screen with the same time-base as that used to scan the probe. The magnification is the ratio of the screen size on the monitor to the amplitude of the probe scan on the specimen. The signal results from secondary electrons, characteristic X-rays, or a wide variety of other excitation phenomena

In the following, we will assume that the reader is familiar with those aspects of microstructure and crystallography that commonly form the basis of introductory courses in materials science and engineering. This includes: a knowledge of *Bravais lattices* and the concept of *crystal symmetry*, common microstructural concepts associated with *polycrystals* and *polyphase materials* (including the significance of microstructural *inhomogeneity* and *anisotropy*), and finally, the thermodynamic basis of *phase stability* and *phase equilibrium* in polyphase materials.

Throughout this book, each chapter will conclude with examples of three specific engineering materials which are characteristic of a wide range of common materials, and we will explore the information which can be obtained about these materials from the different methods of microstructural characterization which are being discussed. The materials which we will consider are:

(a) A low-alloy steel containing *ca* 0.4 %C;
(b) a dense, glass-containing alumina;
(c) a thin-film microelectronic device based on the Al/TiN/Ti system.

There is no good reason why we have not selected an engineering polymer or a composite for these examples. The principles of characterization are the same, even though the details of interpretation may differ.

Our choice of the methods of microstructural characterization which we shall describe is as arbitrary as our selection of these 'typical' materials. Any number of methods of investigation may be, and are, used to characterize the microstructure of engineering materials, but this text is not a compendium of all known techniques. Instead, we have chosen to limit ourselves to those established methods which are commonly found in a well-equipped industrial development department or university teaching laboratory. The methods selected include *optical* and *electron microscopy* (both *scanning* and *transmission*), *X-ray* and *electron diffraction*, and the commoner techniques of *microanalysis* (energy dispersive and wavelength dispersive X-ray analysis, Auger electron spectroscopy, X-ray photospectroscopy and electron energy loss spectroscopy). In each case, a serious attempt is made to describe the *physical principles* of the method, clarify the *limitations*, and explore the extent to which the technique can be used to yield *quantitative* information.

1.1 MICROSTRUCTURAL FEATURES

When sectioned, polished and suitably etched, nearly all engineering materials will be found to exhibit structural features which are characteristic of the material. These may be visible to the unaided eye, or may require a low-powered optical microscope to reveal the detail. The finest sub-structure will only be visible in the electron microscope. Many of the properties of engineering solids are directly and sensitively related to the microstructural features of the material, that is these properties are *structure sensitive*. In such cases, the study of microstructure can reveal a direct *causal relationship* between a particular microstructural feature and a specific property.

In the following, we shall explore some of these structure–property relationships and attempt to clarify further the meaning of the term *microstructure*.

1.1.1 Structure–Property Relationships

It is not enough just to state that 'materials characterization is important'. In practice, it is usually possible to distinguish between those properties of a material which are *structure-sensitive* and those which are *structure-insensitive*. Examples of structure-insensitive properties are the *elastic moduli*, which vary only slowly with composition or grain size. There is little error involved in assuming that all steels have the same tensile (Young's) modulus, irrespective of their composition. In fact, the variation in elastic modulus with temperature (typically less than 10 %) exceeds that associated with alloy chemistry, grain size or degree of cold work. The *thermal expansion coefficient* is another example of a property which is less affected by microstructural variations than it is by temperature or pressure. The same is true of the *specific gravity* of a solid.

In contrast, the *yield strength*, which marks the onset of plastic flow in engineering alloys, is a sensitive function of several microstructural parameters, namely the *grain size* of the material, the *dislocation density*, and the distribution and *volume fraction* of second-phase particles. *Thermal conductivity* and *electrical resistivity* are also structure-sensitive properties, while *heat treating* an alloy may have a large affect on its conductivity, often because both the thermal and the electrical conductivity are drastically reduced by the presence of alloying elements in solid solution in the matrix. Perhaps the most striking example of a structure-sensitive property is the *fracture toughness*, which measures the ability of a structural material to inhibit crack propagation and prevent catastrophic brittle failure. Very small changes in chemistry and highly localized grain boundary segregation, may cause a catastrophic loss of ductility, thus reducing the fracture toughness by an order of magnitude. Although such segregation effects are indeed an example of extreme structure sensitivity, they are also extremely difficult to detect, since the bulk impurity levels associated with the segregation phenomena need only be of the order of 10^{-5} (10 parts per million (p.p.m.)).

A classic example of a structure-sensitive property relationship is the Petch equation, which relates the yield strength of a steel, σ_y, to its grain size D, by introducing the two materials constants, σ_o and k_y, as follows:

$$\sigma_y = \sigma_o + k_y D^{-1/2} \tag{1.1}$$

This relationship presupposes that we are able to determine the grain size of the material quantitatively and unambiguously, and it is worth taking a moment to think about the meaning of the term *grain size* (see Section 1.1.3.1).

The fracture surfaces of engineering components which have failed in service, as well as those of standard specimens which have been tested to failure, are frequently subjected to microscopic examination in order to characterize the failure mechanisms (a procedure which is termed *fractography*). In *brittle*, polycrystalline samples, much of the fracture path often lies along specific low-index planes within the crystal lattice of the individual grains, to yield a *transgranular* or *cleavage* failure. Since neighbouring grains have different orientations in space, the cleavage surfaces are forced to change direction at the grain boundaries. The line of intersection of the cleavage plane in one grain with the grain boundary is very unlikely to lie in an allowed cleavage plane in the second grain, so that cleavage failures in polycrystalline materials must propagate on unfavourable crystal lattice planes, or else link up by *intergranular* grain boundary failure at the grain boundaries. A *fractographic* examination reveals the relative extent of intergranular and transgranular failure (Fig. 1.6). By determining the three-dimensional nature of brittle crack propagation and its dependence on grain size or grain boundary chemistry we are able to explore critical aspects of the failure mechanism.

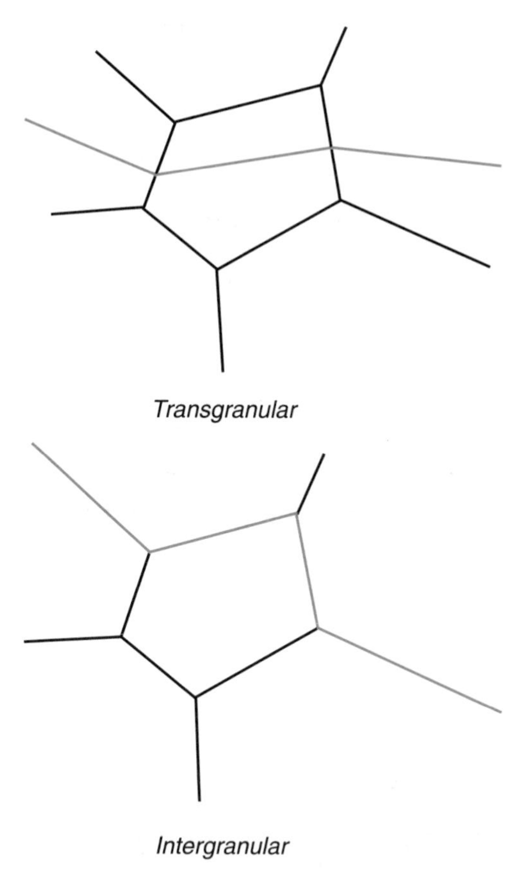

Figure 1.6 Illustrations of transgranular and intergranular brittle failure; in three dimensions some intergranular failure is always present, since transgranular failure occurs by cleavage on specific crystallographic planes

Ductile failures are also three-dimensional in nature. A tensile crack in a ductile material typically propagates by the nucleation of small cavities in the region of hydrostatic tensile stress which is generated ahead of the crack tip. The nucleation sites are often small, hard inclusions, and the distribution of the cavities depends on the distribution of these nucleating sites. The cavities grow by plastic shear at the root of the crack until they join up to form a cellular, ridged surface, termed a *dimpled* fracture (see Fig. 1.7). The complex topology of this dimpled, ductile failure may not be immediately obvious from a two-dimensional micrograph of the fracture surface, and in fractography it is common practice to image the failure from two, slightly tilted, points of view. This allows a rough surface to be viewed and analysed *stereoscopically* (see Section 4.3.6.3). Our eyes give us the same impression of depth by superimposing the two views of the world which we receive separately from each eye (Fig. 1.8).

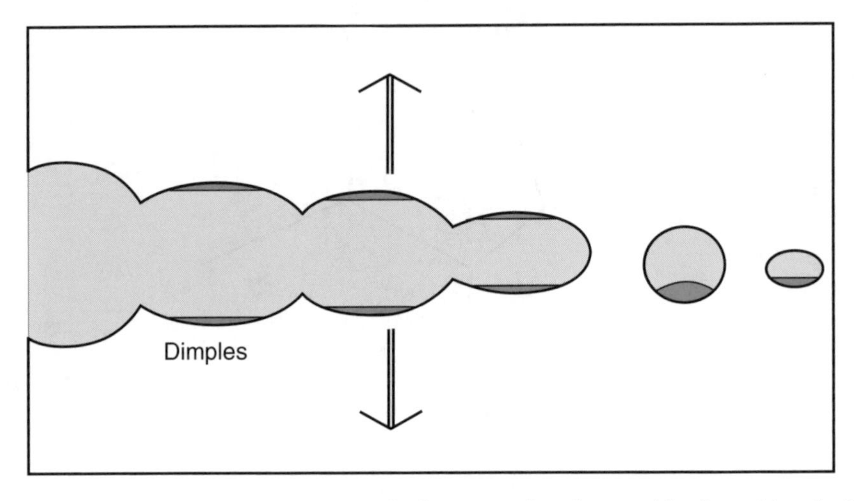

Figure 1.7 Formation of a dimpled ductile fracture surface from cavities formed by plastic flow

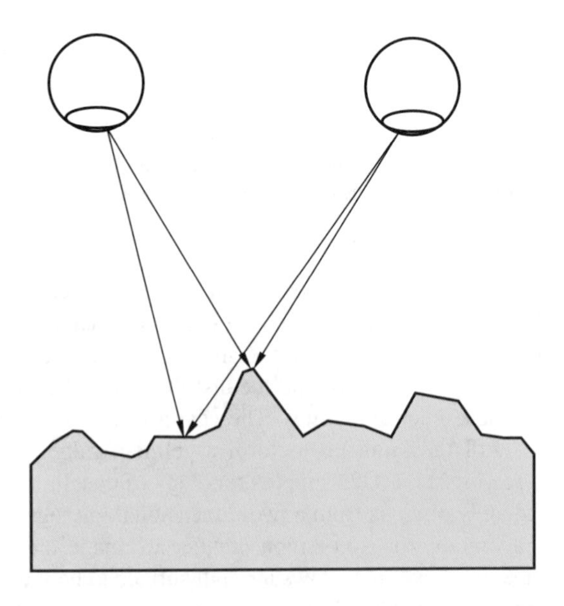

Figure 1.8 Illustration of the principle of stereoscopic imaging. The left and right eyes see the world from two different positions, so that two points at different distances in space subtend different angular separations when viewed by the two eyes

The scale of the microstructure determines many other mechanical properties, in the same way as grain size is related to yield strength. The *fracture strength* of a brittle material, σ_f, is related to the *fracture toughness* K_c by the size c of the defects present in the material (cracks, inclusions or porosity), i.e. $\sigma_f \propto K_c\sqrt{c}$. The contribution of work-hardening to the plastic flow stress, $\Delta\sigma_y$, depends on both the *dislocation density* ρ and the elastic shear modulus G, i.e. $\Delta\sigma_y \propto G\sqrt{\rho}$. Similarly, the effectiveness of precipitation-hardening by a second phase, $\Delta\sigma_\pi$, is largely determined by the average separation of the second-phase precipitates, L, through the relationship $\Delta\sigma_\pi \propto G/(\sqrt{L})$.

1.1.2 Microstructural Scale

Microstructure is a very general term which is used to cover a wide range of structural features, from those visible to the naked eye down to those corresponding to the interatomic distances in the crystal lattice. It is good practice to distinguish between *macrostructure*, *mesostructure*, *microstructure* and *nanostructure*.

Macrostructure refers to those features which approach the scale of the engineering component and are visible to the naked eye, or detectable by common methods of *non-destructive evaluation* (dye penetrant testing, X-ray radiography, or ultrasonic testing). Examples are processing defects such as large pores, foreign inclusions, or shrinkage cracks. Non-destructive evaluation, however, is beyond the scope of this present text.

Mesostructure is a less common term, but is useful for describing those features which are on the borderline of the visible. This is particularly the case with the properties of composite materials, which are dominated by the size, distribution and volume fraction of the reinforcement, the cracking present at the interface or within the matrix, or other forms of defect (such as gas bubbles or de-wetting defects). The mesoscale is also of interest in bonding processes, for example the dimensions of an adhesive or a brazed joint, or the heat-affected zone in a weld.

Microstructure covers the scale of structural phenomena most commonly of concern to the materials scientist and engineer, namely grain and particle sizes, dislocation densities and particle separations, microcracking and microporosity.

The term *nanostructure* is restricted to sub-micron features, that is the lateral dimensions of grain boundaries and interfaces, the initial stages of precipitation, regions of local ordering in amorphous (glassy) solids, and very small particles whose structure is dominated by atoms at the particle surface rather than those within the particle.

Table 1.2 summarizes these different scales of material structure in terms of the magnification required in order to observe the features concerned.

Table 1.2 The scale of microstructural features, the magnification required to reveal the feature, and some common techniques available to the reader for studying the microstructure

Scale	Macrostructure	Mesostructure	Microstructure	Nanostructure
Typical magnification	$\times 1$	$\times 10^2$	$\times 10^4$	$\times 10^6$
Common techniques	Visual inspection	Optical microscopy	Scanning and transmission electron microscopy	X-ray diffraction
	X-ray radiography	Scanning electron microscopy	Atomic force microscopy	Scanning tunnelling microscopy
	Ultrasonic inspection			High-resolution transmission electron microscopy
Characteristic features	Production defects	Grain and particle sizes	Dislocation substructure	Crystal and interface structure
	Porosity, cracks and inclusions	Phase morphology and anisotropy	Grain and phase boundaries	Point defects and point-defect clusters
			Precipitation phenomena	

1.1.2.1 The Visually Observable

The human eye is a remarkably sensitive data collection and imaging system, but it is limited in three respects:

(a) the range of wavelengths in the electromagnetic spectrum which can be detected;
(b) the minimum signal intensity needed to trigger image 'recognition'
(c) the minimum spatial separation which can be resolved without the help of lenses or a microscope.

The eye is sensitive to wavelengths ranging from *ca* 0.4 to 0.7 μm, corresponding to a colour scale from dark red to violet; the peak sensitivity, however, is in the green, and is usually quoted as 0.56 μm, which is a characteristic emission peak in the spectrum obtained from a mercury vapour lamp. As a consequence, optical microscopes are commonly focused by using a green filter, while the phosphors used for the screens of transmission electron microscopes and in monitors for scanning systems are also usually green.

The integration time of the eye is *ca* 0.1 s, and sufficient photons have to be captured by the retina within this time in order to form an image. In absolute darkness, the eye 'sees' isolated flashes of light, which constitute a background of random noise. It requires quite long periods of time for the eye to achieve its maximum sensitivity at low light levels (a process termed *dark adaptation*). Nevertheless, when properly dark-adapted, the eye detects of the order of 50 % of the incident 'green' photons, and a statistically significant image will be formed if an amount of the order of 100 photons can contribute to each picture element (*pixel*). In principle, this is as good as the best available night-viewing systems, but these systems can integrate the image over a much longer period of time than the 0.1 s available to the eye, so that they can operate at much lower light levels.

The ability to identify two separate features depends on the resolution of the eye, which is a function of the aperture of the lens and the distance at which the features are viewed. The concept of resolution was defined by Lord Raleigh in terms of the apparent width of a point source. Abbe showed that if the point source subtends an angle 2α at the lens then its apparent radius δ was given by the expression $\delta = 1.2\lambda/\mu \sin\alpha$, where λ is the wavelength of the radiation from the source and μ is the index of refraction of the intervening medium. Raleigh assumed that two point sources could be distinguished when the peak intensity of one coincided with the first minimum of the other (Fig. 1.9), i.e. the resolution, defined by this Raleigh criterion, is exactly equal to the apparent radius of the point source, δ.

The diameter of the fully dilated pupil (the aperture that controls the amount of light entering the eye) is *ca* 6 mm, while it is impossible to focus on an object if it is too close (the *near point*, which is typically about 150 mm). It follows that $\sin\alpha$ for the eye is of the order of 0.04. Using green light at 0.56 μm and taking $\mu = 1$ (for air), we arrive at an estimate for δ_{eye} of just under 0.2 mm, that is the unaided eye can resolve features which are a few tenths of a millimetre apart. The eye records of the order of 10^6 image features at any one time, corresponding to an object some 20 cm across at the near point (compare the width of this page!).

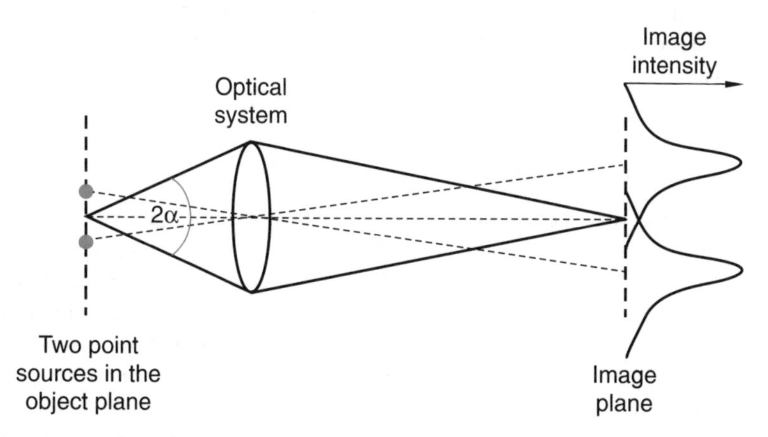

Figure 1.9 The Raleigh criterion defines the resolution in terms of the separation of two identical point sources, which results in the centre of the image of one source falling on the first minimum in the image of the second source

1.1.2.2 'With the Aid of the Optical Microscope'

An image which has been magnified by a factor M will contain resolvable features whose sizes range down to the limit dictated by the resolving power δ of the microscope objective lens. In the image, these 'just resolved' features will have a separation of $M\delta$. If $M\delta < \delta_{eye}$, then the eye will not be able to resolve all of the features recorded in the image. On the other hand, if $M\delta > \delta_{eye}$, then fewer resolvable features will be present in an image of a given size. This means that *less* information will be available to the observer. It follows that there is an 'optimum' magnification, corresponding to the ratio δ_{eye}/δ, at which the eye is able to resolve all of the imaged features and where the potential number of resolvable image points (pixels) is a maximum. Lower magnifications will image larger areas of the specimen, but at the cost of restricting the observable resolution. (In some cases, for example aerial reconnaissance, this may be desirable, and a hand lens is then used to further magnify areas on a photographic negative which are then enlarged on a photographic print.) Higher magnifications than the optimum are seldom justified, since the image then appears blurred and the number of resolvable pixels in the imaged area decreases.

The optical microscope uses visible electromagnetic radiation as the probe, and the best optical lens systems have values of $\mu \sin \alpha$ of the order of unity (by employing a high-refractive index, inert oil as the medium between the objective lens and the specimen). It follows that the best possible resolution is of the order of the wavelength, that is *ca* 0.5 μm. By assuming 0.2 mm for the resolution of the eye, this implies that, at a magnification of $\times 400$, the optical microscope should reveal all of the detail that it is capable of resolving. Higher magnifications are often employed (why strain your eyes?), but there is little point in seeking magnifications greater than 10^3.

Modern, digitized imaging systems are capable of recording image intensity levels at rates of up to 10^6 pixels per second, thus allowing for real-time digital image recording, not only in the optical microscope, but also in any other form of spatially resolved signal collection and processing.

1.1.2.3 Electron Microscopy

Attempts to improve the resolution of the optical microscope by reducing the wavelength of the electromagnetic radiation used to form the image have only been partially successful. Ultraviolet (UV) radiation is invisible to the eye, so that the image must be viewed on a fluorescent screen, and special lenses which do not absorb the radiation are required. The shorter-wavelength radiation is also strongly absorbed by many engineering materials, thus severely limiting the potential applications for a 'UV' microscope. Attempts have also been made to develop an X-ray microscope, by focusing the sub-nanometre wavelength beam with curved crystals, but no practical solutions have been found to the immense technical problems that have been encountered.

Electrons are the only feasible alternative. An electron beam of fixed energy will exhibit wave-like properties, with the wavelength λ being derived from the de Broglie relationship, $\lambda = h/\sqrt{2meV}$, where h is the Planck constant, m is the mass of the electron, e is the electron charge and V is the accelerating voltage. With V in kilovolts and λ in nanometres, the constant $h/\sqrt{2me}$ is equal to 0.037. At an accelerating voltage of only 1 kV, this wavelength is much less than the interplanar spacing in a crystalline solid. However, as we shall see later (Section 4.1.2), it is not that easy to focus an electron beam. Electromagnetic lenses must be used, and the various lens *aberrations* (Section 4.1.2) limit the acceptable values of the collection angle α for scattered electrons to between 10^{-2} and 10^{-3} rad ($360^\circ = 2\pi$ rad). At these small angles, $\alpha \approx \sin\alpha$, and the Raleigh resolution criterion reduces to $\delta = 1.2\lambda/\alpha$ (with $\mu = 1$, since the electron beam will only propagate without energy loss in a vacuum).

Typical interatomic distances in solids are of the order of 0.2 to 0.5 nm, so that, in principle, atomic resolution ought to be achievable at 100 kV. This is indeed the case, but in practice the optimum operating voltage for achieving consistent resolution of the atomic arrays in a crystal lattice is between 200 and 400 kV. Commercial transmission electron microscopes guarantee sub-nanometre resolutions and are capable of detecting essentially all of the microstructural features in engineering materials.

Scanning electron microscopes, in which the resolution depends on the focus of the electron beam into a fine probe, have an additional, that is *statistical*, resolution limit. The beam current available in the probe decreases rapidly as the beam diameter is reduced, and the intensity of the signal which can be detected is proportional to the beam current. For some types of signal (most notably excited X-rays, whose wavelength is characteristic of a chemical species), the statistical limit is reached at probe diameters which are much larger than those set by the electromagnetic performance of the probe lens. X-ray mapping of characteristic radiation in the scanning electron microscope is statistically limited to a spatial resolution of the order of 1 μm.

A secondary electron image, obtained in the same scanning electron microscope, may have a resolution which is only limited by the aperture of the probe-forming lens and the wavelength of the electron beam. Combining the Abbe equation, the Raleigh criterion and the de Broglie relationship linking wavelength to accelerating voltage, we can estimate the probable resolution of a secondary electron image. Since the image is formed from a *secondary* signal, we need to consider the volume of material beneath the impinging beam which generates the signal. As the accelerating voltage is increased, the electrons will penetrate deeper into the sample and will be scattered (elastically and inelastically) over a wider angle. It follows that lower accelerating voltages are desirable in order to limit the spread of the beam in the sample. On the other hand, a finer probe will be formed at higher accelerating voltages, since the wavelength of the electrons is then reduced. Most scanning electron microscopes are designed to operate in the range 1–30 kV, with the lower accelerating voltages being preferred for lower-density (lower-atomic number) specimens. Assuming α to be 10^{-2} and taking 4 kV as a typical lower limit for the accelerating voltage, this yields a potential resolution for the secondary electron image in the scanning electron microscope of *ca* 2 nm.

A great deal will be said in later chapters about the factors limiting the spatial resolution in the different methods of microscopy used to characterize microstructure. For the time being, it will suffice to note that the following four factors should be considered:

(a) the physical characteristics of the probe source;
(b) the optical properties of the system;
(c) the nature of the specimen–probe interaction;
(d) the statistics of data collection and processing.

Each technique has its limitations, and each method should be chosen according to the information required. Modern characterization equipment often combines different techniques on one platform. Thus the limitations of a specific technique do not always mean we have to prepare different samples, or search for another piece of equipment.

1.1.2.4 'Seeing Atoms'

A common, and not in the least foolish, question is 'can we see atoms?'. The answer is, in a certain sense, 'yes', although most of the science and technology behind the imaging of atomic and molecular structure is beyond the scope of this present text.

The story starts in Berlin some sixty years ago, when the electron microscope was developed by Ruske. At that time, Erwin Müller demonstrated that a suitably polished tungsten needle mounted in a vacuum chamber could be made to emit electrons from its tip by applying a positive voltage to the needle. This *field-emission* process was used to image differences in the *work function*, that is the energy required to extract an electron from the surface of the metal, and the image was formed by radial projection of the electrons on to a fluorescent screen. Some twenty years later, Müller admitted a small quantity of gas into the chamber, reversed the voltage on the tungsten needle, and observed ionization of the gas at the tip surface. When the low-pressure gas was cooled to cryogenic temperatures, the radially projected image on a fluorescent screen showed regular arrays of bright spots, and the *field-ion microscope* (FIM) was born (Fig. 1.10). The intensity of these bright spots reflects the electric field distribution over the protruding atoms on the surface of the tip.

At a sufficiently high electric field, the protruding atoms themselves are ionized and are accelerated radially away from the tip. These *field-desorbed* atoms can be detected with a high degree of probability in a time-of-flight (TOF) mass spectrometer, so that not only can the surface of the tip be imaged at 'atomic' resolution, but in many cases the individual atoms can be identified with reasonable certainty in a *field-ion atom probe*.

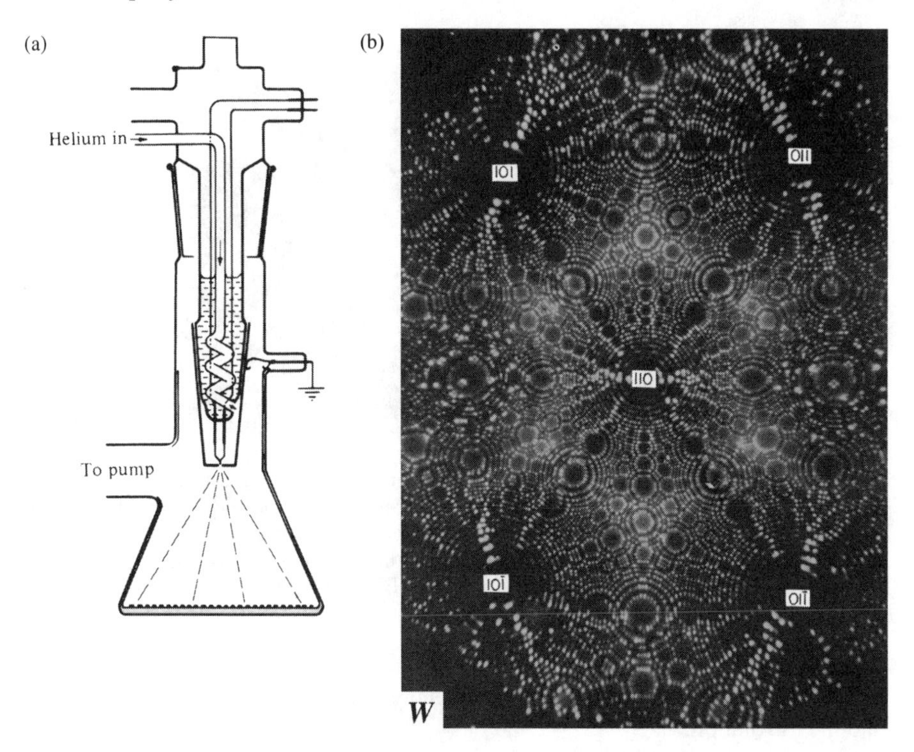

Figure 1.10 The field-ion microscope was the first successful attempt to image 'atoms': (a) a schematic representation of the instrument; (b) a tungsten tip imaged by field-ion microscopy

Unfortunately, most engineering materials are unable to withstand the high electric field strengths needed to generate a field-ion image, while materials containing more than one chemical constituent form rather irregular field-desorbed surfaces. A rather more useful instrument for observing 'atoms' is the *scanning tunnelling microscope* (STM). In this technique, the sharpened tungsten needle is used as a probe rather than as a specimen, and is brought to within a few atomic distances of the specimen surface. The vertical distance z of the tip from the surface is precisely controlled, and the tip can be scanned in the x–y plane while monitoring either the tip current at constant applied voltage or the tip voltage at constant applied current.

The contrast periodicity observed in a scanning tunnelling image reflects the periodicity in the atomic structure of the surface, and some of the first images published demonstrated unequivocally the restructuring of the $\{111\}$ surface of a silicon single crystal to form a 7×7 rhombohedral array (Fig. 1.11). By varying the voltage of the tip with respect to the specimen, the electron density of states can be sampled at varying distances beneath the surface. The resolution of the scanning tunnelling microscope depends primarily on the mechanical stability of the system,

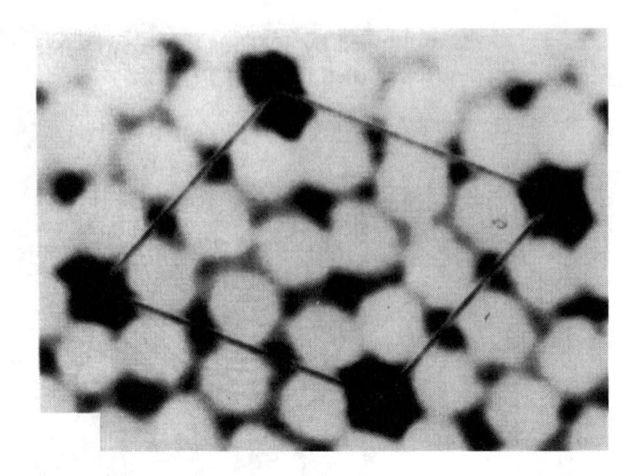

Figure 1.11 The scanning tunnelling microscope provides data on the spatial distribution of the density of states in the electron energy levels at and beneath the surface. In this example, the 7×7 rhombohedral unit cell of the restructured $\{111\}$ surface of a silicon crystal is clearly resolved. Reprinted from I. H. Wilson, p. 162 in Walls and Smith (eds) *Surface Science Techniques*, Copyright 1994, with permission from Elsevier Science

and various commercial instruments are now available, all of which guarantee 'atomic' resolution.

In principle, there is no reason why the same tungsten needle should not be used to monitor the *force* between the needle and a specimen surface over which the needle is scanned. As the needle approaches the surface, it experiences an initial van der Waals attraction (polarization forces), which is replaced by a repulsion as the needle makes physical contact with the specimen. Scanning over the surface at constant displacement (at constant z), and monitoring the changes in the van der Waals force, yields a scanning image in which 'molecular' resolution is readily obtained (Fig. 1.12), and where 'atomic' resolution has also been claimed. This *atomic force microscope* (AFM) is proving to be particularly useful for studying solid surfaces in gaseous or liquid media (areas in which the field-ion microscope and the scanning tunnelling microscope are incapable of operating), and promises to be a powerful method for imaging both organic and polymer membranes. By vibrating the tip (using a piezoelectric transducer), it is possible to monitor the *elastic compliance* of the substrate at atomic resolution, thus providing information on the atomic bonding at the surface and its spatial distribution.

All three of these instruments the FIM, STM and AFM, can provide spatially resolved information on the surface structure of a solid. In all three cases, what we see is one or another aspect of the atomic structure of the surface and the nature of the interatomic bonding. In a sense, we do indeed 'see atoms'. In later chapters we will explore the extent to which transmission electron microscopy and scanning electron microscopy may also allow us to 'see atoms'.

Plate 1

Fig. 1.20 The colour and texture of a porcelain glaze depend on the presence of controlled amounts of impurity cations which introduce *colour centres* into the silicate glass, and may nucleate localized crystallization. The illustration shows a glass jar from the First Century AD, Roman Empire. Reproduced by permission of the Corning Museum of Glass

Plate 2

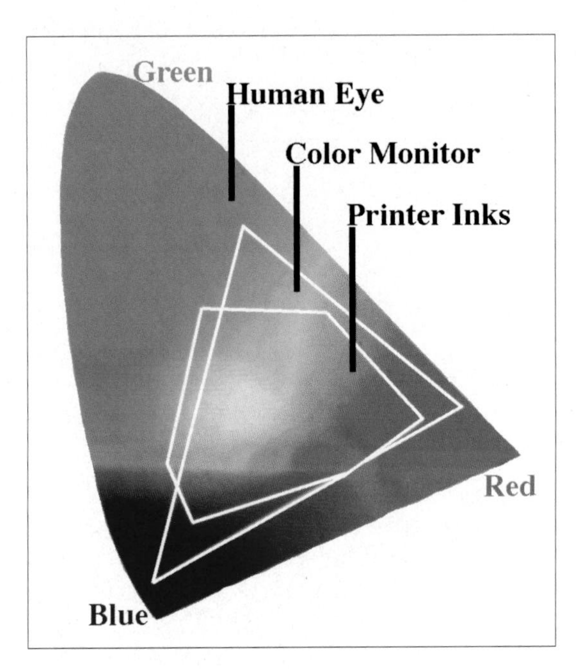

Fig. 7.21 The chromaticity triangle; a chosen colour within the triangle can be selected by combining different intensities of the three primary colours taken from the corners of the triangle

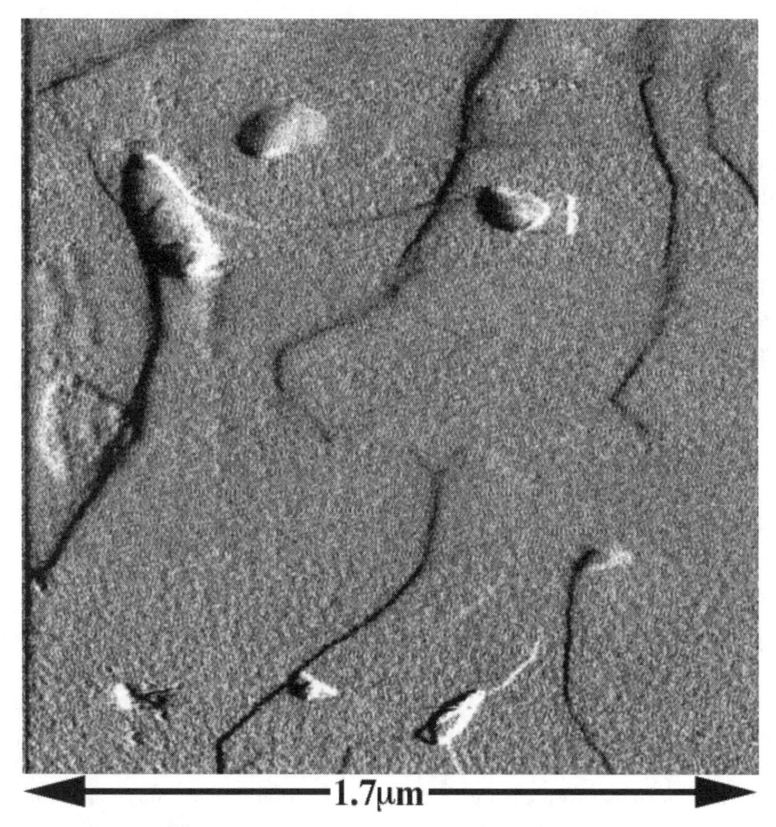

Figure 1.12 Atomic force micrograph showing steps of unit cell height on the surface of GaN (provided by permission of E. Rabkin)

1.1.3 Microstructural Parameters

Microstructural features are commonly described in qualitative terms: the structure may be more or less *equiaxed*, or the particles of a second phase may be *acicular* or *plate-like*. Theories of material properties, on the other hand, frequently include a quantitative dependence on the microstructure, through the introduction of microstructural parameters. These parameters, for example grain size or dislocation density, must be assigned a numerical value if they are to have any predictive value within the framework of the theory. A classic example is the Petch relationship linking the yield strength of a steel σ_y, to its grain size D (see equation (1.1)).

In many cases there is a 'chasm' of uncertainty between a qualitative microstructural observation and its association with a predicted or measured material property. In order to bridge this chasm we need to build two 'support piers'. To construct the first we note that many of the engineering properties of interest are *stochastic*, rather than *deterministic*, the property is not *single-valued*, but is best described by a *probability function*. For example, not all men are 190 cm tall, nor do they weigh 70 kg. Rather, there are empirical functions describing the probability that an individual taken from any well-defined population will have a height or weight which falls within a set interval. Clearly, if an engineering property is stochastic in nature, then the material parameters which determine that property will be stochastic, and we need to understand their *statistical distribution*.

The second support pier is on the microstructural side of the chasm, and is constructed from a knowledge of the *mechanisms of image formation* and the *origin of image contrast*, in combination with an analysis of the spatial relationship between the image observations and the bulk structure of the material. This analysis is termed *stereology*, the science of spatial relationships, and is an important objective of this book. For the time being, we will only consider the terms used for some common microstructural parameters.

1.1.3.1 Grain Size

Most engineering materials are *polycrystalline*, so that they consist of a three-dimensional aggregate of individual grains, each of which is a single crystal whose crystal lattice orientation in space differs from that of its neighbours. The size and shape of these individual grains are as varied as the grains of sand on the sea-shore. If we imagine the polycrystalline aggregate separated out into its individual grains, we might define grain size as the *average* separation of two parallel tangent planes which touch the surfaces of a randomly oriented grain. This is the *caliper* diameter D_C, and it is *not* easy to measure (Fig. 1.13). We could also imagine counting the number of grains in a unit volume of the sample, N_V, and then defining an average grain size as $D_V = N_V^{-1/3}$. This definition is unambiguous and independent of any anisotropy or inhomogeneity in the material. However, most samples are prepared by taking a *planar* section through the three-dimensional microstructure, so we could count the number of grains *sectioned* by the polished surface, and then transform the number of grains intercepted per unit area of the section, N_A, into an average 'grain size' by writing $D_A = N_A^{-1/2}$. Finally, we could lay down a set of lines on our 'random' polished and etched section, and count the number of intercepts which a 'test' line makes with the grain boundary traces seen on the section. If both the test line on the surface and the sample section are truly random, then the test line is a *random intercept* of the boundary array in the bulk material. The number of intercepts per unit length of the test line, N_L, is then related to yet another measure of the grain size, D_L, the *mean linear intercept*, where $D_L = L/N_L$, and is usually the way grain size is defined (with or without some factor of proportionality). However, in industrial quality control a totally different measure is also used. In this case, the sample microstructure is compared with a set of standard microstructures (ASTM Grainsize Charts) and assigned an American Society for Testing Materials (ASTM) grain size, D_{ASTM}, on the basis of the observer's 'best-fit' match with the ASTM chart.

What happens if the grains are not of uniform size? Well, the section through the sample will cut through the individual grains at different positions with respect to their centres of gravity, so that even grains of identical volume will occupy different areas on the plane of a section, depending on how far the plane of the section is from their centre of gravity. This distribution of intercept areas will be convoluted with the 'true' grain size distribution, and it will be difficult (but not impossible) to derive the grain size distribution from the distribution of the areas intercepted by the polished

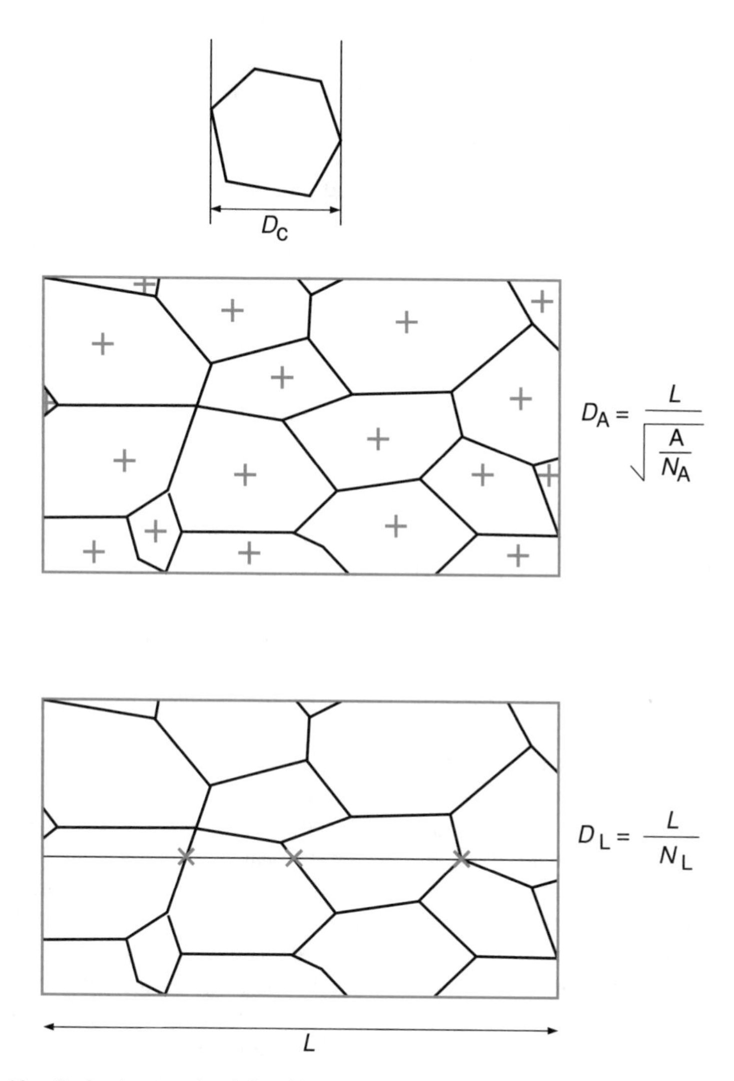

Figure 1.13 Grain size may be defined in several ways which are not directly related to one another, for example the mean caliper diameter, the average section diameter on a planar section, or the average intercept length along a random line

section. If the grains are also elongated (as may happen if the grain boundary energy depends on orientation) and if the elongated grains are partially aligned (as will be the case for a ductile metal which has been plastically deformed), then we shall have to think very carefully indeed about the significance of any grain size parameter which has been measured from a planar surface section. These problems are treated in more detail in Chapter 7.

1.1.3.2 Dislocations and Dislocation Density

Dislocations control many of the properties of engineering materials. A dislocation is a line defect in the crystal lattice which generates a local elastic strain field. Dislocations may interact with the free surface, internal interfaces (such as grain boundaries) and each other, in order to reduce the total elastic strain energy in the lattice. Each dislocation line is characterized by a displacement vector, the *Burger's vector*, which defines the magnitude of the elastic strain field. The angle between the Burger's vector and the dislocation line determines the nature of the strain field (the contributions of the shear, tensile and compressive displacements of the atoms from their equilibrium positions).

In a thin film specimen imaged in transmission electron microscopy, the Burger's vector and line sense of the dislocations can often be determined unambiguously from the diffraction contrast in the region of the dislocation (see Section 4.2.3), but the determination of the *dislocation density* can be difficult. A good theoretical definition of dislocation density is 'the total line length of the dislocations per unit volume of the sample'. This definition has the merit that it is independent of the dislocation distribution (it does not matter if the dislocations are all aligned along a particular crystallographic direction). However, it does not make any allowance for dislocations having different Burger's vectors. This may not matter if the dislocation population is dominated by one specific type of Burger's vector (as is commonly the case in cold-worked metals, where the slip dislocations are usually of only one type). However, it may matter a great deal in assessing the residual dislocation densities in semiconductor single crystals.

A further problem arises when we seek to extend our definition of the term dislocation. For example, when dislocations interact they may form low-energy dislocation networks (Fig. 1.14), which separate regions of the crystal having slightly different orientations in space. The dislocation network is thus a sub-grain boundary. Are we to include the array of dislocations in this sub-grain boundary in our count or not? Plastic deformation frequently leads to the formation of dislocation tangles which form cell structures, and within the cell walls it is not usually possible to resolve the individual dislocations (Fig. 1.15). Small dislocation loops may be formed by the collapse of point defects resulting from plastic deformation, quenching or radiation damage (Fig. 1.16). Do these small loops count as dislocations?

An alternative definition of dislocation density is 'the number of dislocation intersections per unit area of a planar section'. In an anisotropic sample this

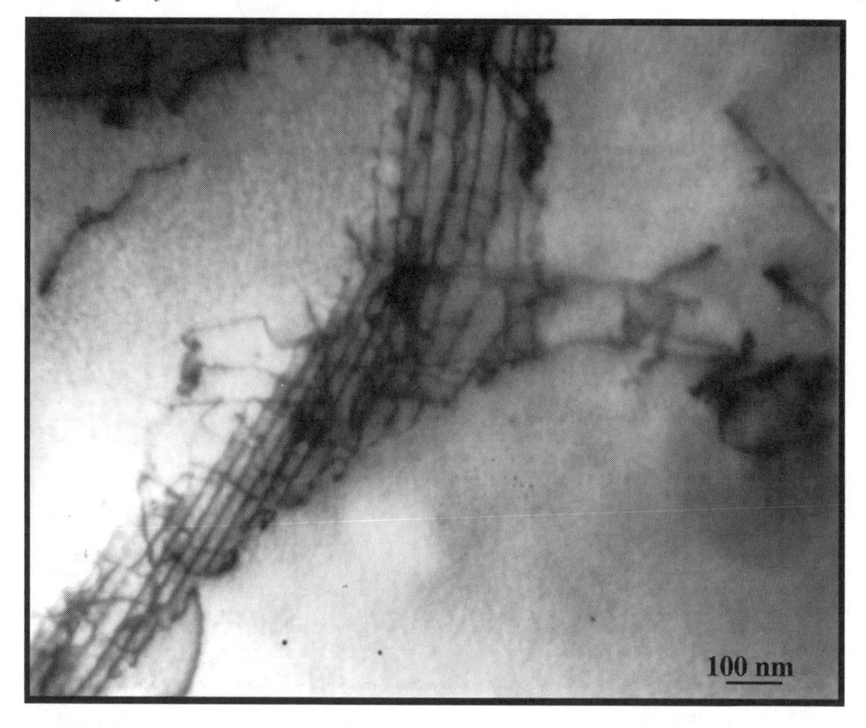

Figure 1.14 A low-energy array of dislocations forms a dislocation network which constitutes a sub-boundary in the crystal and can interact with slip dislocations. Such boundaries separate the crystal into sub-grains of slightly different orientation

definition results in a parameter that would be expected to show a dependence on the plane of the section, so that the two definitions are not equivalent. Furthermore, the dislocation density determined from observations made by using one method need not agree with that derived from other measurements, no matter how the density is defined. After all, the spatial resolution may differ, and the method of specimen preparation could also affect the results.

1.1.3.3 Phase Volume Fraction

Many engineering materials contain more than one phase, and the size, shape and distribution of the particles of a second phase are often dominant factors in determining the effect of the latter on the properties. As with grain size, there are a number of non-equivalent options for defining particle size and shape, which are for the most part analogous to the definitions of grain size, as discussed above. However, there is one microstructural parameter that is independent of the scale of the second phase, and that is f_V the phase volume fraction. Since this is a scale-independent parameter, it can be determined both conveniently and quickly from diffraction data

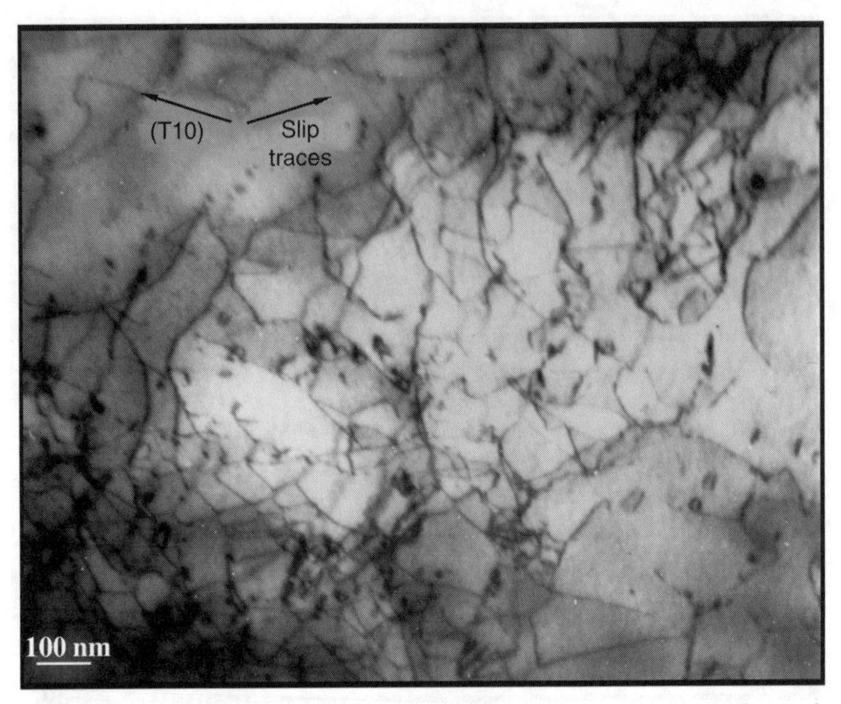

Figure 1.15 Plastic deformation of a ductile metal often results in poorly resolved dislocation tangles which form cells within the grains

(see the Worked Examples in Chapter 2), but it can also be determined from images of a planar section (Fig. 1.17). On a random section, the volume fraction of a second phase can be estimated from the areal fraction of the second phase intercepted by the section, A/A_0. At one time (many years ago) it was accepted practice to cut out the regions of interest from a photographic image and weigh them relative to the 'weight' of the total area sampled. This areal estimate is actually equivalent to a lineal estimate, determined from a random line taken across the section. Providing that the line and section are 'random', the length of the line traversing the second phase relative to the total length of the test line, L/L_0, is also an estimate of the phase volume fraction. Finally, a random grid of test points on the sample section can also provide the same information: the number of points falling on regions of the second phase divided by the total number of points, P/P_0, again estimates the volume of the second phase relative to the total volume of the sample, V/V_0. Thus, for the case of random sampling, we have $f_V = A/A_0 = L/L_0 = P/P_0 = V/V_0$.

1.2 CRYSTALLOGRAPHY AND CRYSTAL STRUCTURE

The arrangements of the atoms in engineering materials are determined by the chemical bonding forces. Some degree of order is always present in solids at the

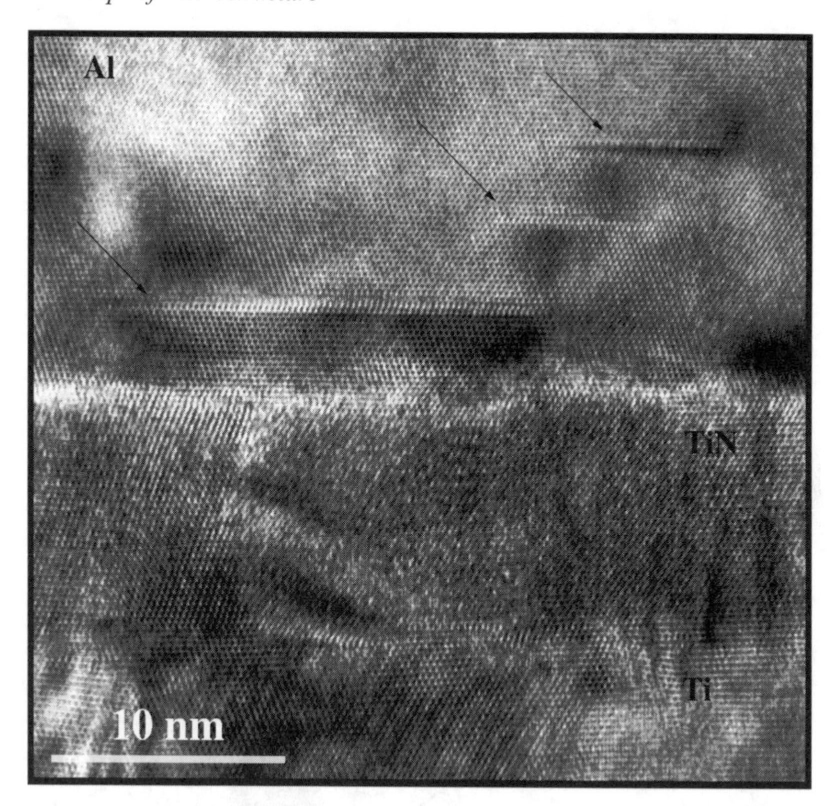

Figure 1.16 Point defects generated by plastic deformation, quenching or radiation damage may condense to form small dislocation loops, seen edge-on in this lattice image

atomic level, even in what appears to be a featureless, structureless glass or polymer. In the following, we will briefly review the nature of the chemical forces and outline the ways in which these chemical forces are related to the engineering properties. We will then discuss some of the tools needed to describe and understand the commonly observed atomic arrangements in ordered, crystalline solids. The body of knowledge that describes and characterizes the structure of crystals is termed crystallography.

1.2.1 Interatomic Bonding in Solids

It is a convenient assumption that atoms in solids are packed together much as one would pack table tennis balls into a box. The atoms (or, if they carry an electrical charge, the ions) are assumed to be spherical, and to have a diameter which depends on their *atomic number* (the number of electrons surrounding the nucleus), their electrical charge (positive, if electrons have been removed to form a cation, or negative if additional electrons have been captured to form an anion), and, to a much

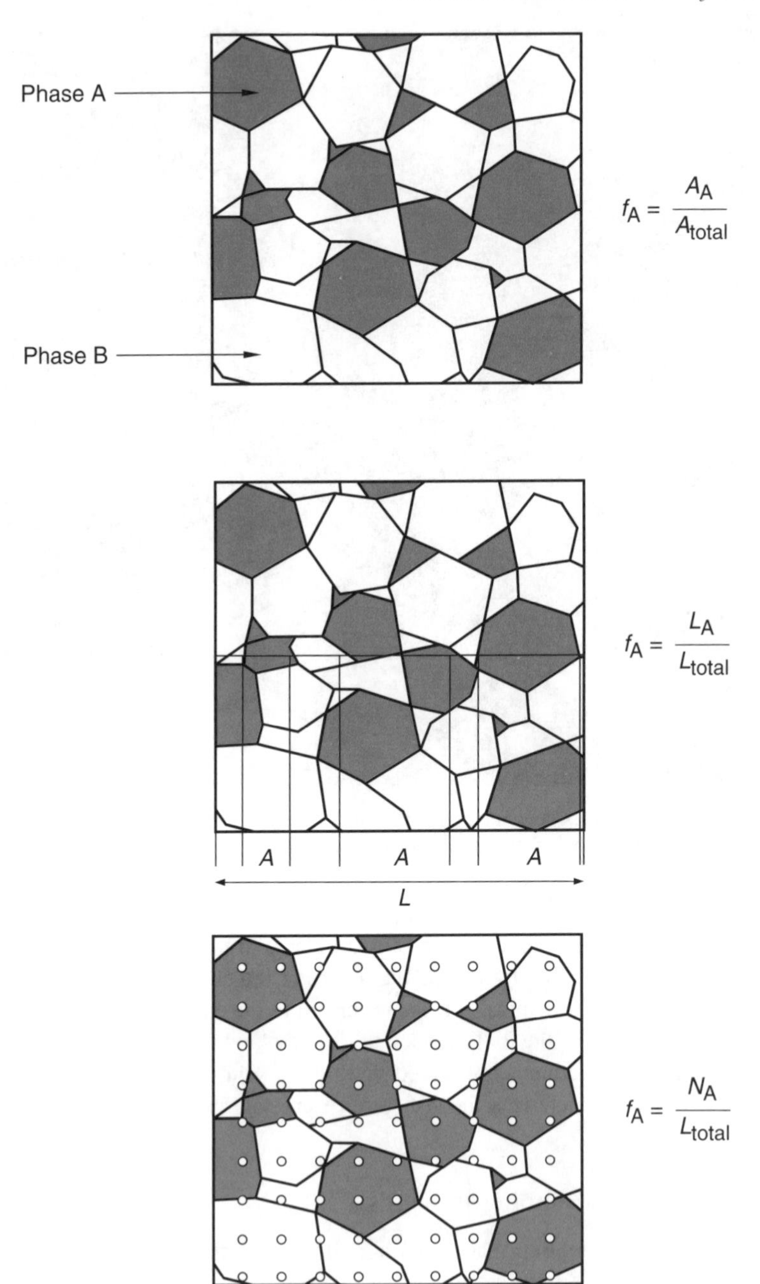

Figure 1.17 The volume fraction of a second phase can be determined from the areal fraction of the phase, seen on a random planar section, or from the fractional length of a random test line which intercepts the second phase particles in the section, or from the fraction of points in a test array which falls within the regions of the second phase

lesser extent, the number of neighbouring atoms surrounding the one being considered (the *coordination number*).

1.2.1.1 IONIC BONDING

In an ionically bonded solid, the outer, valency electron shells of the atoms are completed or emptied by accepting or donating electrons. In cooking salt, NaCl, the sodium atom donates an electron to the chlorine atom to form an ion pair (a positively charged sodium cation and a negatively charged chlorine anion). Both cations and anions now have stable outer electron shells, with the chlorine ion having a full complement of eight electrons in the outer shell, and the sodium ion having donated its lone excess electron to a chlorine atom.

To minimize the electrostatic energy, the cations must be surrounded by anion neighbours and vice versa. At the same time, the outer electron shell of the anions contracts towards the positively charged nucleus as a result of the excess positive charge, while the excess negative charge on the anions causes a net expansion of the outer electron shell. In the case of NaCl and many other ionic crystals, the large anions form an ordered (and closely packed) array, with the smaller cations occupying the interstices. Two opposing factors determine the number of neighbours of opposite charge (the coordination number). The electrostatic (*coulombic*) attraction between ions of opposite charge tends to maximize the density. At the same time, the smaller ion must be larger than the interstices which it occupies in the packing of the larger ion, in order to keep the neighbouring ions of similar charge separated. The smallest possible number of neighbours of opposite charge is 3, and boron ($Z = 5$) is a small, highly charged cation which can have this low coordination number of 3. Coordination numbers of 4, 6 and 8 are found for steadily increasing ratios of the two ionic radii (Fig. 1.18), while the maximum coordination number, 12, corresponds to cations and anions having approximately the same size. The anion is not always the larger of the two ions, and if the cation has a sufficiently large atomic number it may be the anions which occupy the interstices in the cation array. Zirconia, ZrO_2, is a good example of this.

Silicate structures and glasses are also dominated by ionic bonding, but in this case the tightly coordinated cations form a *molecular ion*, most notably the SiO_4 silicate tetrahedron. These tetrahedra may carry a residual negative charge (as in magnesium silicate, Mg_2SiO_4), or they may be covalently linked to form a poly-ion, with the coordination tetrahedra sharing their corner oxygen atoms (as in quartz) to form *oxygen bridges*. Borates, phosphates and sulfates may form similar structures, but it is the silicates that dominate in engineering importance. Since only the corners of the tetrahedra may be shared, and not the edges or faces, the silicates form very open structures which accommodate a wide variety of other cations. In addition, the oxygen corner linkage is flexible, thus allowing the two linked tetrahedra considerable freedom to change their relative orientation. In a glass, the average negative charge on the silicate ion varies inversely with the number of oxygen bridges, and this charge is neutralized by the presence of additional cations

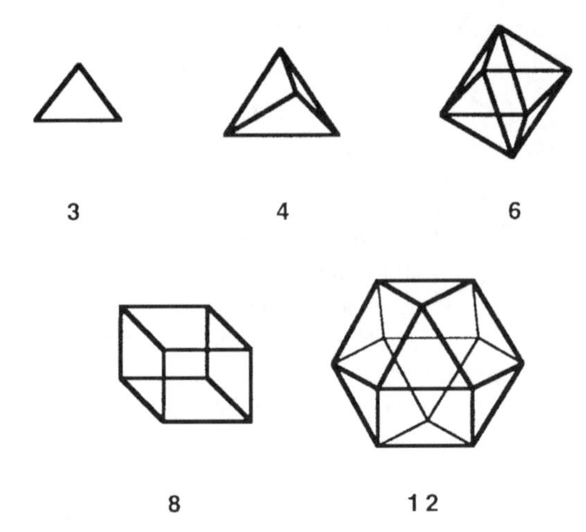

Figure 1.18 The number of neighbours of an ion, its coordination number, is primarily determined by the ratio of the radii of the smaller to the larger ions. The regular coordination polyhedra allow for 3, 4, 6, 8 or 12 nearest neighbours

(modifiers) which occupy the interstices between the tetrahedra. The tetrahedra themselves do not readily change their dimensions (although some substitution of the Si^{4+} ion can occur, most notably by B^{3+} or Al^{3+}), while the oxygen bridges constrain the distance between neighbouring tetrahedra. It follows that glasses possess well-defined *short-range order*, which may extend to distances of the order of 2 nm from the centre of a silicate tetrahedron.

1.2.1.2 Covalent Bonding

Many important engineering materials are based on chemical bonding in which neighbouring atoms share electrons that occupy *molecular orbitals*. In diamond (Fig. 1.19), the carbon atoms all have four valency electrons, and by sharing each of these with four neighbouring carbon atoms, each atom acquires a full complement of eight electrons for the outer valency shell. It is the C–C covalent bond which determines the chemical stability, not only of diamond, but also of most polymer molecules, which are constructed from covalently linked chains of carbon atoms. The oxygen bridges in silicate glasses are also, to a large extent, covalent bonds, and it is no accident that the same oxygen bridges provide the chain linkage in silicone polymers, –(O–SiHR)–.

It is not always possible to describe a bond as being either covalent or ionic. Consider the series, NaCl, MgO, AlN, and SiC. The first two are commonly described as ionic solids, while the third, aluminium nitride, might also be described as being ionically bonded, although the effective charge on the ions is appreciably less than that predicted by their valency (+ or −3). Silicon carbide is tetrahedrally

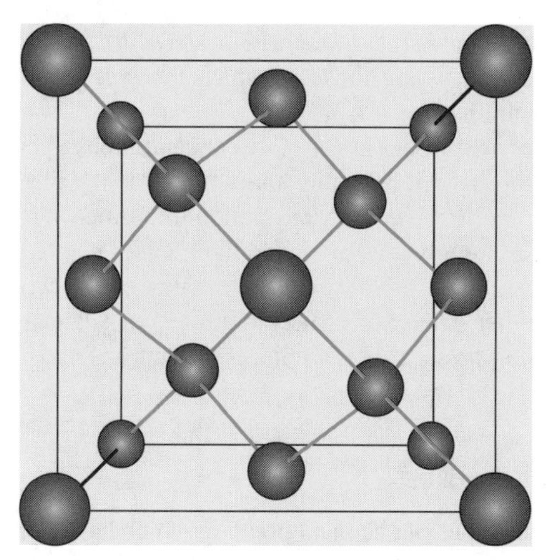

Figure 1.19 In covalently bonded diamond the carbon atoms are tetrahedrally coordinated to each of their nearest neighbours by shared molecular orbitals

coordinated, as is diamond, and, since both constituents are in the same group of the periodic table, one might guess that it is covalently bonded. However, this is not entirely correct, and the physical properties are best simulated by assuming that the silicon atoms carry *some* positive charge, while the carbon atoms carry a corresponding negative charge. To summarize, as the valency increases, so does the contribution from covalent bonding.

1.2.1.3 METALS AND SEMICONDUCTORS

Valency electrons may be shared, not only with a nearest neighbour atom, but quite generally, throughout the solid, that is the molecular orbitals may not be localized to a specific atom pair. Electrons which are free to move throughout the solid are said to occupy a *conduction band*, and to be *free electrons*. The chemical bonding in such a solid is termed *metallic* bonding, and is associated with a balance between two opposing forces, namely the coulombic attraction between the free electrons and the array of positively charged cations, and the repulsive forces between the closed shells of the cations. The properties which are characteristic of the metallic bond are associated with the mobility of the free electrons (thermal and electrical conductivity, and optical reflectivity), and the non-directionality of the bond (mechanical plasticity or *ductility*).

In some cases, small numbers of electrons may be present in the conduction band of a solid, as a result of thermal excitation or the presence of impurities. Such materials are termed *semiconductors*, and their importance has grown, and continues to grow, with the development of the electronics industry. If electrons are thermally excited to occupy the conduction band, then they leave vacant *holes* behind, which may also be mobile. Semiconductors in which *electrons* are the dominant current carriers are termed *n-type*, while those in which the *holes* are responsible for the electronic properties are termed *p-type*.

If certain impurities are present in very low concentrations, then the electrons may not be free to move throughout the solid, and will not confer electrical conductivity, but they will occupy *localized states*. Many cation impurities in ceramics give rise to localized states which strongly absorb visible light, and they are said to form *colour centres*. Such cation additions are of major importance in the ceramics industry, and the effects may be either deleterious (discolouration) or advantageous (a variety of attractive enamels and glazes). An example of the latter is shown in Fig. 1.20 (see Plate 1 facing page 18).

1.2.1.4 Polarization Forces

In addition to the three types of chemical bonding which have been described briefly above, many of the properties of engineering solids are determined by *secondary*, or *van der Waals bonding*, associated with molecular *polarization* forces. In its weakest form, the polarization force arises from the polarizability of an electron orbital. Rare gas atoms will liquefy (and solidify) at cryogenic temperatures as a consequence of the small reduction in potential energy achieved by polarization of an otherwise symmetrical electron orbital. Many molecular gases (H_2, N_2, O_2, CH_4) behave similarly. The properties of a number of engineering polymers are dominated by the polarizability of the molecular chain, most notably polyethylene, $-(CH_2-CH_2)-$. The ductility of the polymer, as well as its softening point and glass transition temperature, are determined by a combination of the molecular weight of the polymer chains and their polarizability. Even this weak bonding is quite sufficient to ensure the mechanical stability of engineering components manufactured from polymers, and to allow them, under suitable circumstances, to crystallize.

Stronger polarization forces exist when the molecular species has a lower symmetry and can posses a permanent *dipole moment*. A good example is carbon dioxide (CO_2), but similar molecular groupings are often present in high-performance engineering polymers. Organic tissues are commonly based on giant *polar molecules* with properties dictated by a combination of the molecular configuration and the position of the polar groups within the molecule.

The strongest polarization forces are associated with a dipole moment due to hydrogen, *the hydrogen bond*. Hydrogen in its ionized form is a proton, with no electrons to screen the nucleus. The ionic radius is therefore the smallest possible, and asymmetric molecular groupings which contain hydrogen can have very high dipole moments. The two compounds that demonstrate this best are water $-(H_2O)$ and ammonia $-(NH_3)$, and the corresponding molecular groups found in engineering and organic polymers are (OH) and (NH_2), where these groups raise both the tensile strength and the softening point of the polymer. The families of polyamides and polyamines (which include the 'nylons', a commercial trade name) depend for their strength and stiffness on the hydrogen bonding between the polymer chains.

1.2.2 Crystalline and Amorphous Phases

We have outlined the nature of the chemical bonding found in engineering solids and the atomic coordination requirements associated with this bonding. In some engineering solids the local atomic packing results in long-range order and the material is crystalline, while in others short-range order has no long-range consequences, and the material is amorphous or glassy.

Single-phase polycrystalline materials are made up of many grains, each of which has identical atomic packing to that of its neighbours, although not in the same relative orientation. In *polyphase* materials the grains of each phase all have the *same* atomic packing, which generally differs from that of the other phases. In thermodynamic equilibrium, the grains of each phase have a unique and fixed composition which depends on the temperature and composition of the material, and can usually be determined from the appropriate phase diagram. In general, solid phases may be either crystalline or amorphous. Amorphous phases may be formed by several quite distinct routes, for example rapid cooling from the liquid phase, condensation from the gaseous phase, or as the result of a chemical reaction. Good examples of the last type are the protective oxide films formed on aluminium alloys and stainless steels by reaction in air at room temperature.

1.2.3 The Crystal Lattice

The well developed facets on many naturally occurring crystals, as well as on those ionic crystals which could be grown from aqueous solution, prompted the early development of the science of crystallography. An analysis of the angles between the facets permitted an exact description of the *symmetry* elements of a crystal, and led to speculation that crystal symmetry was a property of the bulk material. This was confirmed with the discovery that isolated, single crystals would diffract X-rays at very specific angles, to yield a sharp *diffraction pattern* that was a characteristic of *any* crystal of the same material oriented in the same relation to the incident beam, irrespective of its size and shape.

The interpretation of sharp diffraction maxima in terms of a completely ordered and regular atomic array of the chemical elements followed almost immediately, being pioneered by the father and son team of Lawrence and William Bragg. The concept of the *crystal lattice* was an integral part of this interpretation, following on from the realization that the atoms in a crystal must be centred at discrete, essentially fixed, distances from one another, and that these interatomic separations constituted an array of *lattice vectors* that could all be defined in terms of an elementary unit of volume, the *unit cell* of the crystal, which displayed all the symmetry elements observed in the bulk crystal.

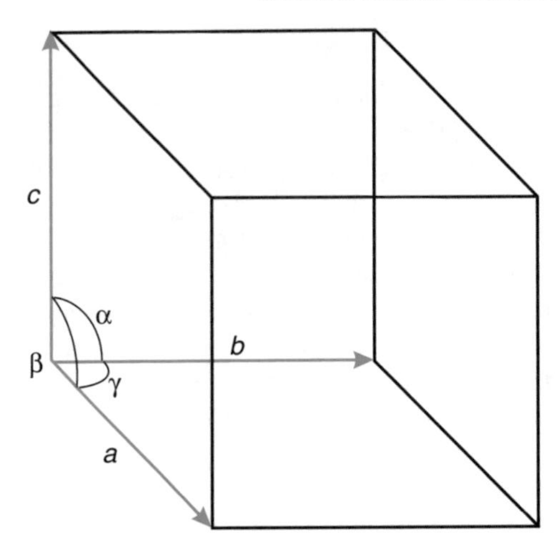

Figure 1.21 Schematic representation of the lattice parameters of a unit cell

1.2.3.1 Unit Cells and Point Lattices

This section introduces some basic aspects of crystallography, and describes how crystalline structure is interpreted and crystallographic data retrieved from the literature.

To understand the structure of crystals, it is convenient at first to ignore the positions of the atoms, and refer to a periodic three-dimensional scaffold within which the atoms are positioned. Such a scaffold is termed a *lattice*, and is defined as a set of periodic points in space. A single lattice cell (unit cell) is a parallelepiped, and each unit cell can be periodically stacked in space, by integer displacements of the unit cell parameters. The unit cell parameters are the three coordinate lengths (a, b, c) determined by placing the origin of the coordinate system at a lattice point, and the three angles (α, β, γ) subtended by the lattice cell axes (Fig. 1.21). Thus a unit cell with $a = b = c$ and $\alpha = \beta = \gamma = 90°$ is a cube. The various unit cells are generated from the different values of a, b, c and α, β and γ. An analysis of how to fill space periodically with lattice points shows that only *seven* different unit cells are required to describe point lattices, and these are termed the seven *crystal systems*, namely triclinic, monoclinic, orthorhombic, tetragonal, rhombohedral, hexagonal and cubic.

The seven crystal systems are defined by using *primitive* unit cells, in which each cell only contains one lattice point at the origin. More complicated arrangements are possible, with each requiring that every lattice point should have an identical surrounding. These permutations were analysed by the French crystallographer Bravais in 1848, who described all possible (14) point lattices (the *Bravais lattices*), as shown in Fig. 1.22.

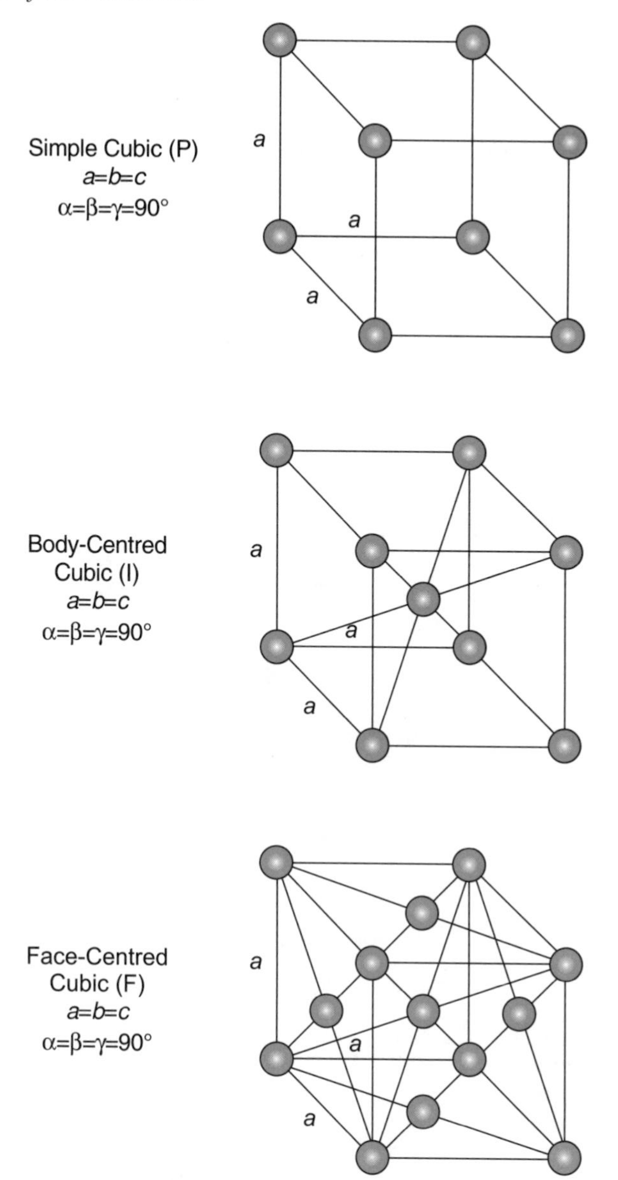

Figure 1.22 The 14 Bravais lattices derived from the seven crystal systems

1.2.3.2 Space Groups

Various degrees of symmetry are possible in the periodic packing of atoms in crystals. For example, if a crystal is built of atoms located only at the corners of a unit cell, then it will have the highest possible symmetry within that particular

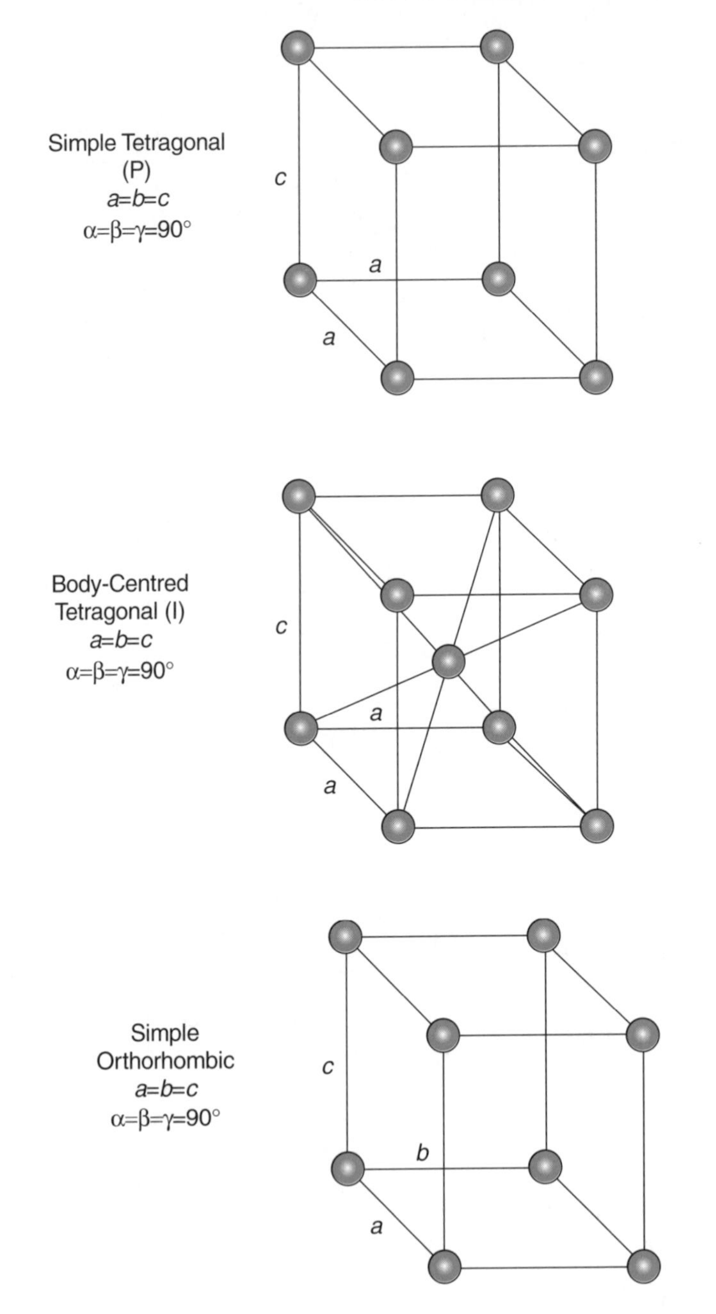

Figure 1.22 (*continued*)

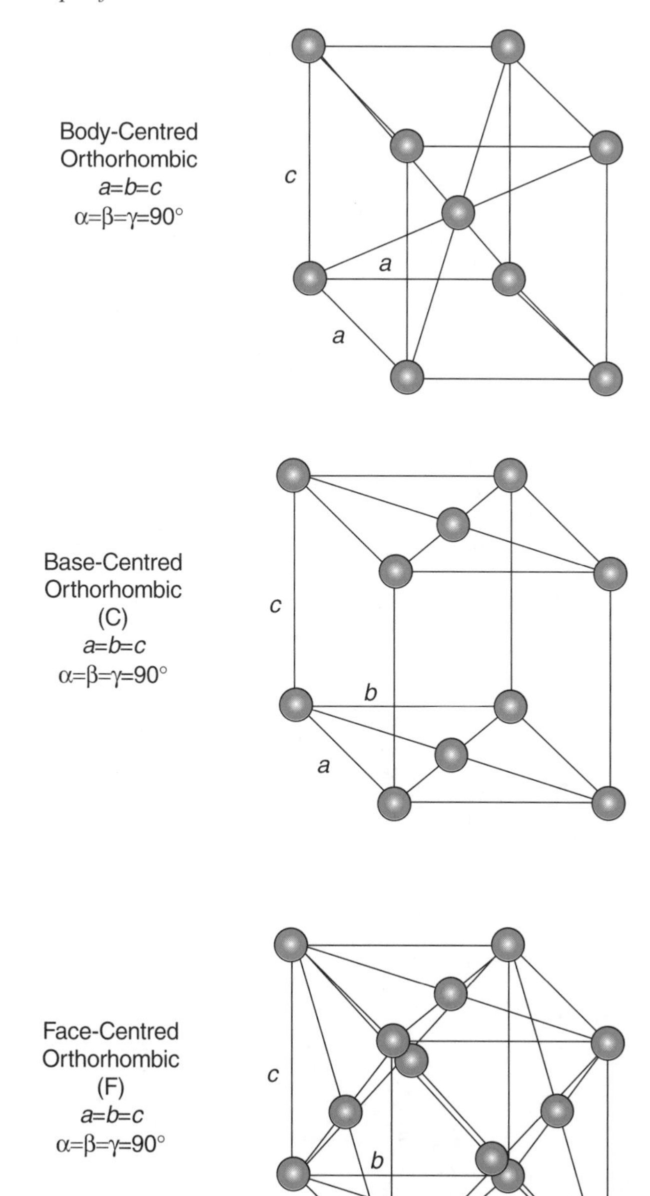

Figure 1.22 (*continued*)

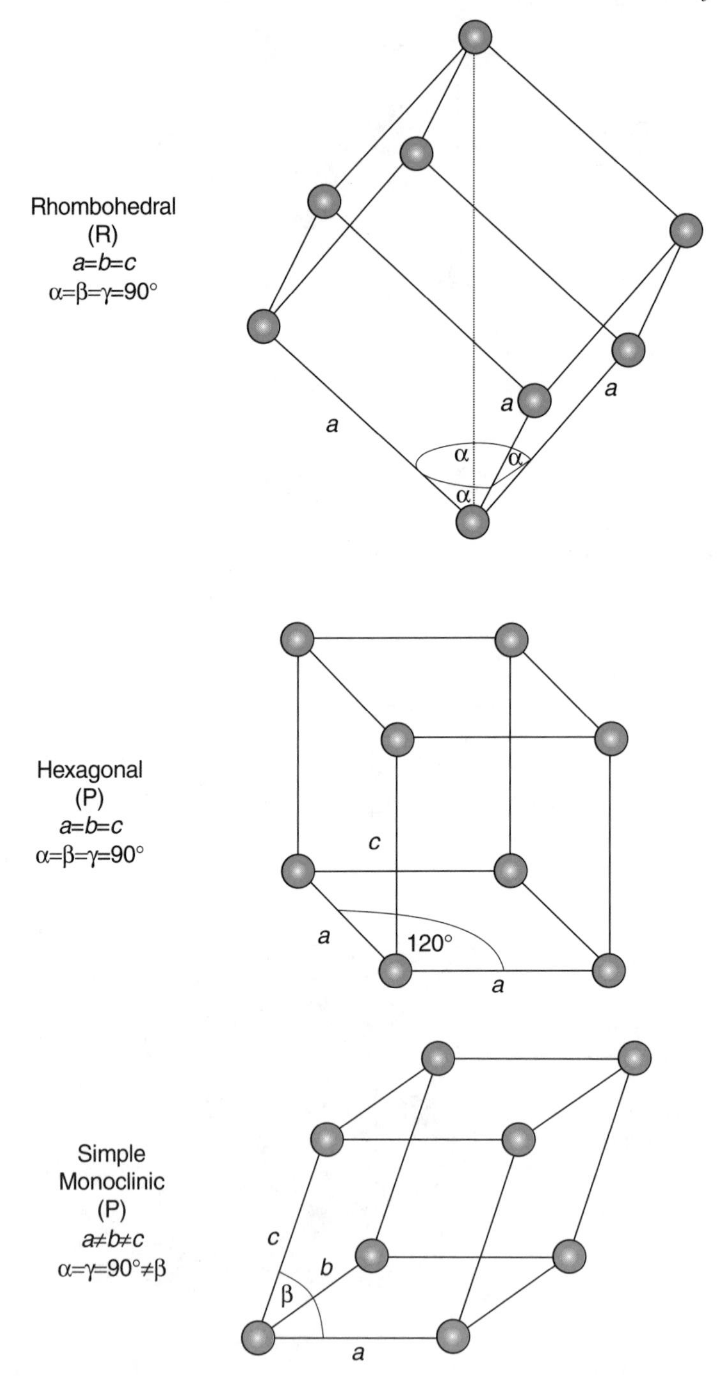

Figure 1.22 (*continued*)

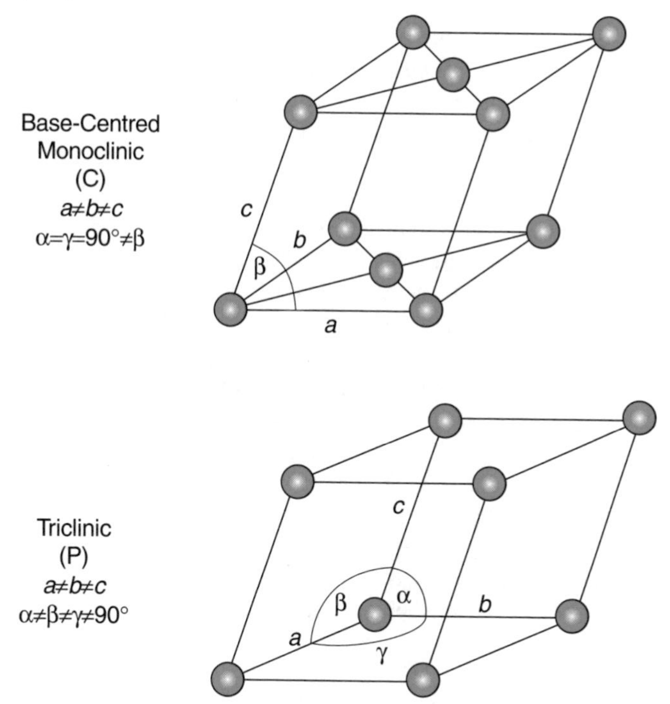

Figure 1.22 (*continued*)

crystal system. This symmetry will be retained if there are symmetrical groups of atoms associated with each lattice point. However, the atomic groups around a lattice point might also pack with *lower* symmetry, thus reducing the symmetry of the crystal, so that it belongs to a different symmetry group within the same crystal class.

These considerations lead to the definition of *space groups*, which provide criteria for filling a Bravais point lattice with atoms in a periodic array. There are a total of 230 different periodic space groups. The structure of a crystal can always be described by using one (or more) space groups, and, given a little practice, this is undoubtedly the most convenient way to visualize a complex crystal structure.

Let us examine the use of space groups to define a crystal structure by using a simple example. Assume that we wish to know the positions of all of the atoms in a copper (Cu) crystal. First, we need a literature source which contains the

Table 1.3 Data for Cu as presented in Pearson's handbook [1]

Phase	Structure type	Pearson symbol space group	a, b, c (*nm*)	α, β, γ (°)	Atoms	Point set	x	y	z	*Occ*
Cu	Cu	cF4 $Fm\bar{3}m$	0.36148		Cu	4a	000	000	000	100

crystallographic data. Probably the best available for materials science is *Pearson's Handbook of Crystallographic Data for Intermetallic Phases'* [1] (the name is misleading, since pure metals and ceramics are also included). The principle data for Cu listed in this book appear in the format shown in Table 1.3.

Following the name of the phase, a *structure type* is given. This is a real material which serves as an example for this particular crystallographic structure, with, in this case, copper being its own structure type. Next the *Pearson symbol* and *space group* are given, which refer to the type of lattice cell and the symmetry of the structure. The first, cF4, means a cubic (c) face-centred lattice (F) with four atoms per unit cell. The symmetry description, $Fm\bar{3}m$, is also the name of the space group and is followed by the lattice parameters. In the case of Cu, $a = b = c = 0.36148$ nm, and $\alpha = \beta = \gamma = 90°$. Since a cubic structure clearly has $\alpha = \beta = \gamma = 90°$, these values are not listed. Finally, the *Wyckoff generating sites* are given. These are the sites of specific atoms within the crystal structure, upon which the space group symmetry operators act. When combined with the space group, the symmetry operators will generate all of the other atomic positions within the unit cell and specify the '*occupancy*'. For Cu, $x = 000$, $y = 000$, $z = 000$ and $Occ = 100$, that is $x = y = z = 0.0$ and the occupancy is $= 1.00$. The last term, the *occupation factor*, indicates the proportion of sites which are occupied by a particular atom species. In the case of Cu, all of the sites (neglecting point defects) are *always* occupied by Cu (that is a value of 1.0).

How do we generate the crystal structure? We first need details of the symmetry operators for the space group $Fm\bar{3}m$. These can be found in the '*International Tables For Crystallography*', Volume A (*Space-Group Symmetry*) [2]. An example of some typical data is given in Fig. 1.23. The data presented in these tables provide all of the symmetry operators for a particular space group. In order to generate the atomic positions from the generating-site data listed for Cu, one has to add the (x, y, z) values of the generating site to the values listed in the tables. Now, returning to our example, Cu has a generating site of type 4a with (x, y, z) equal to $(0, 0, 0)$. In the tables, the Wyckoff generating site for 4a has an operator of $(0, 0, 0)$. This is our first atom site. Now we activate the general operators which are listed. The addition of $(0, 0, 0)$ to our initial generating site of $(0, 0, 0)$ leaves one atom at the origin, while the addition of $(0, 1/2, 1/2)$, $(1/2, 0, 1/2)$ and $(1/2, 1/2, 0)$ to $(0, 0, 0)$ places three more atoms, one at the centre of each face of the unit cell which adjoins the origin. There are therefore four Cu atoms in our unit cell, as shown schematically in Fig. 1.24(a).

Fig. 1.24(a) does not look like a 'complete' face-centred cubic (FCC) structure, since only those atoms are shown which 'belong' to a single unit cell. Additional atoms at sites $(1, 0, 0)$, $(1, 1, 0)$, $(0, 1, 0)$, $(1, 0, 1)$, $(0, 1, 1)$, $(0, 0, 1)$, $(1, 1, 1)$, $(1, 1/2, 1, 2)$, $(1/2, 1, 1/2)$ and $(1/2, 1/2, 1)$ belong to the neighbouring unit cells in the crystal. A more generally accepted (although no more correct) illustration of the unit cell for copper is given in Fig. 1.24(b).

We need a clear definition of when an atom belongs to a specific unit cell. If $0 \leq x < 1$, $0 \leq y < 1$, and $0 \leq z < 1$, then the atom belongs to the unit cell. If not, the atom belongs to a neighbouring cell. This provides a very easy method for counting the number of atoms per cell.

Why is all of this effort required just to define four atoms in an FCC configuration? Well, for this simple structure, the formalism may be unnecessary, but if, for example, you have a structure with 92 atoms per unit cell, the use of these tables

Generators selected (1); $t(1,0,0)$; $t(0,1,0)$; $t(0,0,1)$; $t(0,\frac{1}{2},\frac{1}{2})$; $t(\frac{1}{2},0,\frac{1}{2})$; (2); (3); (5); (13); (25)

Positions

Multiplicity, Wyckoff letter, Site symmetry			Coordinates				Reflection conditions
			$(0,0,0)+$	$(0,\frac{1}{2},\frac{1}{2})+$	$(\frac{1}{2},0,\frac{1}{2})+$	$(\frac{1}{2},\frac{1}{2},0)+$	h,k,l permutable
							General:
192	l	1	(1) x,y,z	(2) $\bar{x},\bar{y},z$	(3) $\bar{x},y,\bar{z}$	(4) $x,\bar{y},\bar{z}$	hkl : $h+k,h+l,k+l=2n$
			(5) z,x,y	(6) $z,\bar{x},\bar{y}$	(7) $\bar{z},\bar{x},y$	(8) $\bar{z},x,\bar{y}$	$0kl$: $k,l=2n$
			(9) y,z,x	(10) $\bar{y},z,\bar{x}$	(11) $y,\bar{z},\bar{x}$	(12) $\bar{y},\bar{z},x$	hhl : $h+l=2n$
			(13) $y,x,\bar{z}$	(14) $\bar{y},\bar{x},\bar{z}$	(15) $y,\bar{x},z$	(16) $\bar{y},x,z$	$h00$: $h=2n$
			(17) $x,z,\bar{y}$	(18) $\bar{x},z,y$	(19) $\bar{x},\bar{z},\bar{y}$	(20) $x,\bar{z},y$	
			(21) $z,y,\bar{x}$	(22) $z,\bar{y},x$	(23) $\bar{z},y,x$	(24) $\bar{z},\bar{y},\bar{x}$	
			(25) $\bar{x},\bar{y},\bar{z}$	(26) $x,y,\bar{z}$	(27) $x,\bar{y},z$	(28) $\bar{x},y,z$	
			(29) $\bar{z},\bar{x},\bar{y}$	(30) $\bar{z},x,y$	(31) $z,x,\bar{y}$	(32) $z,\bar{x},y$	
			(33) $\bar{y},\bar{z},\bar{x}$	(34) $y,\bar{z},x$	(35) $\bar{y},z,x$	(36) $y,z,\bar{x}$	
			(37) $\bar{y},\bar{x},z$	(38) y,x,z	(39) $\bar{y},x,\bar{z}$	(40) $y,\bar{x},\bar{z}$	
			(41) $\bar{x},\bar{z},y$	(42) $x,\bar{z},\bar{y}$	(43) x,z,y	(44) $\bar{x},z,\bar{y}$	
			(45) $\bar{z},\bar{y},x$	(46) $\bar{z},y,\bar{x}$	(47) $z,\bar{y},\bar{x}$	(48) z,y,x	

									Special: as above, plus
96	k	$..m$	x,x,z	$\bar{x},\bar{x},z$	$\bar{x},x,\bar{z}$	$x,\bar{x},\bar{z}$	z,x,x	$z,\bar{x},\bar{x}$	no extra conditions
			$\bar{z},\bar{x},x$	$\bar{z},x,\bar{x}$	x,z,x	$\bar{x},z,\bar{x}$	$x,\bar{z},\bar{x}$	$\bar{x},\bar{z},x$	
			$x,x,\bar{z}$	$\bar{x},\bar{x},\bar{z}$	$x,\bar{x},z$	$\bar{x},x,z$	$x,z,\bar{x}$	$\bar{x},z,x$	
			$\bar{x},\bar{z},\bar{x}$	$x,\bar{z},x$	$z,x,\bar{x}$	$z,\bar{x},x$	$\bar{z},x,x$	$\bar{z},\bar{x},\bar{x}$	
96	j	$m..$	$0,y,z$	$0,\bar{y},z$	$0,y,\bar{z}$	$0,\bar{y},\bar{z}$	$z,0,y$	$z,0,\bar{y}$	no extra conditions
			$\bar{z},0,y$	$\bar{z},0,\bar{y}$	$y,z,0$	$\bar{y},z,0$	$y,\bar{z},0$	$\bar{y},\bar{z},0$	
			$y,0,\bar{z}$	$\bar{y},0,\bar{z}$	$y,0,z$	$\bar{y},0,z$	$0,z,\bar{y}$	$0,z,y$	
			$0,\bar{z},\bar{y}$	$0,\bar{z},y$	$z,y,0$	$z,\bar{y},0$	$\bar{z},y,0$	$\bar{z},\bar{y},0$	
48	i	$m.m2$	$\frac{1}{2},y,y$	$\frac{1}{2},\bar{y},y$	$\frac{1}{2},y,\bar{y}$	$\frac{1}{2},\bar{y},\bar{y}$	$y,\frac{1}{2},y$	$y,\frac{1}{2},\bar{y}$	no extra conditions
			$\bar{y},\frac{1}{2},y$	$\bar{y},\frac{1}{2},\bar{y}$	$y,y,\frac{1}{2}$	$\bar{y},y,\frac{1}{2}$	$y,\bar{y},\frac{1}{2}$	$\bar{y},\bar{y},\frac{1}{2}$	
48	h	$m.m2$	$0,y,y$	$0,\bar{y},y$	$0,y,\bar{y}$	$0,\bar{y},\bar{y}$	$y,0,y$	$y,0,\bar{y}$	no extra conditions
			$\bar{y},0,y$	$\bar{y},0,\bar{y}$	$y,y,0$	$\bar{y},y,0$	$y,\bar{y},0$	$\bar{y},\bar{y},0$	
48	g	$2.mm$	$x,\frac{1}{4},\frac{1}{4}$	$\bar{x},\frac{3}{4},\frac{1}{4}$	$\frac{1}{4},x,\frac{1}{4}$	$\frac{1}{4},\bar{x},\frac{3}{4}$	$\frac{1}{4},\frac{1}{4},x$	$\frac{3}{4},\frac{1}{4},\bar{x}$	hkl : $h=2n$
			$\frac{1}{4},x,\frac{3}{4}$	$\frac{3}{4},\bar{x},\frac{3}{4}$	$x,\frac{1}{4},\frac{3}{4}$	$\bar{x},\frac{1}{4},\frac{1}{4}$	$\frac{1}{4},\frac{1}{4},\bar{x}$	$\frac{1}{4},\frac{3}{4},x$	
32	f	$.3m$	x,x,x	$\bar{x},\bar{x},x$	$\bar{x},x,\bar{x}$	$x,\bar{x},\bar{x}$			no extra conditions
			$x,x,\bar{x}$	$\bar{x},\bar{x},\bar{x}$	$x,\bar{x},x$	$\bar{x},x,x$			
24	e	$4m.m$	$x,0,0$	$\bar{x},0,0$	$0,x,0$	$0,\bar{x},0$	$0,0,x$	$0,0,\bar{x}$	no extra conditions
24	d	$m.mm$	$0,\frac{1}{4},\frac{1}{4}$	$0,\frac{3}{4},\frac{1}{4}$	$\frac{1}{4},0,\frac{1}{4}$	$\frac{1}{4},0,\frac{3}{4}$	$\frac{1}{4},\frac{1}{4},0$	$\frac{3}{4},\frac{1}{4},0$	hkl : $h=2n$
8	c	$\bar{4}3m$	$\frac{1}{4},\frac{1}{4},\frac{1}{4}$	$\frac{1}{4},\frac{1}{4},\frac{3}{4}$					hkl : $h=2n$
4	b	$m\bar{3}m$	$\frac{1}{2},\frac{1}{2},\frac{1}{2}$						no extra conditions
4	a	$m\bar{3}m$	$0,0,0$						no extra conditions

Symmetry of special projections

Along [001] $p4mm$
$\mathbf{a}'=\frac{1}{2}\mathbf{a}$ $\mathbf{b}'=\frac{1}{2}\mathbf{b}$
Origin at $0,0,z$

Along [111] $p6mm$
$\mathbf{a}'=\frac{1}{6}(2\mathbf{a}-\mathbf{b}-\mathbf{c})$ $\mathbf{b}'=\frac{1}{6}(-\mathbf{a}+2\mathbf{b}-\mathbf{c})$
Origin at x,x,x

Along [110] $c2mm$
$\mathbf{a}'=\frac{1}{2}(-\mathbf{a}+\mathbf{b})$ $\mathbf{b}'=\mathbf{c}$
Origin at $x,x,0$

Figure 1.23 The space group Fm$\bar{3}$m which defines the symmetry for Cu. Reproduced by permission of Kluwer Academic Publishers from *International Tables for Crystallography*, Vol. A, *Space Group Symmetry*, T. Hahn, ed. (1992)

is certainly the easiest and least error-prone way to define the atomic positions. It is also convenient if you wish to simulate crystal structures or diffraction spectra: instead of typing in the positions of all of the atoms in the unit cell, you can use the above space group operations.

Now consider an example in which the use of *occupation factors* is important, a Cu–Ni solid solution with 50 atom% Ni. Both Cu and Ni have the simple FCC structure, with complete solid solubility over the entire composition range. The description of the structure in Pearson's handbook would be very similar to that of Cu, but instead of one generating site there are now two, both with the same x, y, and z values, but with each having an occupation factor of 0.5, corresponding to the bulk concentration of 50 atom% Ni. Thus each of the four sites in the cell is occupied half of the time by Cu, and half of the time by Ni. We have just described a *disordered* solid solution! Other than slight changes in the lattice parameters, this is the only difference between pure Cu and the Cu–Ni solid solution.

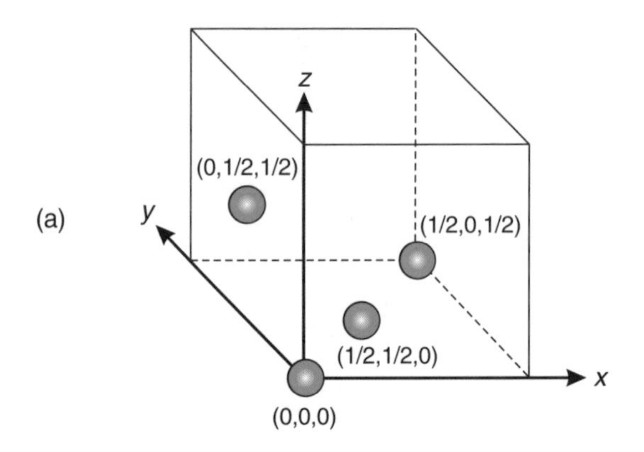

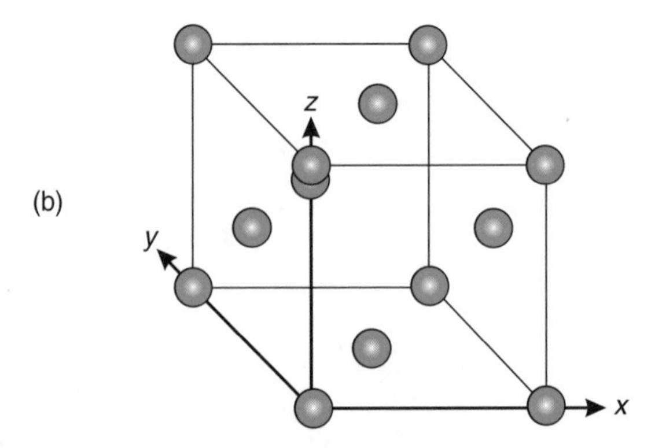

Figure 1.24 Schematic representations of (a) the unit cell of Cu and (b) the same unit cell but with additional atoms from neighbouring unit cells

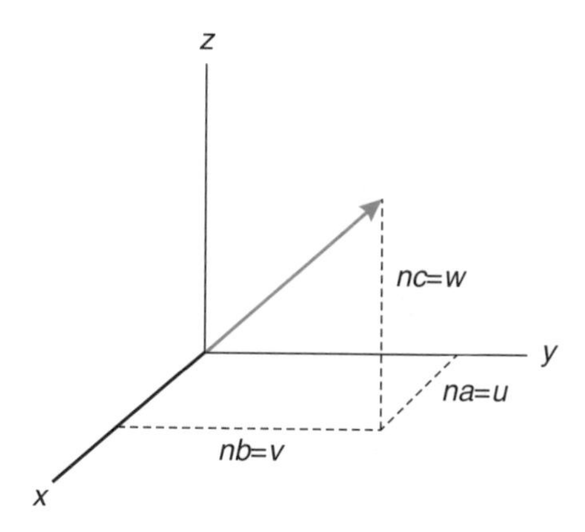

Figure 1.25 Schematic representation of direction indices in a crystal lattice (Reproduced from Hahn, *International Tables for Crystallography*, Vol. A, *Space-Group Symmetry*, with kind permission from Kluwer, Academic Publishers)

1.2.3.3 Miller Indices and Unit Vectors

Crystal planes and crystal directions in the lattice are described by a *vector notation* based on a coordinate system defined by the axes and dimensions of the unit cell. A *direction* is defined by a vector at the origin of the coordinate system and of length sufficient to ensure that the x, y and z coordinates of the vector tip all correspond to an integer number of unit cell coordinates (Fig. 1.25). The length of a *rational* vector always corresponds to an interatomic repeat distance in the lattice. In a crystal, two or more lattice directions may be geometrically equivalent, and it is sometimes useful to distinguish a *family* of crystal directions. For example, in a cubic crystal the x-axis is defined by the direction [100] (in square brackets), but the y and z directions, [010] and [001], respectively, are, by symmetry, geometrically equivalent. Angular brackets are used as a shorthand for the family of $\langle 100 \rangle$ directions. Of course, in lower-symmetry crystals the two directions [100] and [010] may not be equivalent (that is $a \neq b$ in the unit cell), and these two directions do not then belong to the same family. Note that *all* directions which are *parallel* in the crystal lattice are considered equivalent, regardless of their point of origin, and are denoted by the same *direction indices* $[uvw]$. A negative index is perfectly possible, for example $[\bar{u}vw] \neq [uvw]$, and indicates the u coordinate in the negative direction. Since the direction indices define the *shortest* repeat distance in the lattice along the line of the vector, they cannot possess a common factor. It follows that the direction indices [422] and [330] should be written as [211] and [110], respectively. Finally, the direction indices all have the dimension of *length*, with the unit of length being defined by the dimensions of the unit cell.

Crystal planes are described in terms of the reciprocal of their intercepts with the axes of a coordinate system which is defined by the unit cell (Fig. 1.26). If the unit cell parameters are a, b and c, and the crystal plane makes intercepts at x^*, y^* and

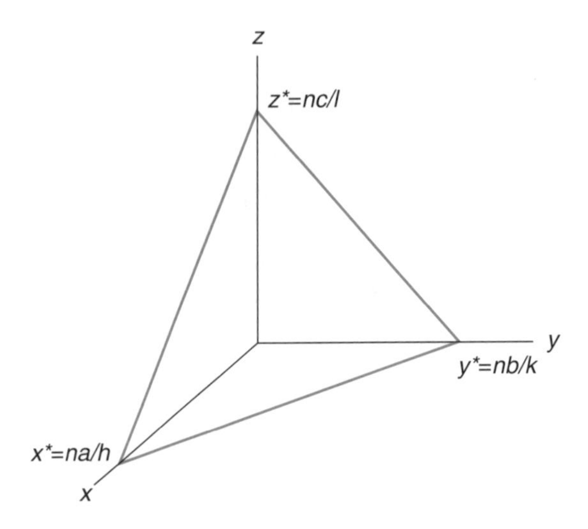

Figure 1.26 The definition of Miller indices which describe the orientation of a crystal plane within the unit cell of the crystal

z^* along the axes defined by the unit cell, then the planar indices, termed the *Miller indices*, are $(na/x^*,\ nb/y^*,\ nc/z^*)$, where the integer n is chosen to clear the indices of fractions. Thus, since a cube plane only intersects one of the axes of a cubic crystal, two of the values x^*, y^* and z^* must be equal to ∞, and the Miller indices are one of the three possibilities (100), (010) and (001) (in round brackets). All planes which are parallel in the lattice are described by the same indices, irrespective of the intercepts that they make with the coordinate axes (although different values of n will be required to clear the fractions), and reversing the sign of all three Miller indices does not define a new plane, so that $(hkl) \equiv (\overline{hkl})$. On the other hand, $(hkl) \neq (\bar{h}kl) \neq (\bar{h}k\bar{l})$, and these indices refer to three crystallographically distinct, non-parallel planes in the lattice. While the letters $[uvw]$ (with square brackets) are used to define a set of direction indices, the letters (hkl) (with round brackets) are used to define a set of lattice planes. If a family of geometrically equivalent (but non-parallel) planes is intended, then this can be indicated by curly brackets, that is $\{hkl\}$. Note that the dimensions of the Miller indices are *inverse lengths*, while the units are the inverse dimensions of the unit cell. Unlike the direction indices, Miller indices having a common factor do have a very specific meaning. In effect they refer to fractional values of the interplanar spacing in the unit cell. Thus the indices (422) are divisible by 2, and correspond to planes which are parallel to, but have half the spacing of the (211) planes.

In *cubic* crystals, and *only* in cubic crystals, any set of direction indices is always *normal* to the crystal planes having the *same* set of Miller indices, so that the [123] direction in a crystal of cubic symmetry is normal to the (123) plane. This is *not* generally true in less symmetric crystals, even though it may be true for *some* directions. If a number of crystallographically distinct crystal planes share a common direction [uvw], then the indices of that shared direction is said to be the *zone axis* of those planes, and the planes are said to lie on a *common zone*. There is a simple way of finding out whether or not a particular plane (hkl) lies on a given zone [uvw]. If it does, then $hu + kv + lw = 0$.

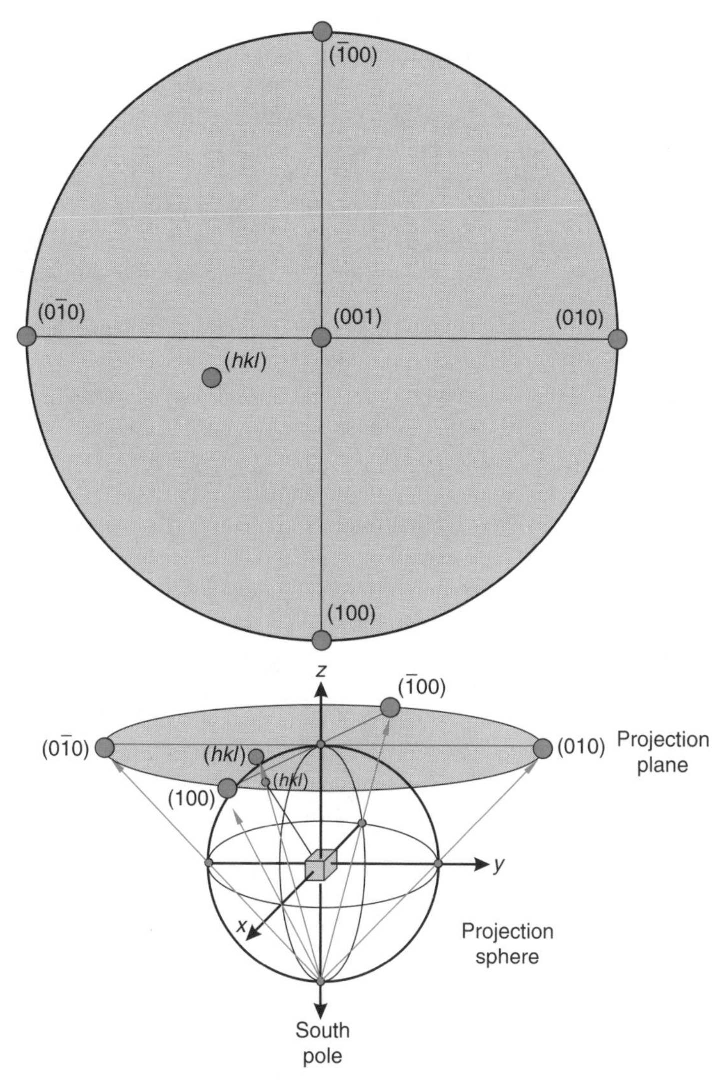

Figure 1.27 Derivation of the stereographic projection, a two-dimensional representation of the angular relationships between crystal planes and directions

1.2.3.4 THE STEREOGRAPHIC PROJECTION

It is a great convenience to be able to plot the prominent crystal planes and directions in a crystal on a two-dimensional projection, similar in principle to the projections used to depict geographical features in map making. By far the most useful of these is the *stereographic projection.*

The crystal is imagined to be at the centre of a sphere and the crystal directions or normals to the prominent crystal planes are then projected from the centre of the sphere to intersect its surface (see Fig. 1.27). Straight lines are then drawn from the south pole of the sphere, through the points of intersection of these crystallographic directions and the crystal plane normals with the sphere surface, to intersect a plane *tangential* to the sphere at the north pole. All points around the equator of the sphere will project on to the tangent plane as a circle with a radius equal to the diameter of the projection sphere. All points on the sphere which lie in the northern hemisphere will project within this circle, while any points lying in the southern hemisphere will project outside the circle. By reversing the direction of projection (from the north pole to a plane tangential to the south pole) we can indicate points lying in the southern hemisphere, using an open circle to distinguish the southern from the

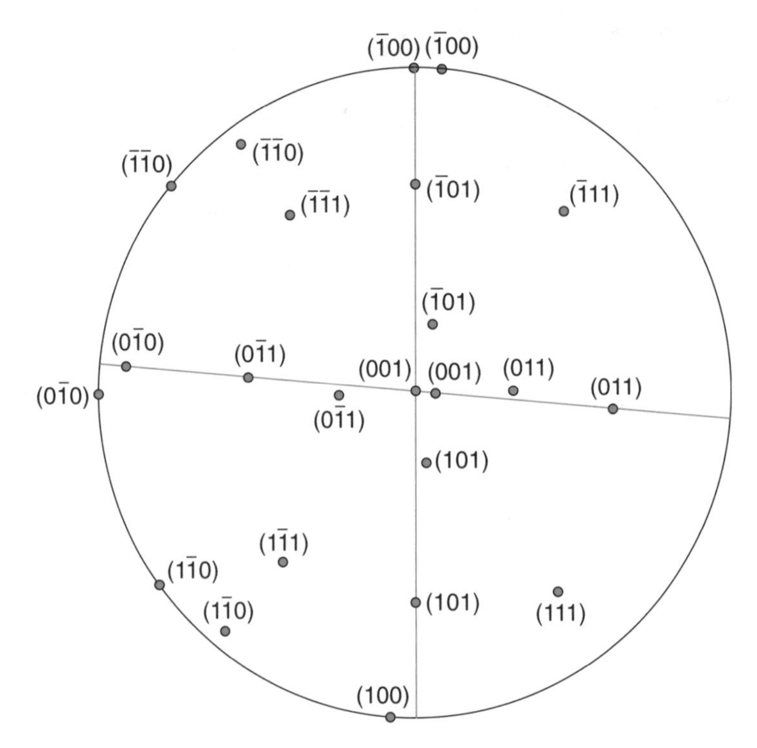

Figure 1.28 A stereographic projection of the lattice of a triclinic crystal. Note that the normals to the faces of the unit cell, denoted by their Miller indices (100), (010) and (001), do not coincide with the axes of the unit cell, the directions [100], [010] and [001]

northern hemisphere points, and thus avoid having to use points outside of the equatorial projection circle.

This two-dimensional plot of the crystal plane normals and crystal directions is the stereographic projection whose radius equals the diameter of the projection sphere. It is conventional to choose a prominent symmetry plane (usually one face of the unit cell) for the plane of the projection. Fig. 1.28 shows the stereographic projection for the *least* symmetrical, *triclinic*, crystal system, with the axes of the unit cell, [100], [010] and [001], and the faces of the unit cell, (100), (010) and (001), plotted on the projection. The (001) plane contains the [100] and [010] directions, while the [001] zone contains the normals to the (100) and (010) planes. However, as noted above, the crystal directions do not coincide with the plane normals having the same indices.

A *stereogram* (stereographic projection) for a cubic crystal, with the plane of the projection parallel to a cube plane, is shown in Fig. 1.29. Plane normals and crystal directions with the same indices coincide, as do the plots of crystal planes and the corresponding zones. The high symmetry of the cubic system divides the stereogram into 24 geometrically equivalent *unit spherical triangles* projected on to the

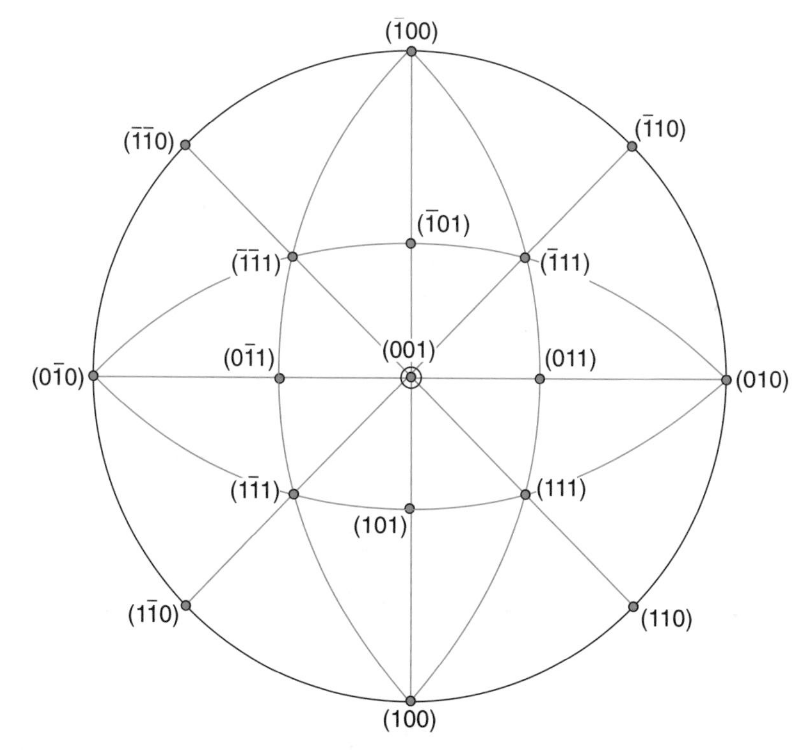

Figure 1.29 A stereographic projection of a cubic crystal with a cube plane parallel to the plane of the projection. The projection consists of 24 unit spherical triangles bounded by great circles, with each triangle containing all of the symmetry elements of the crystal

stereogram from the surface of the projection sphere. (We ignore the southern hemisphere, since reversing the sign of a plane normal does not change the crystal plane.) Each of these unit triangles possesses all of the symmetry elements of the cubic crystal, and is bounded by the traces of one plane from each of the families {100}, {110} and {111}.

The zones defined by coplanar directions and plane normals pass through the centre of the projection sphere (by definition) and intercept this sphere along circles with the diameter of the projection sphere. These circles then project on to the stereogram as traces of larger circles, termed *great circles*, whose *maximum* curvature is equal to that of the stereogram. The *minimum* curvature of a great circle is zero, and corresponds to planes whose trace passes through the centre of the stereographic projection as a straight line. The bounding edges of any unit spherical triangle which defines the symmetry elements are always great circles.

Another property of the stereographic projection is that the cone of those directions which lie at a fixed angle to a given crystallographic direction or crystal

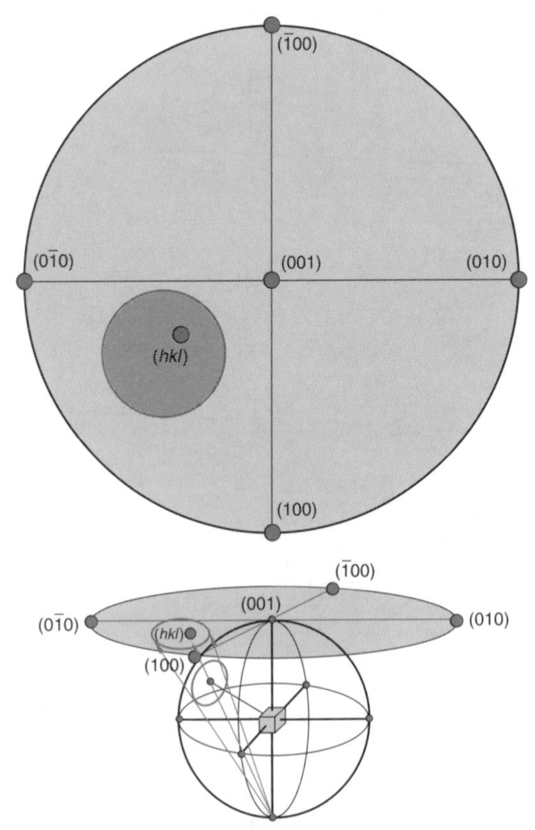

Figure 1.30 A cone defining a constant angle with a direction in the crystal, projects as a small circle on the stereogram

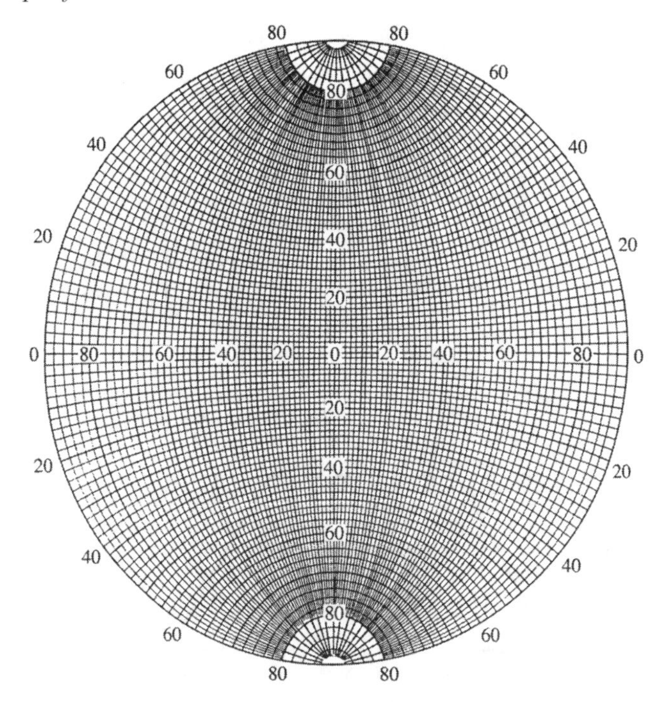

Figure 1.31 The Wulff net gives the angular scale of the stereogram in terms of small circles of latitude and great circles of longitude

plane normal also projects as a circle in the plane of the stereogram, termed a *small circle* (although the axis of the cone does not project to the centre of the circle on the stereogram, see Fig. 1.30). It follows that the angular scale of a stereographic projection is distorted, as can be seen from a standard *Wulff net* (Fig. 1.31), which defines the angular scale in small circles of *latitude* and great circles of *longitude*, identical to those that are commonly used to map the surface of the globe.

Finally, *spherical triangles*, defined by the intersection of great circles, have some geometrical properties which are useful in applied crystallography. The sum of the angles formed at the intersections exceeds 2π, while the sides of the triangle also

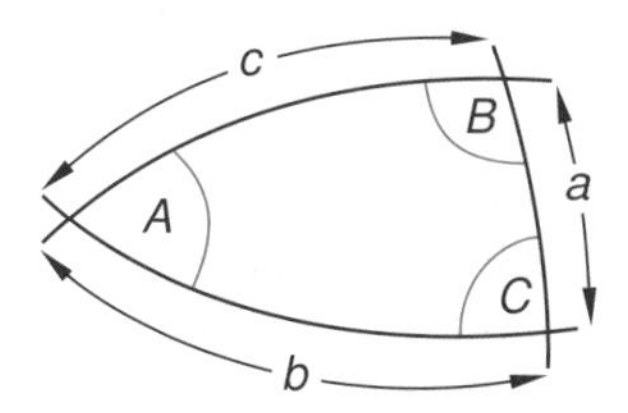

Figure 1.32 All six elements of a spherical triangle represent angles; a simple trigonometrical relationship links all of the latter angles (see text for details)

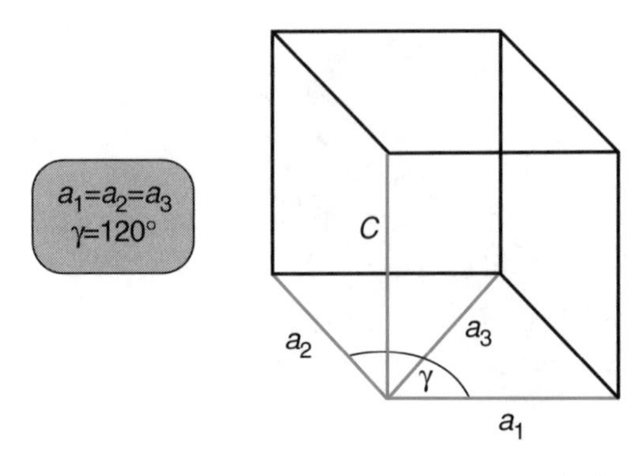

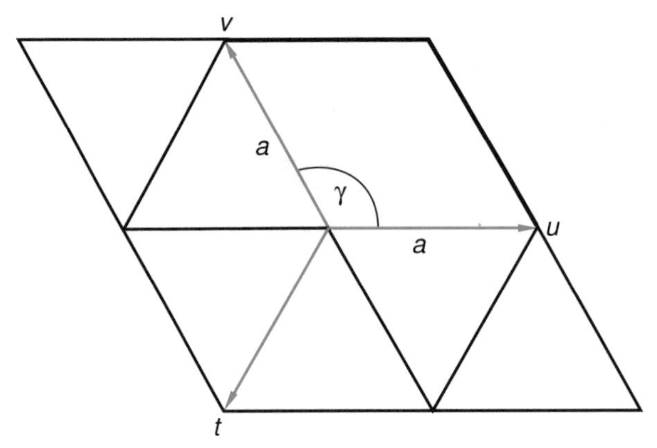

Figure 1.33 The hexagonal unit cell showing the use of a four-axis coordinate system, $x(a_1)$, $y(a_2)$, $u(a_3)$ and $z(c)$

define angles. Great circles that intersect at 90° must each pass through the pole of the other. If more than one of the six elements of a spherical triangle is a right angle, then at least four elements are right angles. If two of the angles are given, then all of the remaining four angles can be derived. If the sides of the spherical triangle are denoted by a, b and c, while the opposing angles are A, B and C, then the relationship $\sin a/\sin A = \sin b/\sin B = \sin c/\sin C$ holds (Fig. 1.32).

Hexagonal (and rhombohedral) crystals present a special problem, since the usual Miller indices and direction indices do not reflect the hexagonal symmetry of the crystal. It is common practice to introduce an additional axis into the basal plane of the hexagonal unit cell. The three axes a_1, a_2 and a_3 are then at angles of 120° to one another, with the c axis mutually perpendicular (see Fig. 1.33). The (redundant) u-axis results in a *fourth* Miller index i when defining a plane, ($hkil$), with the sum $h + k + i = 0$. A stereogram for a hexagonal crystal (for example zinc) is given in Fig. 1.34. The angular distance between the poles lying within the stereogram and its centre, the c-axis [0001], depends on the axial ratio of the unit cell, c/a. Families of crystal planes in a hexagonal lattice have similar indices in the four Miller index

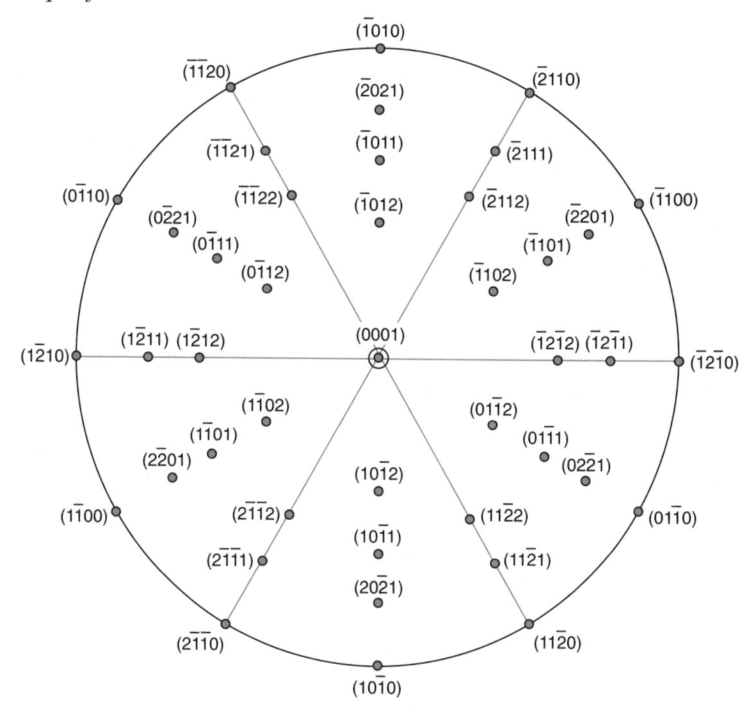

Figure 1.34 A basal plane stereogram for zinc (hexagonal close-packed (HCP) structure), illustrating the use of the four-Miller-index system (*hkil*) for hexagonal crystal symmetry

notation, so that $(10\bar{1}0)$ and $(1\bar{1}00)$ are clearly from the same family, while in the three-index notation, (100) and $(1\bar{1}0)$, this is by no means obvious.

Summary

The term *microstructure* is taken to mean those features of the structure of a material which are revealed by examining a sample, using a suitable *probe*. The term includes the identification of the phases present (crystalline or glassy), the determination of their morphology (the grain or particle sizes and their distribution), and the chemical composition of these phases. Microstructural characterization may be either *qualitative* ('what does the microstructure look like?') or *quantitative* ('what is the grain size?').

The two commonest forms of probe used to characterize microstructure are *electromagnetic radiation* and *energetic electrons*. In the case of electromagnetic radiation, the *optical microscope* and the *X-ray diffractometer* are the two most important tools. The optical microscope uses radiation in the visible range of wavelengths (0.4–0.7 μm) to form an *image* in either reflected or transmitted light. In X-ray diffraction the wavelengths used are of the order of the interatomic and interplanar spacings in crystal structures (0.5–0.05 nm). A wide range of electron energies may be used to probe the microstructure in an *electron microscope*. In *transmission electron microscopy* energies of several hundred kilovolts may be used, while in *scanning electron microscopy* the beam energy may be less than 10 kV.

The interaction of the probe beam with the sample may be either *elastic* or *inelastic*. Elastic interaction involves the scattering of the beam without loss of energy, and is the basis of *diffraction analysis*, using either X-rays or high-energy electrons. Inelastic interactions may result in *contrast* in an image formed from elastically scattered radiation (as when one phase absorbs light while another reflects or transmits). However, inelastic interactions can also be responsible for the generation of a *secondary signal*. In the scanning electron microscope the primary, high-energy beam generates low-energy secondary electrons which are collected to form the image. Inelastic scattering is also the basis of many microanalytical techniques. Both the energy lost by the primary beam and that generated in the secondary signal may be characteristic of the *atomic number* of the chemical elements present in the sample, and the energy dependence of this signal (the energy *spectrum*) provides information on the *chemical composition*.

Many engineering properties of materials are sensitive to the microstructure, which depends on the processing conditions, that is the *microstructure* is determined by the processing route, while the *properties* (and not just the mechanical properties) are determined by the microstructure. This includes not only the microstructural features noted above (grains and particles) but also various *defects* in the microstructure, for example *porosity*, *microcracks* and unwanted *inclusions* (phases associated with contamination).

The ability to distinguish closely spaced features is termed the *resolution* and is determined by the wavelength of the probe radiation, the characteristics of the interaction with the specimen and the nature of the image-forming system. In general, the shorter the wavelength and the wider the acceptance angle of the imaging system, then the better the resolution. Magnifications of the order of 1000 are more than enough to reveal all of the microstructural features accessible to the optical microscope. The wavelength associated with energetic electrons is very much less than the interplanar spacings in crystals, so that the transmission electron microscope is potentially able to resolve the crystal lattice. The resolution of the scanning electron microscope is usually limited by inelastic scattering events occurring under the probe, and is of the order of a few nanometres for secondary electrons, but only of the order of 1 μm for the characteristic X-rays (which are generated much deeper beneath the surface).

Some microscopic methods of materials characterization are capable of resolving individual atoms, in the sense that the image reflects a physical effect associated with individual atoms, but these are beyond the scope of this present text. Many microstructural features may be quantitatively described by *microstructural parameters*. Two important examples are the *volume fraction* of a second phase and the *grain* or *particle size*, both of which have a major effect on the mechanical properties. In many cases, the microstructure within the sample varies, either with respect to direction (*anisotropy*), or with respect to position (*inhomogeneity*).

Crystal structure (or the lack of it, in an amorphous or glassy material) reflects the nature of the chemical bonding, and the four types of bond, *covalent*, *ionic*, *metallic* and *polar* (or van der Waals), combine to give the common classes of engineering materials, namely *metals* and *alloys* (metallic bonding), *ceramics* and *glasses* (covalent and ionic bonding), *polymers* and *plastics* (polar and covalent bonding), and *semiconductors* (primarily covalent bonding).

In a crystal structure, the arrangement of the atoms, ions or molecules is regularly repeated in a spatial array. The smallest unit in this array that contains all the symmetry elements of the bulk crystal is termed the *unit cell* of the crystal lattice, and each cluster of identically arranged atoms in the unit cell can be represented by a single *lattice point*. There are 14 possible ways of arranging the lattice points to give distinctly different lattice symmetries, the 14 *Bravais lattices*.

Characteristic directions in the lattice, corresponding to any particular atomic sequence, can be defined by *direction indices*, while any set of atomic planes can be defined by the normal to these planes, given by *Miller indices*. The direction and Miller indices are conveniently plotted in two dimensions by projecting them on to an imaginary plane. The *stereographic projection* has proved to be the most useful of the geometrically possible projections.

Bibliography

1. P. Villars and L.D. Calvert, *Pearsons's Handbook of Crystallographic Data for Intermetallic Phases*, Vols 1–3, American Society for Metals, Metals Park, OH (1985).
2. T. Hahn (ed.), *International Tables for Crystallography, Volume A: Space-Group Symmetry*, Kluwer Academic, Dordrecht (1992).
3. C. Barrett and T.B. Massalski, *Structure of Metals*, Pergamon, Oxford, (1980).

Worked Examples

We conclude each chapter with examples of characterization for three different material systems, in order to demonstrate the type of information that we can obtain. For this first chapter, we examine the crystallographic structure of the phases we shall encounter in future chapters and examine the basic use of the stereographic projection.

We have seen how the literature data can be used to understand the crystal structure of copper and we now look at two further and slightly more complicated crystal structures which will reappear later. The first is Fe_3C, iron carbide or *cementite*, which exists in equilibrium with α-Fe in most steels (check the Fe–C equilibrium phase diagram). The principle data for Fe_3C, as listed in Pearson's handbook [1], are shown in Table 1.4.

Table 1.4 Data for Fe_3C as presented in Pearson's handbook [1]

Phase	Structure type	Pearson symbol space group	*a, b, c* (*nm*)	α, β, γ (°)	Atoms	Point set	*x*	*y*	*z*	*Occ*
CFe3	CFe3	oP16	0.50890		C	4c	890	250	450	100
		Pnma	0.67433		Fe1	4c	036	250	852	100
			0.45235		Fe2	8d	186	063	328	100

Note that the name of the phase is CFe_3, and not Fe_3C, which is the usual chemical designation for cementite. This is because Pearson's book lists compounds in alphabetical order. We also see that CFe_3 is the structure type for this structure, and that it has a primitive (P) orthorhombic (o) structure with 16 (16) atoms per unit cell. It belongs to the space group Pnma, and has three generating sites, one for carbon, which generates a total of 4 carbon atoms, and two for iron (Fe_1 and Fe_2). The first of these (Fe_1) generates a total of 4 atoms while the second (Fe_2) generates a total of 8 atoms, for an overall total of 12 iron atoms. Since cementite is a stoichiometric phase, the occupation of each site is constant (none or only a limited solid solubility), and the occupation factor for each generating site is 100%. A schematic drawing of the unit cell, containing 16 atoms, is given in Fig. 1.35(a), while four unit cells, using a more conventional representation (showing the atoms belonging to the neighbouring cells but also reflecting the full symmetry of the structure), is shown in Fig. 1.35(b).

The second example that we shall consider is α-Al_2O_3, known as *sapphire* in its single-crystal form and *corundum* in its polycrystalline form, which is the thermodynamically stable form of alumina. Table 1.5 shows the principle data, as given in Pearson's handbook [1], for α-Al_2O_3.

The crystallographic data for α-Al_2O_3 can be quite confusing. We note that the structure is rhombohedral, with 10 atoms per unit cell (hR10), and a space group of $R\bar{3}c$. However, the *lattice parameters* which are given are for a *hexagonal* unit cell ($a = 0.4754$ nm and $c = 1.299$ nm). If you check the *International Tables for Crystallography* [2] under the space group $R\bar{3}c$, you will note that there are *two* ways to describe the unit cell, with the first based on a *rhombohedral* cell with 10 atoms per unit cell, and the second based on a *hexagonal representation*, with 30 atoms per unit cell. The symmetry of the structure does not change when using a hexagonal representation, but it is generally easier to use the hexagonal unit cell, even though not all of the symmetry operations for a hexagonal structure are correct for α-Al_2O_3. The *generating sites* listed in Pearson's book [1] are for the hexagonal unit cell, which is why there are 12 aluminium cations and 18 oxygen anions, thus giving a total of 30 atoms per unit cell. The structure is stoichiometric, so the occupation factors are both 100%. Schematic representations of the rhombohedral and hexagonal unit cells are given in Fig. 1.36. Throughout the rest of this book we will use the *hexagonal* unit cell to describe α-Al_2O_3.

Now let us examine some of the basic uses of the stereographic projection. As described earlier, a stereographic projection is a map which shows us the *angles* between different crystallographic directions and plane normals. Of course, we can

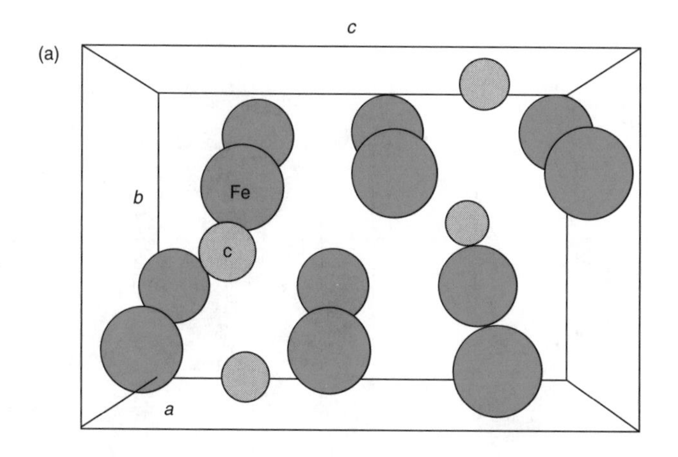

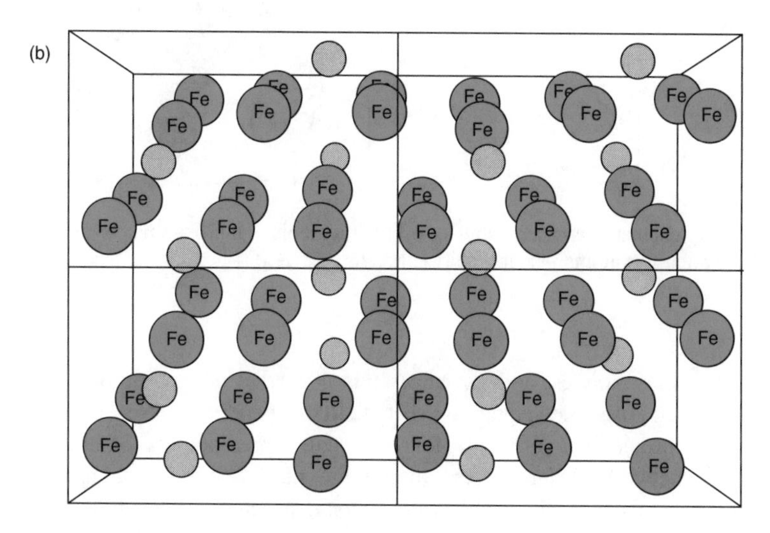

Figure 1.35 Schematic representation of the structure of Fe_3C (cementite) shown for (a) a single unit cell and (b) four unit cells

Table 1.5 Data for α-Al_2O_3 as presented in pearson's handbook [1]

Phase	Structure type	Pearson symbol space group	*a, b, c* (*nm*)	α, β, γ (°)	Atoms	Point set	*x*	*y*	*z*	*Occ*
Al_2O_3	Al_2O_3	hR10	0.4754		Al	12c	0000	0000	3523	100
		$R\bar{3}c$	1.299		O	18c	3064	0000	2500	100

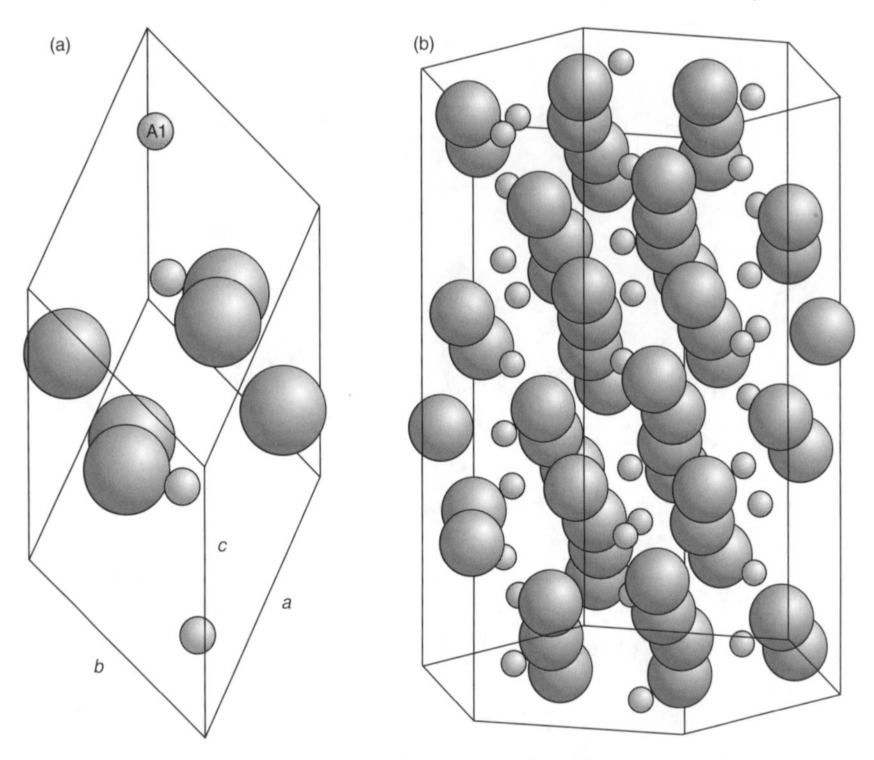

Figure 1.36 Schematic representations of (a) the rhombohedral (10 atoms per unit cell) and (b) the hexagonal (30 atoms per unit cell) unit cells of α-Al_2O_3

also calculate these angles, and nowadays this is carried out by various computer programs which use the equations listed in Appendix I. Nevertheless, the stereographic projection is useful for a visual representation of the angular relationships between different crystallographic planes and directions. We use titanium (Ti) for this example. α-Ti has a *hexagonal* structure, with lattice parameters of $a = 0.29504$ nm and $c = 0.46833$ nm. The stereographic projection (the angular relationships) for *cubic* unit cells does not depend on the lattice parameter, but for all other structures it does (compare the equations at the end of this chapter), so for α-Ti we need the lattice parameters in order to calculate the stereographic projection correctly. Fig. 1.37 shows the stereographic projection for α-Ti, with the (0001) plane at the centre of the projection. (We could in fact, centre the projection on any crystallographic projection or plane normal.) Assume that we are interested in finding which planes lie at an angle of 90° to the basal (0001) plane. We now draw a great circle (using a Wulff net or our computer program), and for this simple case the great circle marks the perimeter of the stereographic projection. We can see that all planes having indices (*hki*0) are 90° from (0001). Simple!

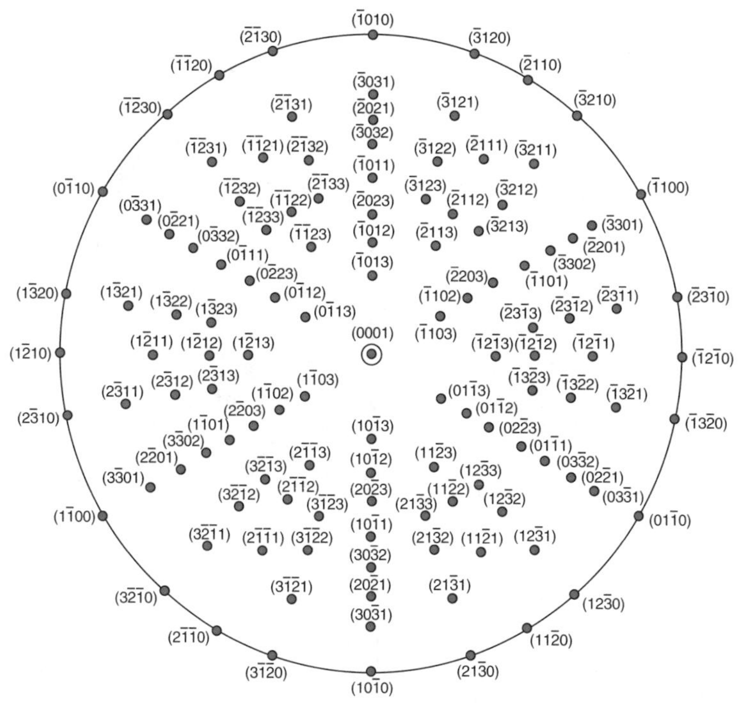

Figure 1.37 The stereographic projection for α-Ti (hexagonal)

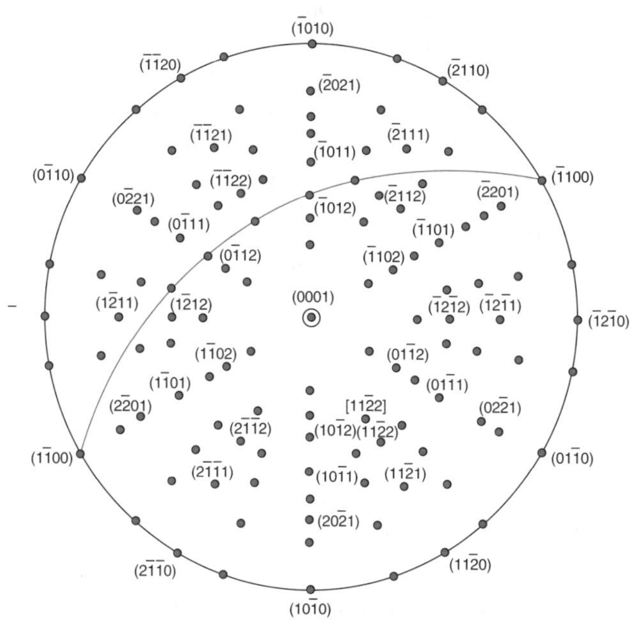

Figure 1.38 The stereographic projection for α-Ti, showing a great circle 90° to the $[11\bar{2}2]$ pole

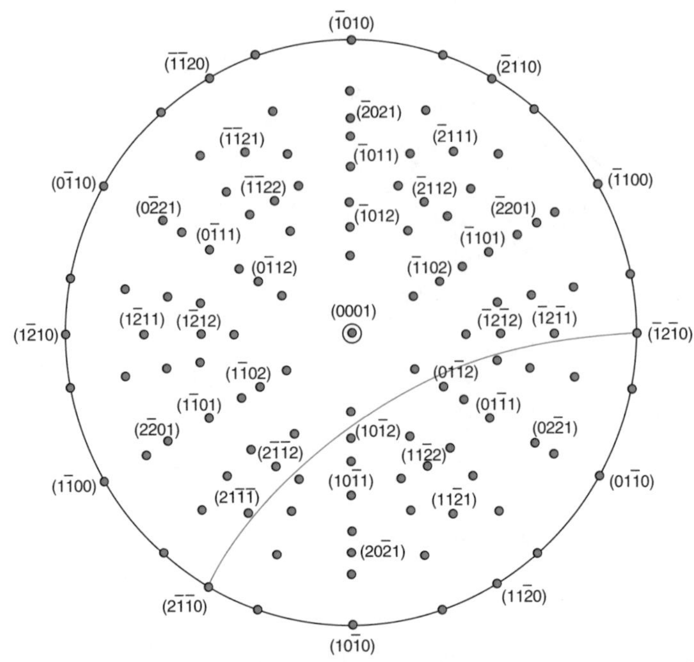

Figure 1.39 The stereographic projection for α-Ti, showing planes at an angle of 60° to the $(11\bar{2}0)$ plane, using a small circle construction

Now let us find the planes which are 90° from the direction $[11\bar{2}2]$. Again, we draw a great circle, this time centred about $[11\bar{2}2]$ (Fig. 1.38). It is important to note that the direction $[11\bar{2}2]$ and the plane $(11\bar{2}2)$ do not coincide on the projection shown in Fig. 1.38, since $[11\bar{2}2]$ is *not* normal to $(11\bar{2}2)$, in the hexagonal lattice. This great circle passes through the poles (normals) of all of the planes which are perpendicular to $[11\bar{2}2]$, and thus contain the $[11\bar{2}2]$ direction.

The stereogram also shows planes whose normals are at an angle of *less* than 90°. An example is given in Fig. 1.39, in which the planes whose poles make an angle of 60° with the normal to the $(11\bar{2}0)$ plane lie on a small circle centred about this plane. (Note that the centre of a small circle on the stereographic projection is not at the geometric centre of the circle.)

Problems

1.1 Give *three* examples of common *microstructural features* in a polycrystalline, polyphase material. In each case, give one example of a physical or mechanical property sensitive to the feature.

1.2 Give *three* examples of small *processing defects* which might be present in a bulk material.

1.3 Distinguish between *elastic* and *inelastic* scattering of a beam of radiation incident on a solid sample.
1.4 What is meant by the term *diffraction spectrum*?
1.5 Give *three* examples of *structure-sensitive* and *three* examples of *structure-insensitive* properties of solids.
1.6 What magnification would be needed to make the following features visible to the eye:
(a) a 1 mm blow-hole in a weld bead;
(b) the 10 μm diameter grains in a copper alloy;
(c) lattice planes separated by 0.15 nm in a ceramic crystal?
1.7 Why is the resolution attainable in the *electron* microscope so much better than that of the *optical* microscope?
1.8 To what extent can we claim to see the 'real' features of microstructure?
1.9 Give *three* examples of *microstructural parameters*, and in *each* case, *one* way in which these parameters are linked quantitatively to material properties.
1.10 Define the terms, *symmetry, crystal lattice* and *lattice point.*
1.11 What features of the crystal lattice are described by the directional indices [*uvw*] and the Miller indices (*hkl*)? (note: be *very* careful and *very* specific!).
1.12 By using literature data, define the *unit cell* and give the *atomic positions* for the following:
(a) Al; (b) α-Ti; (c) α-Fe; (d) TiN.
1.13 Using the crystallographic data for aluminium, which you obtained for the previous question, calculate the *minimum* distance between neighbouring atoms in an aluminium crystal. Compare this value with the diameter of the aluminium atom listed in The Periodic Table. *Which* crystallographic direction (or plane) did you use and *why*?
1.14 Find the total number of equivalent planes in the cubic structure (*multiplicity factor*) belonging to the following families:
(a) {100}; (b) {110}; (c) {111}; (d) {210}; (e) {321}.
1.15 Calculate the distance between the $(\bar{1}012)$ crystal planes in the α-Al_2O_3 structure (their *d*-spacing). Repeat for the $(10\bar{1}2)$ planes. Are these planes crystallographically equivalent?
1.16 Compare the *planar density* (number of atoms per unit area) for the following planes in the *FCC* structure:
(a) {100}; (b) {110}; (c) {111}.
Which of these planes has the highest density?
1.17 Repeat the previous question for the *BCC* structure.

Chapter 2

Diffraction Analysis of Crystal Structure

Radiation striking an object may be scattered or absorbed. When the scattering is entirely *elastic*, no energy is lost in the process, and the wavelength (energy) of the scattered radiation remains unchanged. The regular arrays of atoms in a crystal lattice interact elastically with radiation of sufficiently short radiation, to yield a *diffraction spectrum* in which the radiation is scattered out of the incident beam (Fig. 2.1). Both the diffraction angles and the intensities in the various diffracted beams are a sensitive function of the crystal structure. The diffraction angles depend on the Bravais point lattice and the unit cell dimensions, while the diffracted intensities depend on the atomic numbers of the constituent atoms and their geometrical relationship with respect to the lattice points.

A diffraction pattern or spectrum may be analysed at two levels. A crystalline material may be identified from its diffraction spectrum by comparing the diffraction angles corresponding to the peaks in the spectrum and their relative intensities with a diffraction standard (for example, the JCPDS file,) see worked examples. In this procedure, the diffraction spectrum is treated as a 'fingerprint' of the crystal structure in order to identify the crystalline phases as unambiguously as possible. Alternatively, the diffraction spectrum may be compared with a calculated spectrum, derived from some hypothetical model of the crystal structure. The extent to which the predicted spectrum fits the measured data, the *degree of fit*, then determines the confidence with which the model is judged to represent the crystal structure.

In general, a measured spectrum is first compared with existing data, but if there are serious discrepancies with the known standard spectra then it may be necessary to search for a new model of the crystal lattice in order to explain the results. In recent years, computer procedures have been developed to aid in interpreting crystallographic data, and much of the uncertainty and tedium of earlier procedures has been eliminated.

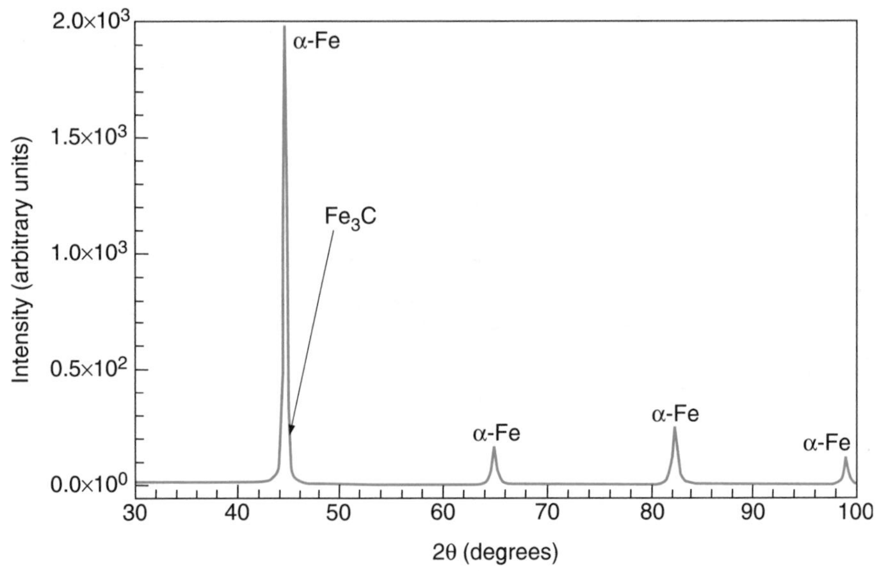

Figure 2.1 The diffraction spectrum from a 0.4%C steel (CuKα radiation, 0.154 nm); most of the peaks are due to body-centred cubic (BCC) α-Fe

2.1 SCATTERING OF RADIATION BY CRYSTALS

The condition for a crystalline material to yield a discrete diffraction pattern is that the wavelength of the radiation should be comparable to, or less than the interatomic spacing in the lattice. In practice, this means that only X-rays, high-energy electrons and neutrons can be used to extract structural information on the crystal lattice. Although suitable sources of neutron radiation are fairly common, few laboratories are equipped with a synchrotron. This present text is therefore limited to a discussion of the elastic scattering of *X-ray* and *electron* beams, although much of the theory is independent of the nature of the radiation.

The specimen dimensions are dictated by the nature of the radiation employed to obtain the diffraction pattern. All materials are highly transparent to *neutrons*, and it is quite common for neutron diffraction specimens to be several centimetres thick. X-rays, on the other hand, especially at the wavelengths normally used (*ca* 0.1 nm), are strongly absorbed by engineering materials and X-ray diffraction data are limited to *submillimetre* surface layers, fine powders or small crystals. *Electron beams* used in transmission electron microscopy may have energies of up to a few hundred kilovolts, and at these energies *inelastic* scattering dominates when the specimen thickness exceeds a tenth of a micron. Electron diffraction data are therefore limited to sub-micron specimen thicknesses. Thus, even though neutrons, X-rays and electrons may be diffracted by the same crystal structure, the information obtained will refer to a very different sample volume, with important implications for the specimen geometry and the procedures used to select and prepare specimens.

2.1.1 The Laue Equations and Bragg's Law

A one-dimensional array of atoms interacting with a parallel beam of radiation of wavelength λ, incident at an angle α_0 will scatter the beam to an angle α and generate a path difference Δ between the incident and scattered beams (Fig. 2.2): $\Delta = (y - x) = a(\cos\alpha - \cos\alpha_0)$, where a is the interatomic spacing. The two beams will be in phase, and hence reinforce each other, if $\Delta = h\lambda$, where h is an integer. Now consider a crystal lattice made up of a *three-dimensional* array of regularly spaced lattice points, which are set at the corners of a primitive unit cell with lattice parameters, a, b, and c. The condition that the scattered (diffracted) beam will be in phase with the incident beam for this three-dimensional array of lattice points is now given by the following set of three equations, known as the *Laue equations*:

$$\Delta = a(\cos\alpha - \cos\alpha_0) = h\lambda \tag{2.1a}$$

$$\Delta = b(\cos\beta - \cos\beta_0) = k\lambda \tag{2.1b}$$

$$\Delta = c(\cos\gamma - \cos\gamma_0) = l\lambda \tag{2.1c}$$

The cosines of the angles α, β and γ, and α_0, β_0 and γ_0, define the directions, respectively, of the incident and the diffracted beams with respect to the *unit cell* of this crystal lattice. The choice of the integers *hkl*, identical to the notation used for *Miller indices*, is, as we shall see below, by no means fortuitous.

A more convenient, and completely equivalent, form of the geometrical relationship determining the angular distribution of the peak intensities in the diffraction spectrum from a regular crystal lattice is the *Bragg equation*

$$n\lambda = 2d\sin\theta \tag{2.2}$$

where n is an integer, λ is the wavelength of the radiation, d is the spacing of the crystal lattice planes responsible for a particular diffracted beam, and θ is the angle

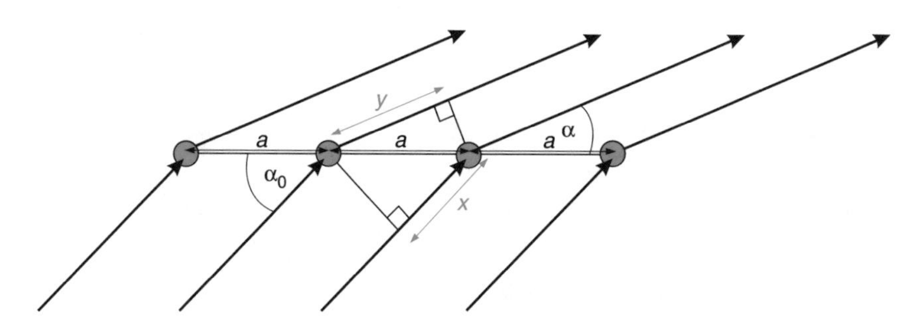

Figure 2.2 When the path difference between the incident and the scattered beams from a row of equidistant point scatterers is equal to an integral number of wavelengths, then the scattered beams are in phase, and the amplitudes scattered by each atom will reinforce each other (see text for details)

the incident beam makes with the lattice planes (Fig. 2.3). The assumption made in deriving the Bragg equation is that the planes of atoms responsible for a diffraction peak behave as a specular mirror, so that the angle of incidence θ is equal to the angle of reflection. The path difference between the incident beam and the beams reflected from two consecutive planes is then $(x-y)$ in Fig. 2.3. The angle between the incident and reflected beams is 2θ, and $y = x\cos 2\theta$. However, $\cos 2\theta = 1 - 2\sin^2\theta$, while $x\sin\theta = d$, the interplanar spacing, so that $(x-y) = 2d\sin\theta$. The distance d between the lattice planes is a function of the Miller indices of the planes and the lattice parameters of the crystal lattice. The general equations (for all Bragg lattices) are as follows:

$$\frac{1}{d^2} = \frac{1}{V^2}(S_{11}h^2 + S_{22}k^2 + S_{33}l^2 + 2S_{12}hk + 2S_{23}kl + 2S_{31}lh) \tag{2.3}$$

where V is the volume of the unit cell:

$$V = abc\sqrt{(1 - \cos^2\alpha - \cos^2\beta - \cos^2\gamma + 2\cos\alpha\cos\beta\cos\gamma)} \tag{2.4}$$

and the constants S_{ij} are given by:

$$S_{11} = b^2c^2\sin^2\alpha \tag{2.5a}$$

$$S_{22} = c^2a^2\sin^2\beta,\ S_{33} = a^2b^2\sin^2\gamma \tag{2.5b}$$

$$S_{12} = abc^2(\cos\alpha\cos\beta - \cos\gamma) \tag{2.5c}$$

$$S_{23} = a^2bc(\cos\beta\cos\gamma - \cos\alpha) \tag{2.5d}$$

$$S_{31} = ab^2c(\cos\gamma\cos\alpha - \cos\beta) \tag{2.5e}$$

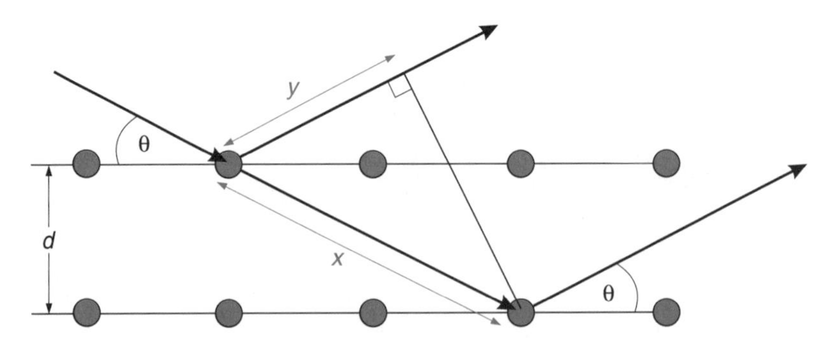

Figure 2.3 If each plane of atoms in the crystal behaves as a mirror, so that the angle of incidence is equal to the angle of reflection, then the condition for the beams reflected from successive planes to be in phase, and hence reinforce each other, is given by Bragg's law (see text for details)

For an orthorhombic lattice, for which $\alpha = \beta = \gamma = 90°$, these equations reduce to

$$\frac{1}{d^2} = \left(\frac{h}{a}\right)^2 + \left(\frac{k}{b}\right)^2 + \left(\frac{l}{c}\right)^2 \tag{2.6}$$

with $V = abc$.

In the Bragg equation, $n\lambda = 2d \sin\theta$, the integer n is referred to as the *order of reflection*. A first-order hkl reflection, ($n = 1$), corresponds to a path difference of a single wavelength between the incident and the reflected beams from the (hkl) planes, while a second-order reflection corresponds to a path difference of two wavelengths. However, from the Bragg equation, this path difference of two wavelengths for the second-order d_{hkl} reflection is exactly equivalent to a single-wavelength path difference from planes of atoms at one half of the d_{hkl} spacing, corresponding to planes with Miller indices ($2h\,2k\,2l$). It is therefore common practice to label the nth-order reflection as coming from planes having a spacing d_{hkl}/n, and having Miller indices ($nh\,nk\,nl$). The Bragg equation is then written as $\lambda = 2d_{hkl} \sin\theta$, where the subscript hkl is now understood to refer to a *specific order* n of the hkl reflection. For example, d_{110}, d_{220} and d_{440} would be the first-, second- and fourth-order reflecting planes for the 110 reflections.

2.1.2 Allowed and Forbidden Reflections

Body-centred (BC) or face-centred (FC) Bravais lattices have planes of lattice points that give rise to destructive (out-of-phase) interference for some orders of reflection. In the body-centred cubic (BCC) lattice (Fig. 2.4), the lattice point at 1/2 1/2 1/2 scatters in phase for all orders of the 110 reflections, but will give rise to destructive interference for *odd* orders of the 100 reflections, that is 100, 300, 500, ..., etc. It is instructive to list the allowed reflections for primitive, body-centred and face-centred cubic Bravais lattices as a function of the integer $(h^2 + k^2 + l^2)$ (see Table 2.1). For cubic symmetry, Bragg's law reduces to $\sqrt{(h^2 + k^2 + l^2)} = 2a \sin\theta/\lambda$, leading to a regular array of diffracted beams. As can be seen, some values of $h^2 + k^2 + l^2$ are always absent, with 7 being the first of these. Other integers may correspond to more than one reflection, and both the 221 and the 300 planes, $h^2 + k^2 + l^2 = 9$, will diffract at the same angle θ.

Those reflections which are disallowed for a particular lattice are referred to as *forbidden reflections*, and there are simple rules to determine which reflections are forbidden. In the FCC lattice, the Miller indices must be either all odd or all even for a reflection to be allowed, and it is the reflecting planes with mixed odd and even indices that are forbidden. In the BCC lattice, the sum $h + k + l$ must be even for an allowed reflection, and if the sum of the Miller indices is odd, then the reflection is forbidden. Occasionally the sequence of the diffraction peaks may be recognized immediately as due to a specific Bravais lattice. For example, the two sets of paired reflections (111 and 200, and then 311 and 222) are characteristic of FCC symmetry.

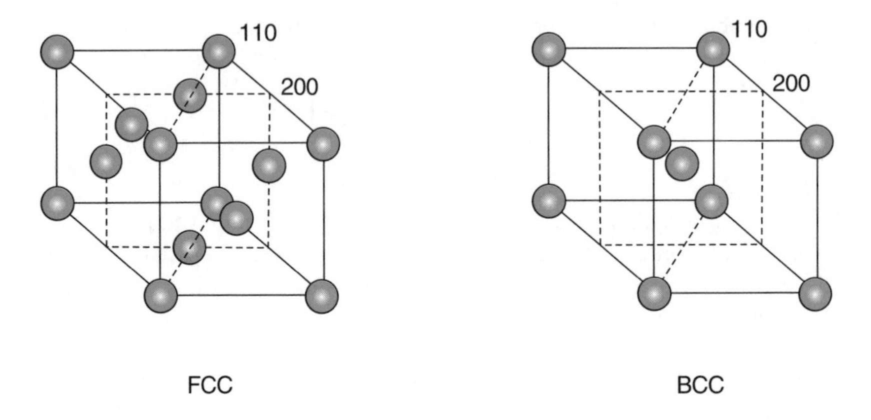

Figure 2.4 The FCC and BCC Bravais lattices contain additional planes of lattice points which lead to some forbidden reflections characteristic of the crystal structure; the darker lattice points are to the front of the unit cells

More often, careful measurements and calculations are needed to identify the crystal symmetry responsible for a given diffraction pattern.

In most crystals the crystal lattice points correspond to *groups* of atoms, rather than individual atoms, while the different atomic species scatter more or less strongly, depending on the atomic number (the number of electrons attached to the atom). In the FCC NaCl structure (Fig. 2.5), the sodium cations are located at the lattice points, while the chlorine anions are displaced from the lattice points by the constant lattice vector 1/2 0 0. Two types of lattice plane exist, those containing both

Table 2.1 Allowed and forbidden reflections in crystals of cubic symmetry

$h^2 + k^2 + l^2$	Primitive cubic	Face-centred cubic	Body-centred cubic
1	100	—	—
2	110	—	110
3	111	111	—
4	200	200	200
5	210	—	—
6	211	—	211
7	—	—	—
8	220	220	220
9	221/300	—	—
10	310	—	310
11	311	311	—
12	222	222	222
13	320	—	—
14	321	—	321
15	—	—	—
16	400	400	400

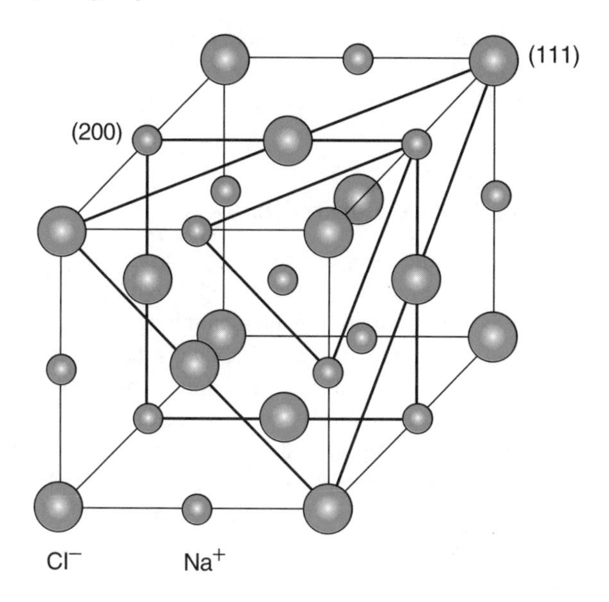

Figure 2.5 In the NaCl structure, each lattice point corresponds to one cation and one anion. These different ions are coplanar for the 200 planes, and therefore scatter in-phase, thus enhancing the 200 diffraction peak. In contrast, the 111 planes of cations and anions are interleaved, and lead to interference, thus reducing the intensity of the 111 diffraction peak. The larger anions are distinguished from the smaller cations

cations and anions (*mixed* planes), such as {200} and {220}, and those consisting of an equi-spaced, alternating sequence of pure anion and cation planes, such as {111} and {311}. Since the cations and anions in the mixed planes are coplanar, the two species always scatter *in-phase*, and the intensities of the diffraction peaks are enhanced by the additional scattering of the second atomic species. On the other hand, the alternating planes of cations and anions ions scatter *out-of-phase* for all *odd-order* reflecting planes, and hence reduce the diffracted intensity, while the same alternating cation and anion planes will scatter *in-phase* for all *even-order* reflections, thus enhancing the scattered intensity. The extent to which the second atomic species will enhance or reduce the diffracted intensity will depend on the difference in scattering power associated with the difference in atomic number of the two atomic species.

2.2 RECIPROCAL SPACE

Bragg's law indicates that the angles of diffraction are *inversely* proportional to the spacing of the reflecting planes in the crystal lattice. In order to analyse a diffraction pattern it is therefore helpful to establish a three-dimensional coordinate system in which the axes have the dimensions of *inverse length* (nm^{-1}). Such a system of coordinates is referred to as *reciprocal space*.

2.2.1 The Limiting-Sphere Construction

The value of $\sin\theta$ is constrained to lie between ± 1, so that, from Bragg's law, the values of $1/d$ *must* fall in the range between 0 and $2/\lambda$ if they are to give rise to a diffracted beam. If the beam of radiation is incident along the x-axis, and the crystal is placed at the origin of the coordinate system, then a sphere of radius $2/\lambda$ encloses the allowed values of $1/d$ in reciprocal space and defines planes which can diffract a wavelength λ. Now imagine a smaller sphere of radius $1/\lambda$ within this limiting sphere, placed so that it touches the limiting sphere on the x-axis, and hence also touches the position of the crystal (Fig. 2.6). A line passing through the centre of this second, reflecting sphere and parallel to the diffracted beam (which makes an angle 2θ with the x-axis) intersects the periphery of the reflecting sphere at the point P.

2.2.2 Vector Representation of Bragg's Law

The two vectors $\mathbf{k}_0$ and $\mathbf{k}$ in Fig. 2.6 define the wave vectors of the *incident* and *diffracted* beams, $|\mathbf{k}_0| = |\mathbf{k}| = 1/\lambda$, and if the vector OP is equal to $\mathbf{g}$, then $|\mathbf{g}| = 1/d$, and Bragg's law, $\lambda = 2d\sin\theta$, can be written as the vector equation, $\mathbf{k}_0 + \mathbf{g} = \mathbf{k}$. Note that the vector $\mathbf{g}$ is *perpendicular* to the diffracting planes, while the wave vectors $\mathbf{k}_0$ and $\mathbf{k}$ are *parallel* to the incident and diffracted beams. As we shall see later, this vector form of the Bragg equation can be very useful indeed.

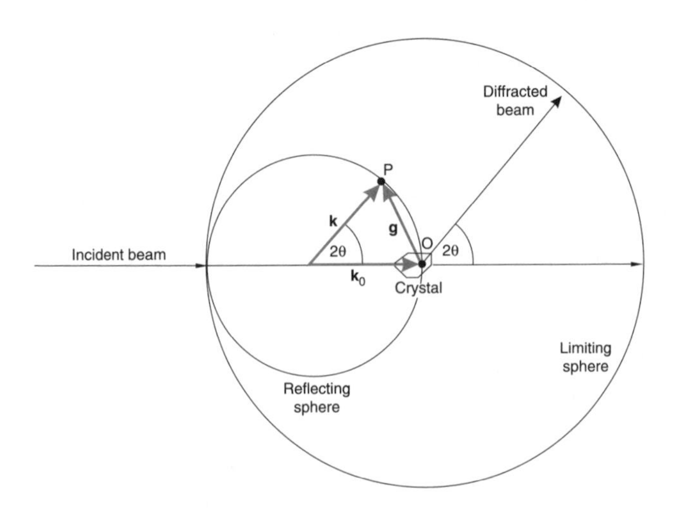

Figure 2.6 The limiting sphere and reflecting sphere constructions (see text for details)

2.2.3 The Reciprocal Lattice

We can now define a *lattice*, in reciprocal space, in some ways analogous to the Bravais lattice in real space (Fig. 2.7). The origin of coordinates in reciprocal space is the point (000) and lies at the position of the crystal in the centre of the limiting-sphere construction. Any reciprocal lattice vector $\mathbf{g}_{hkl}$, drawn from this origin to the point (*hkl*), will be normal to the reflecting planes having Miller indices (*hkl*). The distance from 000 to *hkl*, $|\mathbf{g}_{hkl}|$, is equal to $1/d_{hkl}$, so that the successive orders of reflection, 1 to n, are represented by a series of *equidistant* points along a straight line in reciprocal space which originates at (000). Note that *negative* values of n are perfectly legitimate in this representation.

The condition for Bragg's law to be obeyed is that the reciprocal lattice vector $\mathbf{g}$ should lie on the reflecting sphere. This can be achieved by varying either λ or θ, as we shall see later.

2.2.3.1 The Reciprocal Unit Cell

The dimensions of the *reciprocal lattice unit cell* can be defined in terms of the corresponding Bravais lattice unit cell. The general equations are as follows:

$$a^* = \frac{bc}{V}\sin\alpha; \; b^* = \frac{ca}{V}\sin\beta; \; c^* = \frac{ab}{V}\sin\gamma \tag{2.7}$$

where V is the volume of the unit cell. It is important to note that the axes a^*, b^* and c^* of the unit cell in reciprocal space are, in general, not parallel to the axes of the unit cell a, b and c in real space. This is consistent with the previous treatment

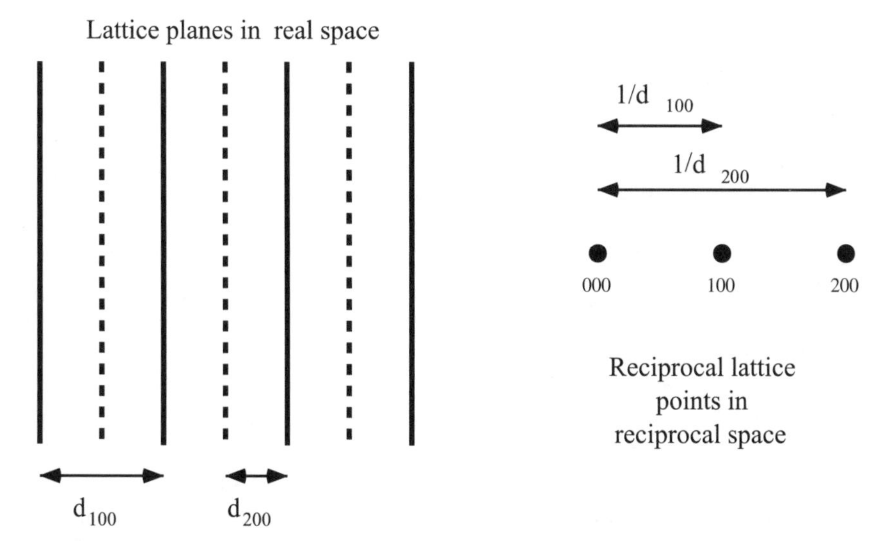

Figure 2.7 The definition of reciprocal lattice points in terms of the lattice planes of a crystal defined by its Miller indices; this example is for the cube planes (*h*00)

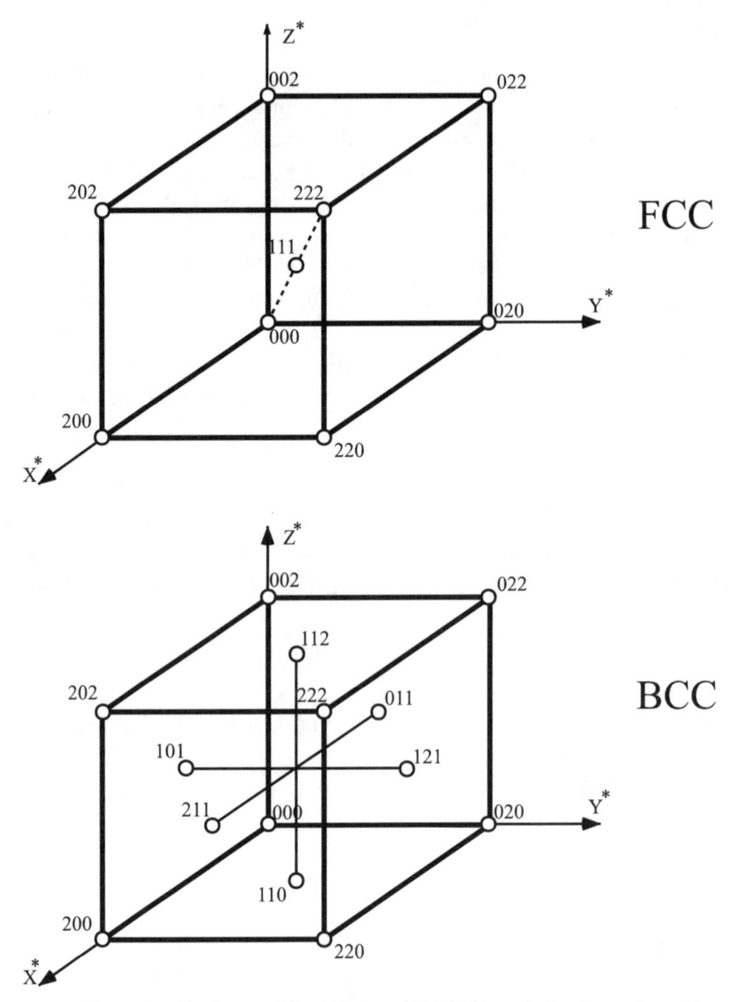

Figure 2.8 Allowed reflections of the FCC and BCC Bravais lattices plotted in reciprocal space; the FCC reciprocal lattice has body-centred cubic symmetry, while the BCC reciprocal lattice has face-centred cubic symmetry

of direction indices and Miller indices. The axes of the unit cell in real space correspond to the direction indices [100], [010] and [001], while the axes of the reciprocal unit cell correspond to the Miller indices (100), (010) and (001).

It should now begin to be clear just why it is useful to define Miller indices in terms of reciprocal lengths, and a comparison of the reciprocal lattices of the FCC and BCC Bravais lattices may help to clarify this even more. Fig. 2.8 compares the two reciprocal lattices. In this figure, all forbidden reflections have been excluded. Both reciprocal lattice unit cell dimensions are defined by **g** vectors of type 200. The FCC reciprocal lattice unit cell contains a *body-centred*-allowed reciprocal lattice point at 111. This reciprocal lattice cell is therefore *body-centred cubic*, while the

BCC reciprocal lattice unit cell contains allowed *face-centred* reciprocal lattice points of the type 110, and is therefore *face-centred cubic*. Reciprocal lattice points in any direction can be derived by simple vector addition and subtraction. The series, 020, 121 and 222, thus constitute a sequence of reflections along a line in BCC reciprocal space which are each separated by the reciprocal lattice vector 101, while the sequence, 200, 111 and 022, in FCC reciprocal space are separated by the vector $\bar{1}11$. Any two *non-parallel* reflections $\mathbf{g}_1$ and $\mathbf{g}_2$ define a plane in reciprocal space. Their common zone $\mathbf{n}$, and another reflection also lying in this zone, $\mathbf{g}_3$, must obey the rules of vector geometry, $\mathbf{g}_1 \times \mathbf{g}_2 = \mathbf{n}$, and $\mathbf{n} \cdot \mathbf{g}_3 = 0$.

2.3 X-RAY DIFFRACTION METHODS

We should now take another look at the *limiting sphere* and *reflecting sphere* construction, but this time with the *reciprocal lattice* superimposed on this construction (Fig. 2.9). The following three factors determine whether or not a particular crystal plane will give a diffraction peak:

(a) *The wavelength of the X-rays in the incident beam*. Reducing the wavelength of the X-rays increases the diameter of the limiting sphere, and therefore brings longer reciprocal lattice vectors (smaller interplanar spacings) within the limiting sphere.
(b) *The angle of the incident beam with respect to the crystal*. As the crystal is rotated about its centre (or, equivalently, as the X-ray beam is rotated about the crystal), the reflecting sphere sweeps through the reciprocal lattice points, thus allowing different diffracting planes to obey the Bragg law in turn. By rotating

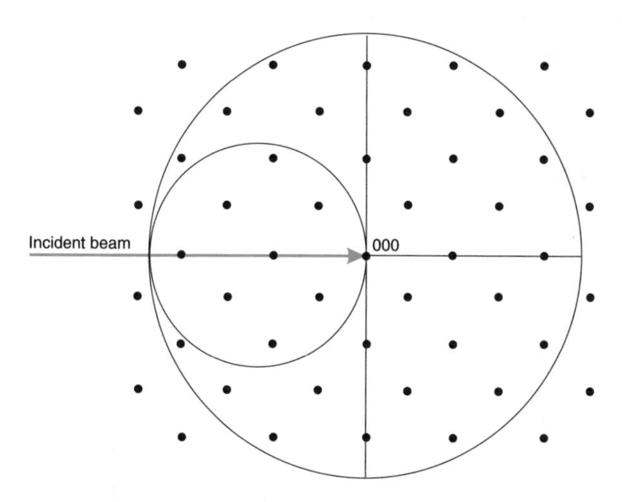

Figure 2.9 Superimposing the reciprocal lattice on the reflecting sphere construction demonstrates the effects of certain experimental variables on diffraction

about two axes at right angles, all of the reciprocal lattice points lying within the limiting sphere can be made to diffract, each at the appropriate Bragg angle, θ_{hkl}.

(c) *The effective size of the reciprocal lattice points*. If the diffraction condition had to be obeyed exactly, then diffraction would only be observed at the precise Bragg angle. In practice, the reciprocal lattice points have a finite size determined by the size and perfection of the crystal. In addition, inelastic absorption of the incident beam limits the pathlength for elastic scattering, and hence limits the crystal dimensions that can contribute to coherent diffraction.

Absorption effects are determined by the value of the *absorption coefficient*. Typical absorption coefficients for X-rays result in pathlengths of the order of tens of microns before coherent elastic scattering phenomena become blurred by inelastic scattering events. Since interplanar spacings are usually in the range of tenths of a nanometre, it follows that, for large, otherwise perfect crystals, the effective diameter of a reciprocal lattice point (the uncertainty in the value of **g**) is of the order of $10^{-5}|\mathbf{g}|$. As a consequence of this, the lattice parameters of crystalline phases can be measured to an accuracy of the order of 10 ppm, which is quite sufficient to allow for accurate determination of changes in crystal dimensions due to *temperature* (thermal expansion), *alloying additions* or *applied stress* (in particular, residual stresses associated with processing and assembly). For small crystallites and heavily deformed crystals it is possible to measure the range of θ over which diffraction from a particular *hkl* plane is observed, and to derive quantitative information on the *crystal size* or degree of perfection.

So far we have assumed that the incident beam is *monochromatic* and accurately parallel. Again, the effect of these assumptions is best understood from the reflecting sphere construction (Fig. 2.10). If there is a spread of wavelengths in the incident beam, then the limiting sphere becomes a shell, and the reflecting sphere generates a 'new moon' crescent, within which reciprocal lattice points satisfy the Bragg law. If the incident beam is not strictly parallel, then the reflecting sphere is rotated about the centre of the limiting sphere, by an angle equal to the divergence or, equivalently, convergence angle of the incident beam, thus generating two crescent volumes within which the Bragg law is satisfied. Both of these effects introduce errors into the determination of lattice spacing by X-ray diffraction which depend on the value of $|\mathbf{g}|$ and the angle between **g** and the incident beam.

2.3.1 The X-Ray Diffractometer

An X-ray diffractometer comprises a source of X-rays, the X-ray generator, a diffractometer assembly, and X-ray data collection and analysis systems. The diffractometer assembly controls the alignment of the beam, as well as the position and orientation of both the specimen and the X-ray detector.

The X-rays are generated by accelerating a beam of electrons on to a pure metal target contained in a vacuum tube. The high-energy electrons eject ground-state electrons from the atoms of the target material, creating holes, and X-rays are

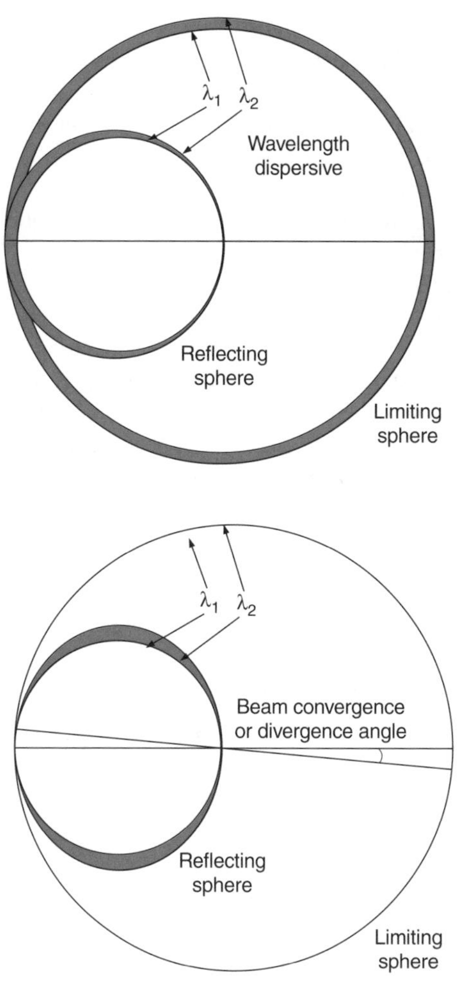

Figure 2.10 The effect of variations in X-ray wavelength or inadequate collimation of the beam can be readily understood from the reflecting sphere construction

emitted during the refilling of these ground states. If all of the electron energy, eV, were to be converted into an X-ray quantum, then the frequency v would be given by the quantum relationship, $eV = hv$, where h is the Planck constant. The X-ray wavelength λ is proportional to the reciprocal of this frequency, so that $\lambda = c/v$, where c is the velocity of light in the medium through which the X-rays propagate. The condition that all of the energy of the exciting electron is used to create a photon sets an upper limit on the frequency of the X-rays generated, and hence a lower limit on the X-ray wavelength. The above relationships lead to an inverse dependence of this minimum wavelength on the accelerating voltage of the X-ray tube, which is given (in vacuum) by $\lambda_{\text{min}} = 1.243/V$, where λ is in nanometres and V is in kilovolts.

Above this minimum wavelength there is a continuous spread of X-ray wavelengths generated by the incident electron beam, whose intensity increases with both incident electron energy and beam current, as well as with the atomic number of the target (that is the density of electrons in the target material). This continuous distribution of photon energies and wavelengths in the X-rays emitted from the target is referred to as *white radiation*.

Superimposed on the continuous spectrum of white radiation are a series of very narrow and intense peaks, the *characteristic radiation* (Fig. 2.11). A characteristic peak corresponds to the energy released when the hole in an inner electron shell, created by a collision event, is filled by an electron which originates in a higher energy shell of the same atom. Thus removal of an electron from the K-shell excites the atom to an energy state E_K, and if the hole in the K-shell is then filled by an electron from the L-shell, then the energy of the atom will decay to E_L, while the decrease in the energy, $(E_K - E_L)$, will appear as an X-ray photon of fixed wavelength which contributes to the K_α line of the *characteristic* target spectrum

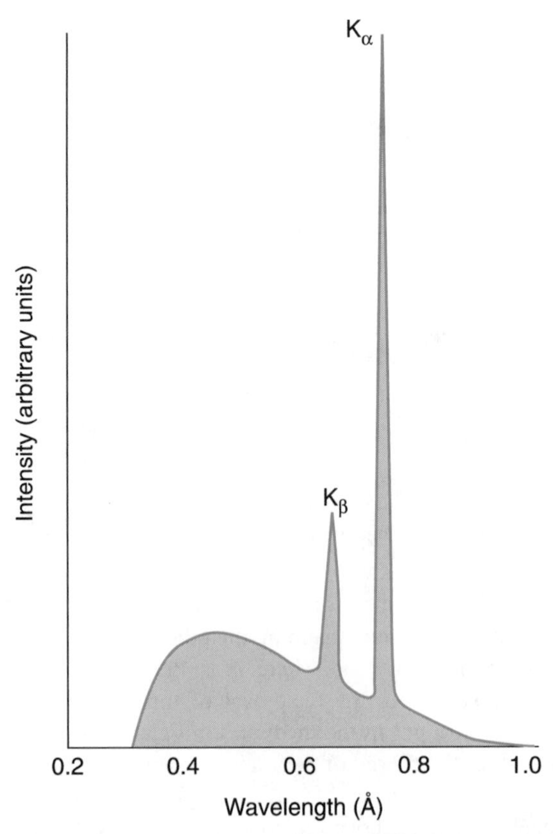

Figure 2.11 An energetic electron beam striking a solid target generates a continuous spectrum of white X-radiation with a sharp cut-off at a minimum wavelength, corresponding to the incident electron energy, together with a discontinuous set of narrow intensity peaks, the characteristic X-radiation

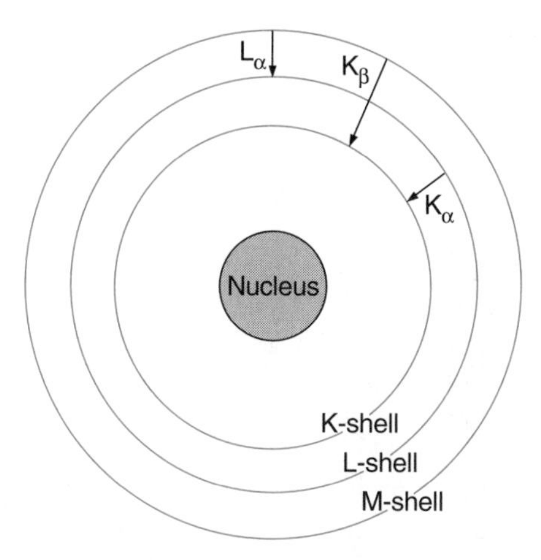

Figure 2.12 Characteristic X-radiation is generated by electron transitions involving the inner shells, with the wavelengths being specific to the atomic species present in the target material

(Fig. 2.12). Filling the hole in the K-shell with an electron from the M-shell reduces the energy state of the atom to E_M, thus leading to a K_β photon, and a *second* line in the K-shell spectrum. Since the residual energy of the atom is lower in the E_M state than it is in the E_L state, this photon has a higher energy, so that the wavelength of the K_β-line in the characteristic spectrum is less than that of the K_α-line.

Further decay of the energy of the excited atom from the E_L and E_M states will result in the generation of L and M characteristic radiation of much longer

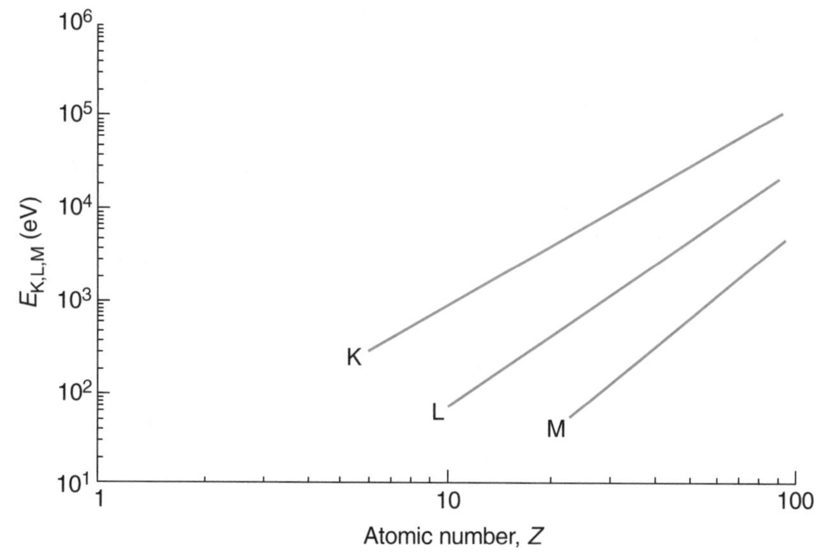

Figure 2.13 Illustration of how the excitation energy required to eject an electron from an inner shell increases with atomic number

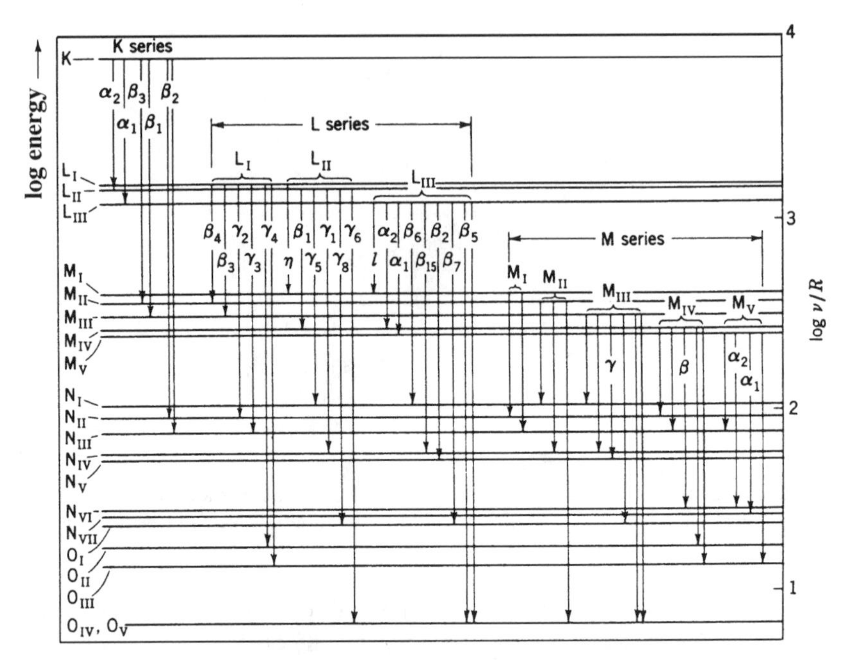

Figure 2.14 The atomic energy levels and characteristic X-ray spectrum for a uranium atom. After Barratt and Massalski, *Structure of Metals*, 3rd revised edition, with permission from Pergamon Press

wavelength, and there are many alternative options for the origin of a donor electron to fill a hole in the L- or M-shells, so that the characteristic L and M spectra consist of several closely spaced lines. Clearly, a characteristic line can only be generated in the target by the incident beam if the electron energy exceeds the excitation energy for that line. The excitation energy increases with the atomic number of the target material (Fig. 2.13), since the electrons in any given shell are more tightly bound to a higher atomic-number nucleus. The low-atomic-number elements in the first row of the Periodic Table only contain electrons in the K-shell, and hence can *only* give K-lines, while only the heaviest elements (of high atomic number) have M- and N-lines in their spectra, which can then be very complex (Fig. 2.14).

If elastic scattering of the X-rays is to dominate their interaction with a sample, then we need to ensure that intensity losses due to inelastic scattering processes are minimized. A monochromatic X-ray beam traversing a thin sample in the x direction loses intensity I at a rate given by $\mathrm{d}I/\mathrm{d}x = -\mu I$, where μ is the linear absorption coefficient for the X-rays. It is the *mass* traversed by the beam, rather than the *sample thickness*, which is important, so that the values tabulated in the literature are generally for the *mass absorption coefficient* μ/ρ, where ρ is the density, rather than the linear absorption coefficient. The transmitted intensity is then given by the following:

$$I/I_0 = \exp\left(-\frac{\mu}{\rho}\rho x\right) \tag{2.8}$$

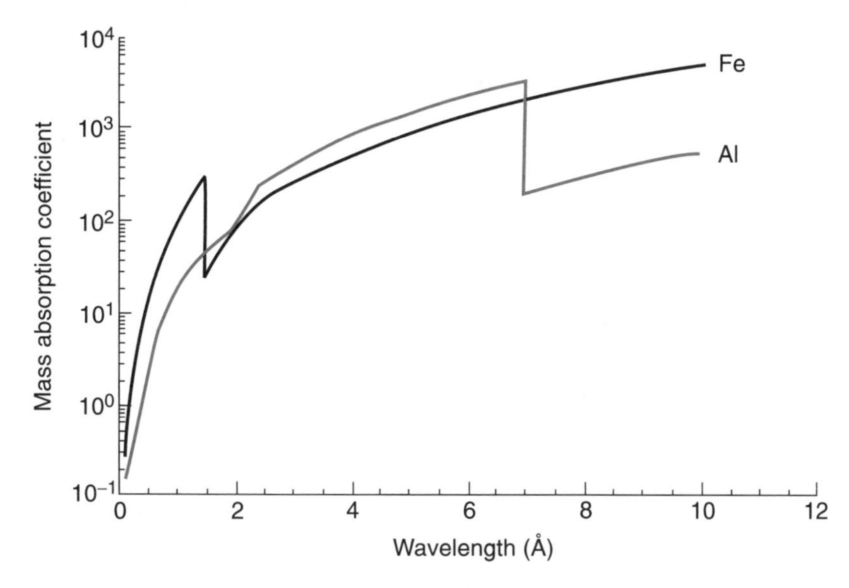

Figure 2.15 The expected dependences of the X-ray mass absorption coefficient on wavelength shown for iron and aluminium

Two plots of the mass absorption coefficient as a function of the X-ray wavelength are shown in Fig. 2.15, for the case of a constructional steel and for an aluminium alloy. The lower-density aluminium alloy has the lower linear absorption coefficient at any given wavelength. All of the materials show a general increase in the mass absorption coefficient with wavelength, but with a sequence of step discontinuities. These are referred to as *absorption edges*, and correspond to the wavelengths at which the incident X-ray photon possesses sufficient energy to eject an inner-shell electron in the specimen, similar to the ejection of an electron by an energetic incident electron. It follows that the absorption edges are the X-ray equivalents of the minimum *excitation energies*, which are involved in the generation of characteristic X-rays, as discussed above. In a similar way to electron excitation, short-wavelength, high-energy X-rays can generate secondary characteristic X-rays of longer wavelength in the specimen target, a process termed *X-ray fluorescence*; however, in X-ray fluorescent excitation, there is no background, white radiation.

In order to avoid fluorescent radiation and minimize absorption of the incident beam, it is important to select radiation for X-ray diffraction measurements that has a wavelength close to the minimum, but on the long-wavelength side of the absorption edge in the specimen. Thus CuK_α radiation ($\lambda = 0.154\,\text{nm}$) is less than ideal for steels and other iron alloys ($E_{\text{FeK}} = 7.109\,\text{keV} \equiv 0.17433\,\text{nm}$). On the other hand, CoK_α radiation ($\lambda = 0.1789\,\text{nm}$) lies just to the long-wavelength side of the K_{Fe}-edge and will give sharp diffraction patterns from steel, which are free of background fluorescence.

Assuming that CoK_α radiation is to be used, then the values of μ/ρ for iron and aluminium are 46 and 67.8, respectively. Inserting the respective densities of the two

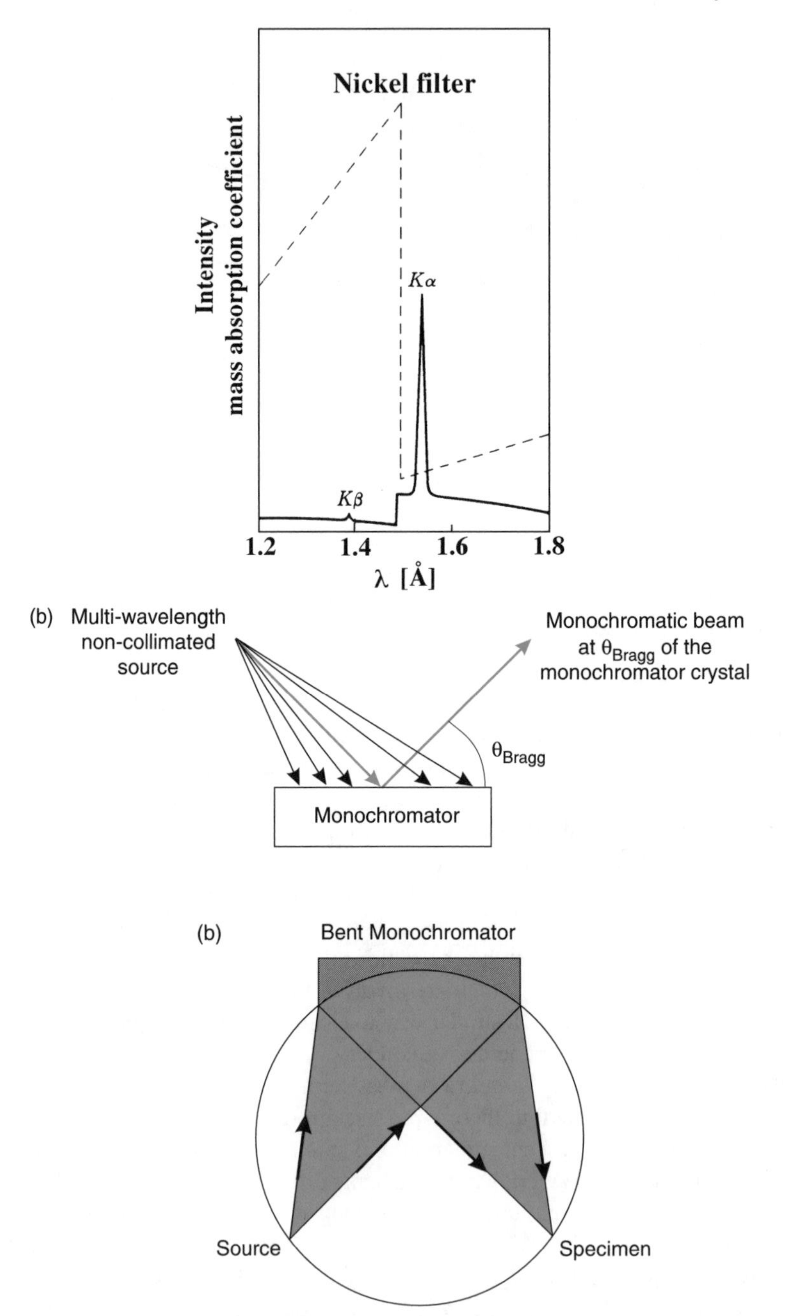

Figure 2.16 Some important spectrometer features: (a) CuK radiation filtered by a nickel foil to remove K_β; (b) schematic representation of a monochromator crystal which allows a specific wavelength to be selected from an X-ray source; (c) schematic representation of a fully focusing spectrometer which maximizes the diffracted intensity collected at the detector

metals, 7.88 and 2.70 g ml^{-3}, we can then derive the thickness of the sample which will reduce the intensity of the incident beam to 1/e of its initial intensity, namely 27.6 μm for iron and 54.6 μm for aluminum. These values effectively determine the thickness of the sample which provides the reflection diffraction signal in each metal when using CoK_α radiation.

X-ray diffraction experiments require either monochromatic or white radiation. Monochromatic radiation is generated by exciting K-radiation from a pure metal target and then filtering the beam by interposing a foil which strongly absorbs the β-component of the K-radiation without any appreciable reduction of the intensity of the α-component. This can be accomplished by choosing a filter which has an absorption edge that falls exactly between the K_α and K_β wavelengths. A good example is the use of a nickel filter ($E_{NiK} \equiv 0.1488$ nm) with a copper target ($E_{CuK} \equiv 0.138$ nm), which transmits the CuK_α beam (0.154 nm), but not the K_β (Fig. 2.16(a)).

More complete selection of a monochromatic beam can be achieved by interposing a single crystal oriented to diffract at the characteristic K_α peak. This monochromatic diffracted beam can then be used as the source of radiation for the actual sample (Fig. 2.16(b)). Finally, the monochromator crystal can be bent into an

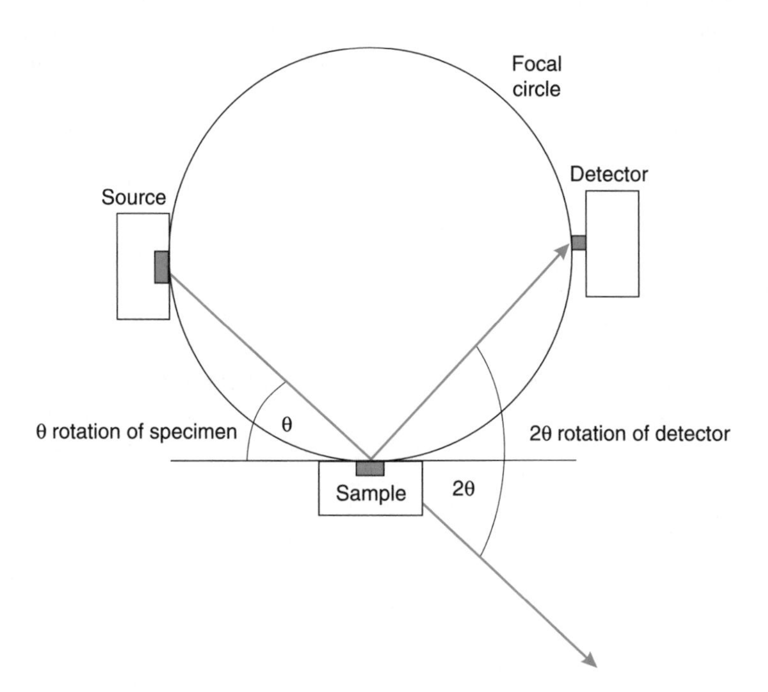

Figure 2.17 Schematic representation of a sample mounted on a goniometer stage which can be rotated about one or more axis, and a detector which travels along the focusing circle in the Bragg–Brentano geometry

arc of a circle (Fig. 2.16(c)), so that radiation from a line source striking any point on the arc of the crystal will satisfy the Bragg condition, thus focusing a diffracted beam to a line at the specimen position.

An X-ray spectrum is usually recorded by rotating an X-ray detector about the sample, the latter being mounted on a diffractometer goniometer stage. The goniometer allows the sample to be rotated about one or more axes (Fig. 2.17). In order to make full use of the potential resolution of the method (determined by the sharpness of the diffraction peaks), the diffractometer must be accurately aligned and calibrated, typically to better than 0.01°. The accurate positioning of the sample is very important, especially in a focusing diffractometer, since any displacement of the plane of the sample will result in a shift in the apparent Bragg angle (see Section 2.4.5). Note that if a given plane in the sample is to remain perpendicular to a radius of the focusing circle, then the detector must rotate around the latter at a rate which is twice that of the sample.

A number of X-ray detectors have been used (including photographic film), but the commonest is the *proportional counter*, in which an incident photon ionizes a low-pressure gas, thus generating a cloud of charged ions which are then collected as a current pulse. In the proportional counter, the charge carried by the current pulse is proportional to the photon energy, and electronic discrimination can be used to ensure that only photons of a selected energy trigger the counting system. The present generation of proportional counters have an energy resolution better than 150 eV, and can be used to eliminate most of the noise associated with white radiation (although they are not able to separate the K_α and K_β peaks). There is a dead-time associated with the current pulse generated in a proportional counter, and a second photon arriving at the counter within a microsecond or so of the first will not be counted. This sets an upper limit to the counting rate and means that peak intensities recorded at high counting rates may be underestimated.

While the thickness of the sample which can be studied by X-ray diffraction is dictated by the mass absorption coefficient for the incident radiation (see above), the lateral dimensions are a function of the diffractometer geometry. For automated powder diffractometers with a 'Bragg–Brentano' geometry, the width of the incident beam is typically of the order of 10 mm, while the length of the illuminated patch depends on the angle of incidence, and is typically in the range 1–7 mm (Fig. 2.18). The size and spacing of *Soller slits* (also called *divergence slits*) determine the area illuminated by the incident beam. For fixed slits, the total illuminated area decreases as the diffraction angle increases (2θ), but to a first approximation, the irradiated volume is independent of 2θ. Some Bragg–Brentano diffractometers include an automatic (compensating) divergence slit, which increases the width of the incident beam as the diffraction angle increases. While the irradiated area then remains constant, the irradiated volume of the sample increases with increasing diffraction angle, and calculated integrated intensities must therefore take into account the dependence of integrated intensity on diffracting volume.

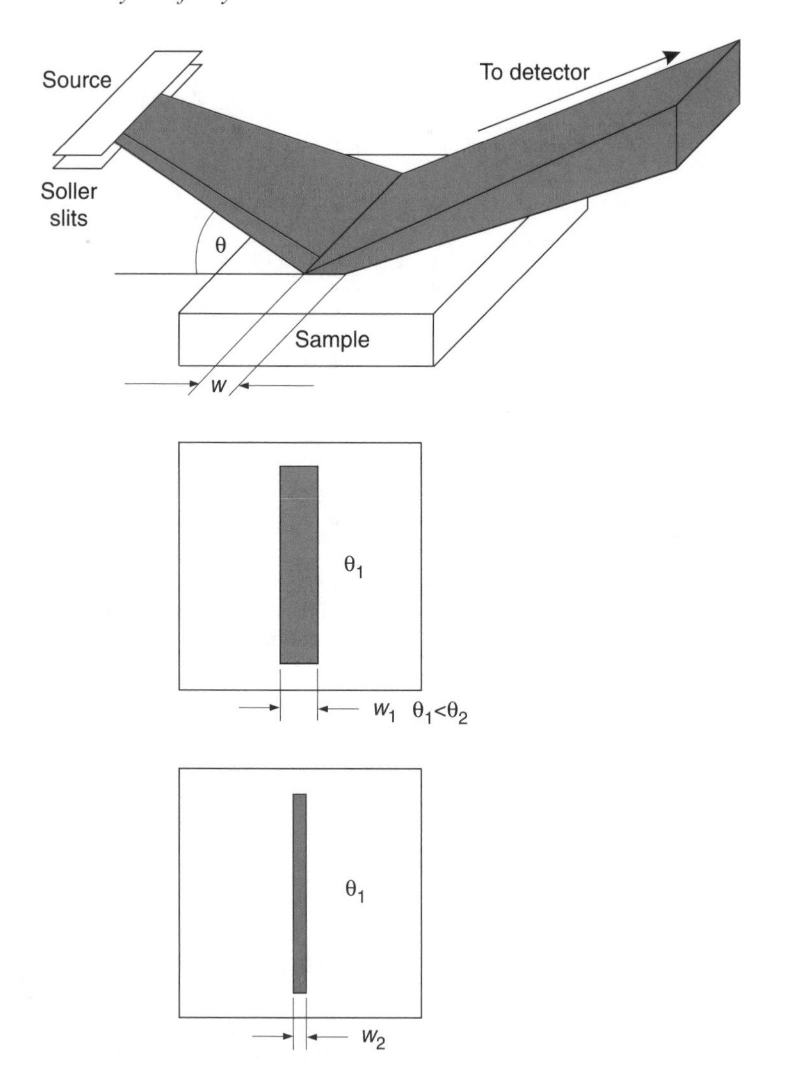

Figure 2.18 Schematic representation of the influence of θ on the exposed surface area for a powder diffractometer which uses the Bragg–Brentano geometry

2.3.2 Powder Diffraction—Particles and Polycrystals

The grain size of engineering materials is usually less than the volume of the material which contributes to X-ray diffraction. This is also true of many powder samples, either compacted or dispersed. The general term *powder diffraction* is used to describe both the nature of the pattern formed and the analysis used to interpret results from these *polycrystalline* samples.

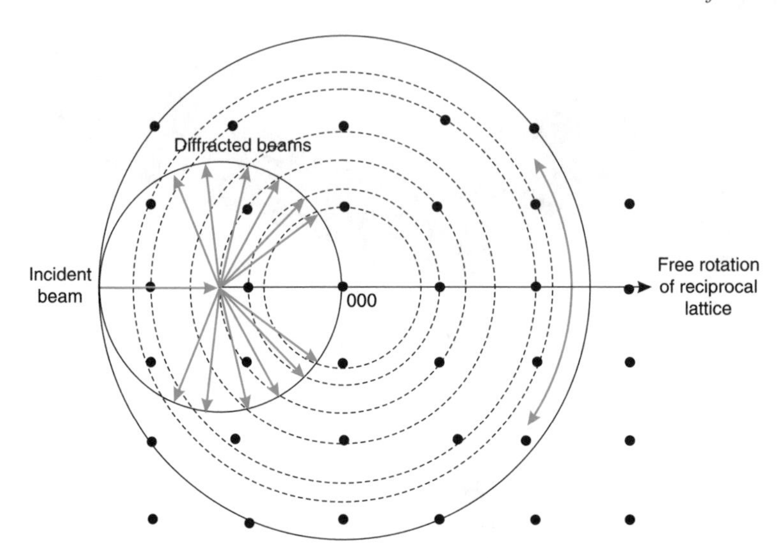

Figure 2.19 In powder diffraction, the case of a randomly oriented polycrystalline sample can be regarded as being equivalent to free rotation of the reciprocal lattice about the centre of the limiting sphere. Each reciprocal lattice vector within the limiting sphere then generates a spherical surface which intersects the reflecting sphere on a cone of allowed reflections from the individual crystals which subtends an angle 2θ with the incident beam

If we assume that the individual grains are randomly oriented and much smaller than the cross-section of the incident beam, then this is equivalent to assuming that the wave vector of the incident beam, $\mathbf{k}_o$, takes all possible orientations in reciprocal space, or equivalently, that the reciprocal lattice is freely rotated about its origin (Fig. 2.19). The grains oriented for Bragg reflection have **g**-vectors which touch the surface of the reflecting sphere, and these grains generate a diffraction cone which subtends a fixed angle 2θ with the incident beam. The detector is rotated about an axis normal to the incident beam and passing through the sample, to give the diffracted intensity as a function of 2θ, the *diffraction spectrum* for the sample.

If the specimen is rotated about an axis normal to the plane of diffraction (the plane containing the incident and diffracted beams) at a constant rate $d\theta/dt$, while the detector is rotated about the same axis at twice this rate, $d(2\theta)/dt$, then the normal to the diffracting planes in the crystals contributing to the spectrum will remain parallel. This is very useful if the grains in the polycrystalline sample are not randomly oriented. Mechanical working (plastic deformation) and directional solidification are two processes which tend to align the grains along specific directions or in certain planes. Such samples are said to contain *crystalline texture*, and the grains are said to be *preferentially oriented*. If the normal to the diffracting planes in a rotating-sample diffraction experiment coincides with specific directions in the bulk material (for example parallel or perpendicular to the direction of mechanical work), then the diffraction spectra may exhibit markedly different intensities from those calculated for a random polycrystal, thus reflecting the non-random probability of finding a grain in a specific orientation.

Preferred orientation plays an important role in many materials applications, and is associated with the *anisotropy* of the physical, chemical or mechanical properties. A classic case is that of the *magnetic hysteresis* of silicon iron, which is markedly different in the ⟨100⟩ and ⟨111⟩ directions. Transformer steels are therefore processed to ensure a strong (and favourable) texture, thus exhibiting low hysteresis losses. In many mechanical applications, texture is considered undesirable, and structural steel sheet is usually cross-rolled (that is rolled in two directions at right angles) in order to reduce the tendency to align the grain orientations.

The *lattice spacings* in polycrystalline samples are dependent on the state of stress of the sample. Stresses may be due to the conditions under which the component performs in service (operating stresses), but they may also result from the way in which the component is assembled into a system (as when a bolt is put under tension when a nut is tightened). However, residual stresses can also result from the *processing history*, for example gradients of plastic work in the component, variations in cooling rate, and some special surface treatments (ion-implantation, chemical surface changes, or mechanical bombardment with hard particles). In all cases, the *residual stresses* are present in the material in a state of *mechanical equilibrium*, even though no external forces are being applied.

One way of determining residual stress is by accurate X-ray measurement of lattice spacings. It is, of course, the lattice strains within the individual grains which are being sampled, and these must be converted to stresses through a knowledge of the elastic constants of the phases present. It is useful to distinguish two types of residual stress, namely *macrostresses* and *microstresses*. Macrostresses are present when large numbers of neighbouring crystals of the same phase experience similar stress levels, and the stresses vary smoothly throughout the sample to generate an equilibrium stress state (for example *compressive stresses* in the surface layers balanced by *tensile stresses* in the bulk). On the other hand, microstresses may also exist, in which the stresses in the individual grains within any element of volume may be widely different and of opposite sign, while the average stress sums to zero. A good example would be stresses due to anisotropy of thermal expansion in a non-cubic polycrystal, leading to constraints on the contraction of neighbouring crystals during cooling. Macrostresses result in a displacement of the diffraction maxima from their equilibrium positions, while microstresses result in a broadening of the diffraction peaks about their equilibrium positions.

Given that lattice spacings are measurable by X-ray diffraction to 1 part in 10^5, it follows that lattice strains are detectable to approximately 10^{-5}. For an aluminium alloy with an elastic modulus of 60 Gpa, this corresponds to a stress of less than 1 MPa, so that accurate determination of residual stress levels which constitute only a few % of the bulk yield stress should be possible. Unfortunately, measurements using X-ray diffraction are confined to the surface layers of a sample, but may nevertheless be extremely helpful, as in controlling the quality of surface coatings.

2.3.3 Single-Crystal Laue Diffraction

The powder method depends on measuring the intensity diffracted from a monochromatic incident beam as a function of the Bragg angle, given by the relationship, $\lambda = 2d \sin \theta$, and identifying the lattice planes responsible for the diffraction peaks in the spectrum. An alternative approach would be to use a beam of white radiation and determine the spatial distribution of the intensity diffracted by a rigidly mounted sample. A particular set of diffracting planes in a crystal will now select that wavelength from the incident beam which satisfies the Bragg criterion for the angle at which the crystal is oriented. In a polycrystalline sample, this will result in sets of reflections for each crystal, many of which will overlap with those from other crystals, thus resulting in a confused pattern which cannot easily be

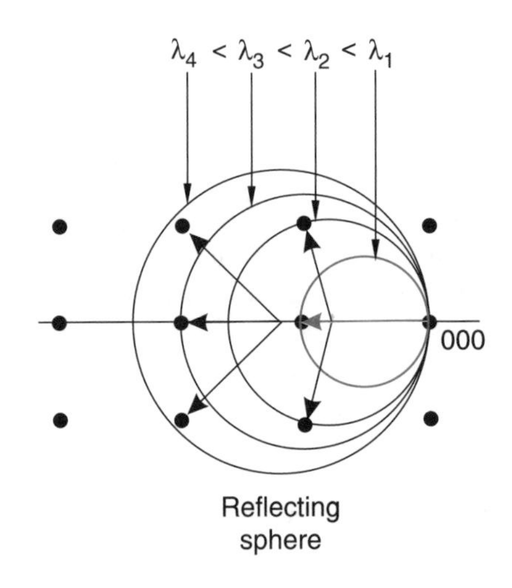

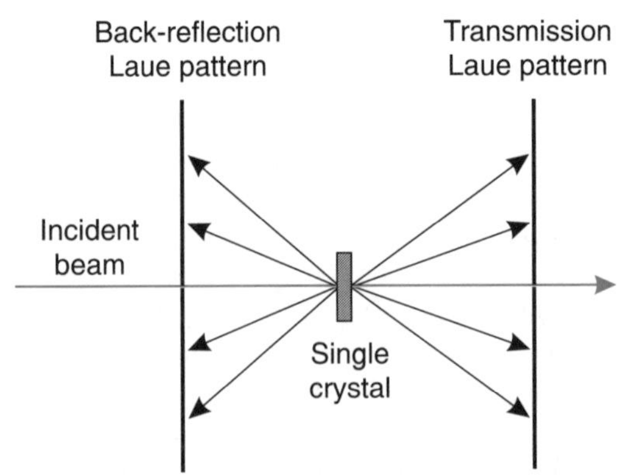

Figure 2.20 Schematic representations of the reflecting sphere construction for single-crystal diffraction using white X-radiation and the experimental configurations used to record Laue diffraction patterns. Each reciprocal lattice point diffracts the specific wavelength which satisfies the Bragg relationship

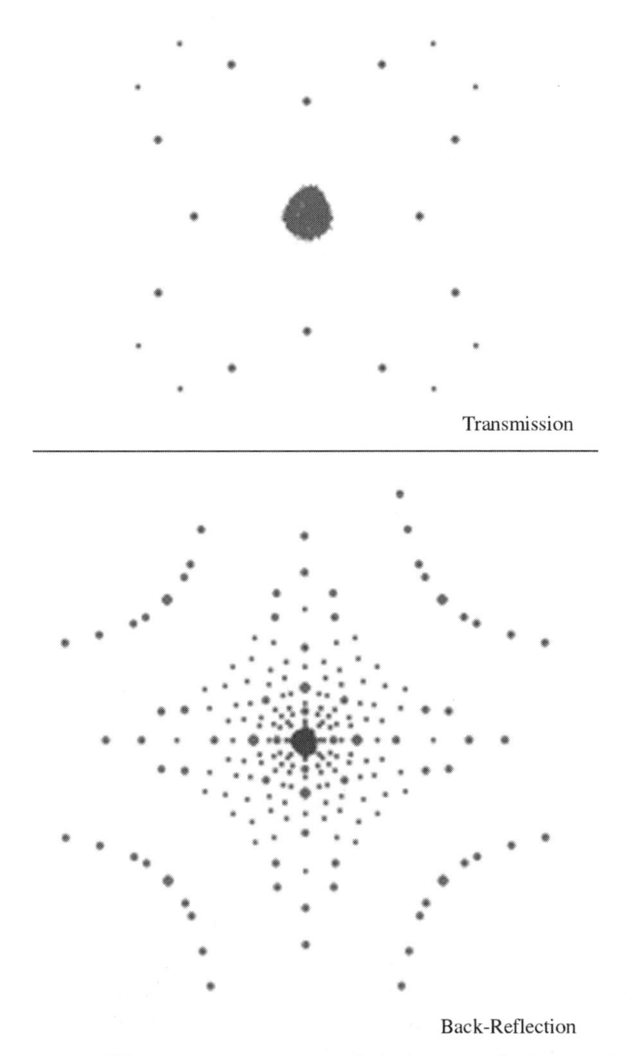

Figure 2.21 Laue diffraction patterns recorded from a single-crystal sample in a symmetrical orientation: (a) in transmission; (b) in reflection

interpreted. However, if the sample is a single crystal, the reflections will form a very distinctive Laue pattern which can be used to determine the orientation of the crystal with respect to the incident beam.

A photographic film can be used to record the single crystal diffraction pattern in a *Laue camera*. Two configurations are possible (see Fig. 2.20). If the specimen is thin enough, a Laue pattern may be recorded in *transmission* on a plane *perpendicular* to the incident beam. The Laue reflections from a set of crystal planes which lie on the same symmetry zone will then intersect the plane of the film on an *ellipse*. More commonly, Laue diffraction patterns are recorded in *reflection*, when there is no

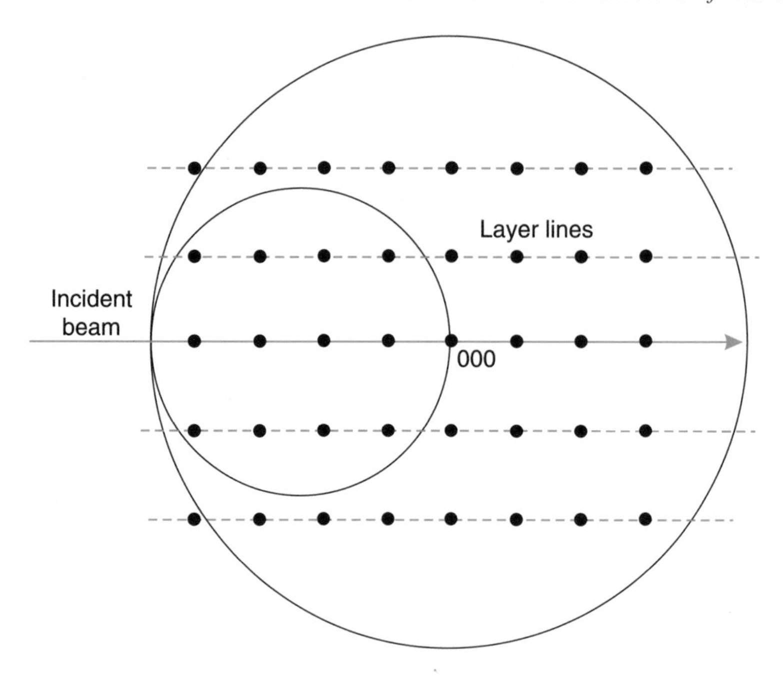

Figure 2.22 Rotating a single crystal about an axis perpendicular to the plane of the spectrometer and using monochromatic radiation brings each lattice point in turn into the diffracting condition, so generating layer lines of reflections

limitation on the thickness of the crystal. The beam passes through a hole in the centre of the recording film, which is again normal to the incident beam. The symmetry zones of the reflections now intersect the film as arcs of *hyperbolae*. Examples of Laue patterns taken in both transmission and reflection are given in Fig. 2.21.

The interpretation of a Laue pattern depends on identifying the symmetry axes of the reflecting zones in order to determine the orientation of the single crystal with respect to an external coordinate system. Information on the perfection of the single crystal can also be derived from a Laue image, primarily the presence of sub-grains or twinned regions which give rise to additional sets of diffracted beams, displaced with respect to those due to the main crystal.

2.3.4 Rotating Single-Crystal Methods

While much crystallographic structure analysis can be achieved when using randomly oriented, powder or polycrystalline samples, single crystals are often required to confirm a crystal lattice model unambiguously. Monochromatic radiation is used and the crystal is mounted at the exact centre of the spectrometer on a goniometer stage. The goniometer allows the crystal axes to be oriented accurately with respect to the spectrometer and permits continuous rotation of the crystal about an axis normal to the plane containing the incident beam and the detector. Rotation brings reciprocal lattice points lying on planes parallel to the plane of the spectrometer into the reflecting condition, thus generating layer lines of reflections (Fig. 2.22).

The single crystal must be large enough to ensure sufficient resolution for structure analysis in reciprocal space, but not so large as to result in geometrical blurring of the reflections in real space. Crystals between 0.1 and 1 mm in size are suitable. The range of the rotation angle is restricted to reduce overlap from multiple reflections, and several spectra must be recorded by rotating the crystal about the prominent symmetry axes.

2.4 DIFFRACTION ANALYSIS

So far we have only discussed the *geometry* of diffraction and shown how a determination of the angular distribution of the diffracted beams can be used to identify the crystal symmetry and determine the lattice parameters to a high degree of accuracy. This information is usually sufficient to identify the crystalline phases present in a solid sample unambiguously, but there is a great deal of information present in the *relative intensities* of the diffracted beams which we have not yet utilized. In order to make use of this information, we will now examine the factors which determine the scattered amplitude.

2.4.1 Atomic Scattering Factors

We shall confine our attention to incident X-rays and electrons, which are scattered by electrons in the solid, and ignore incident neutrons, which are scattered by the atomic nuclei. By taking α as the angle between the scattering direction of the incident beam and the direction in which an interacting electron is accelerated, J.J. Thomson showed that the scattered *intensity* is given by the expression $I = I_0 e^4 \sin^2 \alpha / (r^2 m^2 c^4)$ where e and m are the charge and mass of the electron, respectively, c is the velocity of the electromagnetic radiation, and r is the distance of the accelerated electron from the incident beam (Fig. 2.23).

For an *unpolarized* X-ray beam we need to average the effect of the electric field components which act on the electromagnetic wave (Fig. 2.24). The electric field acts perpendicular to the plane of scattering, x–z, and lies in the y–z plane, with

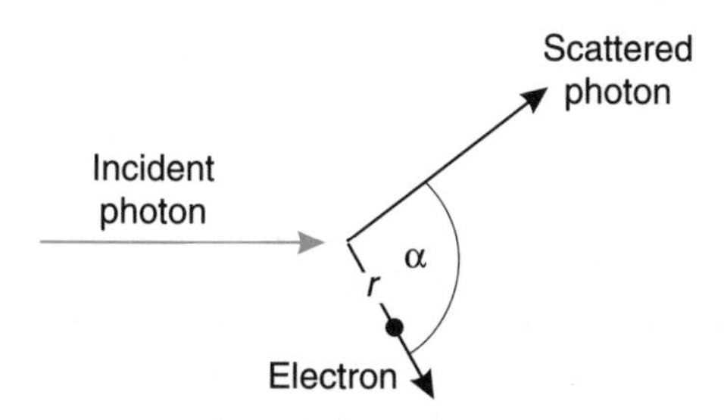

Figure 2.23 Schematic representation of the scattering of an X-ray beam by an electron (see text for details)

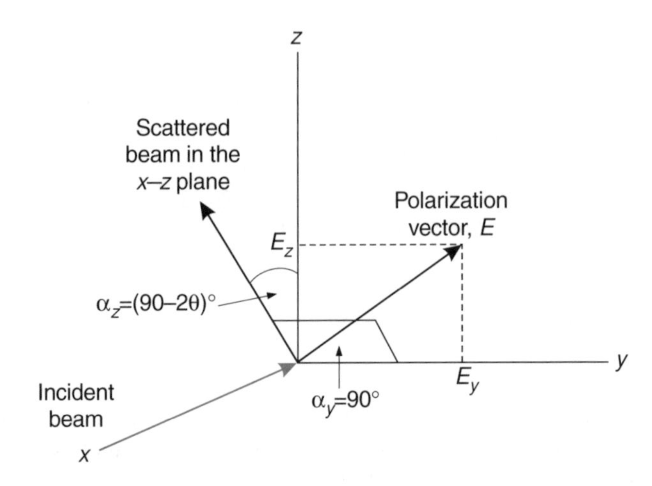

Figure 2.24 If scattering occurs at an angle 2θ in the x–z plane, then the applied electric field is in the y–z plane and the average values of the components E_y and E_z must be equal

components E_y and E_z. For an unpolarized beam, these two components are, on average, equal, while the values of α, the angle between the components of E and the scattering direction, are given by $\alpha_y = \pi/2$ and $\alpha_z = (\pi/2 - 2\theta)$. Assuming that each component contributes half of the total intensity, the factor $\sin^2 \alpha$ should be replaced by $(\sin^2 \alpha_y + \sin^2 \alpha_z)/2$. Substituting for α_y and α_z leads to the following relationship:

$$I = I_0 \frac{e^4}{r^2 m^2 c^4} \cdot \frac{(1 + \cos^2 2\theta)}{2} \tag{2.9}$$

where the term $(1 + \cos^2 2\theta)/2$ is called the *polarization factor.*

Each atom in the sample contains Z electrons, where Z is the atomic number. In the direction of the incident beam, all of the electrons in the atom will scatter in phase. If the *atomic scattering factor*, $f(\theta)$, is defined as the amplitude scattered by a *single atom* divided by that scattered by an *electron*, then it follows that $f(0) = Z$, while for $\theta > 0$, then $f(\theta) < Z$, since at larger scattering angles the electrons around an atom will scatter increasingly out-of-phase. The θ dependence of some representative atomic scattering factors are shown in Fig. 2.25.

2.4.2 Scattering by the Unit Cell

The next step is to derive the amplitude scattered by a unit cell of the crystal structure. If there is a path difference δ between the X-ray beam scattered from an atom at the origin and from any other atom in the unit cell, then this will correspond to a phase difference between the two scattered beams, which is given by $\varphi = \delta 2\pi/\lambda$. If the position of the second atom is defined by the vector $\mathbf{r}$ in the direction $[uvw]$, such that the coordinates of the atom in the unit cell are xyz with $u = x/a$, $v = y/b$ and $w = z/c$, and the atom lies on the plane (hkl), corresponding

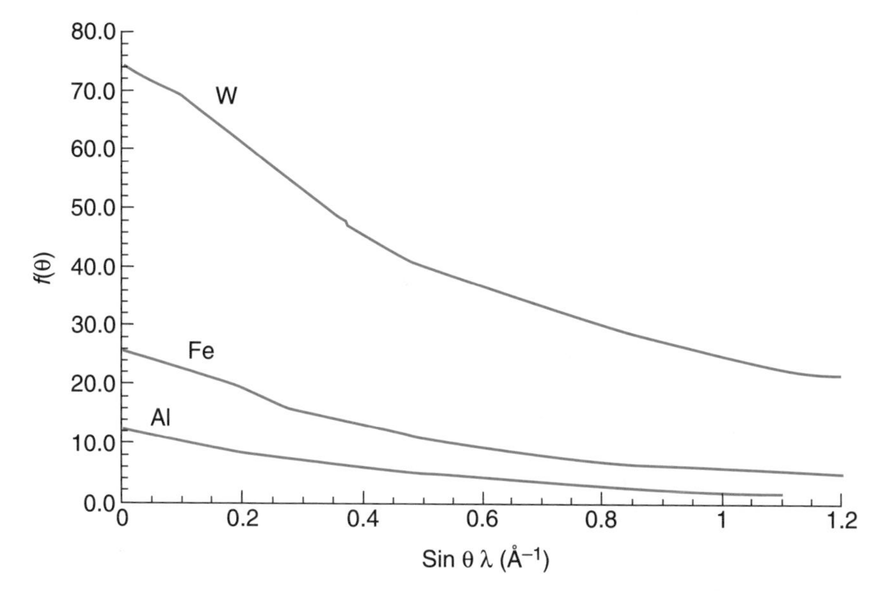

Figure 2.25 The atomic scattering factor as a function of Z and θ, showing curves for aluminium, iron and tungsten

to the reciprocal lattice vector **g** (Fig. 2.26), then the *phase difference* for radiation scattered by the two atoms into the *hkl* reflection is given by the expression, $\varphi_{hkl} = 2\pi(hu + kv + lw) = 2\pi\mathbf{gr}$.

Each atom will scatter an amplitude A, which depends on the atomic scattering factor $f_{Z\theta}$ for that atom, and we can represent both the phase and the amplitude of the scattered wave from each atom by a vector **A**. By using complex notation, $Ae^{i\varphi} = A(\cos\varphi + i\sin\varphi)$, and the contribution to the amplitude scattered into the diffracted beam *hkl* by an atom at *uvw* in the unit cell will be given by $Ae^{i\vartheta} \propto f\exp[2\pi i(hu + kv + lw)] = f\exp[2\pi i\mathbf{gr}]$ (see Fig. 2.27).

2.4.3 The Structure Factor in the Complex Plane

By ignoring a constant of proportionality, corresponding to the scattering due to a single electron, we can define a *normalized* scattering factor for the *hkl* reflection in which we sum the contributions to this reflection from all of the N atoms in the unit cell. This new parameter, the *structure factor* for the *hkl* reflection, is then given by:

$$F_{hkl} = \sum_{1}^{N} f_n \exp[2\pi i(hu_n + kv_n + lw_n)] = \sum_{1}^{N} f_n \exp(2\pi i\mathbf{gr}_n) \tag{2.10}$$

Before proceeding further, we note that, in the complex plane, the following relationships hold:

(a) $e^{i\varphi} = \cos\varphi + i\sin\varphi$, so that the *real* component of the amplitude is resolved along the x-axis and the *imaginary* component along the y-axis of *phase space*.

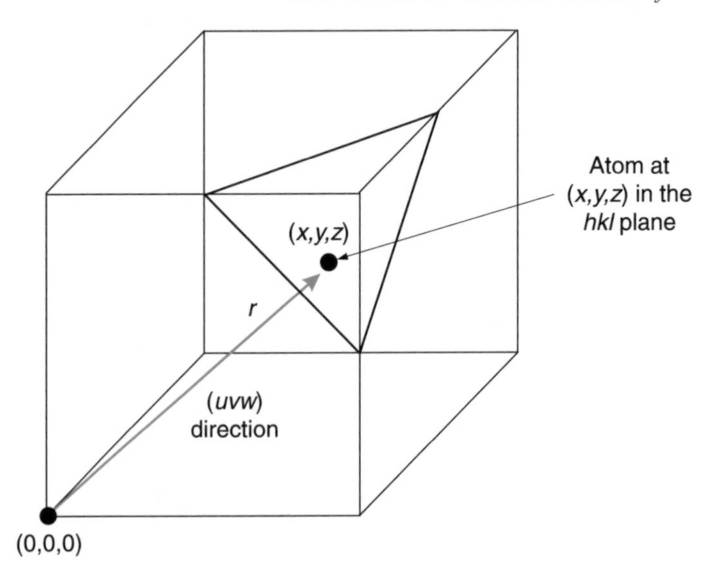

Figure 2.26 Schematic representation of the geometry of atomic positions and scattering planes in the unit cell

(b) $I = |\mathbf{A}||\mathbf{A}^*|$, that is the *intensity* scattered by any combination of atoms is derived from the phase-space vector amplitude by multiplying the *real component* by its *complex conjugate*.

(c) $e^{n\pi i} = (-1)^n$, so that *phase angles* corresponding to even and odd multiples of π have *no* imaginary components of the amplitude, and simply add and subtract, respectively, from the total scattered amplitude.

(d) $e^{ix} + e^{-ix} = 2\cos x$, again a condition for *no* imaginary component.

Copper has an FCC unit cell containing four atoms, with each atom situated at a Bravais lattice point, so that the values of [*uvw*] are [0 0 0], [1/2 1/2 0], [1/2 0 1/2] and [0 1/2 1/2]. It follows that the structure factors for copper are given by the

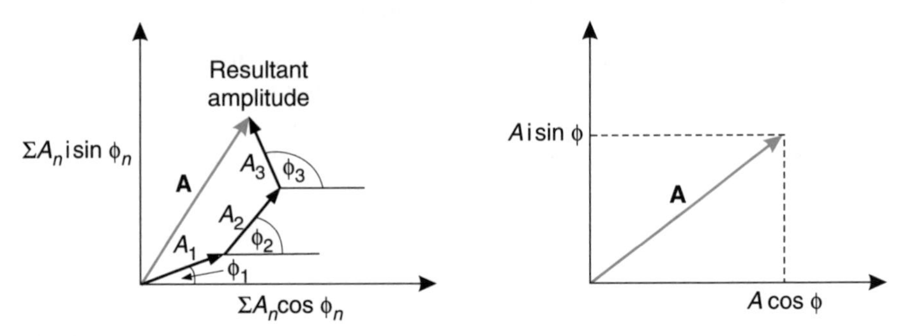

Figure 2.27 The amplitude–phase diagram and a demonstration of its use for summing the scattered amplitudes contributing to a particular reflection by all of the atoms in the unit cell

expression $F_{hkl} = f(1 + e^{\pi i(h+k)} + e^{\pi i(h-l)} + \{e^{\pi i(k+l)})$. If h, k and l are all odd or all even, then $F_{hkl} = 4f$, but if h, k and l are mixed integers, then $F_{hkl} = f(1 + 1 - 2) = 0$, the same result we obtained previously.

In the BCC unit cell, characteristic of α-Fe, the atoms are at the two Bravais lattice points [0 0 0] and [1/2 1/2 1/2], and the structure factors are given by $F_{hkl} = f(1 + e^{\pi i(h+k+l)})$. For $h + k + l$ even, it follows that $F_{hkl} = 2f$, while if $h + k + l$ is odd, then $F_{hkl} = f(1 - 1) = 0$, again as noted previously.

Cubic diamond has an FCC unit cell in which each lattice point corresponds to two atoms, one at the site of the lattice point and the other displaced by a vector [1/4 1/4 1/4]. This is equivalent to two interpenetrating FCC lattices related to one another by this same displacement vector. It follows that the structure factors are given by $F_{hkl} = f(1 + e^{\pi i(h+k)} + e^{\pi i(h+l)} + e^{\pi i(k+l)})(1 + e^{\pi i/2(h+k+l)})$. There are three possibilities here. For $h + k + l$ odd, the amplitude vector for the set of four 'second' atoms has no real component and this vector points either vertically up or vertically down. In both cases, the structure factor for the allowed FCC reflections is increased by $\sqrt{2}$ over that for a single atom at the origin, and a resultant phase angle of $\pi/4$ is introduced. For $h + k + l$ even, only the real component exists, and may be either negative, reducing the structure factor of an allowed FCC reflection to zero, or positive, doubling the structure factor to twice the value for the case of a single atom per lattice point.

Finally, consider the case of common salt (NaCl—FCC). In this example, the cations sit on the Bravais lattice points, while the anions occupy a second FCC lattice displaced by [1/2 1/2 1/2]. The scattering factors of the cations and anions are different, and the structure factors for the different reflections are now given by the relationship $F_{hkl} = (1 + e^{\pi i(h+k)} + e^{\pi i(h+l)} + e^{\pi i(k+l)})(f_{Na} + f_{Cl}e^{\pi i(h+k+l)})$. Even values of $h + k + l$ now result in reinforcement of the intensity for the allowed FCC reflections, $F_{hkl} = 4(f_{Na} + f_{Cl})$, while odd values of $h + k + l$ reduce the intensity, $F_{hkl} = 4(f_{Na} - f_{Cl})$.

2.4.4 Interpretation of Diffracted Intensities

We are now in a position to list all of the physical factors that determine the intensity of an observed diffraction peak in a recorded spectrum. These are as follows:

(a) The *polarization factor* is associated with scattering of unpolarized electromagnetic radiation,

$$\left(\frac{1 + \cos^2 2\theta}{2}\right) \tag{2.11}$$

The exact form of the polarization factor depends on the *geometry* of the diffractometer (see below).

(b) The *structure factors* for the different reflecting planes in the crystal lattice, which include the effect of the *atomic scattering factors* for all of the atoms present in the material,

$$F_{hkl} = \sum_{1}^{N} f_n \exp[2\pi i(hu_n + kv_n + lw_n)] \tag{2.12}$$

(c) The *multiplicity* of the reflecting planes, P, which gives the number of planes belonging to a particular family of Miller indices (determined by the *symmetry* of the crystal). For example, in cubic crystals, planes whose poles fall within the *unit triangle* have a multiplicity of *24*, since there are 24 unit triangles, while those whose poles lie along the *edges* of a unit triangle (and are therefore common to two triangles) have a multiplicity of *12*. On the other hand, there are four $\{111\}$ planes in the stereogram, which correspond to the *apices* of the unit triangles (shared by 6 triangles) and yield a multiplicity of *4*, while the $\{100\}$ reflections correspond to poles on the coordinate axes, shared by 8 triangles, and have a multiplicity of *3*.

(d) The *sampling geometry*. In the powder method, a collector of finite cross-section only samples that proportion of the cone of radiation which is diffracted at the Bragg angle. The fraction of radiation collected is given by the *Lorentz polarization factor*,

$$L = \frac{1 + \cos^2 2\alpha \cos^2 2\theta}{\sin^2 \theta \cos \theta (1 + \cos^2 2\alpha)} \tag{2.13}$$

where θ is the diffracting angle, and α is the diffracting angle of a monochromator placed before the detector.

(e) *Absorption effects*, which depend on the size of the sample and its geometry. In general, absorption can be expected to *increase* at large values of the diffraction angle. The absorption correction can be written as $A(\theta)$, and can be estimated for given geometries and sample densities, by using standard tables of *mass absorption coefficients*. For a *thick* specimen in the *Bragg–Brentano* diffractometer, the absorption factor will *not* be a function of the diffraction angle and is simply $A = 1/2\mu$. Thus this factor can be removed from *normalized* calculated integrated intensities.

(f) *Temperature* is also an important factor, since at high temperatures random atomic vibrations will reduce the coherence of the scattering from the more closely spaced crystal planes. This effect *increases* with $\sin\theta/\lambda$ (that is at small d-values and larger values of hkl), and its influence on the overall integrated intensity can be expressed by the factor

$$e^{-2B\sin^2\theta/\lambda^2} \tag{2.14a}$$

where

$$B = 8\pi^2 \bar{U}^2 \tag{2.14b}$$

with $\bar{U}^2$ being the mean square displacement of each atom.

Summing all the above effects yields a *general relationship* for the diffracted integrated intensity as follows:

$$I = k|F^2|.\frac{1 + \cos^2 2\alpha \cos^2 2\theta}{\sin^2 \theta \cos \theta (1 + \cos^2 2\alpha)} PA(\theta) \exp\left(-2\frac{B \sin^2 \theta}{\lambda}\right) \tag{2.15}$$

where k represents a *scaling factor* and includes I_0. Equation (2.15) can be used to calculate simulated integrated intensities for any given structure by using an appropriate computer program. The *net integrated intensity* is first calculated for each *hkl* reflection, and then all of the calculated intensities are normalized with respect to the *maximum* calculated integrated peak intensity, which is assigned a value of 100 %,

$$I_{hkl}^{n} = \frac{I_{hkl}}{I_{hkl}^{\max}} 100 \tag{2.16}$$

2.4.5 Errors and Assumptions

In the present treatment there is no justification for a quantitative analysis of the errors involved in X-ray diffraction measurements. Some sources of error have been mentioned, while others are minor. It is helpful to distinguish between errors in the measurement of peak positions and errors in the determination of peak intensities. It is especially important to recognize the importance of accurate calibration and alignment if very small changes in lattice parameter (of the order of 10^{-5}) are to be resolved. X-ray diffraction can only sample a small volume of material, but this may be a definite advantage in the analysis of thin films, coatings and solid-state devices. Such applications are critical for a wide range of engineering systems where the important properties are associated with the surface and near-surface regions, for example microelectronic components, opto-electronic devices, wear parts and machine tools. The monitoring of surface stress or chemical change at the surface can often be accomplished by X-ray diffraction, and *in situ* commercial X-ray systems are available for extracting measurements at high temperatures or in a controlled environment.

As an example, we will summarize the errors involved in measuring lattice parameters when using a diffractometer in the Bragg–Brentano geometry. This diffractometer is the commonest automated diffractometer and has a well-defined optical focusing system. The *focal circle* (Fig. 2.28) is tangential to both the source

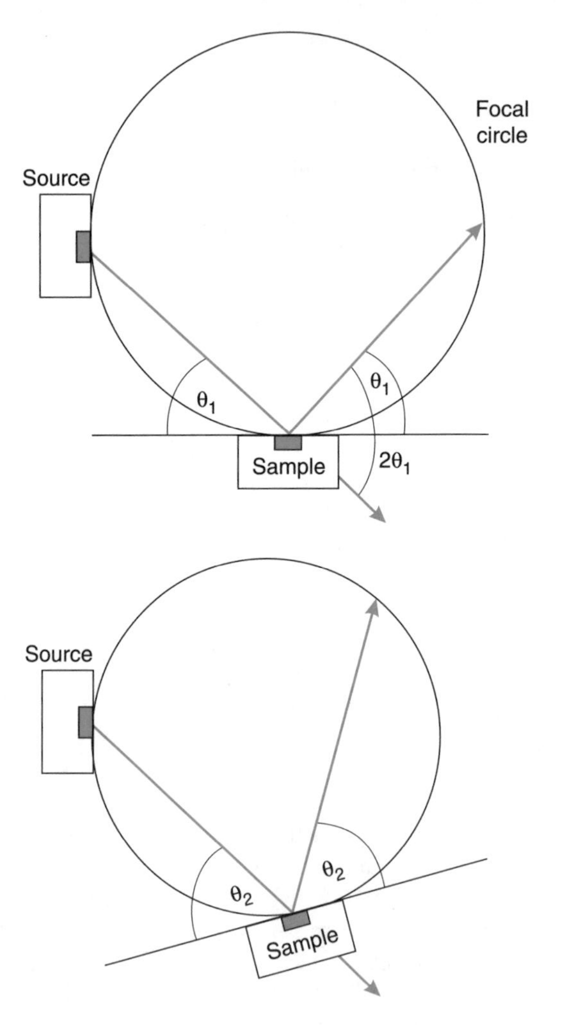

Figure 2.28 The Bragg–Brentano geometry of a diffractometer, shown for two different Bragg angles

and the specimen surface. Since the specimen is rotated in increments of θ, and the detector in increments of 2θ, the radius of the focal circle decreases as θ increases, and the detector must remain on the (changing) focal circle in order to minimize the signal beam spread. These operating conditions determine the three major errors in peak measurement which limit the accuracy in the determination of the lattice spacings and lattice parameters.

The first is *peak broadening*, due either to incorrect alignment of the diffractometer, or misplacement of the specimen on the diffractometer goniometer. This error can be reduced either by better alignment or by the use of smaller

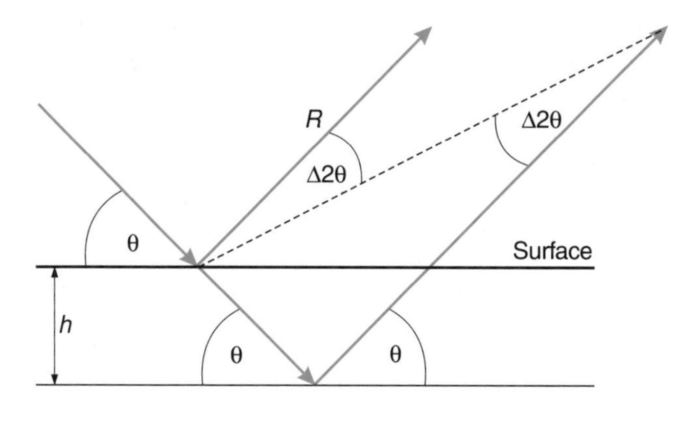

Figure 2.29 Schematic representation of a diffraction signal from below the focusing plane (sample surface) which results in an error in 2θ

receiving slits in front of the detector, but residual broadening will *always* lead to some error in the measurement of the diffraction angle.

The second and third sources of error are both directly related to the specimen itself. The diffraction signal originates from a region *beneath* the specimen surface, and not from the surface itself. This results in an error in the apparent diffraction angle, as shown in Fig. 2.29. A similar error will result if the specimen is placed either above or below the focal point on the goniometer axis. By using Fig. 2.29, we can derive the resulting *defocus error* in θ from the following relationship:

$$\frac{\sin \Delta 2\theta}{h \sin \theta} = \frac{\sin(180 - 2\theta)}{R} \tag{2.17}$$

and thus:

$$\Delta 2\theta \cong \frac{-2h \cos \theta}{R} \tag{2.18}$$

where R is the distance from the sample to the detector.

Finally, we obtain the following:

$$\frac{\Delta d}{d} = \frac{-\Delta 2\theta}{\tan \theta} = \frac{\cos \theta}{\sin \theta} \frac{2h \cos \theta}{R} = k \frac{\cos^2 \theta}{\sin \theta} = \frac{\Delta a}{a} \tag{2.19}$$

It follows that errors in lattice parameter measurements can be reduced by extrapolating the values calculated from each peak to a (hypothetical) value for $\theta = 90^\circ$, that is by plotting the lattice parameters as a function of $\cos^2 \theta / \sin \theta$ and extrapolating to 0. An alternative is to distribute a powder of a diffraction standard (with known lattice parameters) on the surface of the specimen. The peak positions from the standard can then be used to correct for systematic errors.

Some additional factors should be considered in the measurement of peak intensities, especially those determining the efficiency of data collection. There are two considerations here, namely the geometry of the sample in the spectrometer, and the response of the detector to the incident photons. Both of these have been mentioned previously. Some geometrical peak broadening is always associated with the finite diffracting volume, and while the detector response should remain constant over long periods it will always be wavelength-dependent.

Measurement of the comparative intensities from different diffraction peaks (expressed as a percentage of the intensity of the *strongest* diffracted peak) is an important diagnostic tool, both for identifying unknown phases in the sample, and for refining crystal structure models. Such measurements are sensitive to the presence of preferred orientation. The diffraction peak height in an experimental spectrum is an inaccurate measure of the relative peak importance and it is important to determine the integrated peak intensities, that is the total area under the peak, after subtracting background noise, (Fig. 2.30).

In addition to a dependence on preferred orientation, the integrated intensity of a diffraction peak may vary with temperature (see above), or as a result of transmission losses (if the specimen thickness is well below the characteristic absorption

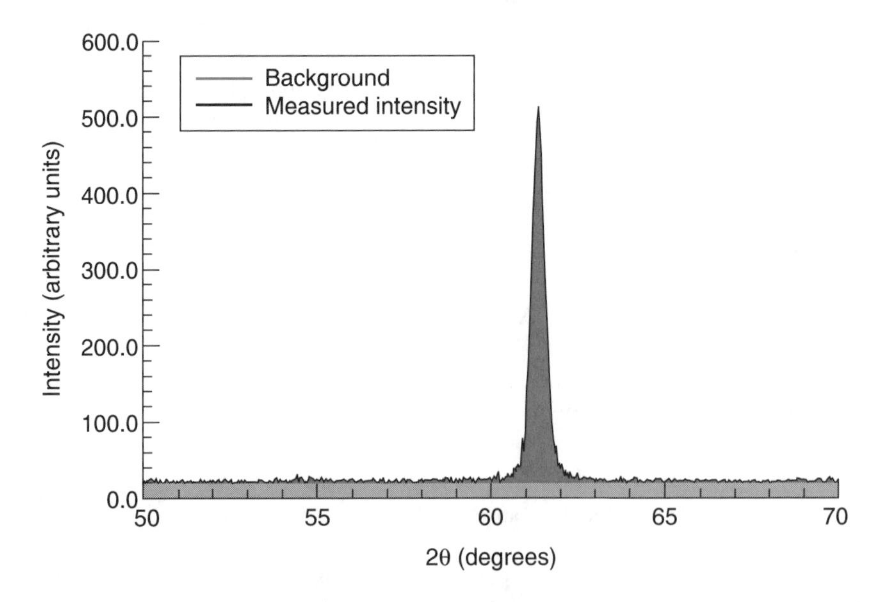

Figure 2.30 A weak diffraction peak, illustrating the calculation of integrated peak intensity from the total area beneath the peak after subtracting background noise

thickness). The reduction in integrated intensity associated with a thin sample can be determined by multiplying the integrated intensity by the following factor:

$$1 - \exp(-2\mu t \operatorname{cosec} \theta) \tag{2.20}$$

For thick specimens, this factor reduces to 1. If μ is known, in principle the thickness t of a thin sample can be measured (not very accurately) by comparing the intensities with those from a bulk sample. For random samples, μt is best determined from relative integrated intensities by a refinement procedure based on the entire diffraction spectrum, but we can also compare the intensity ratio of a single pair of reflections recorded at two Bragg angles θ_1 and θ_2,

$$Y = \frac{I(t)}{I(\infty)} = \frac{1 - \exp(-2\mu t \operatorname{cosec} \theta_1)}{1 - \exp(-2\mu t \operatorname{cosec} \theta_2)} \tag{2.21}$$

This method of determining thickness is generally feasible if μt has values in the range 0.01–0.5.

Preferred orientation changes the relative integrated intensities of the peaks observed in the diffraction pattern, and reflects the presence of important microstructural features which affect the material properties. Accurate methods for determining preferred orientation and in particular the probability of finding a crystal with a specific orientation (*the crystallite orientation distribution factor*) are beyond the scope of this present text. However, there is a fairly simple method which can be used to determine qualitative values of texture. The method (the *Harris method*) is based on a relationship for the volume fraction of a phase with crystal orientations lying within a small solid angle dΩ about an angle (α, β, γ) in an *inverse pole figure*, as follows:

$$P(\alpha, \beta, \gamma)^{\mathrm{d}\Omega}/4\pi \tag{2.22}$$

where $P(\alpha, \beta, \gamma)$ depends only on α, β, and γ. Since every crystal must have some orientation, we can write the following:

$$\frac{1}{4\pi}\iint P(\alpha, \beta, \gamma)\mathrm{d}\Omega = 1 \tag{2.23}$$

For *randomly oriented materials*, P is *independent* of α, β, and γ, and therefore:

$$\frac{P}{4\pi}\iint \mathrm{d}\Omega = 1 \tag{2.24}$$

so that $P = 1$. Values of P greater than one indicate that the corresponding crystallographic direction has a higher probability than would be found in a random polycrystal, while values of P less than one indicate that this direction is less likely to be found. P can be experimentally determined with a Bragg–Brentano diffractometer by using the following equation:

$$P(\alpha, \beta, \gamma) = \frac{I(hkl)}{\sum I(hkl)} \frac{\sum I'(hkl)}{I'(hkl)} \tag{2.25}$$

where $I(hkl)$ is the measured integrated intensity from the plane (hkl) of the sample, and $I'(hkl)$ is the measured or calculated integrated intensity from the same plane in a *randomly* oriented standard of the same material. Values of P are plotted on an *inverse pole figure* (effectively a stereographic projection), and contours of constant P then define the degree of texture relative to the specimen geometry for the hkl reflecting plane.

The measured integrated intensities not only indicate the degree of texture for the diffracting planes, but they can also determine the amount of each phase present in a multiphase material. *Quantitative phase analysis* is important for determining the effect of different processing parameters on the phase content. In order to determine the amount of the phase α in a mixture of α and β, we first redefine the integrated intensity in equation (2.15), by isolating the constants and focusing on the significant variables, as follows:

$$I_\alpha = \frac{K_1 c_\alpha}{\mu_\mathrm{m}} \tag{2.26}$$

where c_α is the concentration of phase α, μ is the linear absorption coefficient of the mixture of phases, and K_1 is a constant. The linear absorption coefficient depends on the amount of each phase present in the mixture,

$$\frac{\mu_\mathrm{m}}{\rho_\mathrm{m}} = \omega_\alpha \left(\frac{\mu_\alpha}{\rho_\alpha} \right) + \omega_\beta \left(\frac{\mu_\beta}{\rho_\beta} \right) \tag{2.27}$$

where ω is the weight fraction of the phase and ρ its density. Rearranging equation (2.27) and combining with equation (2.26) yields the following expression:

$$I_\alpha = \frac{K_1 c_\alpha}{c_\alpha(\mu_\alpha - \mu_\beta) + \mu_\beta} \tag{2.28}$$

Comparing I_α from the mixture to I_α from a pure sample (p) of the α-phase, we obtain:

$$\frac{I_\alpha}{I_{\alpha,\mathrm{p}}} = \frac{\omega_\alpha\left(\dfrac{\mu_\alpha}{\rho_\alpha}\right)}{\omega_\alpha\left(\dfrac{\mu_\alpha}{\rho_\alpha} - \dfrac{\mu_\beta}{\rho_\beta}\right) + \dfrac{\mu_\beta}{\rho_\beta}} \tag{2.29}$$

If we know the mass absorption coefficients of each phase, equation (2.29) is sufficient for quantitative analysis, but if the mass absorption coefficients are unknown we must prepare a set of *standard specimens* (usually mixed powders) and construct a calibration curve. This is usually a good idea in any case, since for most phase mixtures the variation of equation (2.29) with ω_α is nonlinear.

Additional X-ray diffraction techniques, which are beyond the scope of this present book, include the measurement of *particle size*, *residual stress analysis*, *structure refinement* by spectrum fitting (*Rietveld analysis*) and thin-film techniques (particularly important for semiconductor devices). References to these methods can be found in the Bibliography for this chapter.

2.5 ELECTRON DIFFRACTION

The dual wave-particle nature of electrons is expressed by the *de Broglie relationship* for the momentum, $p = mv = h/\lambda$, where m is the mass and v is the velocity of the electron, and λ is the wavelength. Substituting for the electron energy, in electron volts, $eV = 0.5\,mv^2$, yields the (non-relativistic) electron wavelength as follows:

$$\lambda = \frac{h}{\sqrt{2meV}} \tag{2.30}$$

At an accelerating voltage of 100 keV, the electron wavelength is 0.0037 nm, much less than the interplanar spacing in crystals, so that the Bragg angles for electron diffraction are always very small when compared to those for X-ray diffraction, that is the electrons are *elastically* scattered at very small angles. *Electron diffraction* is, as we shall see, both a major source of contrast in thin-film electron microscopy and an important analytical tool in its own right.

2.5.1 Wave Properties of Electrons

The de Broglie relationship is not sufficient to define the wavelength of an electron at high energies, since *relativistic* effects become important. If the rest mass of the electron is m_0, then the relativistic mass is given by $m = m_0 + eV/c^2$, where c is the velocity of the electromagnetic radiation. Substituting in the de Broglie relationship and then rearranging leads to the relativistic equation for the wavelength of the electron λ in terms of the accelerating voltage V:

$$\lambda = \frac{h}{p} \frac{h}{\sqrt{[2m_0eV(1 + eV/m_0c^2)]}} \tag{2.31}$$

The relativistic correction is significant at the accelerating voltages used in electron microscopy, which are typically 100–400 keV.

An electron beam can be focused with the help of electromagnetic lenses, but the focusing mechanism is quite different from that used in the optical microscope, where the latter relies on the refractive index of glass and geometrical optics to achieve a sharp focus (see Chapter 3). A *magnetic field* deflects any electron which has a component of velocity perpendicular to the magnetic field vector, and the deflecting force acts in a direction perpendicular to the plane containing both the velocity and the magnetic field vectors. As a consequence, the electron follows a *spiral path* when passing through a uniform magnetic field.

The magnetic field generated by an electromagnetic lens is *cylindrically symmetric*, and a divergent electron beam passing through such a lens will be brought to a focus, providing that the angular divergence is small. It is fortunate that the elastic scattering of electrons is limited to small angles and permits both *electron diffraction patterns* and *electron microscope images* to be brought sharply into focus in the electron microscope. Assuming a wavelength of 0.0037 nm (100 keV electrons) and interplanar spacings of the order of 0.2 nm, we expect Bragg angles of less than 1°. At these small angles, it is common to quote the Bragg angle in *radians* and to use the approximation, $\sin\theta \approx \theta$. The Bragg relationship can then be written as $\lambda = 2d\theta$, and we will use this form to describe the elastic scattering of electrons by a thin-film specimen in the transmission electron microscope.

Finally, the influence of *inelastic* scattering is important, since electrons which have lost energy by inelastic scattering will have a longer wavelength. The electron beam will no longer be monochromatic, and cannot therefore be brought to a sharp focus. The stopping cross-section for electrons, σ, is defined as $\sigma = (1/N)(\mathrm{d}E/\mathrm{d}x)$, where N is the number of atoms per unit volume and $\mathrm{d}E/\mathrm{d}x$ is the rate of energy loss per unit distance travelled by the electron. As the energy of the electrons is increased (a higher accelerating voltage), σ decreases. (However, with very low electron energies the inner-shell electrons no longer contribute to inelastic scattering of the incident beam and the scattering cross-section decreases.) The general shape of the scattering cross-section curve is shown in Fig. 2.31. The inelastic scattering cross-section increases with Z, so that higher-atomic-number materials can only give sharp electron diffraction patterns for very thin samples. For a 200 keV incident electron beam, tungsten or gold films thicker than *ca* 100 nm will absorb the incident-beam energy and can only give electron diffraction patterns in reflection, at glancing incidence. The maximum thickness for transmission electron diffraction from steel at 200 kV is of the order of 120 nm, while silicon and aluminium specimens must be

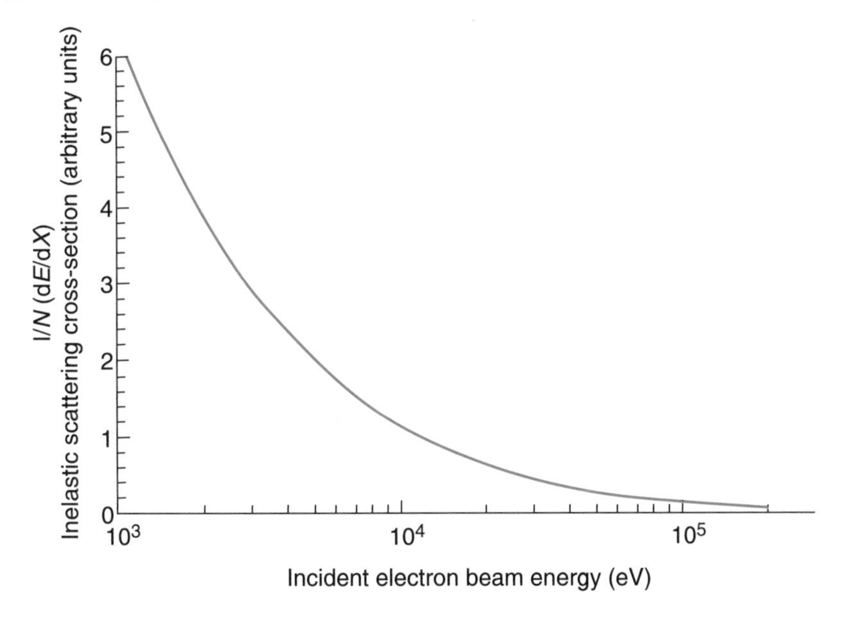

Figure 2.31 The general form of the variation of the inelastic scattering cross-section with incident electron energy

less than *ca* 150 nm in thickness. These values are two orders of magnitude less than the maximum sample thickness in an X-ray diffraction experiment.

2.5.1.1 The Limiting Sphere for an Electron Beam

The limiting and reflecting sphere constructions used to analyse X-ray diffraction phenomena in reciprocal space are equally valid for electron diffraction patterns obtained in transmission electron microscopy, although there are two significant differences.

The first concerns the very short wavelength of the electrons when compared to the interplanar spacings in crystals, $|\mathbf{k}_0| = |\mathbf{k}| \gg |\mathbf{g}|$ (Fig. 2.32), so that, as noted above, $\sin\theta \approx \theta$. In Fig. 2.32, the reflecting sphere construction has been rotated by 90°, a symbolic gesture to the engineering design of the modern electron microscope, in which the beam source (the electron gun) is mounted vertically, thus generating a beam of electrons which is incident on a thin-film specimen mounted in the horizontal plane. This may be compared to the standard design for X-ray diffraction units, in which the X-ray beam is generated in the horizontal plane (for maximum mechanical stability).

The second modification concerns the effective size of the reciprocal lattice points. The elastic scattering cross-section for electrons is much greater than that for X-rays, so that the intensity scattered into the diffracted beam increases rapidly with

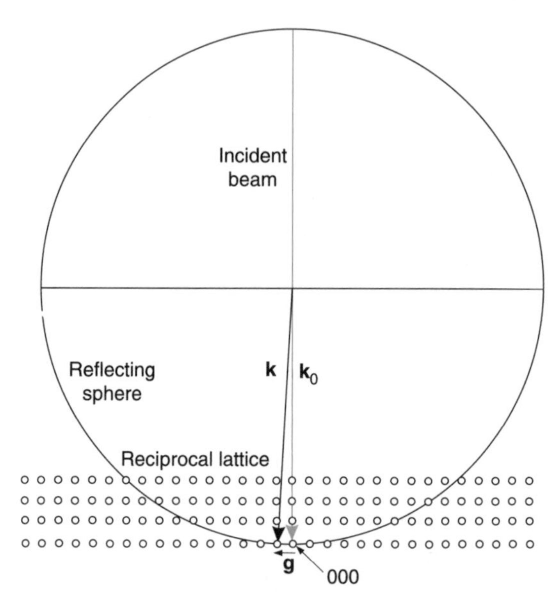

Figure 2.32 The wave vectors in electron diffraction, **k**, are very large when compared to the reciprocal lattice vectors **g**, thus allowing for certain simple geometrical approximations

sample thickness, to the point at which all of the energy in the incident beam may be transferred into the diffracted beam and the diffracted beam starts to be rediffracted back into the incident beam (Fig. 2.33). This process, termed *double diffraction*, leads to oscillations in the diffracted intensity with increasing specimen thickness. These oscillations have a periodicity t_0, the *extinction thickness*, which is characteristic of the electron energy and the structure factor for the actively diffracting planes, and is given by:

$$t_0 = \frac{\pi V_c}{\lambda |F(hkl)|} \tag{2.32}$$

where V_c is the volume of the lattice unit cell. Typical values for t_0 at the accelerating voltages used in transmission electron microscopy are less than 100 nm. It is the *extinction thickness* that limits the effective size of the reciprocal lattice points in electron diffraction, to *ca* $1/t_0$, equivalent to the order of $10^{-2}d$ (compared to values of $10^{-4}d$ or less in X-ray diffraction). An additional 'small crystal' effect may also dominate the Bragg condition in electron diffraction, thus increasing the size of the reciprocal lattice points for very thin films and small crystals (see Chapter 4).

In electron diffraction, both the radius of the reflecting sphere and the size of the reciprocal lattice points are large when compared to X-ray diffraction, therefore relaxing the diffracting condition set by Bragg's law to allow several diffracted beams of electrons to be scattered simultaneously from a thin sample with a

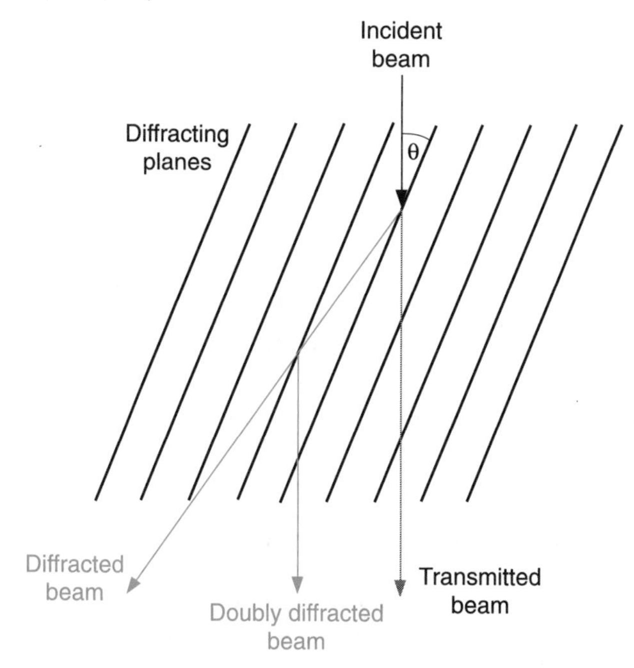

Figure 2.33 Schematic representation of double diffraction of the electron beam, which leads to diffracted intensities which oscillate with film thickness

symmetry axis in the plane of the film (Fig. 2.34). This situation would not be possible in X-ray diffraction.

2.5.2 Ring Patterns, Spot Patterns and Laue Zones

The electron diffraction pattern from a thin film of a single crystal, which is oriented with a major zone axis parallel to the electron beam, will contain all of the reciprocal lattice points which are intersected by the reflecting sphere. This will include those reciprocal lattice points which surround the 000 spot (corresponding to the directly transmitted beam), in so far as the (slight) curvature of the reflecting sphere permits them to fulfil the Bragg condition, but it may also include *additional* points which lie in a layer of the reciprocal lattice *above* that containing the origin (Fig. 2.35). The first set of points is referred to as the *zero-order Laue zone*, while all subsequent rings of diffracting spots are termed *higher-order zones*.

Adequate calibration of the electron microscope enables distances in the diffractogram to be accurately interpreted as distances in reciprocal space, although there are several sources of calibration error. These include not only the physical limitations of the technique, in particular the finite diameter of the reciprocal lattice points, but also experimental limitations associated with electromagnetic lens aberrations in the microscope, curvature of the thin-film specimen, and the

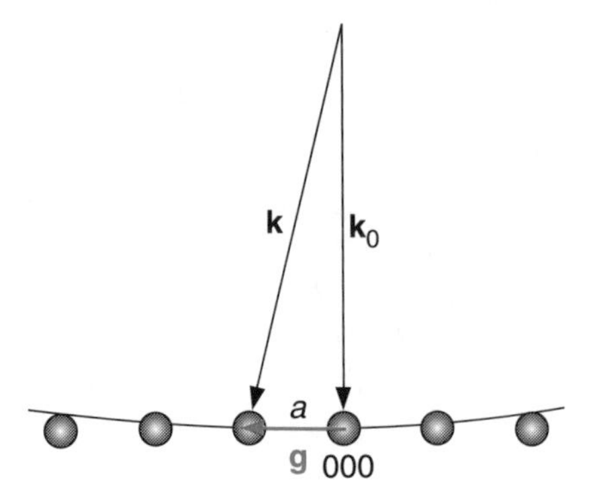

Figure 2.34 Since the radius of curvature of the reflecting sphere is large and the reciprocal lattice points have a finite diameter, Bragg diffraction occurs even though the Bragg condition is not exactly satisfied

photographic response of the recording film. In general, it is not possible to specify lattice spacings derived from electron diffraction measurements to much better than 2 % of the lattice parameter, which is at least two orders of magnitude worse than can be achieved by X-ray diffraction. Although this is true for standard electron diffraction techniques, usually called *selected-area diffraction*, an alternative technique, namely *convergent-beam electron diffraction*, can, for some crystals, be very much more accurate and can be used to determine localized lattice strains, as well as to solve the crystal structure. Convergent-beam electron diffraction is, however, beyond the scope of this present book. The interested reader should consult the texts by Spence and Zuo [5] or Williams and Carter [6], listed in the Bibliography for this chapter.

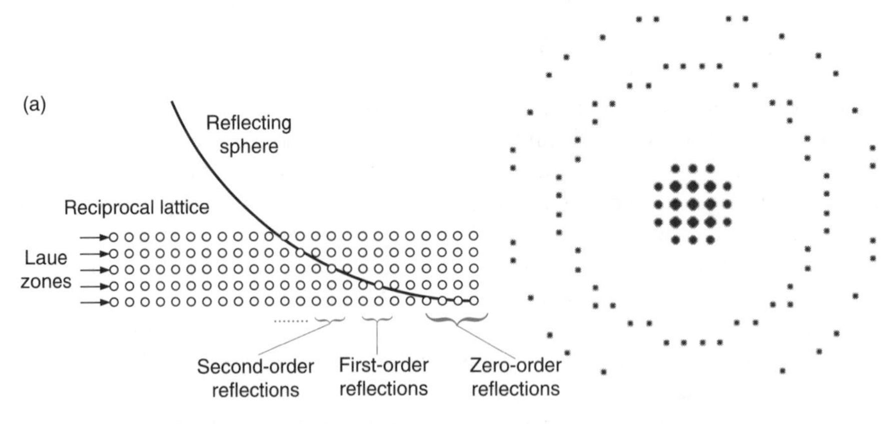

Figure 2.35 Single-crystal electron diffraction from more than one Laue zone: (a) mechanism of formation; (b) a diffraction pattern from a [100] oriented aluminium single-crystal film

The electron beam can also be focused electromagnetically to a fine probe, or small apertures can be used to limit the diameter of a parallel electron beam, thus enabling selected-area diffraction from very small specimen areas, of the order of 20 nm. The volume sampled is then a minute fraction of that sampled by X-rays, yielding phase information on individual crystallites and precipitates in a polycrystalline material.

If the area illuminated by the electron beam includes a large number of crystallites, then a *powder* pattern is generated, analogous to an X-ray powder pattern. In the case of electron diffraction, the fluorescent screen or photographic film is positioned normal to the incident beam, and records successive rings of reflections from each family of reflecting planes (Fig. 2.36). The radius R of a specific ring on the powder pattern is related to the d-spacing of the reflection and the wavelength of the electron beam λ by the following relationship:

$$d = \frac{2\lambda L}{2R} \tag{2.33}$$

where L is the *effective camera length* of the electron microscope when used as a diffraction camera. Equation (2.33) is written with a factor of 2 in both the numerator and denominator since it is good practice to measure the distances $2R$ between two diffraction spots hkl and $\overline{hkl}$ in an electron diffraction pattern, in order to avoid errors associated with determining the position of the directly transmitted 000 beam. The parameter L can be varied in most microscopes, in order to select a value suitable for the lattice parameters of the phases being studied, and good calibration of the microscope should give the term λL, the *camera constant* of the microscope, with an accuracy of $\sim 1\,\%$.

2.5.3 Kikuchi Patterns and their Interpretation

For moderately thick transmission electron microscope specimens, a proportion of the incident electrons will undergo inelastic scattering. These electrons, having lost some energy, are deflected out of the path of the incident beam, but before exiting the specimen these same electrons may also be elastically scattered. If the specimen is a sufficiently perfect single crystal, this secondary elastic scattering will lead to a characteristic *Kikuchi line pattern*, superimposed on the single-crystal spot pattern associated with Bragg diffraction of the primary incident beam. The Kikuchi line pattern arises because the angular distribution of diffuse, inelastically scattered electrons falls off rapidly with angle, typically obeying an $I = I_0 \cos^2 \alpha$ law, where α is the scattering angle. The crystal lattice planes will then elastically scatter any electrons incident at the Bragg angle (since little energy is lost in an initial inelastic scattering event, the Bragg angle is almost unchanged), but more electrons will be elastically scattered away from the diffuse distribution in the region closer to the incident beam. This leads to a *dark* line in the diffuse scattering pattern close to the centre and simultaneously generates a parallel *white* line at a distance from the dark

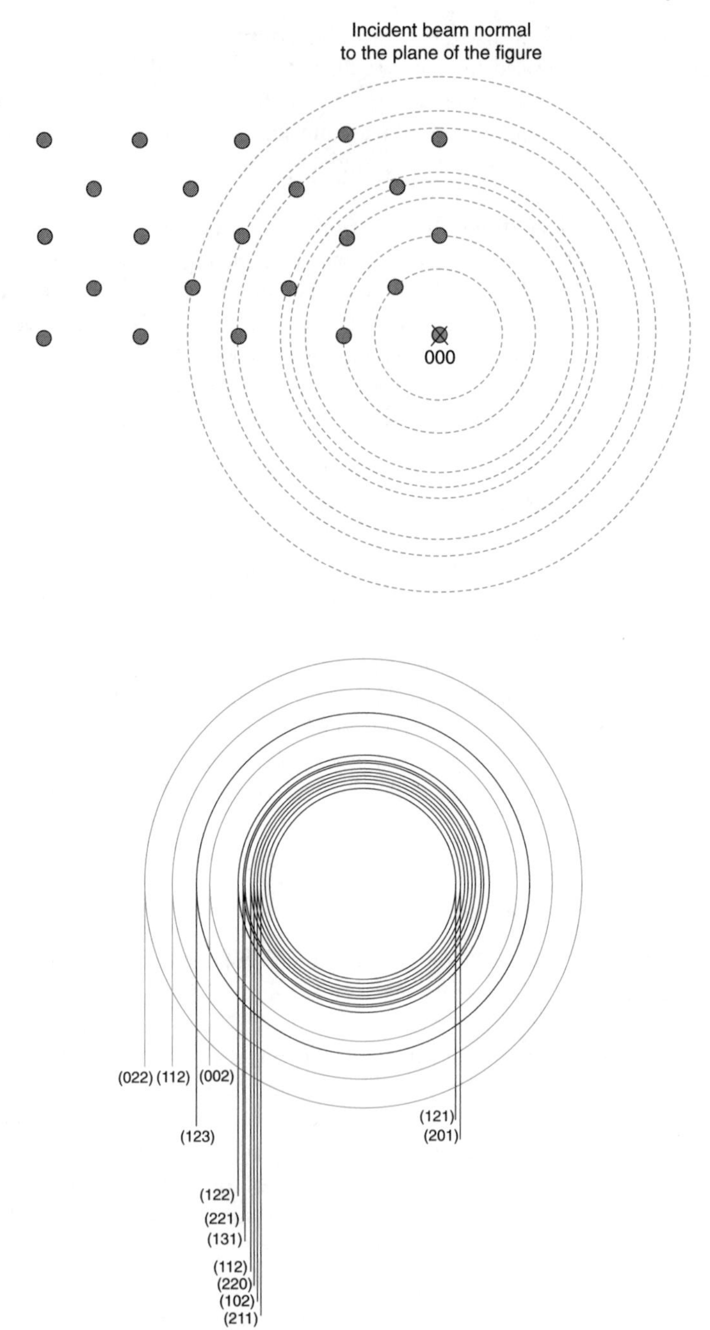

Figure 2.36 Powder patterns in electron diffraction: (a) mechanism of generation; (b) a powder ring pattern from a tempered carbon steel

line determined by the interplanar spacing (Fig. 2.37). (These Kikuchi lines are actually sections of hyperbolae, but since the **g**-vectors are so much smaller in magnitude than the electron wave vectors **k**, they appear on the diffraction pattern as straight lines.) From their geometry, the *spacing* of a light/dark pair of Kikuchi lines is proportional to the value of |**g**|, while the *displacement* of the *mid-point* between the lines is an accurate measure of the angle which the reflecting lattice planes make with the primary incident beam.

When the Kikuchi pattern is symmetrically aligned with respect to the incident beam (Fig. 2.38), the incident beam is then accurately parallel to a symmetry zone of the crystal, and this zone can be readily identified from the Kikuchi pattern. A series

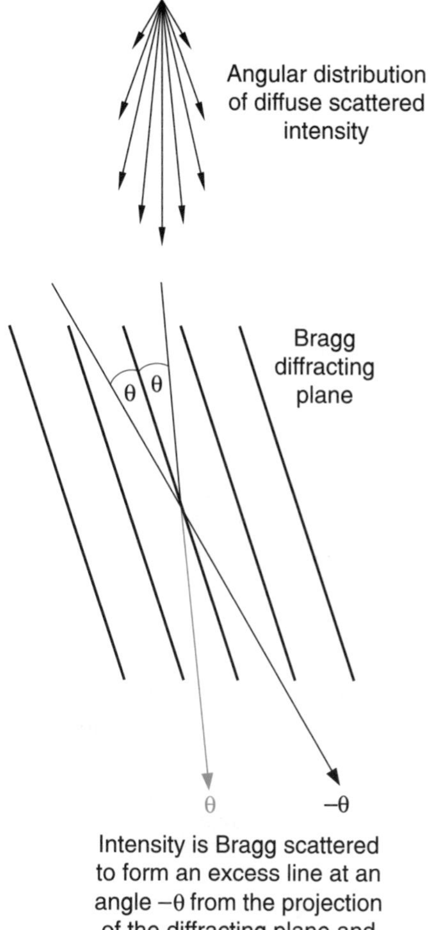

Figure 2.37 Mechanism of formation of Kikuchi line diffraction patterns (see text for details)

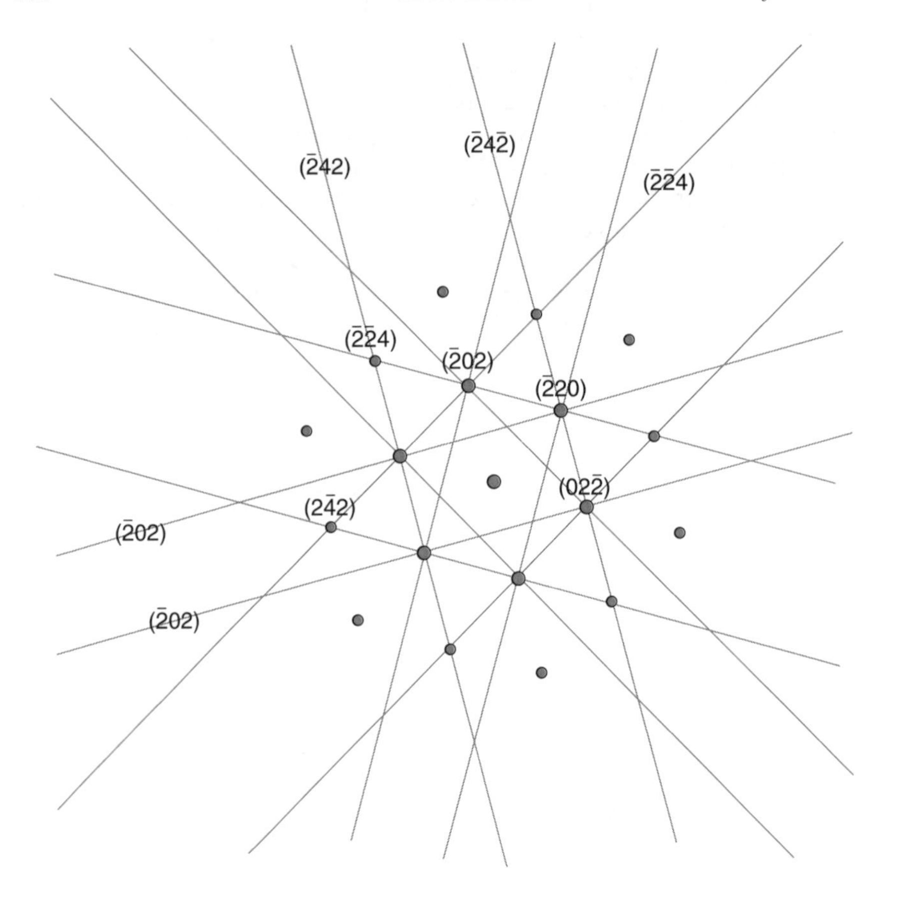

Figure 2.38 Kikuchi diffraction pattern from a [111] oriented aluminium crystal superimposed on the single-crystal spot pattern

of Kikuchi patterns, taken by tilting the specimen about two axes at right angles in the plane of the specimen, can be used to generate a *Kikuchi map* (Fig. 2.39). The Kikuchi map accurately reflects the crystal symmetry and can be used to identify the orientations of specific grains in the electron microscope, almost by inspection. Kikuchi patterns are used to align a crystal exactly on a zone axis, or to shift a crystal off a zone axis by a given angle. With the help of a Kikuchi map, you can tilt the crystal in a controlled way from one zone axis to another.

Summary

Regular arrays of atoms in a crystal elastically scatter short-wavelength radiation (either *X-rays*, *electrons*, or *neutrons*) at well-defined angles to the incident beam.

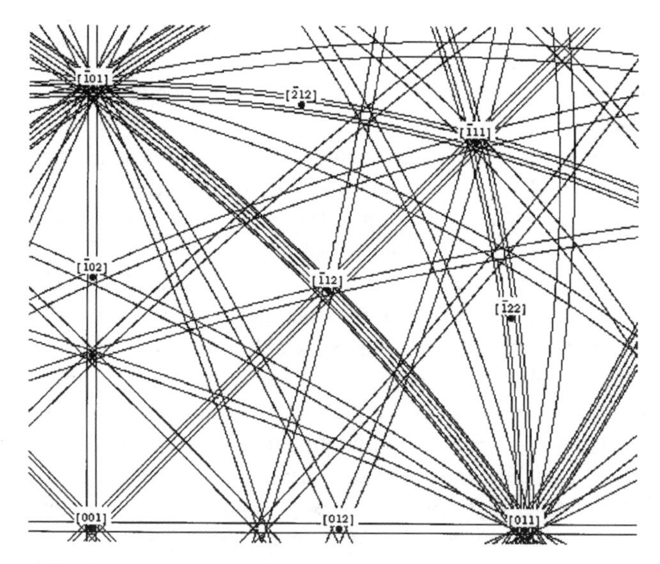

Figure 2.39 Kikuchi map of a cubic silicon crystal with the principle reflecting planes identified

The scattering angle and the scattered intensity are functions of the radiation, the wavelength and the crystal structure, and the process is termed *diffraction*. A *diffraction pattern* is the *angular distribution* of the scattered intensity in space, and is recorded by 'intercepting' the scattered radiation, for example by a photographic emulsion. A *diffraction spectrum* is the *intensity* of the diffracted radiation recorded as a function of the *scattering angle*.

Laue showed that the allowed angles of scattering for a regular crystal were a simple function of the *lattice parameters* of the unit cell, and Bragg simplified the scattering relationships to yield the *Bragg law*, $\lambda = 2d \sin \theta$, which relates the *angle of scattering*, 2θ (relative to the direction of the incident beam), to the *spacing* of the planes in the crystal lattice, d, and the *wavelength* of the incident radiation, λ. The allowed angles of scattering (the *diffraction angles*) always correspond to integer values of the *Miller indices*, but not all possible combinations of the Miller indices give rise to diffraction peaks, and some reflections (*forbidden* reflections) are disallowed because the *Bravais lattice* contains face-centred or body-centred lattice points.

The *intensities* of the diffraction peaks are determined by the atomic number and position of the atoms associated with the lattice points, that is the *space group*, the *Wyckoff positions* of the atoms and the atomic species. Thus some 'allowed' peaks may be enhanced (the atoms corresponding to each lattice point scattering *in-phase*), while others may be reduced, or even absent (*destructive interference*).

A convenient representation of the angular positions of the allowed diffraction peaks is given in *reciprocal space*, in which the diffracting planes are represented by points (*reciprocal lattice points*) whose positions are determined by the *reciprocal* of the interplanar spacing and the *direction vector* normal to the reflecting planes. In reciprocal space, the Bragg equation defines a sphere (the *reflecting sphere*) and any reciprocal lattice point which can be made to intersect the surface of this sphere should give rise to a diffracted beam.

Reducing the wavelength of the incident radiation *increases* the radius of the reflecting sphere, and may allow reciprocal lattice points further from the origin (smaller interplanar spacings) to diffract. *Rotating* the crystal (or reducing the grain size of a polycrystal) *increases* the probability that reciprocal lattice points will intersect the sphere and give rise to diffraction. The volume of the specimen sampled by the incident beam depends on the radiation. In *neutron diffraction*, elastic scattering generally occurs over distances of the order of centimetres before the probability of an inelastic scattering event becomes significant, while even very-high-energy (MeV) *electrons* will be inelastically scattered if the sample thickness exceeds a few microns. *X-rays* are an intermediate case. Most engineering materials will generate an elastically scattered signal from a region 20 to 100 microns in depth when irradiated with X-rays whose wavelengths are of atomic dimensions.

An *X-ray diffractometer* consists of a *source* of X-rays, a sample *goniometer* (which positions the sample accurately in space) and a *detector*. The detector can be rotated about the sample to select or sample specific diffraction angles. (In a few cases, a photographic emulsion may be used to record a diffraction spectrum, or an array of detectors may replace the single detector.) *White* X-radiation may be used for some purposes, for example to record a diffraction pattern from a single crystal in a *Laue camera*. Many X-ray diffraction studies are based on *polycrystalline* samples which give *powder diffraction* spectra from *monochromatic radiation*. In powder diffraction, fine grains of crystalline phases diffract in a series of cones which are intercepted in turn by the rotating detector. If the crystal grains are not randomly oriented in space, but possess some *preferred orientation* (or crystalline texture), then the diffraction pattern will show intensity anomalies which can be analysed, for example in a *texture goniometer*, to derive a *pole figure* (the distribution of diffracted intensity for a particular plane with respect to the sample coordinates) or a *crystallite orientation distribution function* (the probability of finding a crystal having a particular orientation in space).

In many cases, an accurate determination of the *geometry* of the diffraction pattern is sufficient to deduce the phases present and the orientation distribution of the individual crystals, but far more information can be derived from a measurement of the *relative intensities* of the diffraction peaks. Each unit cell in a crystal scatters a proportion of the incident beam into each diffraction peak (associated with a specific *hkl* reflection). The *structure factor* gives the relative scattering power of the different *hkl* planes in the crystal, and can be calculated from a suitable model structure in order to predict the relative peak intensities for a random polycrystal. The *structure factor* is not the *only* parameter determining the relative diffracted intensities, and *diffraction analysis* must also take into account several other effects: the Lorentz or *polarization factor*, the *multiplicity* of the reflecting planes, the specimen *geometry*, the angular dependence of *absorption* losses and the effect of *thermal vibration* at elevated temperature.

Electron diffraction differs from X-ray diffraction in many significant respects. The electron wavelengths which are of practical importance in an electron microscope are very small when compared to the interplanar spacings in crystals, and the effects of *inelastic scattering* are pronounced (leading to the formation of *Kikuchi patterns*, for example). Useful sample thicknesses are *always* sub-micron. The short wavelengths of energetic electrons increases the diameter of the reflection sphere, which is now very large when compared to the spacing of the reciprocal lattice points. Finally, the small volume of the region responsible for electron diffraction significantly relaxes the conditions for diffraction, thus *broadening* the diffraction peak width and, equivalently, smearing the reciprocal lattice points over a region of reciprocal space which is no longer small in comparison to the spacing between these points.

Bibliography

1. C.S. Barrett and T.B. Massalski, *Structure of Metals*, Pergamon, Oxford (1980).
2. B.D. Cullity, *Elements of X-Ray Diffraction*, Addison-Wesley (1956).
3. J.B. Cohen, *Diffraction Methods in Materials Science*, Macmillan, New York, (1966).
4. I.D. Noyan and J.B. Cohen, *Residual Stress: Measurement by Diffraction and Interpretation*, Springer-Verlag, Berlin (1987).
5. J.C.H. Spence and J.M. Zuo, *Electron Microdiffraction*, Plenum, New York (1992).
6. D.B. Williams and C.B. Carter, *Transmission Electron Microscopy*, Plenum Press, New York (1996).

Worked Examples

Let us now use the techniques that we have discussed for the characterization of our selected materials. We will start with a simple example, an *automated diffractometer* (Bragg–Brentano) used to verify the crystal structure of a metal powder and measure its lattice parameters. Armed with a good diffraction spectrum from the sample, accurately mounted in the diffractometer, we use a literature database, the *Joint Committee of Powder Diffraction Standards* (JCPDS, now called the International Centre for Diffraction Data). JCPDS is a database of *experimentally observed* and *calculated* diffraction spectra (both d-spacings and relative intensities), which can be compared to the measured spectrum in order to identify the phases present in the sample. This database is available in two formats:

(a) *Tabulated cards* for the different spectra, which can be accessed either from the *names* or *chemistry* of the compounds, or by the *d-spacings* of the strongest reflections.

(b) *A computerized database* which can be accessed by using a computer program which *automatically* compares the major d-spacings derived from a spectrum with those in the database.

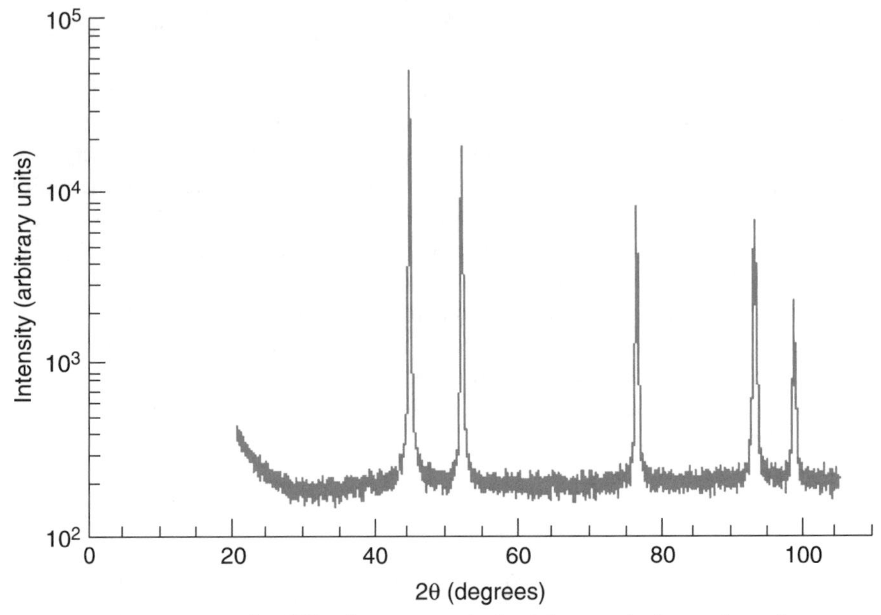

Figure 2.40 X-ray powder diffraction pattern obtained from an 'unknown' sample

The X-ray diffraction spectrum of the powder we wish to identify is shown in Fig. 2.40. We first generate a table of the *d*-spacings derived from the prominent reflections, by using Bragg's law and the known wavelength for CuK_α radiation ($\lambda = 0.1540598$ nm). We now input these *d*-spacings into the computerized JCPDS database, and the output gives us the possible phases which best match the experimentally observed *d*-spacings and relative intensities. Now we extract the cards for each of the options from the JCPDS database in order to compare the measured values of the *d*-spacing and intensity (since this is a powder sample, texture should *not* be a problem). The 'unknown' powder in this case is nickel (Ni), whose JCPDS card is shown in Fig. 2.41. Of course, our sample could have contained several different phases of the metal, but in such cases additional information is usually available on the phases that might exist or the elements that might be present in the material.

From the same powder sample diffraction spectrum of Ni, we can also determine the exact lattice parameter. Any variations in lattice parameter would indicate either the presence of residual stress (very unlikely in a powder specimen), or a nickel-based alloy (one or more alloying elements in solid solution). By careful calibration of the spectrometer and assuming a linear dependence of lattice parameter on composition (*Vegard's Law*), we could determine the concentration of the alloy from an exact determination of the lattice parameters.

Fig. 2.42 shows the results for our sample, in which the apparent lattice parameter for nickel, determined from the individual reflections by using the relationship between *hkl* and the lattice parameter for a cubic structure, has been plotted as a function of $\cos 2\theta / \sin \theta$. The systematic errors are quite small, and the best value for

4-0850 MINOR CORRECTION

d	2.03	1.76	1.25	2.034	Ni
4-0854					
I/I_1	100	42	21	100	NICKEL
4-0850					

Rad. CuKα λ 1.5405 Filter Ni
Dia. Cut off Coll.
I/I_1 G. C. DIFFRACTOMETER d corr. abs.?
Ref. SWANSON AND TATGE, JC FEL. REPORTS, NBS 1951

Sys. CUBIC S.G. O_H^5 – Fm3m
a_0 3.5238 b_0 c_0 A C
α β γ Z 4
Ref. IBID.

εα nωβ εγ Sign
2V D 8.907 mp Color
Ref.

SPECTROGRAPHIC ANALYSIS SHOWS <0.01% EACH OF MG, SI AND CA.
AT 26°C
TO REPLACE 1-1258, 1-1260, 1-1266, 1-1272, 3-1043, 3-1051

d Å	I/I_1	hkl	d Å	I/I_1	hkl
2.034	100	111			
1.762	42	200			
1.246	21	220			
1.0624	20	311			
1.0172	7	222			
0.8810	4	400			
.8084	14	331			
.7880	15	420			

Figure 2.41 The JCPDS card for nickel. Reproduced by permission of the International Center for Diffraction Data

the measured lattice parameter obtained by extrapolating to $\theta = \pi/2$ is $a = 3.5226$ Å. This is close to the values listed by the JCPDS ($a = 3.5238$ Å), and in *Pearson's Handbook of Crystallographic Data for Intermetallic Phases*† ($a = 3.5232$ Å). In order to confirm that a solid solution is present, we would need to prepare a calibration curve and then use analytical techniques to determine the chemistry of the sample.

The same approach can be used to determine the lattice parameters of a polycrystalline sample of α-Al_2O_3 (Fig. 2.43). For alumina, the situation is more complicated, since we have to refine two lattice parameters for the hexagonal unit cell. This can be achieved by using a simple computer program. We could also determine a *correction factor* to account for any systematic errors in our experiment by using a standard sample, and then correcting the measured lattice parameters for the phases of interest accordingly.

Fig. 2.44 shows a diffraction spectrum from a two-phase mixture of α-Fe and Fe_3C in 1040 steel. The reflections from the carbide are weak, since the volume fraction of carbide in the steel is low. Overlapping of some carbide reflections with those of iron also occurs, and this complicates the analysis of the spectrum. Careful inspection of the diffraction spectrum, and comparison with simulated diffraction patterns, is important in order to identify each reflection correctly.

Fig. 2.45 show diffraction spectra from a different type of sample, namely a thin polycrystalline film of aluminium deposited on a thin film of TiN formed on an even thinner film of titanium. These films were deposited sequentially on a single-crystal silicon substrate. In order to detect such thin films in a Bragg–Brentano

† See the Bibliography in Chapter 1.

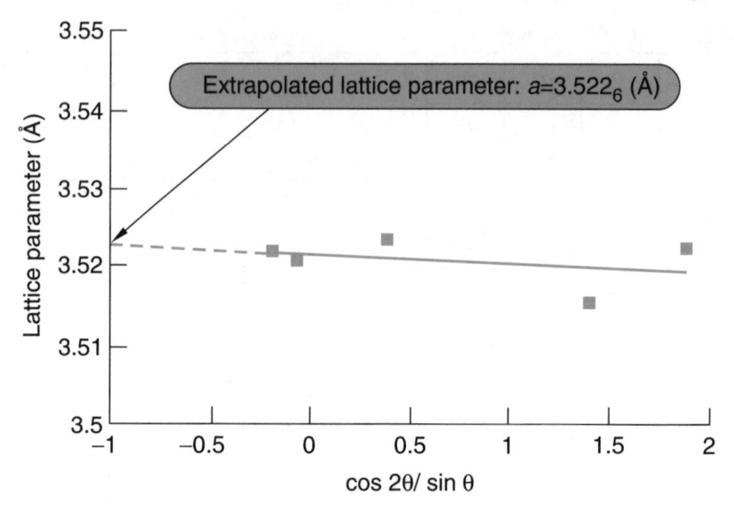

Figure 2.42 The lattice parameter of nickel determined from the *d*-spacings taken from Fig. 2.40, showing the method of extrapolating the data to $2\theta = 180°$ as a way to minimize errors

diffractometer, long counting times are required for each value of 2θ, of the order of 20 s for the patterns shown in Fig. 2.45. Specialized thin-film diffractometers significantly reduce the counting time required for such specimens, but here we shall only consider the Bragg–Brentano diffractometer.

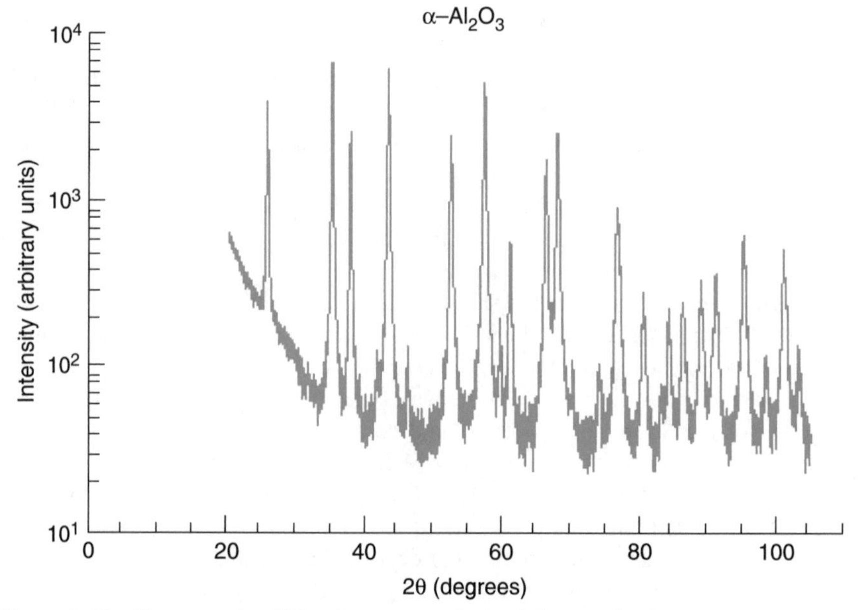

Figure 2.43 X-ray powder diffraction pattern obtained from α-alumina

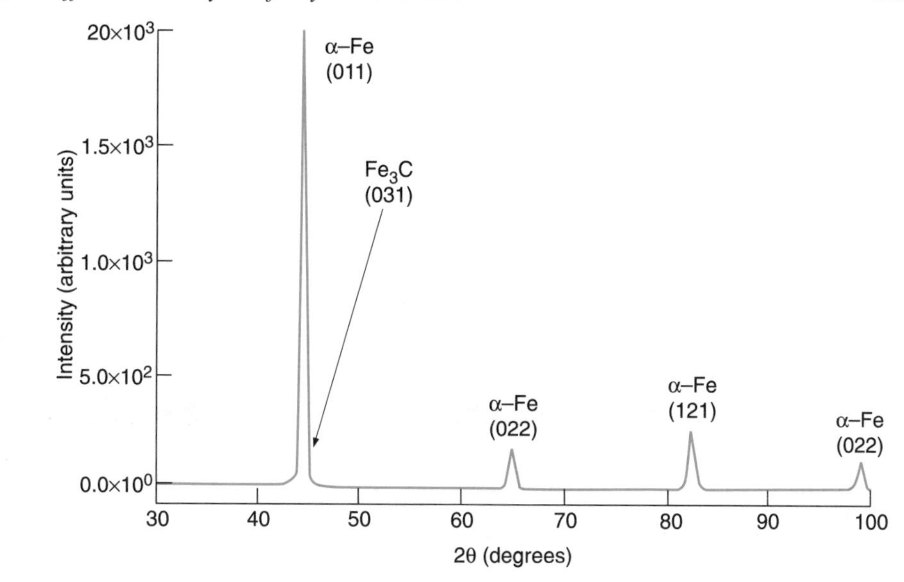

Figure 2.44 X-ray powder diffraction pattern obtained from 1040 steel. (Note that this appears to be the same pattern as that shown in Fig. 2.1)

The reflections shown in Fig. 2.45 have all been indexed. The high-intensity peak from silicon reflects the fact that the silicon single crystal has been oriented in the diffractometer to diffract from the $\{400\}$ planes. Even a slight misalignment of the silicon crystal would remove this reflection.

Consider the reflections due to the deposited thin films. Only a few reflections are detectable, reflections which, according to the JCPDS, are not necessarily the strongest reflecting planes. It follows that the deposited films must have a preferred orientation with respect to the plane of the silicon substrate crystal. Qualitatively, the titanium film has a texture in which the [0001] direction is normal to the silicon substrate surface, while the TiN and aluminium films both have a texture with a [111] direction normal to this surface.

Now, consider a calibration curve prepared for quantitative phase analysis of a sample containing a mixture of alumina and nickel (Fig. 2.46). The calibration curve was prepared by mixing known amounts of the two phases in a powder form. The strong deviation from linearity demonstrates the need for such a calibration curve.

Let us move on to electron diffraction. As stated earlier, selected area electron diffraction does not have the precision which is available in X-ray diffraction, but we can obtain diffraction patterns from single grains in a polycrystalline sample, or from selected regions within a single grain. In Chapter 4, we will use selected area diffraction to correlate between the crystallographic orientation of a grain, and the defects observed in an electron micrograph. It is important to be able to solve a selected area diffraction pattern in order to identify the reflections responsible for contrast in the image and to determine the zone axis and principle directions in the thin-film sample.

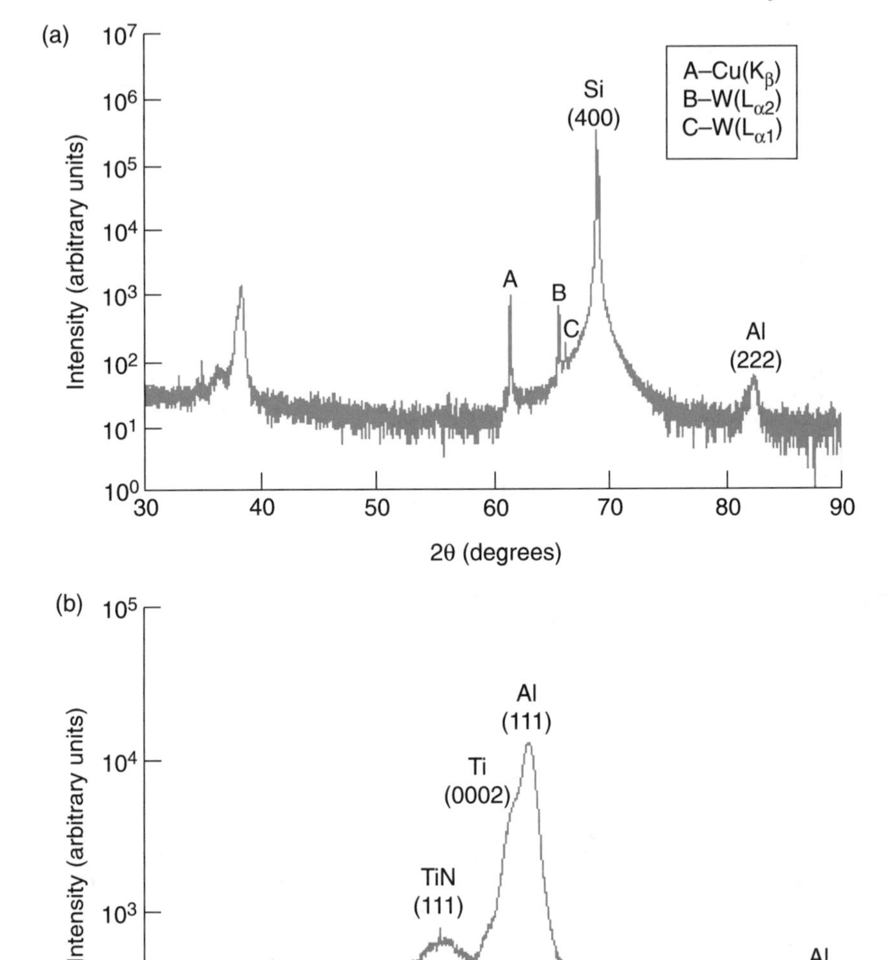

Figure 2.45 X-ray powder diffraction patterns obtained from a thin polycrystalline film of Al/Ti/TiN deposited on a single crystal of Si: (a) full range of 2θ (30–90°); (b) enlarged area for $2\theta = 30$–45°

Our first example is a selected-area electron diffraction pattern from a randomly oriented aluminium polycrystal (Fig. 2.47). The ring pattern results from the intersection of the reciprocal lattices from a large number of grains in the polycrystalline sample with the reflection sphere. The *d*-spacing for each ring is determined by measuring the diameter of the ring and inserting the known value of

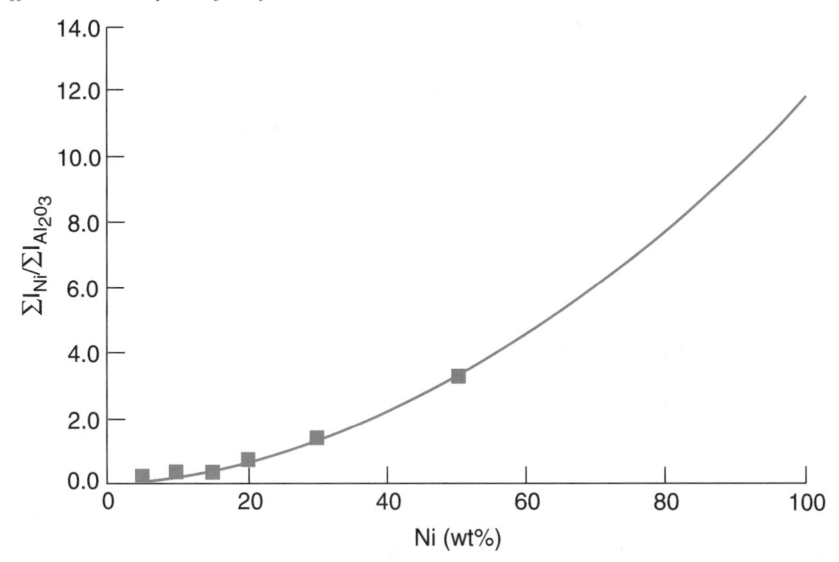

Figure 2.46 Calibration curve for quantitative analysis of a two-phase mixture of nickel and alumina, showing the ratio of the integrated peak intensities, summed for nickel and alumina, as a function of the nickel content

λL (the microscope camera constant) in equation (2.33). In this present case, λL is equal to 2.43 Å cm (10 Å = 1 nm), and the corresponding *d*-spacing of each ring is given. We can either use the information for aluminium from the JCPDS to insert the *hkl* values of each ring, or we can calculate a list of the *d*-spacings for every possible *hkl* in aluminium by using the relationship between the *d*-spacing and the lattice parameter for a cubic crystal. This calculation will not tell us if the reflection has a non-zero structure factor, but calculation of the structure factor for each reflection is also quite simple, being achieved by using an appropriate computer program.

The first step in solving any selected-area diffraction pattern is to calibrate the values of λL for the different camera length settings of the microscope. Although most modern transmission electron microscopes display a value of the camera length L, this value is only approximate, and calibration is necessary by using a standard specimen of known lattice parameter. We use aluminium (not a good choice, however, since the lattice parameter of aluminium is sensitive to dissolved impurities). A single-crystal region of an aluminium foil is oriented perpendicular to a low-index zone axis and a series of diffraction patterns is recorded for various values of the camera length L at an accelerating voltage of 200 kV (Fig. 2.48).

Can we index an electron diffraction pattern without knowing either λL or the zone axis of the crystal? Actually, there is a rather simple solution to this problem. Aluminium has an FCC structure, for which the following holds:

$$\frac{1}{d^2} = \frac{h^2 + k^2 + l^2}{a^2} \tag{2.34}$$

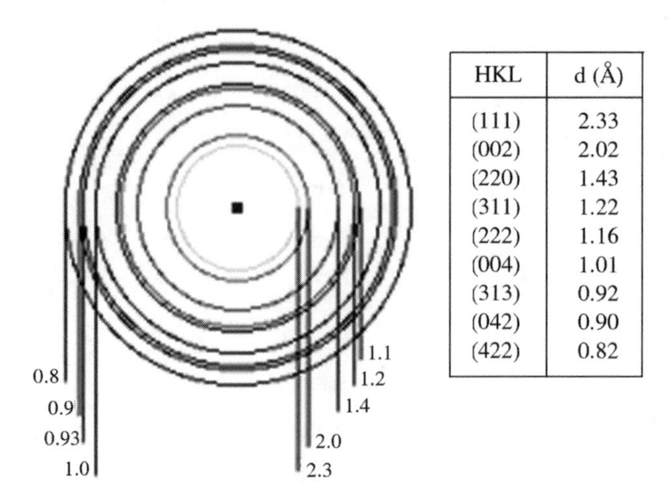

HKL	d (Å)
(111)	2.33
(002)	2.02
(220)	1.43
(311)	1.22
(222)	1.16
(004)	1.01
(313)	0.92
(042)	0.90
(422)	0.82

Figure 2.47 Selected-area electron diffraction pattern obtained from a polycrystalline aluminium specimen. Since the average aluminium grain size is much smaller than the selected area, a 'ring' pattern is formed. The measured *d*-spacings of the rings are indicated on the pattern, and a table of *d*-spacings for different planes in aluminium is also given

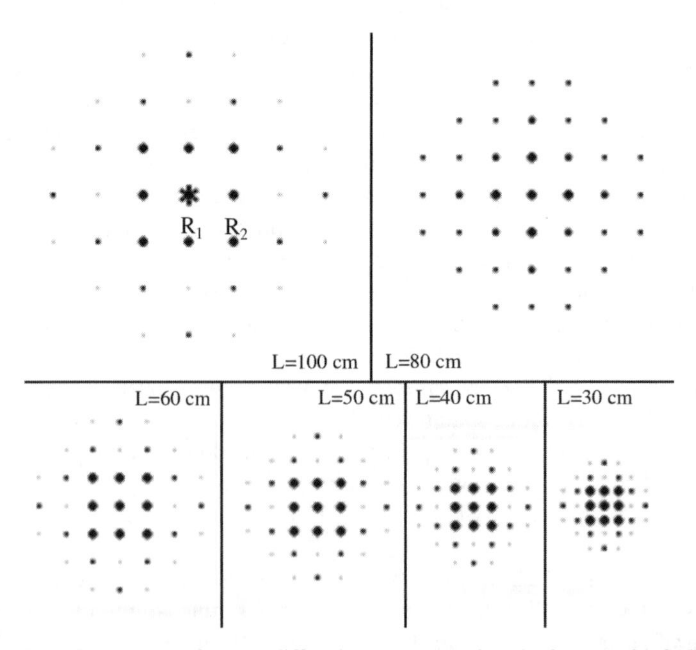

Figure 2.48 Selected-area electron diffraction patterns of a single crystal of aluminium, recorded at different nominal camera lengths L (cm): (a) 100; (b) 80; (c) 60; (d) 50; (e) 40; (f) 30

Putting $(h^2 + k^2 + l^2) = N^2$, and using the ratio of the d-spacings for two reflections, we can write the following:

$$\frac{d_2^2}{d_1^2} = \frac{N_1^2}{N_2^2} \tag{2.35}$$

In order to relate this equation between d and N to the values of R, that is the distance from the centre spot to the reflections of interest, we substitute $d = \lambda L/R$ into equation (2.35) as follows:

$$\frac{d_2^2}{d_1^2} = \frac{N_1^2}{N_2^2} = \frac{R_1^2}{R_2^2} \tag{2.36}$$

This equation now shows the relationship between the ratio of the measured R values, and the corresponding d-spacings. This would be difficult to solve without a computer, but we can use the data from Table 2.1. We measure the distance from the central spot to each of two diffraction spots of interest (Fig. 2.48(a)) and find that $R_1^2/R_2^2 = 0.5$. Simple examination of Table 2.1 immediately identifies the reflecting planes as being (200) and (220). Since the lattice parameter for pure Al is known ($a = 4.05$ Å), the relevant d-spacings can be calculated, $d_{(200)} = 2.02$ Å and $d_{(220)} = 1.43$ Å, and we can now return to our original relationship between d and λ, $d = 2\lambda L/2R$, in order to calculate λL. This same procedure is possible for any diffraction pattern taken from aluminium at any arbitrary camera length, thus allowing for an accurate calibration of the camera constant λL for the microscope (Fig. 2.49).

In principle, standard statistical methods should be used to calculate the errors in λL, so it is better to index the complete diffraction pattern. We now know that point 1 in Fig. 2.48(a) corresponds to (200), while point 2 corresponds to (220), and the angle subtended by these two vectors at the origin (000) is measured to be 45°. We now take the *vector cross-product*† between the two directions [200] and [220] (the direction vectors are always perpendicular to planes for the cubic system) in order to find the zone axis of the diffraction pattern, [001]. We check this by examining a stereographic projection for the cubic crystal structure, and note that the angle between (200) and (220) should be 45°, with both planes on a great circle whose zone axis is [001] (Fig. 2.50). Now continue to move along the great circle defined by [001] (the perimeter of the stereographic projection), taking angles between plane normals from the stereogram and correlating these to the measured angles between reflections on the diffraction pattern. Of course, λL and $2R$ should also be used to measure the d-spacing for each reflection, and thus confirm that the indices we

† It is an accepted convention that the zone axis points *up* the microscope column, from the specimen to the electron source, and *normal* to the emulsion side of a negative. It is important to follow this convention when relating crystallographic directions indexed from a diffraction pattern to specific features shown in electron micrographs (see Chapter 4).

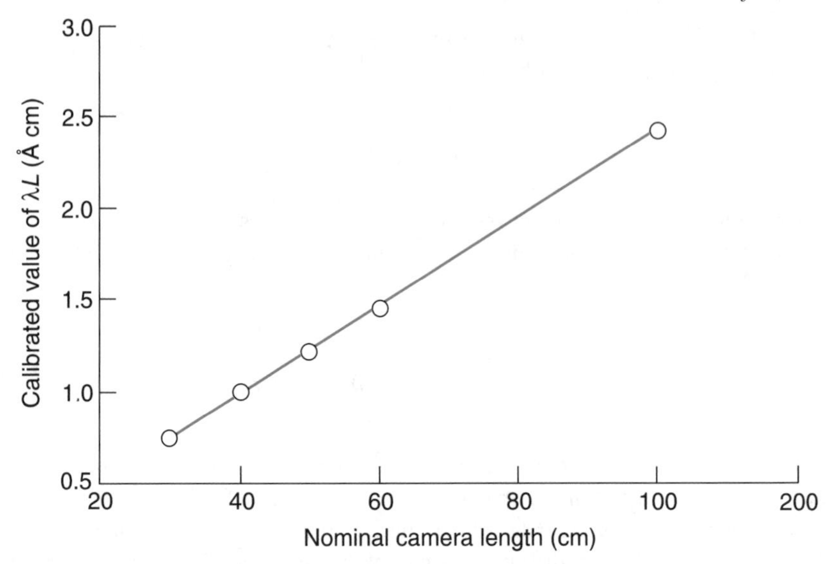

Figure 2.49 Calibration curve for the value of the camera constant (λL) as a function of the camera length indicated by the microscope

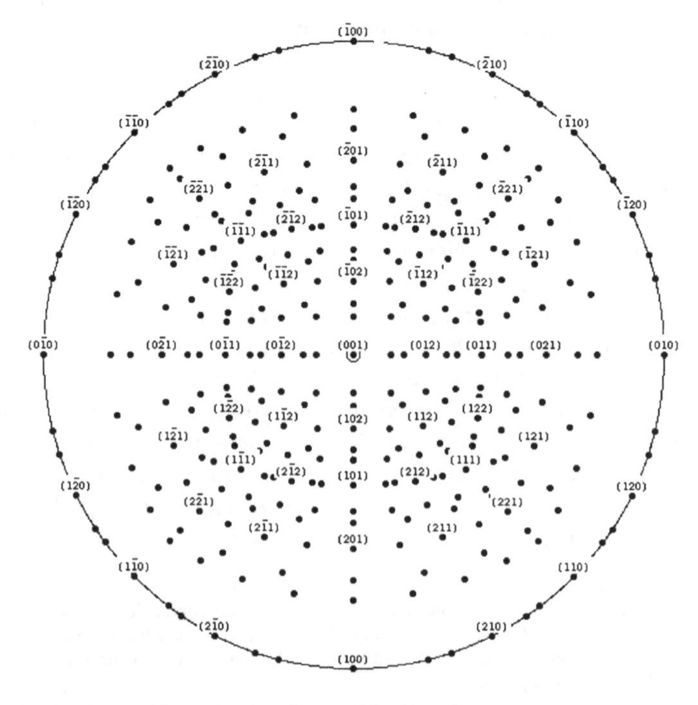

Figure 2.50 Stereographic projection for a cubic crystal

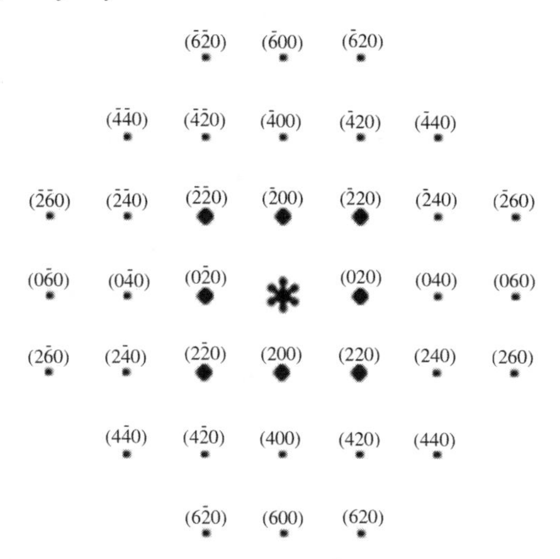

Figure 2.51 A fully indexed selected-area electron diffraction pattern obtained from aluminium for the same zone axis as the patterns shown in Fig. 2.48

assign to each reflection are correct. The result is a fully indexed selected-area diffraction pattern, as shown in Fig. 2.51.

For the FCC structure, this procedure is straightforward, but things become more complicated for non-cubic structures, for example the diffraction pattern from alumina shown in Fig. 2.52. Remember that directions are NOT usually perpendicular to planes, so in order to determine a zone axis (which is a direction) from a set of planes, we need a stereographic projection which includes *both* planes

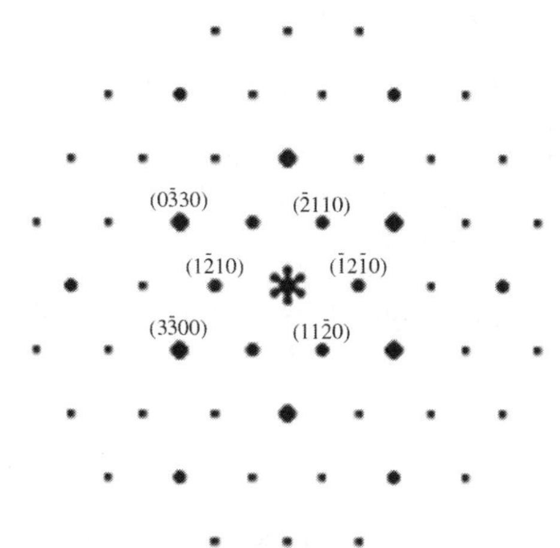

Figure 2.52 Partially indexed selected-area electron diffraction pattern obtained from α-alumina

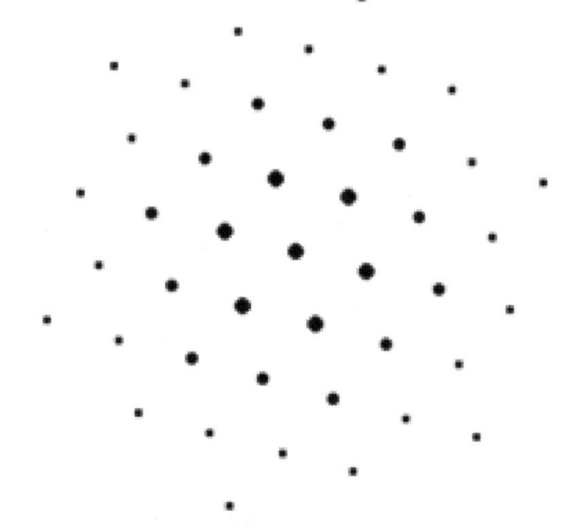

Figure 2.53 Selected-area electron diffraction pattern obtained from α-Fe (Problem 2.15)

and directions (you will need to find a computer program to generate your own stereographic projections!). The expressions listed in Appendix I can be used to determine a zone from any two indexed planes lying on the zone, as well as the angle between the two planes, but the answer should be checked against a computer simulation. In addition, of course, you will still have to solve your own patterns!

You may have noticed that the indices of all of the diffraction spots are related by a simple rule of vector addition or subtraction. This rule is valid for electron diffraction patterns corresponding to any lattice symmetry, since the lattice planes are directions in reciprocal space, while the diffraction pattern is a two-dimensional section through reciprocal space. This vector addition rule should *always* be used to check a solution.

Finally, many electron microscopists use the JCPDS data to help solve electron diffraction patterns. However, this can be misleading. Although the structure factor calculation is the same for both X-ray and electron diffraction, the atomic scattering factors are very different, while the other coefficients in the calculation of total intensity also differ for the two cases. Furthermore, the JCPDS *always* refers to a random powder specimen, for which weakly scattering planes may remain undetected. Such a plane may diffract strongly in electron diffraction, whenever its reciprocal lattice vector touches the reflecting sphere.

Problems

2.1 The minimum lattice spacing which can be detected by diffraction of an incident beam is just *half* of the wavelength of the incident radiation. Why is this?

2.2 When a first-order reflection is *forbidden* (e.g. 110 in the FCC lattice), the second-order reflection (e.g. 220 in FCC) is generally *allowed*. Why is this?

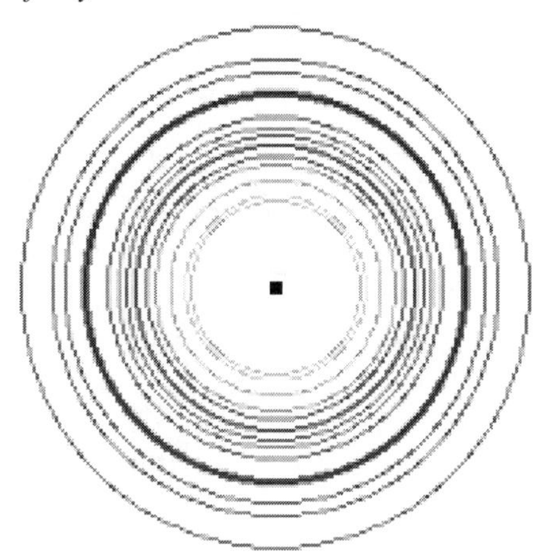

Figure 2.54 Selected-area electron diffraction pattern obtained from a cast metal block (Problem 2.16)

2.3 In a primitive cubic (PC) lattice, the reflections 221 and 300 diffract at the *same* Bragg angle. Find another pair of reflections in this lattice with the same property.
2.4 Which reciprocal lattice vector separates the lattice points 110 and $\bar{1}11$ in reciprocal space? Write down the Miller indices of *two* other reflections that lie on the same zone.
2.5 In general, reciprocal lattice vectors are parallel to lattice directions having the same indices *only* in the case of a cubic crystal. Nevertheless, this equivalence *is* found for some zones of high symmetry in other Bravais lattices. Give *two* examples of these.
2.6 Name *three* factors which may relax the exact diffraction condition, so that *some* diffracted intensity is measured for crystal orientations which deviate from the Bragg condition.
2.7 Distinguish between *white* and *characteristic* X-rays and give *one* application for each type of radiation in X-ray diffraction.
2.8 Define the term *mass absorption coefficient*. By using literature data, estimate the thickness of an iron foil that will ensure 90 % transmission for CuK_α radiation and FeK_α radiation. How do you account for the large difference in the two calculated thicknesses, given that the two wavelengths are quite close?
2.9 Estimate the *minimum* level of residual macrostress detectable by an X-ray line shift in steel (elastic modulus 220 GPa). (Justify any necessary assumptions that you may need to make.)
2.10 Diamond has an FCC structure, but with additional *forbidden* reflections. Determine the first three forbidden reflections and explain their origin.
2.11 Will the measurement of lattice spacing be more accurate for thin films or for bulk specimens? Based on your answer, suggest a suitable specimen holder for the accurate measurement of lattice spacings in a *powder*.
2.12 What, if any, differences would you expect to observe in the diffracted intensity from the same phase in a *thin film* as opposed to a *bulk specimen* (assume that you are using a Bragg–Brentano diffractometer).
2.13 Index the diffraction pattern obtained from Ni (shown in Fig. 2.40) by using the JCPDS data given in Fig. 2.41.
2.14 Index the diffraction pattern obtained from alumina shown in Fig. 2.43. To index the pattern, write a computer program to calculate *d*-spacings for a sequence of planes (*hkil*).
2.15 Fig. 2.53 is a selected-area electron diffraction pattern of α-Fe. Find λL for the diffraction pattern, index the pattern, and determine the zone axis. Mark both the zone axis and the

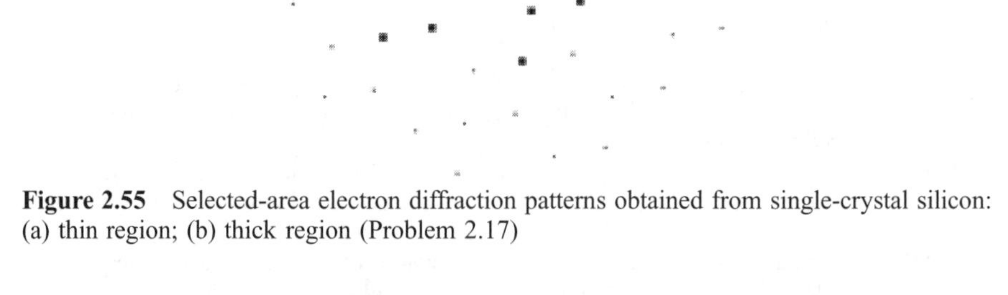

Figure 2.55 Selected-area electron diffraction patterns obtained from single-crystal silicon: (a) thin region; (b) thick region (Problem 2.17)

great circle containing the diffracting planes on a standard stereographic projection for a cubic crystal.

2.16 Fig. 2.54 is a selected-area electron diffraction pattern obtained from a *polycrystalline* region, taken from a cast metal block. There is concern that silicon may have been introduced into the cast as an impurity. Solve the pattern and index the rings. Is there evidence for the presence of crystalline silicon?

2.17 Fig. 2.55 shows two selected-area electron diffraction patterns taken from single-crystal silicon about the same zone axis. One pattern was taken from a thin region of the specimen, while the other was taken from a relatively thick area. Index the reflections and solve for the zone axis. Note and explain the thickness-dependent difference between the two patterns.

Chapter 3

Optical Microscopy

The optical microscope is the primary tool for the morphological characterization of microstructure in science, engineering and medicine. In the medical sciences, thin slices of biological tissue and other preparations are prepared for *transmission* optical microscopy, with or without staining, and frequent use is made of additional image contrast available by the use of *dark-field* optical microscopy or *phase contrast*. The geologist also works primarily in transmission, polishing his mineralogical specimens down to a thickness of less than 50 μm and mounting them on transparent glass slides, but in his case *polarized light* is the most frequent source of contrast, and provides information on the optical properties and spatial orientation of the crystalline phases which are present in the sample.

Metallurgical samples for metallographic examination were originally prepared by Henry Sorby in 1864 as thin slices, by using the methodology developed earlier for mineralogical specimens, and his specimens have survived intact and are still available for examination (Fig. 3.1). However, the presence of the conduction electrons renders metals *opaque* to visible light and all metallurgical samples must be examined in *reflection*. It follows that only the *surface* of the sample is imaged, and that it is the *topology* and the optical properties of the surface that are responsible for contrast in specimens of metals and their alloys examined in the optical microscope.

Polymers and plastics can be imaged in either reflection or transmission, but the amorphous, glassy phases give poor contrast. However, the development of crystalline polymer phases, by slow cooling from a viscous liquid state, has been very successfully studied in transmission by casting thin films of the molten polymer onto a glass slide. In polarized light, the growing crystals show a contrast which is characteristic of the orientation of the optically anisotropic crystal lattice with respect to the polarization vector of the incident beam (Fig. 3.2). Filled plastics and polymer matrix composites can be examined in reflection, although the extreme differences in mechanical response between a low-elastic-modulus polymer and a high-modulus filler or reinforcement makes specimen preparation difficult. Furthermore, the ready availability of *scanning electron microscopes* has reduced

Figure 3.1 A 'Widmanstätten' microstructure in steel (specimen prepared by Henry Sorby, 1864. Reproduced with permission from Smith, *A History of Metallography*, p. 166. Published by The University of Chicago Press

Figure 3.2 'Spherulites' in polyethylene, where arrays of crystallites growing from a common nucleus can be observed in polarized light (magnification ×525). Reproduced by permission of John Wiley & Sons, Inc.

the importance of studying the morphology of these materials in the optical microscope. Nevertheless, the scanning electron microscope is not very sensitive to material *anisotropy*, which is the principle source of contrast in polarized-light microscopy. For example, elastomeric (rubbery) polymers exhibit molecular alignment at high elastic strains which gives rise to a readily observable optical anisotropy.

Ceramics and semiconductors are usually prepared for *reflection* microscopy, despite their obvious similarity to mineralogical samples, although in some cases it may actually be easier (and more informative) to prepare a thin slice for transmission examination. Poor reflectivity, coupled with strong absorption of visible light, makes for poor optical contrast in many ceramic samples when viewed in reflection, while their resistance to chemical attack often makes it difficult to find a suitable etchant to reveal the surface topology of a polished ceramic sample. In addition, the presence of very small quantities of impurities and dopants can alter the response of the sample to surface preparation, often through strong segregation of the dopant to grain boundaries and interfaces.

In this present chapter, we will emphasize the contrast mechanisms which are typical of optical microscopy, together with the basic theory of interaction between a specimen sample and the incident beam of visible light. The emphasis will be on *reflection* microscopy (the *metallurgical* microscope), although much of the discussion is equally applicable in transmission.

3.1 GEOMETRICAL OPTICS

Rapid developments in the physical sciences, and the associated explosion in technology associated with semiconductors, microelectronics and communications, has led to dramatic changes in the syllabi of modern science courses. In particular, there is little place in the modern syllabus for the mundane topic of *geometrical optics*, and professional scientist's understanding of image formation in the telescope or the optical microscope can no longer be taken for granted. Nonetheless, some appreciation of what happens inside an optical microscope is desirable.

3.1.1 Optical Image Formation

The lens of an optical magnifying glass forms an image of an object because the *refractive index* of glass is much greater than that of the atmosphere, and reduces the wavelength of the light in the glass. A parallel beam of light incident at an angle on a polished block of glass is therefore deflected, and the ratio of the angle of incidence to the angle of transmission is determined by the refractive index of the glass (Fig. 3.3).

In the case of a *convex* glass lens (a lens of *positive* curvature), the spherical curvatures of the front and back surfaces of the lens result in the angle of deflection

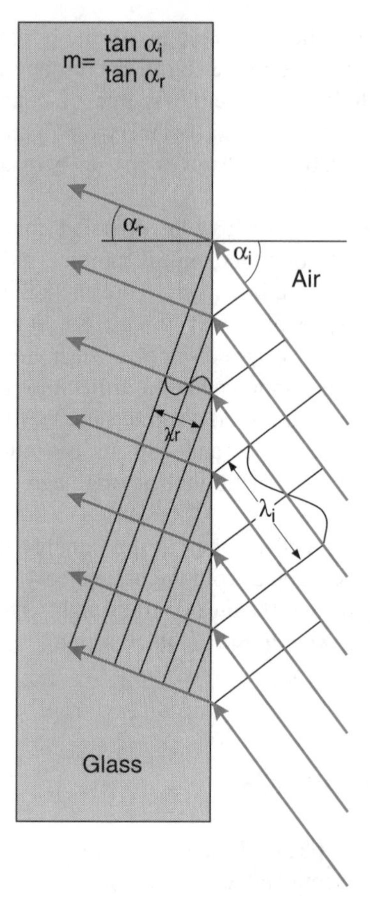

Figure 3.3 A beam of parallel light (a planar wave-front) is deflected on entering a block of glass because of the change in wavelength associated with the refractive index μ of the glass

varying with distance from the axis of the lens, thus bringing a *parallel* beam of light to a focus at a distance f which, for a given wavelength, is a characteristic of the lens, and is termed its *focal length* (Fig. 3.4). If the lens curvature is *negative*, then the lens is *concave* and a parallel incident beam diverges, and will appear to originate at a point in front of the lens, at an *imaginary* focus corresponding to a *negative* focal length.

There is no reason why the front and back surfaces of the lens should have the same curvature, nor even why one surface should not have a curvature of opposite sign to the other. It is the *net* curvature of the two surfaces taken together which determine whether the lens is convex or concave. Similarly, there is no reason why the refractive index should be the same for all of the lenses in an optical system, and different grades of optical glass possess different refractive indices. The 'lenses'

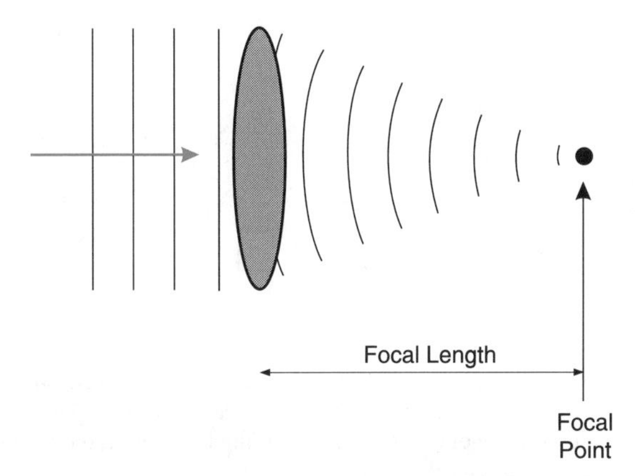

Figure 3.4 A parallel beam transmitted through a convex lens along its axis converges to a focus at a fixed distance from the plane of the lens, known as the focal length

used in the optical microscope are assemblies of convex- and concave-lens components with appropriate refractive indices. Depending on their position in the microscope, these lens assemblies are referred to as the *objective* lens, the *intermediate* lens or the *eyepiece*. It is also quite common for the medium on one side of a lens to be a liquid, with that on the other being air. An *immersion* lens is one that is designed to be used as an objective with an inert *high-refractive-index liquid* between the sample and the lens.

The assumption that a parallel beam of light will be brought to a point focus at a distance determined by the focal length of the lens is only an approximation. The refractive index of the glass varies with the wavelength of light, so that the focal length is only a constant for *monochromatic* light (light of fixed wavelength). Visible (white) light is not brought to a single focus, but rather it is *dispersed*, with the *shorter* wavelengths (blue light) being focused at a *greater* distance from the lens (Fig. 3.5(a)). A parallel beam of white light is not brought to a sharp focus by a glass lens, but rather to a region of finite size (the 'disc of least confusion'), a condition which is referred to as *chromatic aberration*. For thicker, larger-diameter lenses, even monochromatic light is not fully focused, and the outermost regions of the lens (corresponding to the largest angles of deflection) have a shorter focal length than the region near the axis (Fig. 3.5(b)), again resulting in a finite size for the focused beam, a *disc of least confusion*, rather than a point. This latter condition is termed *spherical aberration*. The lens systems used for the lens assemblies in an optical microscope will correct for these aberrations as far as possible, but it is important to check the manufacturers' recommendations to ensure that the lens combinations (for example eyepiece and objective) are fully compatible and appropriate for the type of specimen being examined. For example, objective lenses for biological samples usually compensate for the thickness of a *glass cover slip* (0.1 mm) which protects the sample.

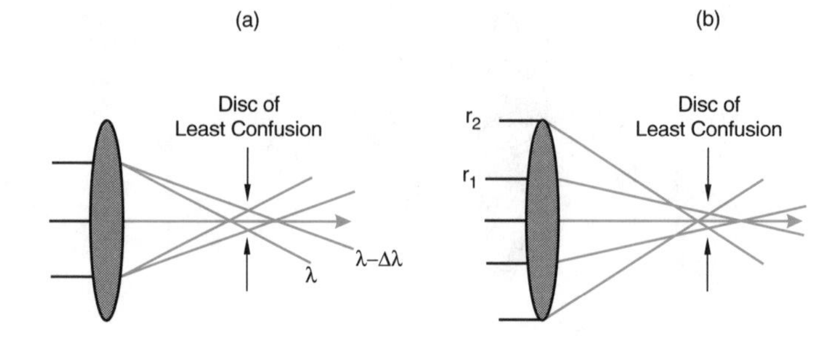

Figure 3.5 (a) The focal length of a thin lens depends on the wavelength of the radiation, thus giving rise to chromatic aberration. (b) When the diameter of the lens is no longer small compared to the focal length, the outermost regions of the lens have a reduced focal length and thus result in spherical aberration

The human eye (Fig. 3.6) forms an image of the visible world on the *retina* by focusing light transmitted through the *lens* of the eye. In order to prevent too much light entering the eye, the *iris* acts as a variable aperture, reducing the effective diameter of the lens in bright light. The retina is covered in a dense array of optical receptors, the *rods* and *cones*, which respond to the incident optical signal with

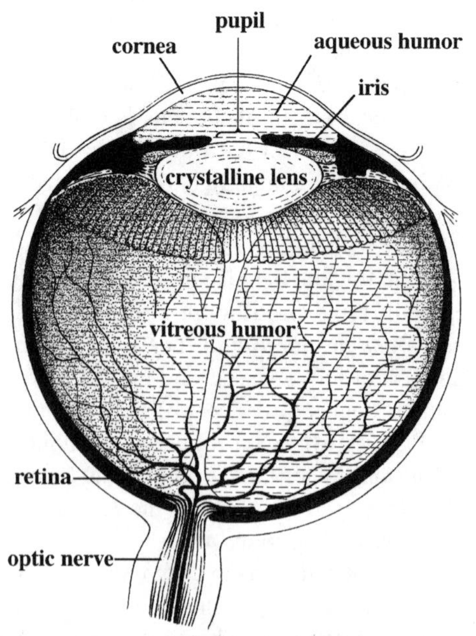

Figure 3.6 In the human eye, the lens focuses an image on to the retina, while the iris acts as a variable aperture to limit the amount of light admitted. The space between the lens and the retina is filled with liquid, so that this is an immersion-lens system

remarkable sensitivity. In many animal species, including, of course, man, the retina responds not only to the *intensity* but also to the *wavelength* of the light, thus resulting in *colour vision*. However, some 20% of the human race have defective colour vision in one form or another, while the response of the eye to colour gradually fades as the intensity is reduced, so that as dusk falls the world becomes grey.

The image formed by a simple lens can be analysed by using a *ray diagram* (see Fig. 3.7). A point in the image which is not on the lens axis and at a distance $-u$ in front of the lens, generates a wave signal *parallel* to the axis which will be deflected by the lens to pass through the focal point at a distance f behind the lens. On the other hand, a wave travelling from the same point through the *conjugate focal point* at $-f$ will be *deflected* parallel to the axis on passing through the lens, while a wave travelling through the *centre* of a thin lens will be *undeflected*. All three rays meet at a point which is a distance v behind the lens, to form the image. It is an exercise in trigonometry to show that the magnification is given by $M = v/u$, while the object and image distances are related by the relationship $1/u + 1/v = 1/f$.

3.1.2 Resolution in the Optical Microscope

The wavelength of the electromagnetic radiation transmitted through the earth's atmosphere from the sun varies from the infrared to the ultraviolet, but the peak intensity is in the *green* region of visible light, close to the peak sensitivity of the eye, *ca* 0.56 μm. The *resolution* of a lens is defined in terms of the *spatial distribution* of intensity which is observed through the lens at its focus for a point source of light situated at infinity.

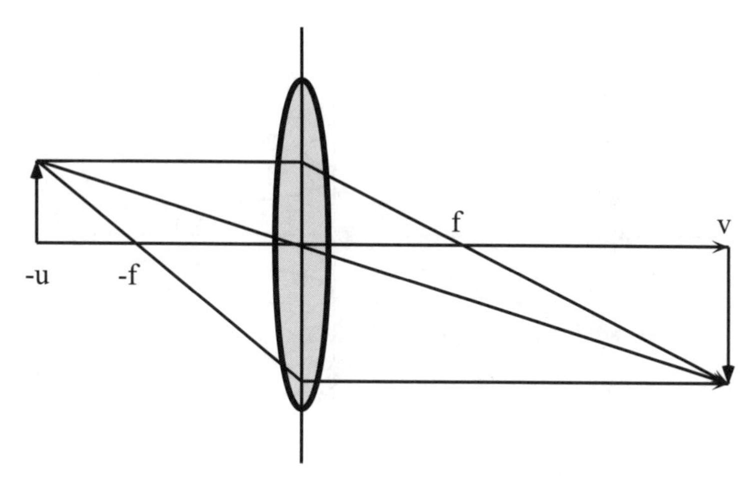

Figure 3.7 A simple ray diagram relates the distance of the lens from the object, u, to both the focal length of the lens, f, and the position of the image plane, v, and determines the magnification in the image, $M = v/u$

3.1.2.1 POINT-SOURCE ABBE IMAGE

The calculated intensity distribution assumes a parallel beam of light travelling along the axis of a thin lens and brought to a focus at the focal distance (Fig. 3.8). For the *cylindrically symmetric* case, the ratio of the peak intensities for the primary and secondary peaks in the intensity distribution is *ca* 9 : 1, while the width of the primary peak is given by the *Abbe equation* as follows:

$$\delta = 0.61 \frac{\lambda}{\mu \sin \alpha} \tag{3.1}$$

where λ is the wavelength of the radiation, α is the aperture (half-angle) of the lens (determined by the ratio of the lens radius to its focal length), and μ is the refractive index of the medium between the lens and the focal point ($\mu \approx 1$ for air).

3.1.2.2 IMAGING A DIFFRACTION GRATING

A *diffraction grating* consists of an array of parallel lines. When illuminated by a normally incident, parallel beam, a cylindrical wave-front is generated from each of the lines in the grating, and these wave-fronts interfere to generate both zero-order and diffracted transmitted beams (Fig. 3.9). When the spacing of the grating is large compared to the wavelength, the angle of diffraction for the *n*th order beam is given by $\sin \theta = n\lambda/d$. It follows that a lens can only be used to image a diffraction grating if the angular aperture of the lens α is large enough to accept both the zero-order and the first-order beams, so that $\sin \alpha \geqslant \sin \theta = \lambda/d$.

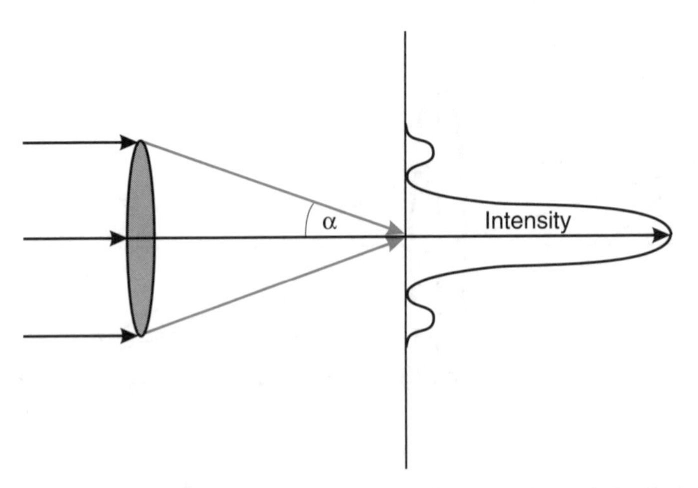

Figure 3.8 The Abbe equation gives the width of the first intensity peak for the image of a point object at infinity in terms of the angular aperture of the lens α and the wavelength of the radiation λ

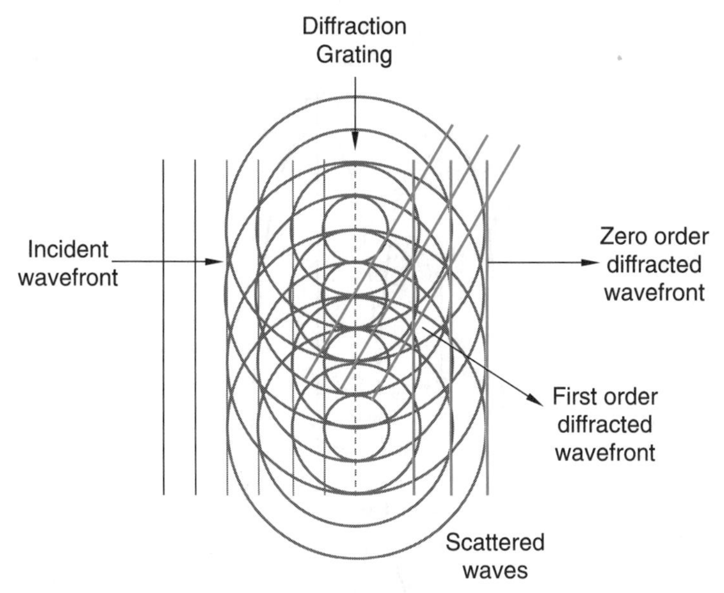

Figure 3.9 The diffraction pattern from a grating generates a series of diffracted beams. In order to image the grating, at least the zero- and first-order beams must be admitted to the aperture of the lens

We can compare this condition both with the Bragg equation for diffraction from a crystal lattice, $\lambda = 2d \sin\theta$, and with the Abbe relationship which defines the Raleigh resolution criterion (see below), $\delta = 0.61\lambda/\mu \sin\alpha$. All three cases give the *limiting conditions* for transmitting information about an object when using electromagnetic radiation as the information carrier. The key parameter is the ratio λ/d, and all three criteria state that the resolution limit on the microstructural information available is determined by the wavelength of the radiation used.

3.1.2.3 Resolution and Numerical Aperture

Raleigh defined resolution in terms of the ability of a lens to distinguish between two point sources at infinity when they are viewed in the image plane. His criterion for resolution was that the angular separation of the two sources (of equal intensity) should ensure that the maximum of the primary image peak of one source should fall on the first minimum of the image of the second source. This condition implies that the combined image of the two sources will show a small (actually 13%), but detectable intensity minimum at the centre (Fig. 3.10). It follows that the Raleigh criterion corresponds exactly to the width of the primary intensity peak given by the Abbe equation, $\delta = 0.61\lambda/\mu \sin\alpha$.

It is important to recognize the fundamental significance of the Abbe relationship, which tells us that, for *any* imaging system based on wave optics, *no* image detail can be transmitted which is much below the wavelength used to transmit the

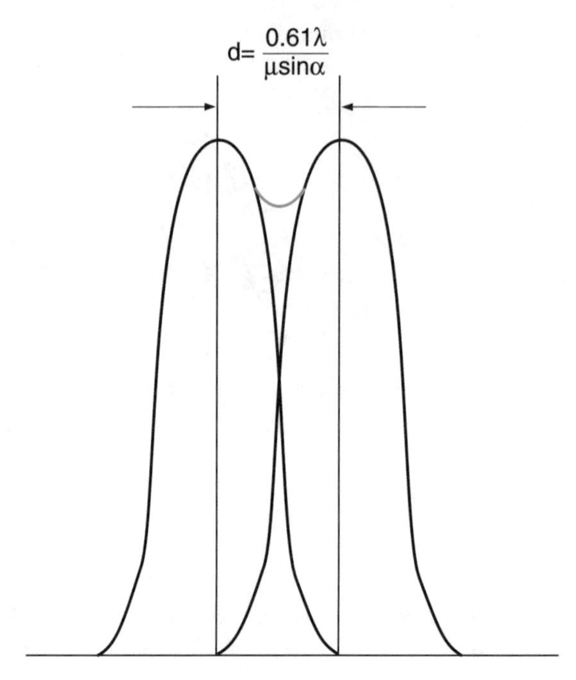

Figure 3.10 The Raleigh resolution criterion requires that two point sources at infinity have an angular separation which is sufficient to place the maximum intensity of the primary image peak of one source at the position of the first minimum of the second

information. It also tells us that if we wish to *maximize* the information in the image, we should collect as much as possible of the signal generated by the object, so that we should increase the aperture of the lens. Actually, it is possible to make use of the particle properties of phonons in order to achieve a resolution which is below the wave-optics limit. This is the basis of *confocal microscopy*, in which a light pipe is scanned across the specimen surface. The resolution in the confocal microscope is limited by the diameter of the light pipe and its distance from the surface and can be appreciably better than that set by the Abbe relationship, but a fuller discussion of this technique goes well beyond this present text.

As noted previously, the lens aberrations limit what can be achieved in the optical microscope. The parameter $\mu \sin \alpha$ is termed the *numerical aperture* of the lens (*NA*) and is an important characteristic of the objective lens systems. The maximum values of *NA* are of the order of 1.3 for an immersion lens system and 0.95 for lenses operating in air. The value of *NA* is usually marked on the side of the objective lens by the manufacturer.

It is important to distinguish between *resolution* and *detection*. As the signal generated by a point source decreases, it becomes increasingly difficult to detect against the background noise of the detection system. In the reflection microscope,

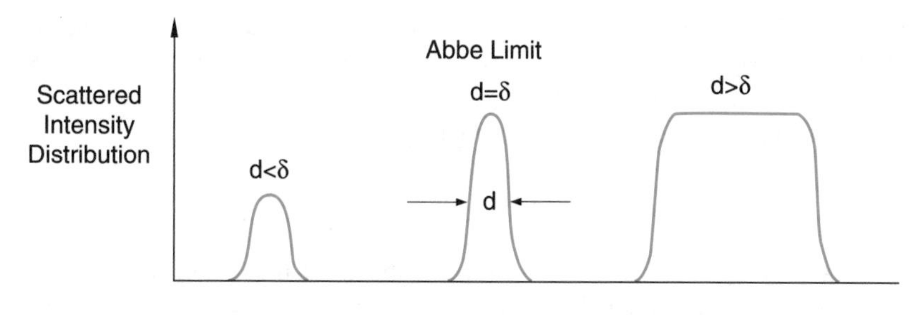

Figure 3.11 Large objects of diameter d are blurred by the diffraction limit δ derived from the Abbe relationship, but objects smaller than the Abbe width are still detectable in the microscope, although the intensity is reduced and they have an apparent width given by the Abbe equation

small objects generally scatter light outside the lens aperture, thus resulting in an intensity deficit in the image field. As the object becomes very small, the contrast decreases while the apparent size of the object (the Abbe width) remains unchanged. At some limiting size, the object will cease to be detectable, but this limit is well below the resolution limit dictated by the wavelength of the light and the *NA* of the lens. The situation is illustrated schematically in Fig. 3.11. One reason for using dark-field illumination is that it is easier to detect small features which scatter light into the imaging field, rather than relying on the small amount of light which is scattered out of the objective aperture in a bright-field image.

3.1.3 Depth of Field and Depth of Focus

Since the resolution available for an object in focus in the image plane is limited by the numerical aperture of the objective lens, it follows that the object need not be at the exact object distance from the lens u, but may be displaced from this plane without sacrificing any resolution (Fig. 3.12). The distance over which the object remains in focus is defined as the *depth of field*,

$$d = \delta \tan \alpha \tag{3.2}$$

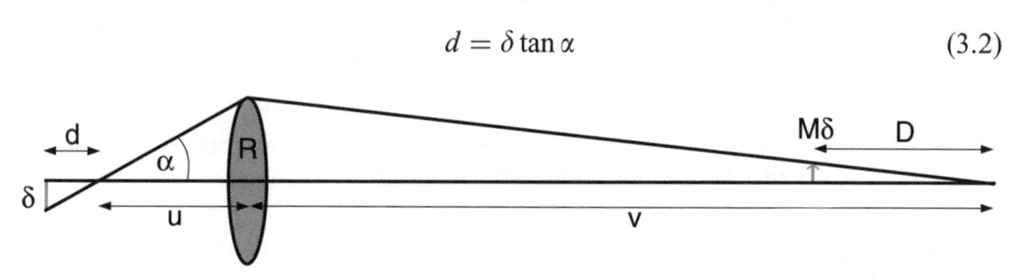

Figure 3.12 Since the resolution is finite, the object need not be in the exact object-plane position in order to remain in focus, and there is an allowed depth of field d. Similarly, the image may be observed without loss of resolution if the image plane is slightly displaced, so that there is an allowed depth of focus D

where α is half the angle subtended by the objective aperture at the focal point. Similarly, the image will remain in focus if it is displaced from its geometrically defined position at a distance v from the lens. The distance over which the image remains in focus is termed the *depth of focus*, as follows:

$$D = M^2 d \tag{3.3}$$

where M is the magnification. (Both of these expressions (equations (3.2) and (3.3)) are approximate and assume that the objective can be treated as a 'thin lens', which is never the case in a commercial instrument.) Since the resolution is given by $\delta = 0.61\lambda/\mu \sin\alpha = 0.61\lambda/NA$, it follows that the depth of field decreases as the numerical aperture increases. For the highest image resolution, the specimen should be positioned to an accuracy of better than 0.5 μm, which determines the required mechanical stability of the specimen stage.

The depth of focus is considerably less critical. Bearing in mind that a magnification of the order of 100 is necessary if all of the resolved detail is to be recorded, displacements of the order a millimetre are acceptable.

3.2 CONSTRUCTION OF THE MICROSCOPE

A simplified design for a reflection optical microscope is shown in Fig. 3.13. The microscope is an assembly of three separate systems, namely the *illuminating system*, the *specimen stage* and the *imaging system*. We will now discuss each of these components in turn.

3.2.1 Light Sources and Condenser Systems

There are two conflicting requirements for the *light source*. On the one hand, the area of the specimen being examined beneath the objective lens should be uniformly flooded with light, in order to ensure that all of the microstructural features experience the same illuminating conditions, while on the other hand the incident light needs to be focused on to the specimen to ensure that the reflected intensity is high enough for comfortable viewing and photography.

The source of light should be as bright as possible. At one time this was achieved by striking a carbon arc, which gave an excellent, if somewhat unstable, source of white light, while the mercury arc lamp generated intense monochromatic emission in the green ($\lambda = 0.546$ mm), which reduced the need for *achromatic* objective lenses (see below). The commonest light source for small microscopes is a resistance-heated carbon filament, while the more expensive microscopes are equipped with a xenon discharge tube, which is a stable intense source of white light.

In addition to the source, there are other important components in the illuminating system (Fig. 3.13). The *condenser lens* focuses an image of the source, and for some

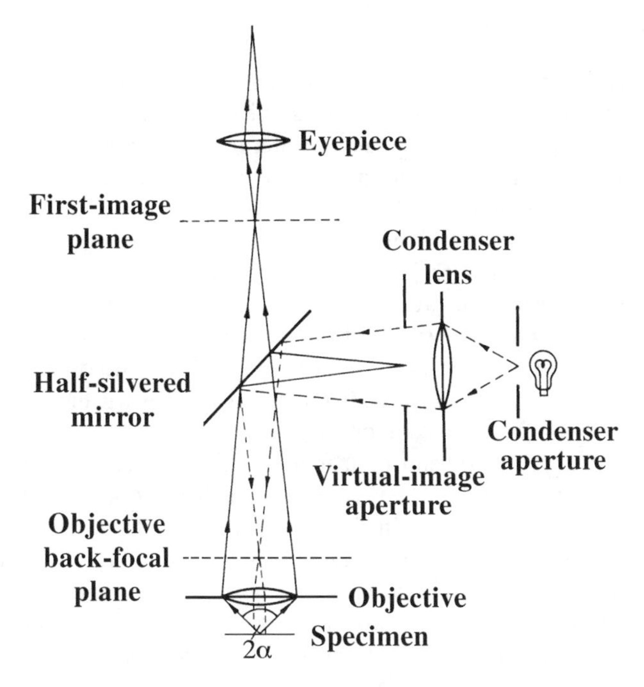

Figure 3.13 The principle components of the reflection optical microscope and their geometrical relationship to each other

viewing conditions the image plane of the source should be close to the *back-focal plane* of the objective lens, so that the specimen is illuminated by a near-parallel beam. The *condenser aperture* limits the amount of light from the source which is admitted into the microscope. Contrast in the image can often be improved by using a very small condenser aperture, but at the cost of drastically reducing the image intensity and introducing artifacts, associated with the diffraction pattern of the 'point' source. A second aperture, the *objective* or *virtual-image aperture*, is placed in the image plane of the objective (see Fig. 3.13) and serves the important function of ensuring that light is not internally reflected within the microscope, leading to unwanted background intensity. The size of the virtual-image aperture should be adjusted to the field of view of the microscope at the magnification used. Both the condenser and virtual-image apertures are continuously variable *irises* which can be adjusted to the required size.

Many reflection microscopes also permit the illuminating system to be repositioned so that optically transparent specimens can be viewed in transmission. This is important not only for examining the thin tissue samples of biology and medicine, but also for mineralogical samples, partially crystalline polymers and thin-film semiconductor materials.

3.2.2 The Specimen Stage

The primary requirement for the specimen stage is *mechanical stability*, and given an expected *ca* 0.3 μm resolution for the instrument, it is clearly essential that the positioning of the specimen be accurate to better than this limit. The position of the specimen in the x–y plane is only one aspect of the stability required, since the image is brought into focus by adjusting the vertical location of the specimen. The accuracy of this z-adjustment must be greater than the depth of focus for the largest *NA* objective lens, and so is just as critical as the x–y stability.

The necessary mechanical precision is commonly achieved by coarse and fine micrometre screws for all three (x, y and z) coordinates, and both the time-dependent drift of the stage and the mechanical 'slack' in the system need to be minimized. (The 'slack' is the difference in the micrometre reading when the same feature is brought into position from opposite directions.)

In general, it is the z-adjustment that presents the most problems, since the necessary stage rigidity implies a fairly massive and hence heavy construction. Two possibilities exist, depending on whether the specimen is to be placed beneath or above the objective lens. In the former case (as illustrated in Fig. 3.13), the plane of the prepared sample surface must be accurately positioned *normal* to the microscope axis, and this is commonly achieved by supporting the specimen from below using plasticine (Fig. 3.14). If the microscope design places the specimen above the

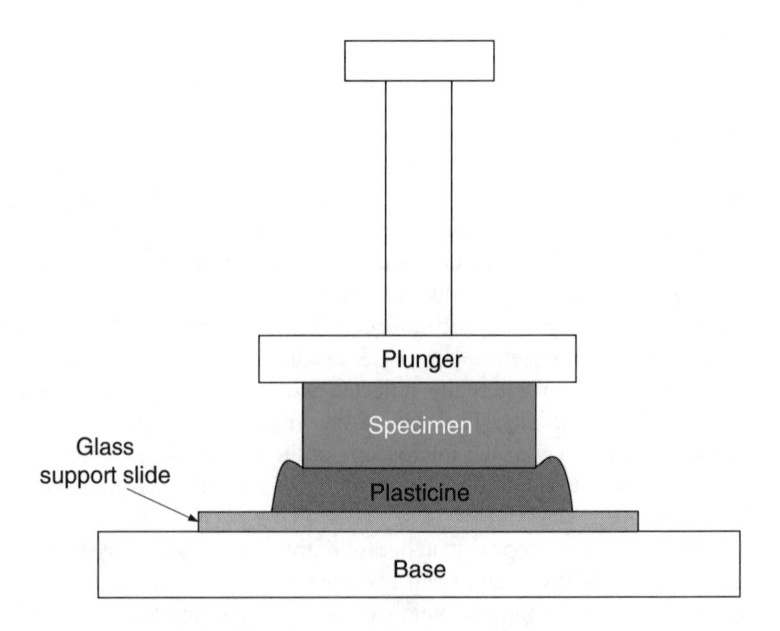

Figure 3.14 If the specimen is to be placed beneath the objective lens, then it must be mounted with the plane of the specimen surface accurately normal to the microscope axis

objective, then the prepared surface rests on an annular support and no mounting device is necessary.

3.2.3 Selection of Objective Lenses

A very wide range of *objective lenses* are available, depending on the nature of the specimen and the desired imaging mode. The performance of the objective lens is primarily dependent on its numerical aperture *NA*, and this is almost always found inscribed on the side of the objective lens assembly, together with the magnifying power for that lens. Most objective lenses are achromatic, so that they are not limited to one wavelength of light, but can be used for recording *colour* images of the microstructure.

Histological examination of soft tissues, which accounts for the major proportion of the work of the optical microscope in the life sciences, requires that the specimen be protected from the environment by mounting a thin tissue slice on a glass slide and then protecting it with a thin coverslip. Similar techniques are used for many polymer specimens, particularly those which are partially crystalline. These materials can be cast on to the slide and the specimen thickness controlled by applying pressure to the coverslip. Objective lenses designed for use with such specimens have been corrected for the refractive index and thickness (0.1 mm) of the optically flat coverslip.

Not only the resolution, but also the *brightness* of the image depends on the *NA* of the objective lens. The brightness (the amount of light per unit area of the image) decreases as the square of the magnification, but the larger the numerical aperture (cone angle) of the lens, the more light is collected. The *NA* may vary by an order of magnitude in going from a low-power (low-magnification) lens to a high-power (high-magnification) immersion objective, requiring a similar order of magnitude increase in the magnification to observe all of the image detail. It follows that the *overall* reduction in brightness occurs by a factor of $10^2/10 = 10$.

The working distance of the objective lens from the specimen surface decreases dramatically as the *NA* is increased, down to of the order of 0.1 mm for the highest-powered lenses. It is only too easy to damage a lens by 'driving' the specimen 'through the focus' and into the glass lens, and good lenses are expensive to replace.

Specially designed long-working-distance lenses are available. These usually create an intermediate image at unit magnification by reflection (Fig. 3.15). The primary purpose of a long-working-distance objective is to allow a specimen to be imaged in a hostile environment. This may be a corrosive medium or it may be an environment at elevated or cryogenic temperatures. Despite the apparent attraction of *in situ* experiments, remarkably little optical microscopy has been carried out under *dynamic* conditions. The difficulties are formidable. The *dimensional stability* of the specimen and its support structure is one major problem, while another is the need to ensure that the optical path between the specimen and the objective is not obscured by a condensate or by chemical attack of any component.

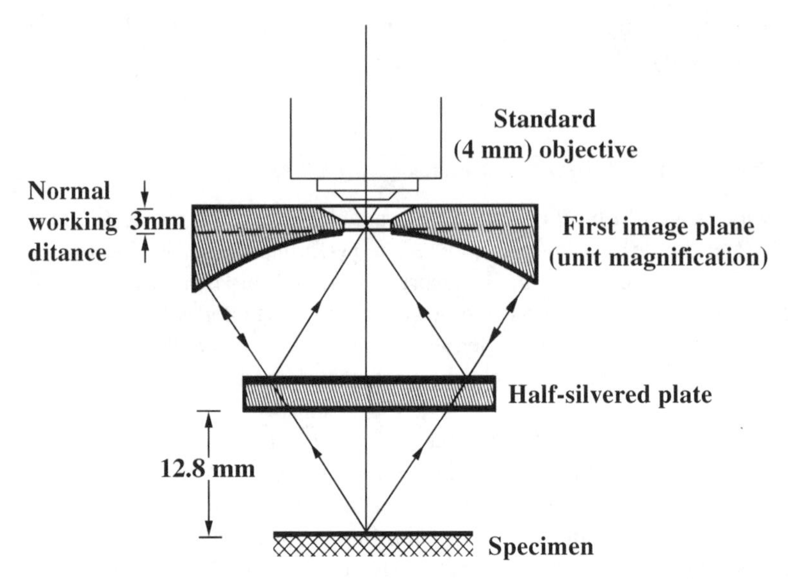

Figure 3.15 Schematic representation of a long-working-distance attachment for studying specimens at high temperatures or in a hostile environment

Cryomicroscopy is subject to the formation of ice crystals, while a *high-temperature stage* may form opaque deposits which originate from the heating elements, the specimen or the supporting structure. In order to be successful, an *in situ* stage must combine a rapid response time with experimental stability. In the case of a heating stage, a compromise is required between the large heat capacity needed to ensure thermal stability at the temperature of the experiment, and the small heat capacity necessary to allow the experimental conditions to be reached in a reasonable time.

Many other specialized objective-lens assemblies are also available. One of the most useful, for both reflection and transmission work, is the dark-field objective which illuminates the specimen with a cone of light surrounding the lens aperture. The light scattered by the specimen into the lens aperture is then used to form a dark-field image in which the intensity is the inverse of that observed in normal illumination (Fig. 3.16).

3.2.4 Image Observation and Recording

The image magnification provided by the objective is limited, and insufficient if the image is to be fully resolvable by the human eye. There are three acceptable options available. The first is to insert an eyepiece, and often an additional intermediate lens, in order to view the image at a final, working magnification which is comfortable for the eye. The second is to use the additional lenses to focus the image on to a light-

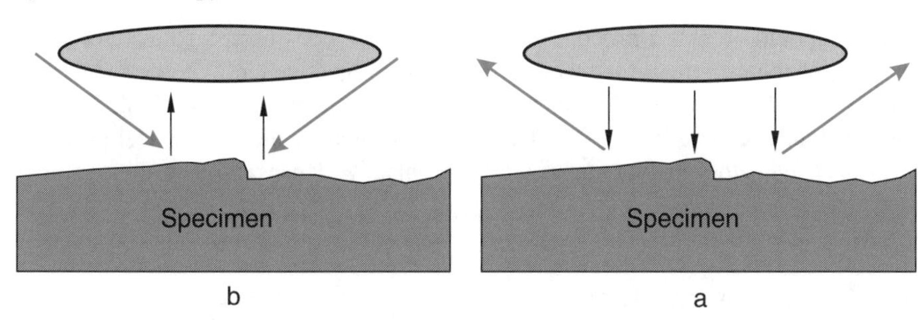

Figure 3.16 A dark-field image (a) is formed by collecting the diffusely scattered light into the objective lens aperture, and is the inverse of the normal, bright-field image (b), where the scattered light falls outside of the objective aperture and is lost to the objective lens

sensitive, photographic emulsion, usually for further photographic enlargement. The third option is to scan the image in a television raster and display it on a monitor. In recent years, advances in the availability of high-quality *charge-coupled device* (CCD) cameras has made it technically possible to record a digital image from an objective lens without any additional lenses. At the time of writing, commercial exploitation of this last option is still at an early stage.

3.2.4.1 MONOCULAR AND BINOCULAR VIEWING

Visual observation is most commonly performed with a *monocular* eyepiece, which enlarges the primary image by a factor of 3–15. A typical (non-immersion) objective ($NA = 0.95$) may have a primary magnification of 40 and a resolution of 0.4 μm, so that, in order to ensure that all resolved features are readily visible to the eye (0.2 mm), some further magnification is required, typically $(0.2 \times 10^3)/(0.4 \times 40) = 12.5$.

Many microscopes have an additional *intermediate* lens (×4, for example), so that a ×3 or ×5 eyepiece should then be sufficient to resolve all of the image detail. Even without an intermediate lens there is no real reason to use a ×15 eyepiece, since the superfluous additional magnification will just reduce the field of view, enlarging the resolved features to the point where they appear blurred to the eye.

Some microscopes are equipped with a beam splitter and a binocular viewer. For those who have difficulty viewing comfortably through one eye this is undoubtedly a convenience, but the microscopist should be aware that there are some disadvantages associated with its use. It is unusual for the focal planes of both eyes to be identical, so that one eyepiece needs to be independently focused. The procedure is then to focus on the plane of the specimen, by using one eye and a fixed-focus eyepiece, and then to adjust the variable focus of the second eyepiece (without changing the specimen plane) until the images observed by both eyes merge into a single, simultaneously focused, image. This procedure is further complicated by the need to adjust the separation of the two eyepieces to match the separation of the observer's eyes. It should be noted that the binocular eyepiece does not provide stereoscopic viewing of the sample, which would require two objective lenses to be focused on the same field of view. Stereobinoculars, with twin objectives, are available and may have magnifications of up to 50. This limit is dictated by the geometrical problems associated with the positioning of the twin objective lenses close to the specimen surface.

3.2.4.2 Photographic Recording

A photographic emulsion and the human eye react very differently to light. The emulsions have a maximum sensitivity to the ultraviolet (*ca* 0.35 μm) and both black and white, and colour films rely on dyes to extend the photosensitivity of the silver halide emulsions beyond the green (Fig. 3.17). Orthochromatic emulsions are not sensitive to red light, which is a convenience in processing, and are a common choice for recording monochromatic microscope images in green light. Panchromatic film is the common choice for black and white photographic recording in daylight, but here too the sensitivity falls steadily with increasing wavelength in the visible range.

The speed of an emulsion is its response to a fixed radiation dose at a standard wavelength and depends on three factors, the *exposure time*, the '*grain*' of the emulsion and the *development process*. A photosensitive silver halide grain will react to subsequent development only if it is excited by a pair of photons. The time interval between the arrival of the two photons is important, since the grain may decay from its initial excited state in the interim. As the incident intensity decreases,

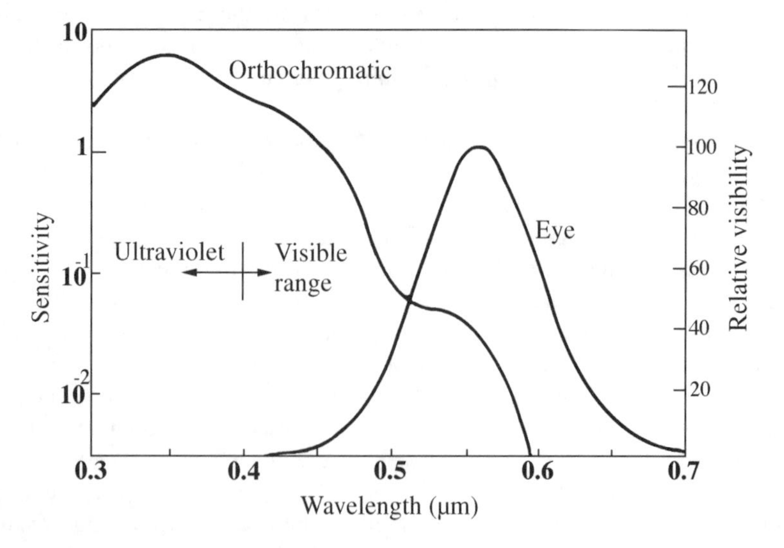

Figure 3.17 The eye is most sensitive to green light, whereas photographic emulsions have sensitivities which decrease steadily with increasing wavelength

the interval between photon excitations of the same grain increases, and the response of the emulsion is reduced, a phenomenon which is termed *reprocity failure*. Larger halide grain sizes increase the photon collision cross-section and improve the photosensitivity, but at the cost of a 'grainier' image with poorer inherent resolution. Some compromise is usually required between *fine-grained, slow-speed emulsions* and *coarse-grained, high-speed* emulsions.

During development, a grain of silver, which is nucleated at an activated halide crystal, grows into a cluster of silver grains which encompasses a much larger volume than that associated with the original halide crystal, so that the resolution available in the recorded image is affected by the growth of the silver grains during the development process. An emulsion designed for photomicroscopy, and developed according to the recommendations of the manufacturer, should have a resolution of the order of 10 to 20 μm and be capable of enlargement by a factor of 10. It should therefore be possible to photograph an image without any loss of information at an appreciably lower magnification than that required to view the resolved microstructure.

The contrast attainable in a given emulsion is defined in terms of the dose dependence of the blackening of the developed emulsion. The dose is the amount of light per unit area, E, multiplied by the time of exposure, t, while the blackening, D, is the logarithm of the ratio of the incident intensity, I_0, to the intensity of light transmitted through the exposed and developed emulsion, I, as follows:

$$D = \log(I_0/I) \tag{3.4}$$

The contrast γ is defined as the maximum slope of the curve of D versus $\log Et$. Emulsions with a high γ lose image detail because they tend to register as black or white with few intermediate grey levels, while low-γ emulsions lack contrast because the grey levels are close together. A major disadvantage of photographic recording is the non-linearity of the response of the emulsion and the difficulty of controlling the many parameters involved in *exposing*, *developing*, *enlarging* and *printing* the emulsion. It is very difficult to make quantitative measurements of image intensity or contrast based on photographic recording.

3.2.4.3 Television Cameras and Digital Recording

Television monitors, for some considerable time, have been used for the presentation of microscope observations in 'real time' to large groups of people, with heat-sensitive paper being used to record images taken from the scanned signal of a television camera. Both accessories are a convenience, but no more than that, primarily because the number of scan lines in a standard television display does not allow anything like the same number of pixel points (image elements) to be resolved as the eye is capable of detecting in a single field of view. In addition, the contrast response of heat-sensitive paper is poor and the life of the recorded image limited. Photographic emulsions are still required if a permanent record is needed.

The new generation of charge-couple-device (CCD) cameras is free of these limitations. The CCD camera can 'grab' an image frame in a few seconds and record a digitized image of 10^6 or more pixel points. Furthermore, the response (below saturation) is essentially linear, so that there is a one-to-one correlation between the recorded 'brightness' of a pixel point and the original intensity of the signal from the object.

Moreover, the digitized image can be processed quickly and efficiently, by using standard computer programs to enhance contrast, remove background or extract quantitative image data. Individual image frames can be combined to extract comparative information, for example on dynamic changes taking place in the sample.

Finally, the diameter of the CCD camera is typically 25 mm or less, so that the image can be recorded by using an objective lens alone, without the need for any intermediate lenses. There is little doubt that within the next few years we will see a major change in the design of optical microscopes, with computer control replacing manual control of the focus, and the CCD camera and computer monitor replacing the photographic system and the eyepiece.

3.3 SPECIMEN PREPARATION

For many students, good specimen preparation is a major obstacle to successful optical microscopy. It is unfortunate that every material presents its own individual and unique problems of specimen preparation. For example, the *elastic modulus* and the *hardness* determine the response of the sample to sectioning, grinding and polishing, while the *chemical activity* determines the response to electrolytic attack and chemical etching. In the following discussion, we will generalize as far as possible, while still recognizing that each metal alloy, every ceramic material and all plastic compositions are almost certain to respond differently.

3.3.1 Sampling and Sectioning

The problem with all microscopes is that they lose the 'larger picture' by focusing on the details. It is only too easy to lose track of the relationship between a microscope image (recorded from a particular position and in a specific orientation) and the engineering component from which the image was taken.

Engineering systems are assembled from components, which frequently have complex geometries and come in a wide range of shapes and sizes. They are produced by a variety of processing routes, and are unlikely to be of uniform microstructure. They often have a 'right-way-up', and the materials from which they are made are often inhomogeneous and anisotropic. Both the chemical composition and the microstructural morphology may vary across a section, even if it is only the surface layers which are 'different'. Preferred orientation may be restricted to the microstructural morphology (elongated inclusions, aligned fibres or flattened grains), or it may be associated with crystalline texture in which certain directions in the crystals are preferentially aligned along specific directions in the component (the axis of a copper wire or the rolling direction in a steel plate). Crystalline texture may exist in the absence of morphological texture, with the grains appearing equiaxed even though they all share a common crystallographic axis.

As a consequence, it may not be easy to decide how to section a component for microscopic examination, although it is always helpful to define the principle axes of the component and to ensure that the plane of any section is aligned with at least one of these axes. In even the simplest cases, it is usually desirable to examine two sections, for example those perpendicular and parallel to a significant symmetry axis of the component. In the case of rolled sheet, sections taken perpendicular to all three principle directions are desirable, the *rolling direction*, the *transverse direction* and the *through-thickness direction*. In a large casting, the microstructure will vary due to differences both in the cooling rate and the effects of segregation. Sections taken from the first portion of a casting to solidify may have totally different microstructures from the portion which solidified last.

If a section has been taken perpendicular to a principle direction of the component, then it is important to identify one other principle direction lying in the plane of the section, for example the trace of a free surface, or a growth or rolling direction. It is only too easy to confuse structurally significant directions, either during mounting and preparation of the section, or as the result of image inversion, either in the microscope or during photographic processing. Newspapers frequently publish *inverted* images of right-handed individuals engaged in apparently left-handed activities!

3.3.2 Mounting and Grinding

For convenience during surface preparation most samples need to be mounted for ease of handling. A polymer resin or moulding compound is the commonest form of sample holder, and can be *die-cast* or *hot-pressed* around the specimen without distorting the sample or damaging the microstructure (Fig. 3.18). Clearly, the sample must be small enough to fit into the die cavity. Very small samples can be supported in the correct orientation by using a coiled spring or other mechanical device.

Once the sample is securely mounted, the surface section can be ground flat and polished. Rough grinding requires care, since it is easy to introduce sub-surface damage. It is helpful to ensure that the surface section is planar, even before the sample is mounted. A wide range of grinding media are available, either cemented, metal-bonded or resin-bonded. The commonest grinding media are alumina, silicon carbide and diamond, and all three are available in a wide range of grit sizes. The latter is defined by the sieve size which will just collect the grit in question, with the quoted sieve size referring to the number of apertures per inch so that the grit size is an inverse function of the particle size, e.g. a #320 grit will be collected by a #320 sieve, having passed though a #220 sieve. Beyond #600, grit sieving is no longer practical since the grit particles aggregate, but the same definitions are used and very

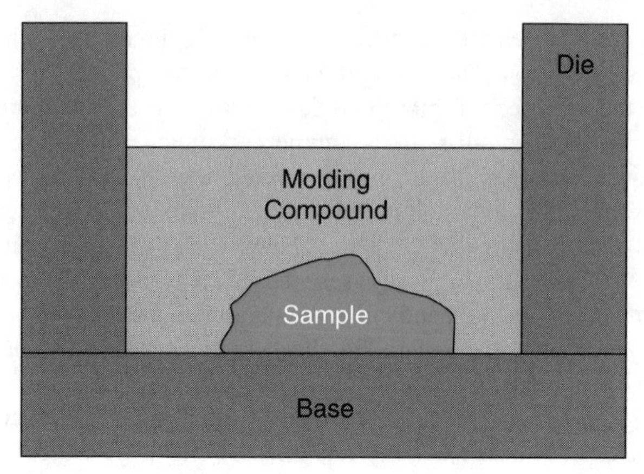

Figure 3.18 Samples are commonly mounted face down in a plastic holder for ease of handling during surface preparation

fine, sub-micron grits are available. In general, a #80 grit is used for coarse grinding, corresponding to particles of a few tenths of a millimetre in diameter.

Grinding is a complex machining process in which the sharp edges of the grinding medium cut parallel to the surface of the sample. The rate of material removal depends on the number of particle contacts, and the depth of cut (both functions of the applied pressure), plus the shear velocity at the interface between the grinding medium and the work piece. Heat generated at the interface and the debris resulting from material removal are two major obstacles to effective grinding. The former increases the ductility of the work piece, and hence the work required to remove material, while the latter clogs the cutting surfaces. These effects can be inhibited by flushing the surface with a suitable coolant to remove both heat and debris from the grinding zone.

The cutting edges of grinding media particles are blunted during the grinding process. New cutting surfaces can be exposed through wear of the surrounding matrix, which releases the blunted grit particles and exposes fresh cutting surfaces. This may occur in the course of grinding of the work piece, but is more often accomplished by dressing the grinding wheel, using an alumina 'dressing stone' to remove the debris and expose new cutting surfaces. The choice of matrix is important. Metal-bonded grits are most frequently used in cutting discs for sectioning hard and brittle samples. They are also used for many coarse-grinding operations, for which diamond is the preferred medium. Resin-bonded discs give a much coarser cut in sectioning operations, but are generally preferred for grinding, since they are less liable to lose cutting efficiency due to the build-up of debris.

The grinding direction may be important. For example, it is usually undesirable to grind a region near a free surface perpendicular to that surface, since the cutting particles are almost certain to introduce extensive damage as they bite into the edge of the sample. Cutting in the reverse direction, so that the grit particles are exiting from the free surface, will result in much less damage. The extent of the sub-surface damage depends on the elastic rigidity and hardness of the material, so that soft metals are extremely difficult to grind. On the other hand, brittle materials are prone to sub-surface cracking. It is important to recognize that sub-surface damage is introduced during grinding and ensure that subsequent polishing removes the damaged layer.

3.3.3 Polishing and Etching Methods

The primary aim of polishing is to prepare a surface which is both flat and devoid of topographical features unrelated to the bulk microstructure of the sample. Each polishing stage is designed to remove a layer of damaged material resulting from the previous stage of surface preparation. There are *three* accepted methods of polishing a sample, *mechanical*, *chemical* and *electrochemical*. Of these three, mechanical polishing is by far the most important.

In mechanical polishing, the mechanical damage of the earlier stages of preparation are removed by resorting to finer and finer grit sizes. The number of polishing steps necessary to reduce topographical roughness to below the wavelength of light and eliminate sub-surface mechanical damage may be as few as three or as many as ten. The carrier for the polishing grit may be paper-backing (for a silicon carbide grit, down to perhaps a #600 grit SiC paper), followed by a cloth polishing wheel (down to perhaps a 0.25 μm diamond grit).

In a chemical-polishing solution, the products of chemical attack form a viscous barrier film at the surface of the sample which inhibits further attack (Fig. 3.19). Regions of negative curvature (surface grooves and pits) develop thicker layers of the viscous semi-protective film, while ridges and protrusions (with positive curvature) are protected by much thinner layers and are attacked faster. As a result, the sample is rapidly smoothed to develop a topographically flat surface. Electrolytic polishing is somewhat similar, but the sample must be an electrical conductor, and is

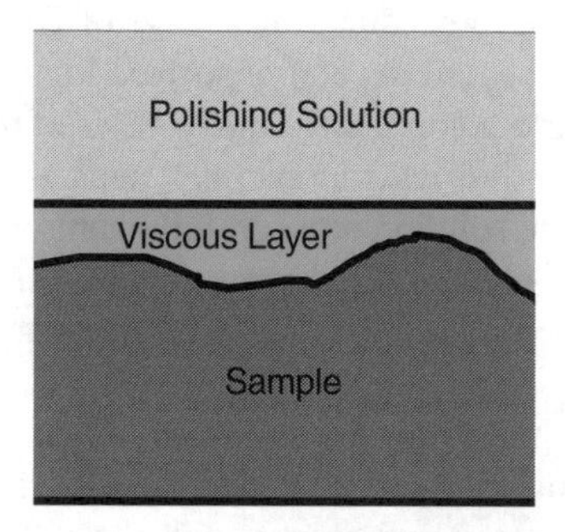

Figure 3.19 Both chemical and electrolytic polishing rely on a viscous liquid layer to enhance the attack of protruding regions and inhibit attack at grooves and recesses, thus ultimately forming a mirror-like, polished surface

almost always a metal. Positively charged cations are dissolved in the electrolyte and form a viscous anodic film of high electrical resistance. For successful electrolytic polishing, most of the voltage drop across the electrolytic cell should be across this anodic film, so that the rate of attack is again dependent on film thickness. External adjustment of the voltage across the cell ensures better control for electrolytic polishing when compared to chemical polishing, and soft materials such as lead alloys, which are difficult to prepare by mechanical polishing, can be successfully electropolished. Nevertheless, the development of increasingly sophisticated mechanical polishing methods, suitable for even the softest engineering materials, has led to a steady reduction in the importance of chemical and electrochemical methods of surface preparation.

Etching of the sample refers to the selective removal of material from the surface in order to develop surface features which are related to the microstructure of the bulk material. If the different phases that are present differentially reflect and absorb incident light, then etching may be unnecessary. Most non-metallic inclusions in engineering alloys are visible without etching, since the metallic matrix reflects incident light, while the inclusion absorbs it and appears dark. Optically anisotropic samples observed in polarized light also show contrast, without etching, which is associated with differences in crystal orientation (see Section 3.4.3).

Etching may develop the surface topography, for example by grooving grain boundaries or giving rise to differences in the height of neighbouring grain surfaces. Etching may also form thin surface films whose thickness reflects the underlying phase and grain structure. Such films may either absorb light or give rise to interference effects which depend sensitively on the film thickness.

Most etching methods involve some form of chemical attack, which is more pronounced in those regions of the surface which have a higher energy (such as grain boundaries). Thermal etching is an exception. In thermal etching, short-range surface diffusion reduces the surface energy of the polished surface and affects the local topology. At grain boundaries, thermal grooves are formed. In many materials, the surface energy of a single crystal is highly anisotropic, and thermal etching reduces the total surface energy by forming surface facets.

The commonest etching procedures make use of chemically active solutions to chemically etch the surface and develop a surface topology which is visible in the microscope. The solvents are commonly one or other of the alcohols, but molten salt baths may also be used for less reactive samples, such as ceramics. In most cases, the sample is immersed in the solution at a carefully controlled temperature for a given time, then rinsed thoroughly and dried (typically, by using alcohol). In some cases, electrolytic etching is used to promote the localized attack (for example on stainless steels).

Chemical staining is used to form a surface film whose thickness depends on the surface features of the microstructure. A steel sample will develop such a film when oxidized in air at moderate temperatures, with the different grains appearing in a rainbow of colours, depending on the thickness of the film formed on each grain (see Section 3.4.5.4).

3.3.3.1 Steels and Non-Ferrous Alloys

Engineering alloys are perhaps the commonest materials to be prepared by mechanical polishing for microscope investigation. Ferrous alloys cover a very wide range of hardness, from brittle martensites to low-yield-strength transformer steels. There are few polishing problems which cannot be solved by using simple 'rules of thumb'. The commonest forms of polishing defect are surface relief (differential polishing), the rounding of edges, and scratches or plastic deformation. Assuming that the polishing media are free of contamination (primarily debris from the earlier stages of preparation), then scratches and plastic deformation can be prevented by selecting a more compliant support for the polishing media, in order to reduce the forces applied to the individual particles and increase the number of particle contacts per unit area of the sample. Conversely, surface relief and edge rounding can be inhibited by selecting a less compliant support.

3.3.3.2 Pure Metals and Soft Alloys

The softest materials are the hardest to polish mechanically, because they are so susceptible to mechanical damage. If excessive force is applied to the polishing media, then wear debris will be embedded in the soft surface of the sample and dragged in the direction of shear. Even if no obvious signs of damage are visible on the polished surface, subsequent etching (see below) is liable to reveal sub-surface traces of plastic deformation. Successful preparation of these materials is accomplished by polishing at low shear velocity and applied pressure, thus inhibiting polishing debris from smearing and adhering to the sample surface.

3.3.3.3 Semiconductors, Ceramics and Intermetallics

Brittle materials are usually easier to polish mechanically than soft materials. In particular, there are unlikely to be problems associated with polishing debris adhering to the surface. On the other hand, microcracking of these materials is very possible, especially at the interfaces between two layers of different compliance, where there is a danger of adhesive failure.

The selection of the polishing medium is important, and its hardness should exceed the hardness of the sample. Neither alumina nor silicon carbide can be successfully polished with SiC grit, and diamond is the medium of preference. The same is true, to a lesser extent, for silicon nitride samples. Cubic boron nitride (CBN) has some advantages as a grinding medium. The hardness exceeds that of SiC but the oxidation resistance is better than that of diamond. It is currently available only in the larger grit sizes ($>3\ \mu m$).

An important point to realize is that even the most brittle of materials can deform plastically in the high-pressure zone beneath a point of contact with a grit particle. Plastic flow in this region will generate internal stresses which, when the contact stresses are removed, may cause cracking and chipping around the original contact, so that damage to the surface and near-surface region develops adjacent to the original contact area. Again, the solution is to limit the applied pressure and select a larger compliance for the carrier of the grinding media.

3.3.3.4 Composite Materials

Some of the most difficult materials to prepare for optical microscopy are engineering composites in which a soft, compliant matrix is reinforced with a stiff and brittle fibre. While such materials have been successfully prepared in cross-section, very few optical micrographs have been published showing the distribution of the reinforcement parallel to the fibres. Any attempt to prepare such a sample is liable to result in loss of fibre adhesion, fracture of loose fibres, and damage to the soft supporting matrix by fragments of the hard reinforcement.

In addition to an awareness of the problems and artifacts (features which are associated with preparation defects) involved in specimen preparation for the optical microscope, it is important to recognize that some samples may just be *unsuitable* for optical microscopy!

3.4 IMAGE CONTRAST

Image contrast in the optical microscope may be developed by several alternative routes, most of which require some surface preparation. A brief description of the various etching procedures commonly used to develop contrast in alloy and ceramic samples has been given previously (Section 3.3.2). A quantitative definition of contrast is best given in terms of the intensity difference between neighbouring resolved image features, $C = \ln(I_1/I_2)$. For small intensity differences this reduces to:

$$\Delta C = \Delta I / I \tag{3.5}$$

and a comparison with the Raleigh criterion for resolution (see Section 3.1.2.3) suggests that ΔC should be *at least* 0.14 if features separated by a distance equal to the resolution are to be visible.

At larger separations, very much smaller contrast variations are distinguishable, and many computerized image-processing systems operate with 256 grey levels, corresponding to intervals of $\Delta C = 0.004$. However, the largest *NA* objectives accept light scattered from the surface at much higher angles than those of lower-*NA* lenses, so that the contrast obtained in the image from any given feature is reduced. In general, higher magnification images always show lower contrast.

3.4.1 Reflection and Absorption of Light

Electromagnetic radiation incident on a polished solid surface may be *reflected, transmitted* or *absorbed.* Specular surfaces, highly reflecting to visible light, are characteristic of the presence of free conduction electrons in the solids, and hence of metallic materials. However, most metals absorb a significant proportion of the incident light. Thus copper and gold absorb in the blue, so that the reflected light appears reddish or yellow. Silver and aluminium reflect over 90 % of normally incident visible light, and both are used for mirror surfaces.

The high reflectivity of polished aluminium is not affected by the presence of the thin, amorphous oxide protective film on the surface, since the thickness of this film is well below the wavelength of visible light. Indeed most samples, and certainly all metals and alloys (with the partial exception of gold), are normally covered by some kind of surface film, either as a result of surface preparation (polishing and etching), or due to reactions in the atmosphere. As long as these films are uniform, coherent and of a thickness less than the wavelength of the incident light, they do not interfere with the reflectivity.

The relationship between the fraction of the incident light which is reflected and that transmitted or absorbed depends on the angle of incidence. The refractive index of the solid, that is the ratio of the wavelength in free space to that in the solid, determines the critical angle beyond which no light can be transmitted, and in dark-field illumination the critical angle may be exceeded, thus increasing the light signal scattered into the objective lens (Fig. 3.20). The fraction of the incident light reflected from the surface is sensibly independent of the sample thickness for all thicknesses exceeding the wavelength, and depends only on the material and the angle of incidence, but the transmitted light depends sensitively on the thickness,

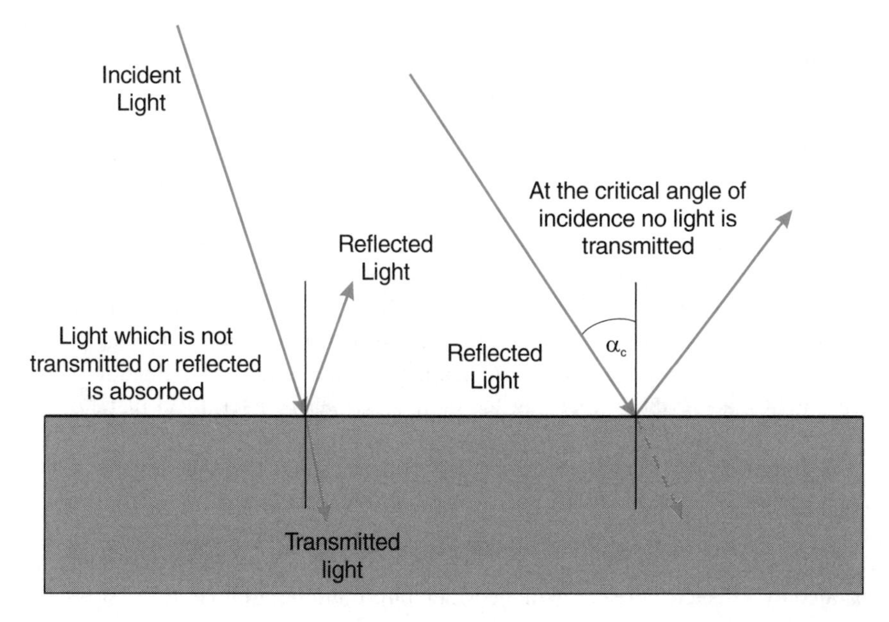

Figure 3.20 Schematic representation of the relationship between the intensity reflected, transmitted and absorbed, and the 'critical' angle for a specularly reflecting surface

decreasing exponentially as the thickness increases, in a manner exactly analogous to the case of X-ray absorption (see Section 2.3.1).

Mineralogical samples are commonly prepared as thin sections and examined in polarized light (see Section 3.4.3). It is important that they should be sufficiently thin to allow adequate transmission and ensure that significant features in the object do not overlap in the projected image of the thin slice. In general, samples of a thickness of 50 μm or less are suitable.

Ceramic and polymer samples generally transmit or absorb an appreciable fraction of the incident light, giving poor contrast, not only because of the weak reflected signal, but also due to light scattered back into the objective from sub-surface features which are below the focal plane in the image. An evaporated or sputtered surface coating of aluminium highlights the surface topography of such samples, but at the cost of losing information associated with variations in reflectivity and absorption. Some of these materials (in particular crystalline polymers and glass-ceramic materials) can be studied in transmission, by using polarized light. Convincing separation of the topological features from effects uniquely associated with the bulk microstructure is best accomplished by imaging the same area both before and after coating with the aluminium reflecting film.

3.4.2 Bright-Field and Dark-Field Image Contrast

In normal, bright-field illumination, only a proportion of the incident light is reflected or scattered back into the objective lens. There are two limiting contrast conditions in bright-field illumination. The first is that discussed previously, in which the light source is focused at the back focal plane of the objective lens, so that the light from a point source would be incident normally on the sample surface, while that from an extended source is incident over a range of angles determined by the size of the source image in the back focal plane of the objective (Fig. 3.21(a)). If, on the other hand, the light source is imaged in the plane of the sample, then the distribution of the intensity over the sample surface will reflect that in the source. In fact, if the sample is a specular reflector, then an image of the source will be visible in the microscope (Fig. 3.21(b)). For high-quality light sources, which emit uniformly, there is an advantage in maximizing the incident intensity by focusing the condenser system so that the source and sample planes coincide, but with poorer quality light sources, especially at low magnifications, uniform illumination of the sample is best achieved by focusing the light source at the back focal plane of the objective.

Topographical features which scatter the incident light outside the objective lens will appear dark in the image. This is true of both steps and grain boundary grooves (Fig. 3.22). The features themselves may have dimensions considerably less than the limiting resolution of the objective, but two features will only be observed if they are separated by a distance greater than the Raleigh resolution **and** give rise to sufficient contrast. In most cases, the size of a topological feature at the surface reflects the surface preparation as much as it does the bulk microstructure. An obvious example

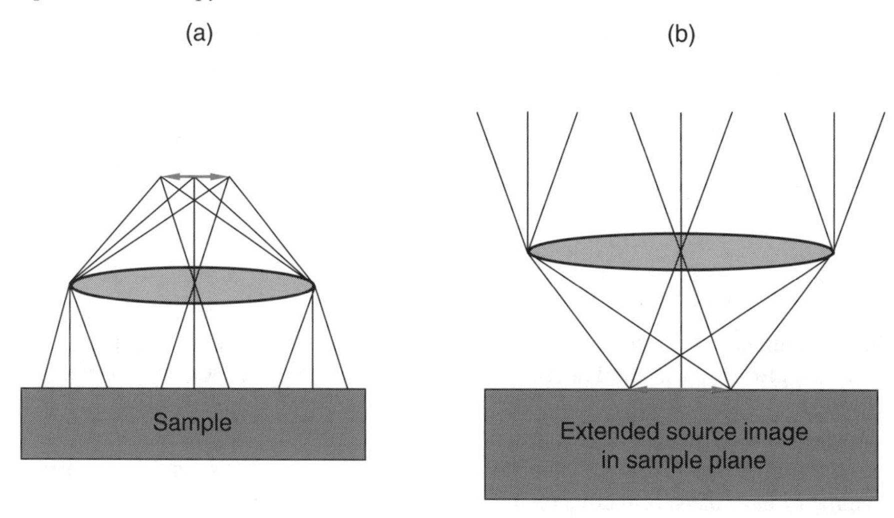

Figure 3.21 Schematic representation of the two limiting conditions for bright-field illumination: (a) the source is imaged in the back focal plane of the objective; (b) the source is imaged in the focal plane of the object

is the 'grooving' at grain and phase boundaries associated with chemical or thermal etching. It follows that considerable care may need to be exercised when measuring a microstructural parameter, such as particle size or porosity. We will return to this in Chapter 7.

Since contrast is mainly determined by comparing the intensity of the signal from some feature with that of the background, features which appear faint in the image can often be enhanced by using a dark-field objective (see Fig. 3.16). In some cases, topographical information can be enhanced by deflecting the condenser system,

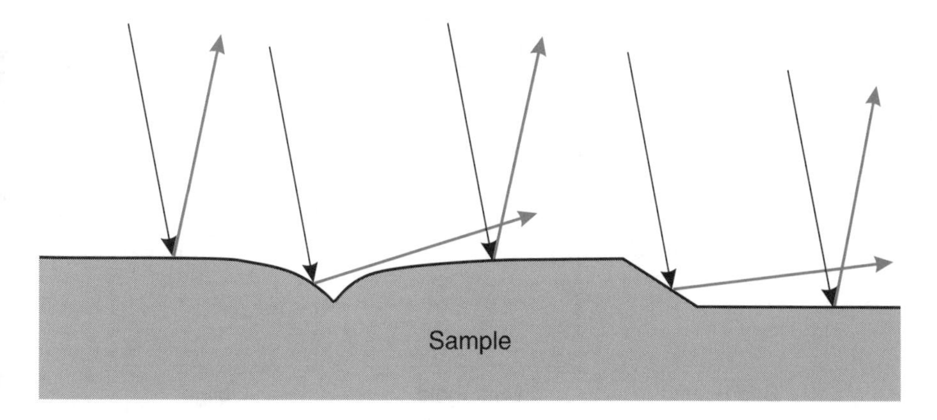

Figure 3.22 In bright-field illumination, topographical microstructural features are revealed by the scattering of light outside of the objective lens aperture

so that the specimen is illuminated from one side only. Such oblique illumination is sometimes available as a standard attachment. The apparent shadowing of topological features in oblique illumination helps to bring out the three-dimensional nature of the surface, but usually with some loss of resolution. The contrast obtained from a single feature by using bright-field, dark-field and oblique illumination are compared in Fig. 3.23.

3.4.3 Optical Anisotropy and Polarized Light

Many samples are *optically anisotropic*, i.e. the refractive index, and hence the wavelength of light in the material, are a function of the direction of propagation. Crystals of cubic symmetry are optically *isotropic*, while those with tetragonal, rhombohedral or hexagonal symmetry are characterized by *two* refractive indices, *parallel* and *perpendicular* to the primary symmetry axis. Crystals of even lower symmetry have *three* refractive indices. If a beam of light is incident on an optically anisotropic, transparent crystal, the beam will be transmitted through the crystal as two beams whose electromagnetic vectors are aligned parallel to the principle optical axes of the crystal (Fig. 3.24). On exiting from an anisotropic crystal, the two beams recombine, but any object viewed through a slice of the crystal will appear doubled, due to the different deflections of the two beams travelling through the slice.

A beam of light is said to be polarized when the electromagnetic wave vectors are not randomly oriented perpendicular to the direction of propagation, but are instead aligned in a specific direction. Sunlight reflected at a shallow angle from a car roof is partially polarized, with the direction of polarization in the plane of the roof. 'Polaroid' sunglasses only transmit light which is vertically polarized, so the glare

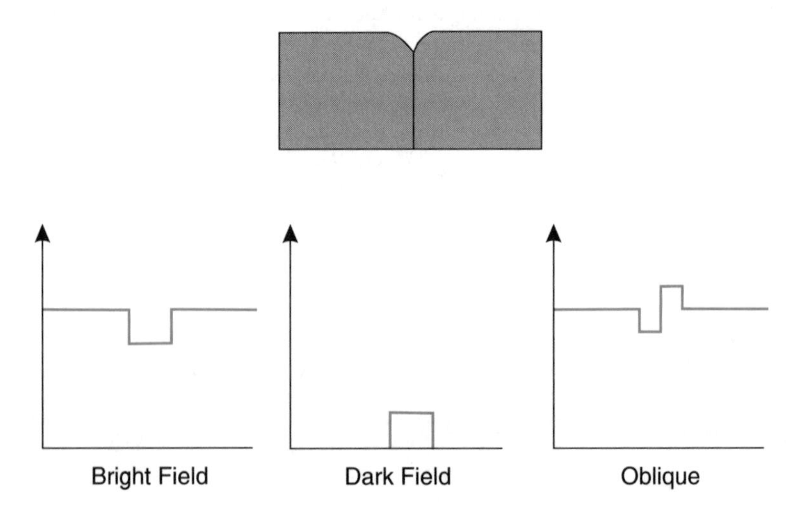

Figure 3.23 A schematic comparison of contrast in bright-field, dark-field and oblique illumination

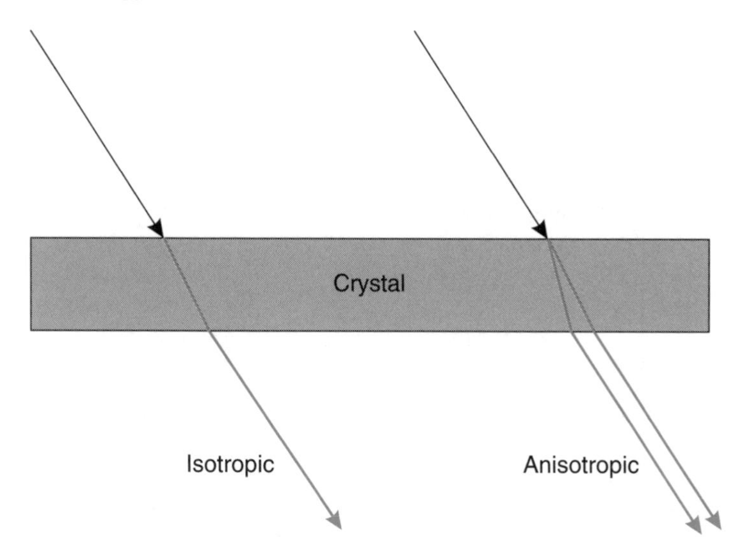

Figure 3.24 A monochromatic beam of light incident on a transparent crystal will propagate as two beams of different refractive indices (different wavelengths) if the crystal is optically anisotropic

from the (approximately horizontal) car roof is cut out from the transmitted beam by the sunglasses. If the sunglasses are removed and rotated through 90°, the glare will be found to be transmitted, since the plane of polarization of the lens is now parallel to that of the reflected glare.

3.4.3.1 Polarization of Light and its Analysis

In the polarizing microscope, the incident illumination is plane-polarized by inserting a 'polarizer' into the path of the condenser assembly. The plane of polarization is chosen to be perpendicular to the plane of assembly of the microscope components, in order to avoid any spurious changes in polarization in transmission through the system. A second polarizing element, termed the 'analyser', is placed in the path of the imaging lenses and can be rotated so that the plane of polarization of the analyser may make any angle of between 0 and 90° with that of the polarizer. When the planes of polarization of the polarizer and the analyser are set at 90°, the sample is said to be viewed through crossed polars and no light reflected from an isotropic sample can be transmitted through the analyser.

When a beam of plane-polarized light is reflected from an optically anisotropic surface, the components of the electromagnetic wave vector are resolved into two components, with both being parallel to the principle optic axes of the surface (Fig. 3.25). The two components are now slightly out-of-phase, due to the optical anisotropy, and the combined reflected beam therefore has a wave vector which rotates as the beam propagates, varying in amplitude as it does so. The tip of the

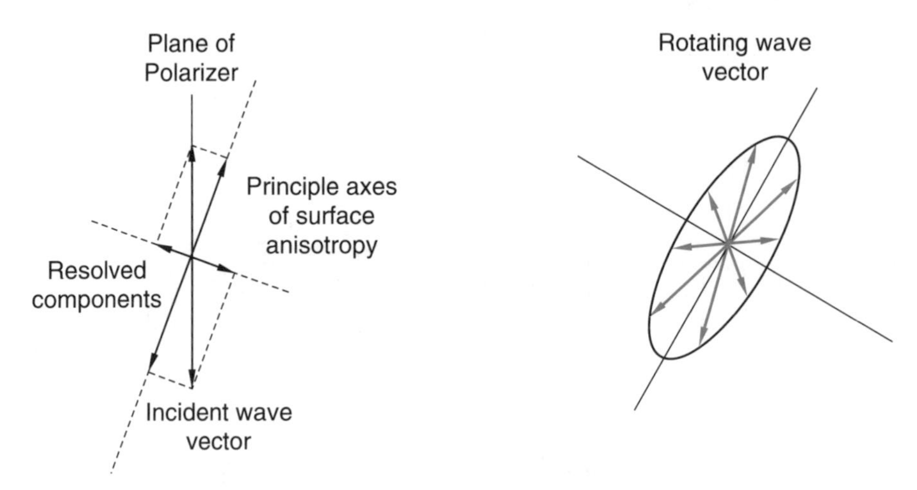

Figure 3.25 (a) Plane polarized light is resolved into two components when it interacts with an anisotropic sample. (b) The out-of-phase component then recombines to form an elliptically polarized wave

amplitude vector in the plane normal to the direction of propagation describes an ellipse, and the beam is said to be elliptically polarized. If an analyser with a plane of polarization at 90° to that of the polarizer intercepts this reflected beam, then only that component of the amplitude parallel to the plane of polarization of the analyser will be transmitted. The situation is summarized in Fig. 3.26.

From this figure, it is clear that the maximum amplitude will be transmitted through the analyser when the principal axes of the sample are set at 45° to the crossed polars. Rotating the analyser will result in more light being accepted, and when the two polars are set parallel all of the light reflected is admitted to the final image.

In order to maximize the contrast in polarized light, the condenser aperture should be stopped down and the source focused on the back focal plane of the objective, so that nearly all of the incident light is normal to the sample surface. The *NA* of the objective lens has a pronounced effect on the contrast, since only at low values of *NA* will the path of the reflected light be sensibly normal to the specimen surface. At high values of the NA, the light will be collected from the specimen surface over a wide range of angles, and the effects of optical anisotropy will be considerably reduced.

3.4.3.2 The 45° Optical Wedge

If a thin slice of a transparent, optically active, *birefringent* crystal, such as quartz, is inserted into the optical path between the polarizer and the analyser with the optical axes of the quartz slice set at 45° to those of the crossed polars, then the amplitudes of the two components of the beam travelling through the crystal will be identical,

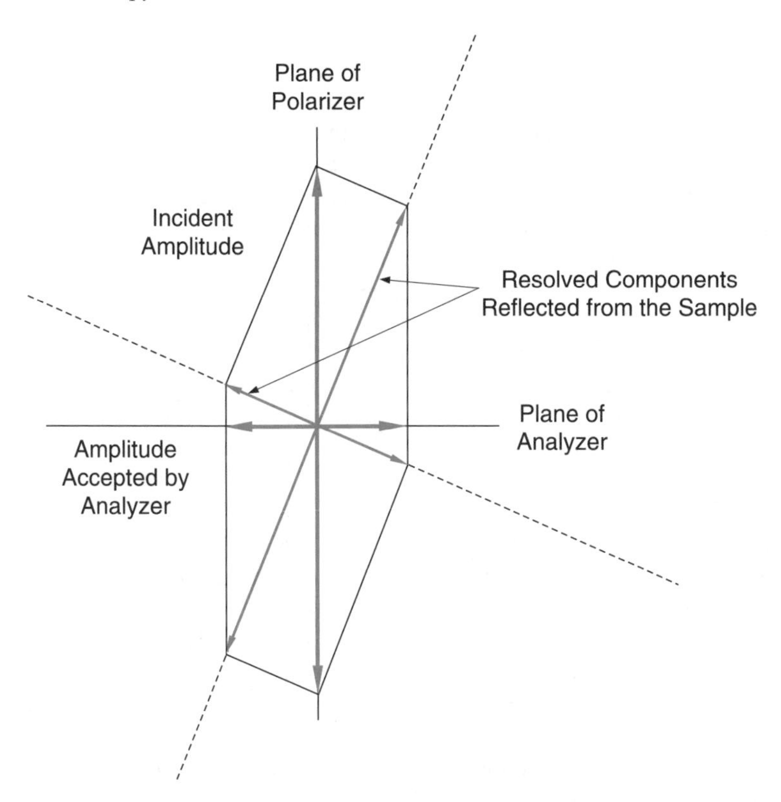

Figure 3.26 A plane-polarized beam of light incident on an optically anisotropic surface will be reflected as two beams with wave vectors parallel to the principal optic axes of the surface. These combine to form an elliptically polarized reflected beam, which is resolved by the analyser, whose plane of polarization is set at 90° to that of the polarizer in this representation

while the phase difference between the two beams at the exit surface will depend on the thickness of the crystal. If the crystal is cut as a wedge and a white-light source is used, then the two beams exiting the wedge will interfere if the phase difference $\Delta\phi$ is equal to π, or more generally, whenever $\Delta\phi = (2n + 1)\pi$. The path difference between the two waves travelling through the quartz crystal, Δd, is related to the difference in the refractive index in the two principal directions and the wavelength of the incident light. If the refractive index η is given by $\eta = \lambda_0/\lambda = c/v$, where λ_0 is the wavelength in space, λ is that in the crystal, c is the velocity of light in space and v that in the crystal, the following expression applies:

$$\Delta d = t(\eta_1 - \eta_2) = (\lambda_0/2\pi)\Delta\phi \tag{3.6}$$

where t is the thickness of the crystal.

For each thickness of the crystal along the wedge, there is a specific wavelength in an incident white-light source which will interfere destructively and be removed from the transmitted beam. The shortest wavelengths of the visible spectrum are removed first (violet), leaving the transmitted light, a pale yellow; this is followed by 'blue' interference, leading to mauve transmitted light, then green followed by red (transmitting cyan). For thick crystals, $\Delta d = (2n + 1)\lambda_0/2$, where n is an integer, and the order of the 'interference' colours will be repeated, although with decreasing brilliance for successive orders, since for the thicker regions the interference condition is fulfilled for more than one wavelength in the visible range at each thickness.

This sequence of interference colours was first noticed by Newton, and is referred to as *Newton's colours*. Using parallel, rather than crossed, polars the colours are reversed (see Fig. 3.27).

3.4.3.3 White Light and the Sensitive Tint Plate

Instead of a quartz wedge, it is possible to insert a thin, transparent slice of quartz of uniform thickness, selected to introduce a phase difference equivalent to destructive interference of blue light, tinting the light accepted into the imaging system by the analyser to mauve. If the sample is optically isotropic, reflection from the surface introduces no further phase shift and the microstructure then appears mauve when viewed between crossed-polars in white light when using this sensitive tint plate. However, if the surface is anisotropic, and thus introduces an additional phase shift into the reflected beam, then this will result in a colour shift along the sequence of Newton's colours. A positive phase shift will increase the wavelength for destructive interference, and that area of the specimen surface will appear more blue, while a negative phase shift will result in destructive interference at a shorter wavelength and a colour shift towards the red. Rotating the sample by 90° will reverse the sign of the phase shifts introduced by the different optically anisotropic features present in the microstructure.

The sensitive tint plate is a very powerful tool for exploring anisotropy in reflection, and is especially important when studying highly anisotropic crystalline polymers in transmission (see Fig. 3.28). Considerable detail is visible in this figure, with the very high optical anisotropy ensuring excellent contrast, even when working at high resolution and with large *NA* objectives.

3.4.3.4 Reflection of Polarized Light

A few further words are necessary concerning the interaction of polarized light with a reflecting surface. As noted earlier (Section 3.4.3), unpolarized light is partially polarized when it is reflected at an angle, with the direction of polarization perpendicular to the plane containing the incident and reflected beams. Furthermore, most surfaces prepared for microscopic examination are covered by a surface film. It follows that *linearly polarized light* incident on the surface is likely to undergo a

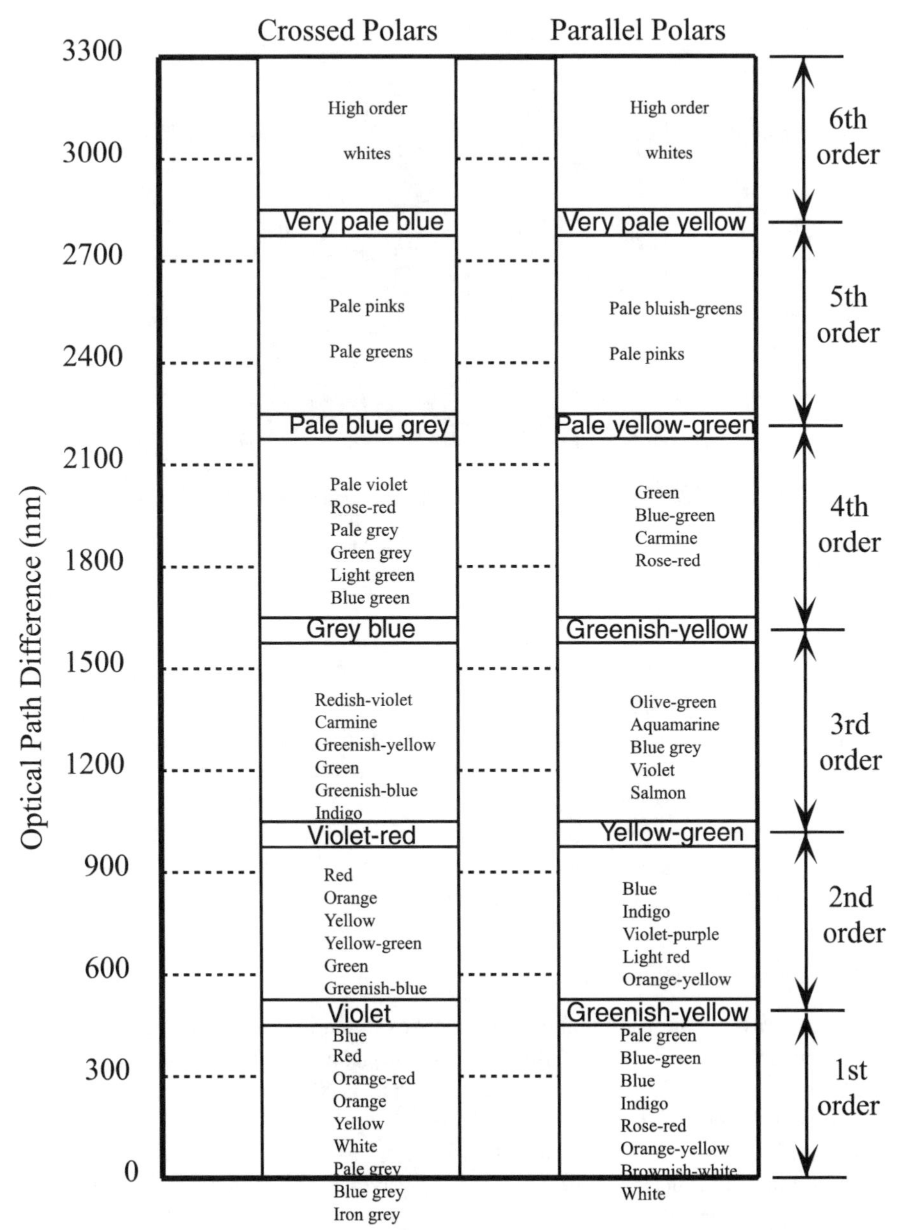

Figure 3.27 Newton's thickness interference colours in crossed and parallel polars

phase change and, to some extent, become *elliptically polarized*. The cause may be *surface topography* (fine surface facets or aligned grooves which impart optical anisotropy to an otherwise isotropic surface), or it may be associated with the conditions of *illumination* (a primary beam incident at a glancing angle, as in dark-

Figure 3.28 Crystalline polymers viewed in transmission by polarized light reveal a wealth of detail associated with the crystallization process. Reproduced by permission of John Wiley & Sons, Inc.

field illumination) or the conditions of collection of the *reflected* light (a large-*NA* objective accepting a wide cone angle).

In many cases, complete extinction is not obtained in crossed polars, and does not necessarily mean that the sample is anisotropic. Alternative explanations are loss of linear polarization resulting from an oblique angle of incidence or the existence of 'anisotropy' due to surface topology. Changing the plane of focus of the source by adjusting the condenser-lens setting should give an indication of whether or not the illuminating conditions are responsible for the effect.

3.4.4 Phase Contrast Microscopy

Phase contrast is important in the histological examination of organic tissues viewed in transmission, when small differences in the scattering from features of weak contrast need to be amplified, but the technique is also useful for revealing fine topological features in reflection microscopy.

If light is reflected from two neighbouring regions which differ slightly in height by an amount h (which is much less than the wavelength), then the reflected beams will be out of phase by a small *phase angle*, given by:

$$\delta\phi = 2\pi(h/\lambda) \tag{3.7}$$

If h is sufficiently small, the *phase shift* between the two specularly reflected wave vectors is equivalent to a small *elastically scattered* amplitude whose wave vector is sensibly out of phase with the specularly reflected beams by $\pi/2$ (Fig. 3.29). The

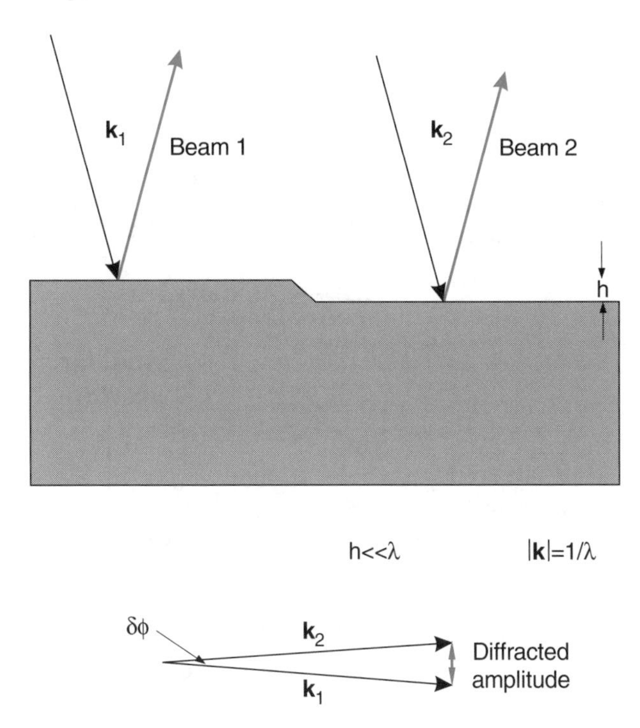

Figure 3.29 Phase contrast is based on the light scattered from small features associated with a difference in height, h. The scattered amplitude is *ca* $\pi/2$ out-of-phase with the specularly reflected beams

problem is to convert this very weak scattered signal into an observable intensity difference. This requires that the scattered amplitude be comparable to the amplitude of the specularly reflected signal and the phase difference be shifted by an additional $\pi/2$, so that constructive or destructive *interference* may occur.

One solution to this problem is illustrated in Fig. 3.30. The source of light is an *annular condenser aperture*, ensuring that the path length to the specimen is sensibly identical for all rays, and this aperture is focused on the *back focal plane* of the objective. An intermediate, *auxiliary* lens is used to bring the image of the source, reflected by the sample surface, into focus on a *grooved phase plate* which is placed behind the final imaging lens. The diameter and width of the groove in the phase plate exactly matches the annular source aperture, so that *specularly reflected* light passes through the groove, while *scattered* light is transmitted through the ungrooved portion. The depth of the groove is selected to introduce the additional required phase shift of $\pi/2$, so that the scattered signal now either reinforces ($+\pi$) or interferes ($-\pi$) with the specularly reflected signal, that is 'bumps' and 'hollows' in the surface topology will appear brighter or darker than the background in the image. In order to ensure that the amplitude of the specularly reflected signal is comparable

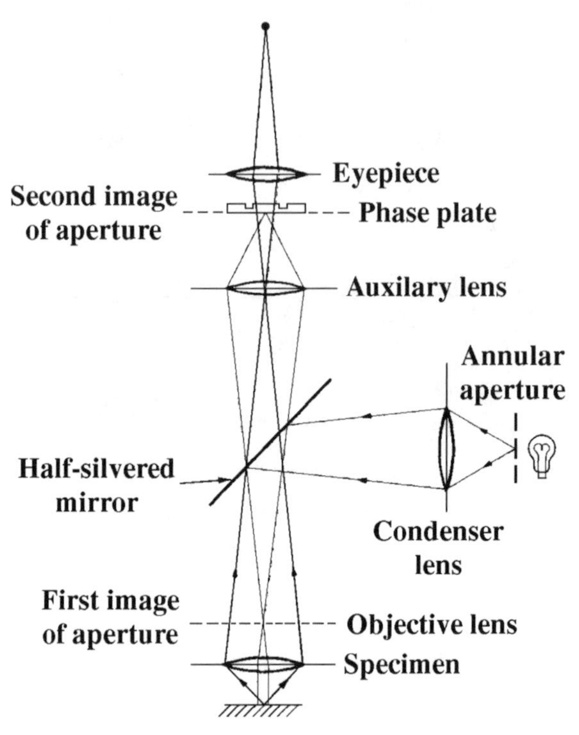

Figure 3.30 Schematic representation of an optical system for phase contrast reflection microscopy

to that of the scattered signal, the grooved region is coated with an absorbing layer that reduces the directly transmitted amplitude by about 90 %.

A series of phase plates are available, especially for use with transmission samples, and provide phase shifts of $(2n + 1)\pi/2$. These are particularly useful in studying biological tissue samples in which small differences in refractive index can be characterized by using the *phase contrast microscope*. Bearing in mind that the height (or phase) differences must be small, and assuming that a reasonable value of h/λ to achieve this is 0.1, it follows that the phase contrast microscope will have no difficulty in picking up variations in surface topology of the order of 10 % of the wavelength (or of the order of 50 nm). In practice, surface steps of the order of 20 nm are readily detectable.

3.4.5 Interference Contrast and Interference Microscopy

In phase contrast, intensity variations are due to interference between a specularly reflected beam and an elastically scattered beam from the same sample surface. In *interference microscopy*, the light reflected from the sample interferes with light reflected from a *standard reference surface*. In order to achieve this, the two beams must be *coherent* (have a *fixed phase relation*), and this is achieved by ensuring that both beams originate from the *same* source, by using a *beam splitter* to first separate and then recombine the two signal amplitudes.

3.4.5.1 Two-Beam Interference

The simplest arrangement for achieving interference is to coat an optical-quality glass coverslip with a thin layer of silver or aluminium, such that rather more than half the incident light will be transmitted by the coverslip (see below). The light reflected from the thin metal film now constitutes the reference beam, while that transmitted through the coating, and subsequently reflected from the sample, is the interfering beam (Fig. 3.31). We shall ignore *multiple reflections* for the time being, noting only that these will affect both the resolution and the contrast. We also assume that there is no absorption, so that the incident beam is either reflected or transmitted, but *not* absorbed.

If the *reflection coefficient* of the metal film is R, then, in the absence of absorption, the *transmission coefficient* is $(1–R)$. If we assume that the reflection coefficient of the sample is 1, then the intensity reflected from the sample is $(1–R)$. The reflected beam from the sample is transmitted back through the metal film, with a transmission coefficient of $(1–R)$, and the remaining intensity is *re-reflected*. Therefore, the intensity of the second beam is $(1–R)^2$, as compared to the intensity of the reference beam, R. For the two beams to be of comparable intensity, it follows that the required value of R is $(3–\sqrt{5})/2$, or approximately 0.38.

The condition for *destructive interference* is that the two beams differ in path by $(2n + 1)\lambda/2$, so that the phase difference is equal to π. However, the *path difference* is just *twice* the separation h of the partially reflecting reference surface from the sample, that is $2h = (2n + 1)\lambda/2$, with the first destructive peak occurring when the two surfaces are separated by $\lambda/4$. Successive *interference fringes* correspond to

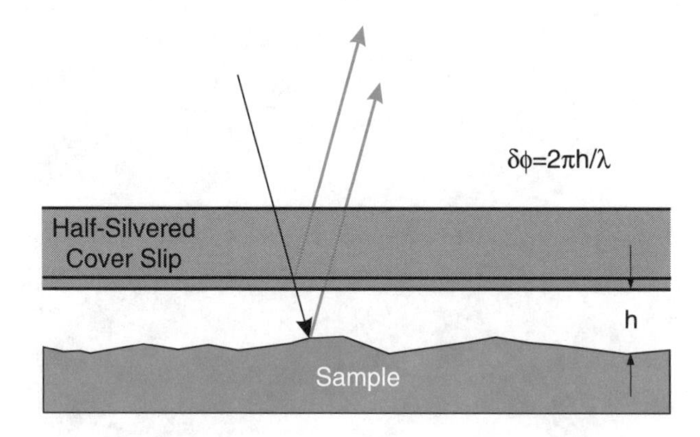

Figure 3.31 Schematic representation of the condition for two-beam interference when using a half-silvered reference surface

height contours separated by $\Delta h = \lambda/2$. By assuming that shifts in an interference fringe are detectable to *ca* 10% of the fringe separation, and that the wavelength used is in the green, simple *two-beam interferometry* should then be capable of detecting topological height differences to an accuracy of ± 20 nm, which is roughly the same sensitivity as that achieved in phase contrast. An example of this is shown in Fig. 3.32.

Interference contrast and sensitivity can be improved by placing a drop of immersion oil between the coverslip and the sample, thus reducing the effective wavelength by a factor μ, the *refractive index* of the oil. The objective lens numerical aperture for good two-beam interference is limited, since for an *NA* greater than *ca* 0.3, the difference in path length for light passing through the periphery of the lens and along the axis is sufficient to destroy the coherency of the beam. As in phase contrast microscopy, the condenser system should be focused on the back focal plane of the objective. It is useful to view the specimen in white light, when the orders of the interference fringes correspond to Newton's colours. Those regions of the specimen which are in contact with the reference plate then appear white, and it is a sobering experience to note *how few* contact points there are in the absence of any appreciable external force.

3.4.5.2 Systems for Interference Microscopy

The Mercedes (or Rolls Royce) of microinterferometers is that designed by Linnik (Fig. 3.33), although this is seldom used. Two identical objective lenses are used, in order to ensure that no path differences are introduced by the optical system. The position of the *reference surface* can be adjusted along the optic axis and can also be tilted about two axes at right angles and in the plane of the surface, in order to adjust the spatial separation and orientation of the fringes.

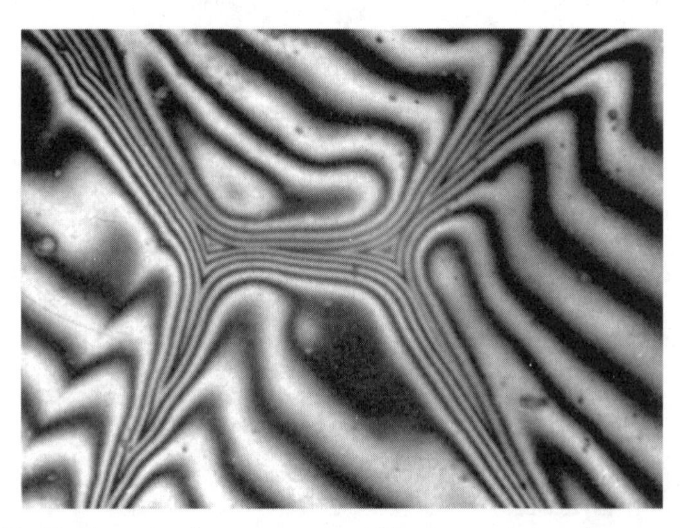

Figure 3.32 Two-beam interference pattern resulting from grain-boundary grooving

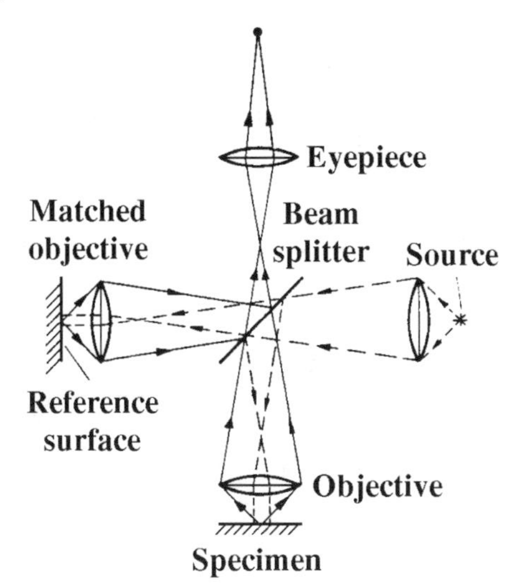

Figure 3.33 Schematic representation of the optical system used in a Linnik micro-interferometer

The position of the reference surface is adjusted in white light so that, at a small angle of tilt, Newton's colours can be observed on either side of the white contour marking the line of coincidence of the reference image with the image of the sample.

Far simpler systems are commercially available as standard attachments, but most of these have a rather limited life, since they rely on half-silvered reflecting surfaces which are easily damaged if brought into contact with a sample. The only real disadvantage of a metal-coated coverslip (which is cheap and can be regarded as a consumable) is the inability to adjust the separation between the reference surface and the sample, or the distance between the interference fringes, but for many applications this is not critical.

Nomarski interference microscopy is based on a rather different concept, and does not set out to create interference fringes, but rather to provide a simple alternative to reflection phase contrast microscopy. A double quartz wedge is inserted into the optical system in the 45° position and creates twin images of the sample viewed in polarized light, which are separated in the image plane by a small lateral shift. The nominal path length of the twin beams is *identical*, so that any difference in phase must be due to a microstructural feature on the scale of the separation of the twin images, usually a height change. The contrast obtained in Nomarski interference reflects these small topological differences, and is very similar to phase contrast.

3.4.5.3 Multi-Beam Interference Methods

As noted previously, in simple two-beam interference a proportion of the light is multiply reflected. With the previous assumptions of no absorption losses, and an optimal reflection coefficient for the half-silvered coverslip of 0.38, these multiply reflected intensity losses amount to 24 % of the incident light and result in unwanted background. By increasing the reflection coefficient of the reference surface, the proportion of light which is multiply reflected can be increased, until the value is close to 1. With the geometry shown in Fig. 3.34, the dependence of the total reflected intensity, I, is given by the relation:

$$I = \left[\frac{T^2}{(1-R)^2}\right]\left[\frac{1}{1+\left(\frac{4R}{(1-R)^2}\right)\sin^2 \delta/2}\right] \tag{3.8}$$

where T and R are the transmission and reflection coefficients, respectively, and the reflection coefficient of the sample is now assumed equal to that of the reference surface. The parameter δ is given by the relation:

$$\delta/2 = (2\pi t \cos\theta)/\lambda \tag{3.9}$$

where t is the separation of the reference and sample surfaces. Once again, we will assume that there is no absorption in the reference film, so that $T + R = 1$, and that the intensity collected falls to zero when $2t\cos\theta = n\lambda$. In this case, the fringes are now localized at the reference surface, and the number of beams which contribute is determined by the angular tilt of the reference surface with respect to the optic axis and the separation between the reference surface and the sample. The width of the black interference fringes depends on the number of beams involved and can be very narrow when compared to the $\cos^2$ intensity dependence obtained in the two-beam case.

For multiple beam interferometry to be effective, the incident beam should be parallel and the surfaces separated by no more than a few wavelengths. The best patterns have been produced by coating the sample with a thin plastic film and evaporating silver on to the atomically smooth surface of the plastic. Under these conditions, the very sharp interference fringes show topological changes in the

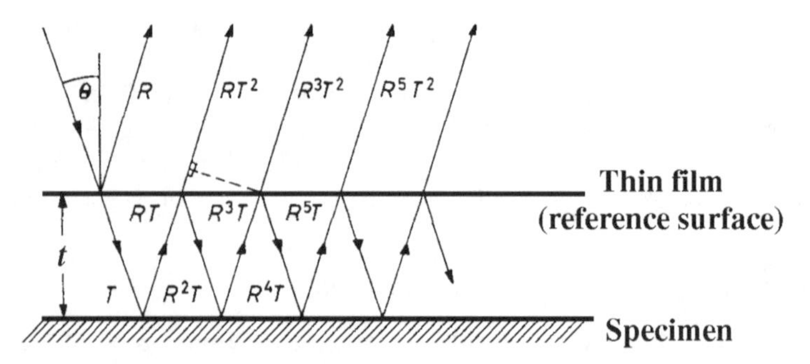

Figure 3.34 Schematic representation of the geometry required for multiple beam interference

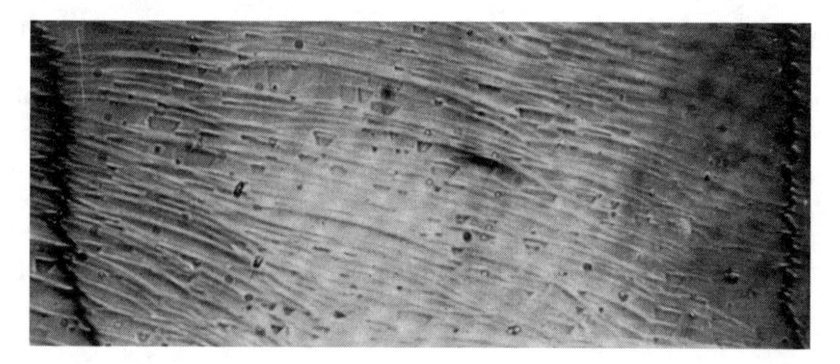

Figure 3.35 Multiple beam interference fringes obtained from a polished quartz specimen

surface at the nanometre level, and it is quite possible to image growth steps which are a few atoms in height (see Fig. 3.35).

Of course, there is a penalty to be paid: *multiple beam interferometry* is only possible with a low-*NA* objective, and the lateral resolution is typically no better than 1 or 2 μm.

3.4.5.4 Surface Topology and Interference Fringes

A few more words are in order concerning the information which can be derived using one or other of the above interference techniques. The small *depth of field* of the optical microscope means that the region of the object in focus is a thin slice of thickness of the order of 1 μm or less. The image is then a planar projection of this slice of material, and contains information on both the topology and the physical properties of the surface. The *lateral* resolution is not only limited by the wavelength of the light and the *NA* of the objective lens, but also by the method of surface preparation and the means used to record the image.

Interference micrographs contain image contrast based on variations in the phase of the light reflected from the surface. These phase variations may be associated with three quite distinct effects:

(a) If they are associated with *surface anisotropy*, then they may be revealed by *polarized light*, and the phase shift arises from the difference in wavelength of the light polarized parallel to the two optic axes of the sample surface.
(b) If they are due to small *topological features* or *spatial inhomogeneities* in the optical properties (variations in refractive index), then they may be detected by *phase contrast* (*or Normanski*) *microscopy*, which identifies the sign of the phase shift and its approximate magnitude.
(c) If they are due to variations in *surface topology*, then *interference microscopy* is often the most appropriate tool, quantitatively monitoring the separation between an optically flat reference surface and the sample surface.

The vertical resolution for topological features can be much better than the lateral resolution, being of the order of 20 nm for *two-beam interference* or *phase contrast* and as little as 2 nm for *multiple beam interference*, but this vertical resolution is only available at the expense of lateral resolution, since the requirements for good interference images limit the numerical aperture of the objective lens to of the order of 0.3. In spite of this, quantitative measurements using interference microscopy have proved extremely valuable, and applications range from measurements of thermal grooving at grain boundaries, to the height of facets on the surfaces of a growing crystal, and the size of the slip steps in ductile materials.

3.5 RESOLUTION, CONTRAST AND IMAGE INTERPRETATION

Finally, we summarize some of the problems and pitfalls associated with the interpretation of optical images.

It should be clear by now that the image of a sample only contains information on the microstructure of the material if the observer understands the separate processes used to obtain the image, that is

(a) *preparation* of the sample;
(b) *imaging* in the microscope;
(c) *recording* the image.

The ultimate *resolution* of the features observed in the recorded image includes contributions from all of these three processes. The width of the grain boundary grooves obtained during etching may limit the resolution just as much as the *NA* of the objective lens or the wavelength of the light used. In addition, the grain size of the photographic emulsion used to record the micrograph, plus the photographic process used to develop the negative and print the final image, can have a similar effect.

Contrast can be due to the surface topology, the optical properties of the sample (reflection coefficient and/or optical anisotropy), or the presence of a surface film. Contrast may be associated with variations in image intensity, but can also result from a difference in phase or wavelength, as in the changes in colour associated with phase shifts and observed with a sensitive tint plate. The enhancement of contrast by photographic means (increasing the γ of the recording film) reduces the number of grey levels which can be recorded and may result in a loss of information. Contrast enhancement using phase contrast or Normarski interference will also result in some loss of lateral resolution. Contrast is generally better for a low-*NA* objective, but is again at the cost of reduced resolution.

Finally, resolution is *useless* without image contrast, but poor resolution for the sake of good contrast is *equally* self-defeating. The eye is still the judge of what constitutes the best compromise.

Summary

The *optical microscope* is the tool of preference for the microstructural characterization of engineering materials, both because of the wealth of information available in the magnified image and the ready availability of high-quality microscopes, specimen preparation facilities and inexpensive methods of image recording. The visual impact of the magnified image is immediate, and its interpretation is in terms of spatial relationships which are familiar to the casual observer of the macroscopic world.

The *geometrical optics* determine the relationship between the object placed on the microscope stage and its magnified image. The ability to resolve detail in the image is limited by the wavelength of the light used to form the image and the angle subtended by a point in the object plane at the objective aperture. The aperture of the eye and the range of wavelengths associated with visible light limit the resolution of the eye to *ca* 0.2 mm, so that the unaided eye can distinguish two features at a comfortable reading distance (30 cm) if they are separated by 0.2 mm.

The *numerical aperture* (*NA*) of an objective lens is the product $\mu \sin \alpha$, where μ is the refractive index of the medium between the lens and the object and α is the half-angle subtended at the objective (the angular aperture of the lens). Values of *NA* vary from of the order of 0.15 for a low-magnification objective used in air to *ca* 1.3 for a high-power, oil-immersion objective, leading to a limiting (best possible) resolution of the order of half the wavelength of visible light, *ca* 0.3 μm.

A sharply focused image will only be obtained if the features to be imaged are all in the plane of focus, and high-*NA* lenses do not allow for much latitude. It follows that specimens to be imaged at the best resolution must be accurately planar. This limited *depth of field* is the primary reason why samples of opaque materials must by polished optically flat, while those of transparent materials must be prepared as thin, parallel-sided sections.

The components of the optical microscope include the light source and condenser system, the specimen stage, the imaging system and the image recording device. Each of these components has its own engineering requirements, namely the intensity and uniformity of the *light source*, the mechanical stability and positional accuracy of the *specimen stage*, the optical precision and alignment of the *imaging system*, and the sensitivity and reproducibility of the *recording system*. While photographic recording is still the primary option, the rapid development of charge-coupled devices is leading to the development of a new range of microscopes in which digital recording and computer-aided image enhancement are well on the way to making the photographic darkroom obsolete.

A major consideration in the application of the optical microscope is the selection and preparation of a suitable sample. In the first place, the sample must be representative of the features which are to be observed, and the sampling procedure must take account of both *inhomogeneity* (spatial variations in the features and their distribution) and *morphological anisotropy* (orientational variations, as in fibrous or lamellar structures). Secondly, preparation of the sample surface must reveal the intersection of bulk features with the plane of the section without introducing *artifacts* (such as scratches or stains). In most cases, surface preparation is a two-stage process. The first step is to prepare a flat, polished, mirror-like surface, while the second is to develop contrast by the use of suitable chemical etchants, solvents or differential staining agents.

Image contrast is a sensitive function of the mode of operation of the microscope, but most engineering materials are examined by reflection microscopy. The contrast then depends on local differences in the absorption and scattering of the incident light, and the optics of image formation. *Chemical etchants* commonly develop topographic features which scatter the incident beam outside the objective aperture, but they may also differentially stain the surface, so that some features absorb more light than others. In many cases, different phases and impurities ('inclusions') will also give contrast which is associated with differences in their ability to reflect incident light, quite independent of the action of an etchant.

The optical imaging conditions can also be controlled in order to enhance contrast. A '*dark-field*' image is formed by the scattered light from the object, rather than by the specularly reflected light used to form a '*bright-field*' image. If the specimen is illuminated with *plane-polarized* light, then the changes in polarization which accompany the interaction with the specimen can be analysed with '*crossed-polars*', which convert rotation of the plane of polarization into variations in image intensity. An *optical wedge* or *sensitive tint plate* can be used to further enhance the contrast in polarized light, thus yielding quantitative information on the optical properties of the sample.

In *phase contrast* reflection microscopy, small (>20 nm) topological differences can be converted into variations in image intensity, while in transmission the differences in refractive index can be similarly imaged. By combining the light reflected from the sample surface with that reflected from an optically flat surface, it is possible to obtain *optical interference*, in which the interference fringes again reflect the topology of the sample. *Multiple-beam interference* allows very small differences in the height of surface features to be detected with a sensitivity of a few nanometres. This vertical 'resolution' is therefore several orders of magnitude better than the image resolution, where the latter is limited by the numerical aperture of the objective and the wavelength of the incident light.

It is important to remember that the quality of the recorded image depends on three independent factors, namely the preparation of the sample, the optical imaging in the microscope, and the system used to record the final image. The eye is the best judge of that particular combination of resolution and contrast which yields the most information.

Bibliography

1. *Metals Handbook, Volume 7: Atlas of Microstructures of Industrial Alloys*, American Society for Metals, Metals Park, OH (1972).
2. J.L. McCall and P.M. French (eds), *Metallography in Failure Analysis*, Plenum, New York (1978).
3. J.B. Wachtman, *Characterization of Materials*, Butterworth-Heinemann, Oxford (1993).

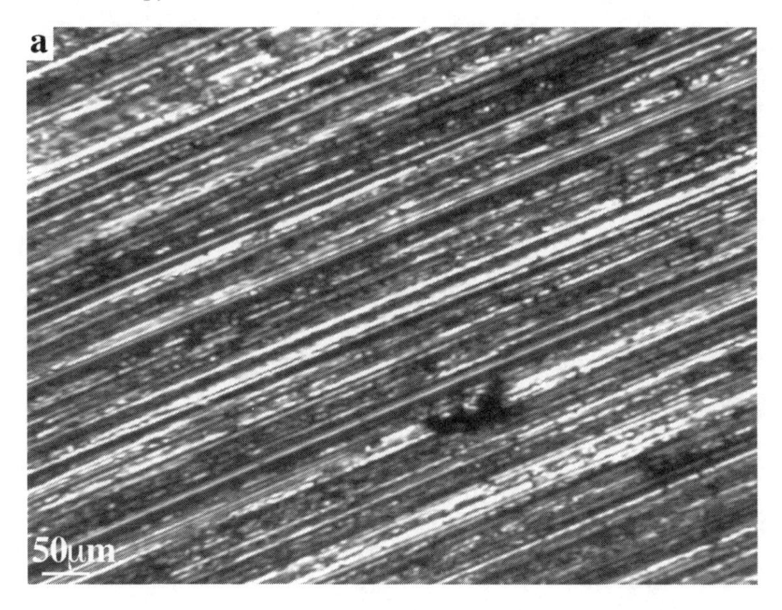

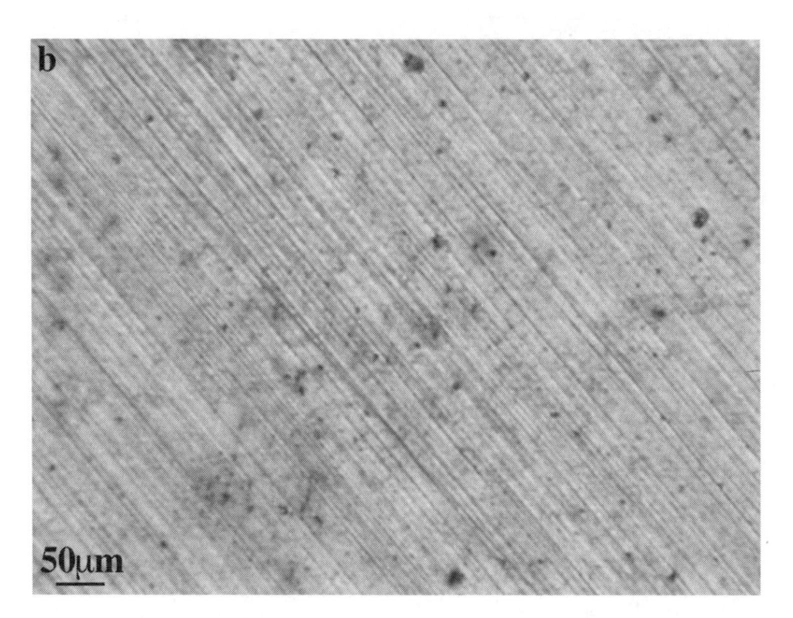

Figure 3.36 Optical micrographs of 1040 steel obtained after polishing with a sequence of diamond grits: (a) after rough grinding the surface to achieve a planar surface; (b) after polishing with a 6 μm diamond grit; (c) after polishing with a 1 μm diamond grit; (d) after polishing with a 0.25 μm diamond grit

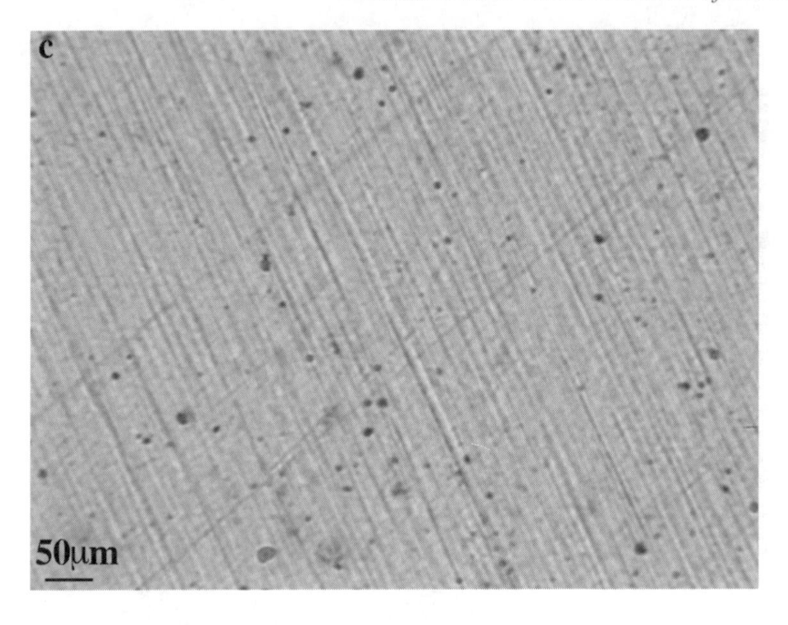

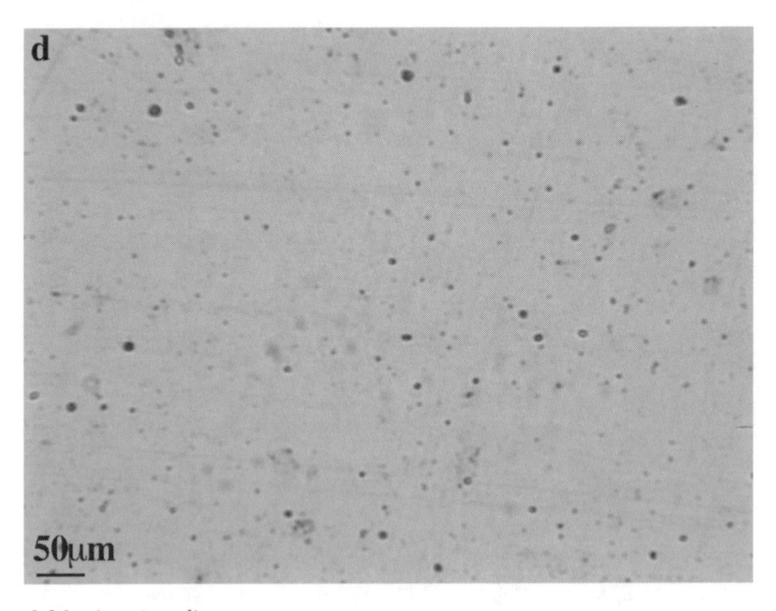

Figure 3.36 *(continued)*

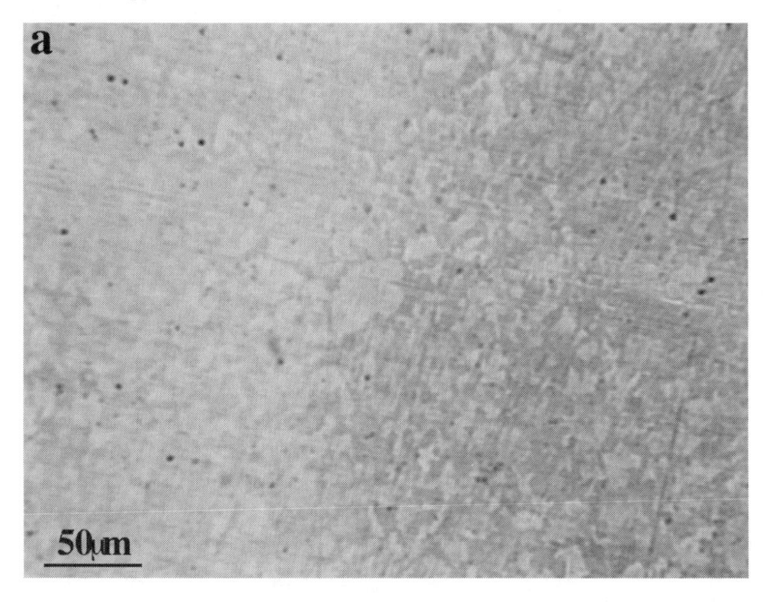

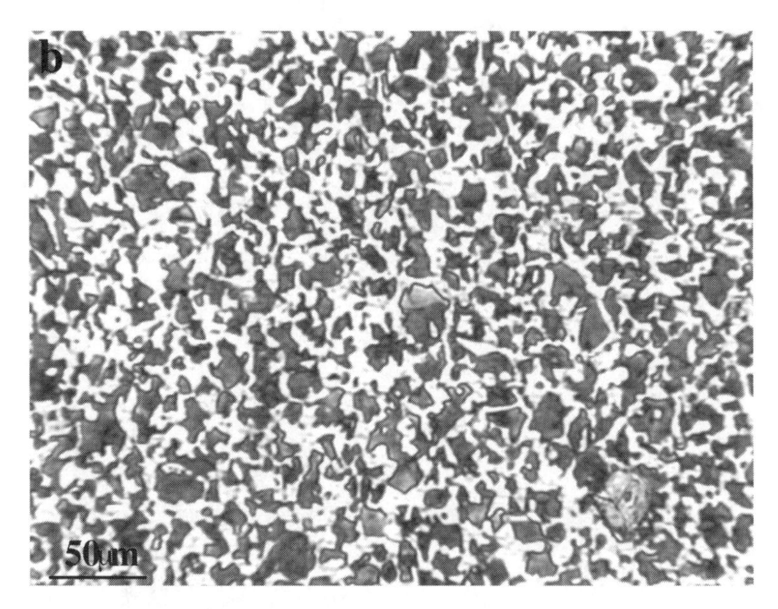

Figure 3.37 The same 1040 steel as shown in Fig. 3.36 after etching for different lengths of time in a very dilute nitric acid solution: (a) under-etched; (b) a good etch; (c) over-etched

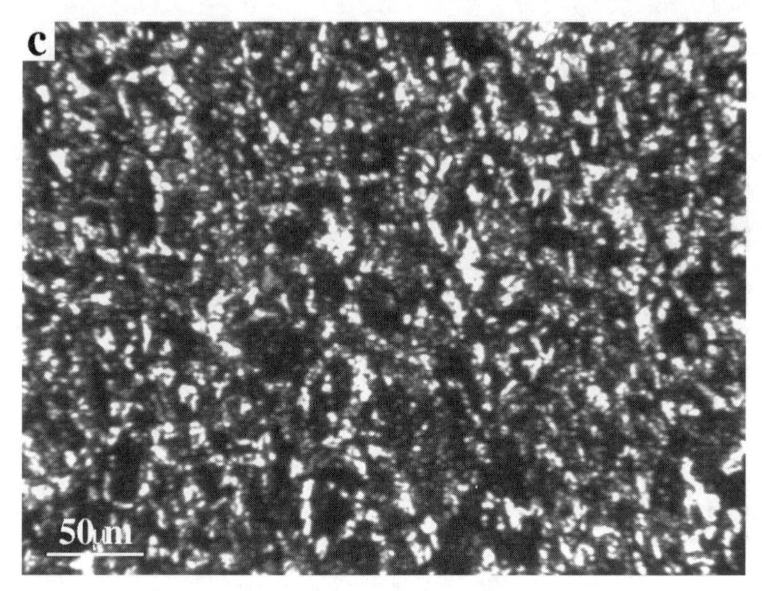

Figure 3.37 (*continued*)

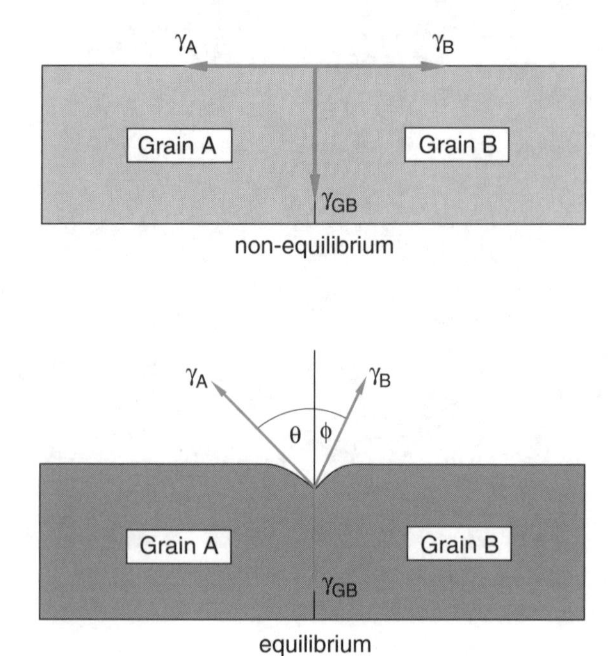

Figure 3.38 A grain boundary intersecting a polished surface is not in equilibrium (a). At elevated temperatures (b), surface diffusion forms a grain-boundary groove in order to balance the surface tension forces

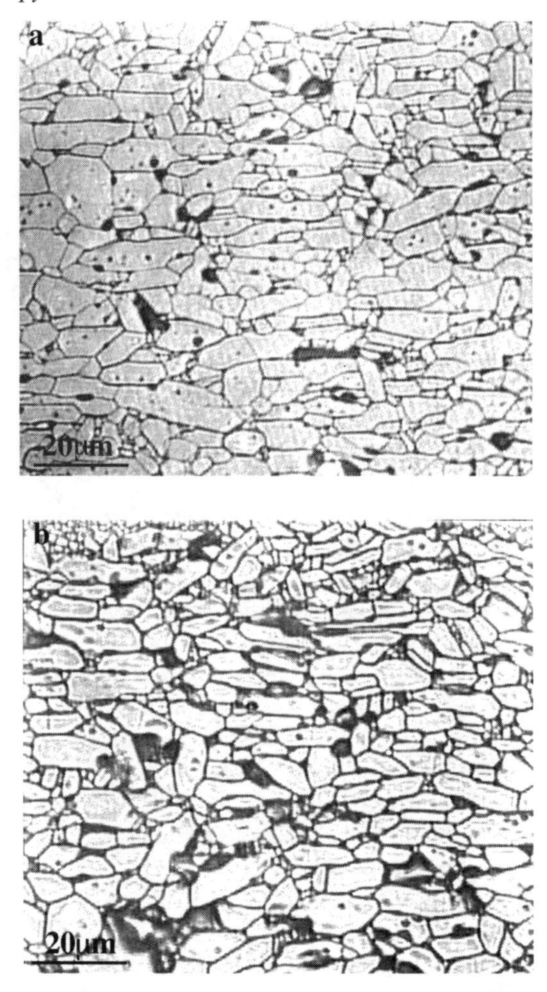

Figure 3.39 Polished alumina, thermally etched at 1200°C for 30 min (a) is adequate, but etching for 2 h at the same temperature (b) clearly improves the visibility of the grain boundaries

Worked Examples

We begin this section by considering the influence of the polishing process on the surface finish of a 1040 steel. Small samples are cut from a large block and then embedded in a *thermoplastic* mount. Granules of the thermoplastic are poured around the specimen in a metal die and compacted at a moderate pressure and temperature. Alternatively, and in order to avoid any damage to the specimen, a *thermosetting plastic resin* can be cast into a mould containing the specimen and then cured at room temperature.

Fig. 3.36 compares the optical micrographs of a 1040 steel surface obtained at different stages of *polishing*. A 'good' polish results in a planar surface with no scratches visible under the optical microscope. A final polish with 0.25 μm diamond particles removes most surface artifacts, and provides a specimen ready for *etching* (Fig. 3.36d).

The choice of etchant must be consistent with the information sought from the sample. A good compendium of etchants for metals and alloys is *The Metals Handbook* [1], which both describes the experimental sample preparation procedures and also gives examples. For our 1040 steel, we wish to determine the average grain size and shape, so we chose a very dilute nitric acid solution. Fig. 3.37 shows optical micrographs of the polished steel after etching in such a dilute acid solution for increasing lengths of time. The two main variables for a given etchant are *time and temperature*—the higher the temperature, then the faster the etching rate. Too light an etch gives poor contrast and fails to reveal the microstructure adequately, while over-etching results in pitting and loss of resolution.

Alumina is an inert oxide, and difficult to prepare for optical microscopy by chemical etching, even though some chemical etchants exist (based on concentrated acids) which can be used at elevated temperatures. Instead, microstructural contrast in alumina and other ceramics is commonly developed by *thermal etching* in which a polished specimen is heated to a temperature sufficient for surface diffusion to occur.

The intersection of a grain boundary with a *polished* surface is unstable since the surface-tension forces are not in equilibrium (Fig. 3.38(a)). If surface diffusion can take place, then grooves will develop along the line of intersection of the grain boundary with the free surface in order to balance the component of the surface forces perpendicular to the boundary (Fig. 3.38(b)). The '*thermally etched*' grain-boundary grooves scatter light outside the objective lens aperture, and provide the contrast needed to identify the grain boundaries.

Mass-transfer by diffusion is a function of both temperature and time, and *insufficient* thermal etching will result in poor contrast (Fig. 3.39(a)). *Increased* thermal etching increases the depth and width of the grooves, thus improving the contrast (Fig. 3.39(b)). However, the same diffusion processes promote grain-boundary migration and grain growth, while 'over-etching' results in 'wide' grain boundaries and loss of resolution.

Problems

3.1 Which optical properties of an engineering material are important in determining the preparation of a sample for optical microscopy?
3.2 Are short-sighted individuals blessed with 'better' resolution? Discuss!
3.3 The Raleigh resolution criterion assumes two point sources of light, and unlimited contrast in the image. Which factors are likely to prevent attainment of the Raleigh resolution?
3.4 The better resolution obtained with a high-numerical-aperture objective is accompanied by reduced contrast and depth of field. Why is this?

3.5 *Mechanical stability* is a necessity for the specimen stage of an optical microscope. Compare (quantitatively) the stability required parallel to the optic axis of the microscope with that required in the plane of focus.
3.6 Photographic recording of the optical image is now gradually being replaced (at least in part) by digital recording. What are the reasons for this?
3.7 Which sections from the following components would you select for examination in an optical microscope: (a) a multilayer capacitor; (b) steel wire; (c) a nylon sheet; (d) a sea shell?
3.8 *Chemical etching* is often used to develop image contrast on a polished sample surface. Give *three* examples of this.
3.9 Give *one* example where you consider that *dark-field* illumination would give more information in optical microscopy than *bright-field* illumination, and justify your choice.
3.10 What is the orientation of the principal axes of an optically anisotropic sample with respect to the axis of polarization of the *incident* light when the observed intensity is a maximum using crossed polars? Explain your answer.
3.11 What is a *sensitive tint plate* and when would you consider using it?
3.12 *Two-beam interference* images have been used to analyse the shape of grain-boundary grooves formed by thermal etching. How does the spacing of the fringes and the angle that they make with the boundary affect the vertical resolution?

Chapter 4

Electron Microscopy

The *electron microscope* has extended the resolution available for morphological studies from that dictated by the wavelength of visible light to dimensions which are well into the range required to image the lattice planes in any crystal structure, that is from of the order of 0.3 μm to of the order of 0.15 nm. The first attempts to focus a beam of electrons by using electrostatic and electromagnetic lenses were made in the 1920s, with the first electron microscopes appearing in the 1930s, pioneered primarily by I. Ruska, working in Berlin. These were *transmission* instruments, intended for samples of powders, thin films and sections prepared from bulk materials. *Reflection electron microscopes*, capable of imaging the surfaces of solid samples at glancing incidence, made their appearance after the 2nd World War but these were almost immediately superseded by the first *scanning electron microscopes*, combined with the microanalytical facilities available in the microprobe (see Chapter 5).

Sub-micron resolution was demonstrated on the earliest *transmission electron microscopes*, manufactured in Europe, and later in Japan and the United States. The early developments in electron microscopy are an international success story: in the immediate post-war period, commercial *transmission* instruments were manufactured in Germany, Holland, Japan, the United Kingdom and the United States. The first *scanning* instruments were made in the United Kingdom, while the first *microprobe* was a French development.

In this chapter, we will outline the basic principles of focusing a beam of electrons before discussing the factors that limit resolution in electron microscopy. We will then compare the requirements for both transmission and scanning electron microscopy before discussing in detail the specimen preparation procedures, and the origin and interpretation of contrast for both instruments.

First, however, we need to distinguish clearly between the imaging modes in the two types of instrument.

The *transmission electron microscope* is in many ways analogous to a transmission *optical* microscope—but usually 'upside down', in the sense that the source of the electron beam is at the top of the microscope 'column' while the recording system is at the bottom (Fig. 4.1). The *electron gun* replaces the light source and is maintained at a high voltage (typically 100–400 kV). A heated tungsten filament is capable of generating electron beam current densities of the order of 5×10^4 A m^{-2}, while lanthanum hexaboride (LaB_6) crystals generate *ca* 1×10^6 A m^{-2}. A field-emission source, by comparison, can generate current densities of the order of 1×10^{10} A m^{-2}. The high-energy electrons from the gun are focused by an electromagnetic *condenser lens* system, whose focus is adjusted by controlling the lens currents (not the lens position, as would be the case in the optical microscope). The *specimen stage* is much more complex than indicated in the diagram, and allows for specimen *tilt* as well as some *z*-adjustment along the optic axis. An important point to note is that fine focusing of the image in the transmission electron microscope is not achieved by adjusting the position of the specimen along the *z*-axis (to alter its distance from the objective lens), but rather by adjusting the *electromagnetic objective lens current* in order to focus a first image from the elastically scattered electrons transmitted through the thin film specimen. The *final imaging system* also employs electromagnetic lenses, and the final image is observed on a *fluorescent screen*. Typical screen current densities are of the order of 10^{-10} to 10^{-11} A m^{-2}, but may be even lower at the highest magnifications. Photographic emulsions are commonly used to record the final image, but as in optical microscopy, advances in the development of *charge-coupled devices* combined with computerized image processing are making digital-image recording increasingly competitive.

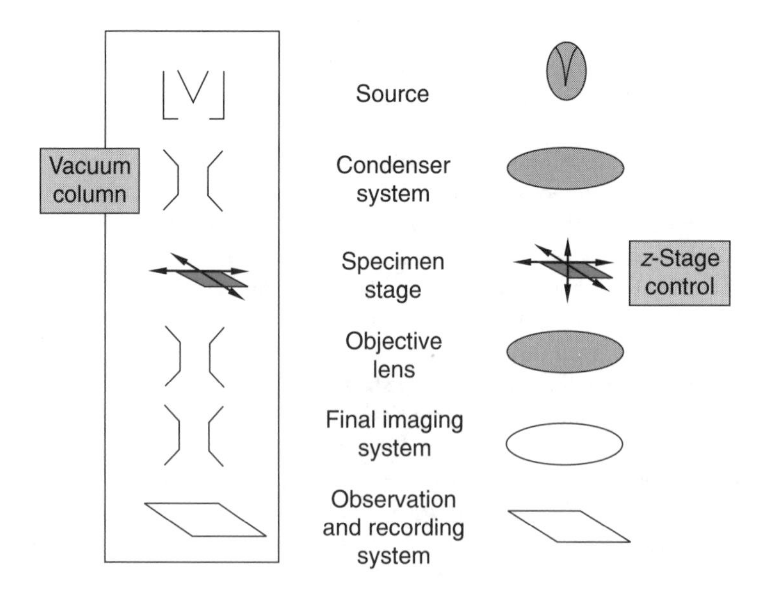

Figure 4.1 Similar to the transmission optical microscope, the transmission electron microscope has a source, a condenser system, a specimen stage, an objective lens and an imaging system, as well as a method for observing and recording the image

The high-energy electron beam has a very limited path length in air, so that the whole electron microscope column must be under *vacuum*. *Specimen contamination* under the beam (the development of a carbonaceous layer on the specimen surface) is a serious problem which may restrict viewing time for a particular area and limit the achievable resolution. In general, the vacuum should be better than 10^{-6} torr, while for the highest resolution a vacuum of 10^{-7} torr is desirable, and all sources of contamination should then be trapped, usually by *cryogenic cooling* of the specimen surroundings.

The *scanning electron microscope* (Fig. 4.2), again has a source of high-energy electrons and a condenser system, but now employs a *probe lens* to focus the electron beam into a fine probe that impinges on the specimen. The electromagnetic probe lens in the scanning electron microscope fulfils a similar function to the *objective lens* in the transmission electron microscope, since it determines the ultimate resolution attainable in the microscope. However, the probe lens is placed *above* the specimen, and has no part in collecting the image signal from the latter. Indeed, the *elastically scattered* electrons are of no real importance in scanning electron microscopy, and it is the variety of possible *inelastic scattering* processes that occur when the electron probe interacts with the sample, which dominate the information collected in this instrument. The electron energy of the beam used in scanning electron microscopy is appreciably less than that used in transmission,

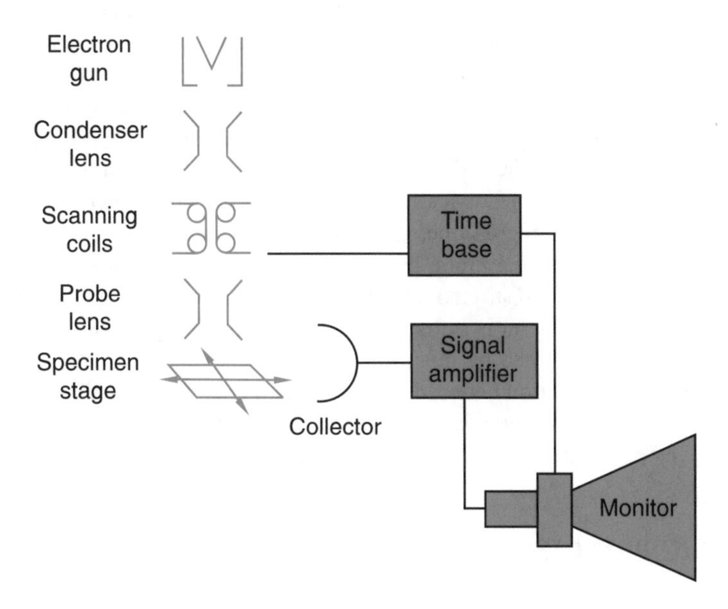

Figure 4.2 In the scanning electron microscope, a fine probe of high-energy electrons is focused on to the sample surface and then scanned across the surface in a television raster. A signal generated by the interaction of the probe with the sample is collected, amplified and displayed on a monitor with the same time base as the raster used to scan the sample

being usually of the order of 5–30 keV, although much lower energies (down as low as 200 eV) are often useful.

The 'image' in scanning electron microscopy is obtained by scanning the electron probe over the sample surface in a *television raster*, and then collecting an image signal from the surface and displaying it, after suitable amplification and processing, on a monitor with the same time base as that used to scan the probe (see Fig. 4.2). This is very similar to the principle employed in a *video camera*, where the image focused by the camera is scanned with a beam of electrons and the modulated signal transmitted, recorded and finally displayed on a monitor—the *television screen*.

The power of the scanning electron microscope derives from the very wide range of signals that may result from the interaction of the electron probe with the sample surface. These include *characteristic X-rays*, generated by excitation of inner-shell electrons, *cathodoluminescence*, in the range of visible light and associated with valency electron excitation, *specimen current* due to the net absorption of electric charge, and *back-scattered electrons*, inelastically scattered out of the surface from the probe beam. However, the most commonly used signal is that derived from low-energy *secondary electrons* emitted from the surface of the target by the inelastic interaction of the sample with the primary beam. The secondary electrons are emitted in large numbers from a highly localized region at the point of impact of the probe. They are therefore readily detected and are capable of forming an image whose resolution is limited only by the probe diameter.

The basic difference in the operation of the *transmission* and the *scanning* instruments can be summarized in terms of the two modes of data collection which are employed to form the image. In both the *optical microscope* and in *transmission electron microscopy*, information is collected continuously over the full field of view and focused by suitable lenses to form a magnified image. In other words, the data from all image points are collected simultaneously, and the time required is the same as that necessary to obtain a statistically significant signal from a *single image point*. In the scanning electron microscope, the information is collected *sequentially*, for each data point in turn, as the probe beam is scanned across the field of view. The *scanning speed* must be restricted in order to ensure that the specimen signal recorded for each image point is statistically adequate, and the total time required to form the image is determined by this minimum scanning speed and the number of image points. This distinction between an *optical image*, in which the image data are acquired for all image points *simultaneously*, and a *scanning image*, in which the image is developed *sequentially* (one point at a time) cannot be overemphasized.

4.1 BASIC PRINCIPLES

While the design and construction of an electron microscope is outside the scope of this present text, it is important to have some appreciation of the basic physical principles which determine the behaviour of electrons in a magnetic field and their interaction with matter.

4.1.1 Wave Properties of Electrons

The focusing of an electron beam is possible because of the dual, *wave–particle* character of electrons. This wave–particle duality is expressed in the *de Broglie relationship* for the wavelength of a particle:

$$\lambda = h/mv \tag{4.1}$$

where m is the mass of the particle, v is its velocity and h is the Planck constant. Assuming that the accelerating voltage in the electron gun is V, then the electron energy is given by:

$$mv^2/2 = eV \tag{4.2}$$

where e is the charge on the electron. It follows that $\lambda = h/(2meV)^{0.5}$, or $\lambda = (1.5/V)^{0.5}$ nm when V is in volts. This numerical value is approximate, since at the accelerating voltages commonly used in the electron microscope, the *rest mass* of the electron, m_0, is appreciably less than the *relativistic mass*, m, and a correction term should be included, in the equation:

$$\lambda = \frac{h}{\sqrt{\left[2m_0eV\left(1 + \frac{eV}{2m_0c^2}\right)\right]}} \tag{4.3}$$

where c is the velocity of light. The relativistic correction amounts to *ca* 5 % at 100 kV, rising to 30 % at 1 MV. The electron wavelength at 100 kV is 0.00370 nm, which is nearly two orders of magnitude less than the interatomic spacings typical of the solid state. At 10 keV, which is typical of many applications of scanning electron microscopy, the wavelength is only 0.012 nm, still appreciably less than the interatomic distances in solids.

4.1.1.1 Electrostatic and Electromagnetic Focusing

Electrons are deflected by both electrostatic and magnetic fields, and can be brought to a focus by suitably engineering the field geometry. In the region of the electron gun, the beam is influenced by the electrostatic field created by the anode and bias cylinder, which usually create a first-focus, 'virtual' source. All subsequent focusing in the electron microscope is magnetic and is achieved by *electromagnetic lenses* equipped with soft iron (essentially zero hysteresis) *pole pieces*. The focal length of an electromagnetic lens is controlled by varying the lens current.

An electron travelling in a magnetic field is deflected in a direction at right angles to the plane containing the magnetic field vector and the original direction of travel of the electron (its momentum vector). In a *uniform* magnetic field, an electron travelling off-axis follows a *helical path* (Fig. 4.3(a)), while electrons originating at

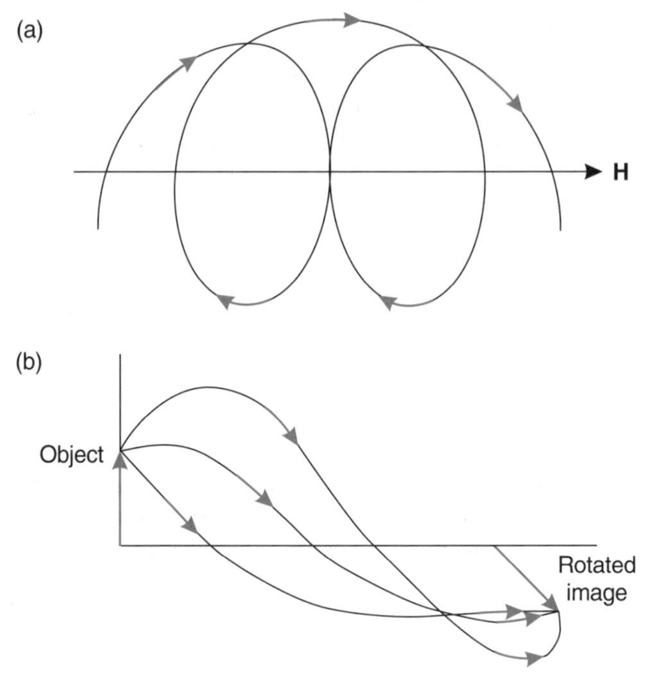

Figure 4.3 An electron in a magnetic field is deflected at right angles to both the momentum and magnetic field vectors: (a) an off-axis electron follows a spiral path; (b) electrons originating at a point off the axis are brought to a rotated focus

any point within the field will be focused at a second point after spiralling along the axis (Fig. 4.3(b)).

The image formed by focusing electrons using an electromagnetic lens differs in several important respects from that formed by focusing light using a glass lens. In the first place, the image is rotated, so that the focusing of the objective lens by adjusting the lens current is accompanied by *rotation* of the image about the optic axis. It follows that two images of the *same* object taken at *different magnifications* will also be rotated with respect to one another. This rotation can be compensated by *reversing* the magnetic field vector over a proportion of the optical path, and it is now common practice to design an electromagnetic lens such that the windings of the lens coils in the upper and lower halves of the lens carry the current in opposite directions. Even with a compensated lens, the electrons are not confined to travel in a plane, which is quite different from the behaviour of light.

In the light-optical microscope, there is an abrupt change in refractive index when the light is deflected as it enters a glass lens, while the refractive index is constant within the glass lens. With an electromagnetic lens, the deflection of the electrons is *continuous*, and the magnetic field created by the lens pole pieces varies continuously over the optic path within the lens.

Finally, the angle subtended by the electron path with respect to the optic axis is always very small ($<1^{\circ}$) so that the path length through the magnetic field of the lens is very long when compared to the spread of the beam. This may be compared to the case of the optical microscope, for which the numerical aperture of the objective may correspond to an acceptance angle for scattered light of between 45 and 90°. In effect, the 'numerical aperture' (this term is *not* used in electron microscopy) of an electromagnetic lens never exceeds 10^{-2}!

4.1.1.2 Thick and Thin Electromagnetic Lenses

The physics of electromagnetic focusing means that the simple geometrical optics which we applied to the light-optical microscope (Section 3.1) is a very inadequate approximation to the optics of image formation in the electron microscope. In particular, the simple relationships between the *focal length*, the *magnification*, and the relative positions of object and image along the optic axis no longer hold because they are based on the assumption that the lens is thin when compared to the total optical path between object and image. Actually, this *thin-lens approximation* is also insufficient for high-powered objectives in the light microscope, and is replaced by much more complex *thick*-lens calculations. In electron microscopy, *all* of the lenses are 'thick lenses'.

Nevertheless, we can illustrate the imaging modes in the electron microscope by using *ray diagrams*. These will be presented in two dimensions, as though the electrons were not deflected out of the plane, and we approximate the electron path changes by sharp deflections at the lens positions, as though the thin lens approximation were still valid. *No* quantitative calculations are possible in this approximation and none will be given.

4.1.1.3 Resolution and Focusing

Given that the maximum beam divergence in the electron microscope is less than 1°, the Raleigh criterion for the image of a point source can be reduced to the following, $\delta = 0.61\lambda/\mu \sin\alpha \approx 0.61\lambda/\alpha < 60\lambda$. By inserting the value for the wavelength at 100 kV, 0.0037 nm, the potential resolution of the electron microscope should be of the order of 0.2 nm. At higher operating voltages, an even better resolution should be possible (see below).

As a gross approximation, we can use the light-optical expression for depth of field, $d \approx \delta/\alpha$, so that the thin-film specimens used in transmission electron microscopy should be of the order of 20–200 nm in thickness if both the top and bottom of the film are to be in focus simultaneously. A similar evaluation can be used for the depth of focus, $D = M^2 d$, so that at a magnification M of 10000 the expected depth of focus is of the order of metres! There is no problem whatsoever in focusing an image on a fluorescent screen and subsequently recording the same image on a photographic emulsion placed some distance *beneath* the focusing screen.

4.1.2 Resolution Limitations and Lens Aberrations

At this point, we will consider the optical performance of the electron microscope in more detail, in particular the reasons why the angular divergence of the electron beam is limited to such small values.

4.1.2.1 Diffraction Limited Resolution

The diffraction limit on resolution is, as in light-optical microscopy, that given by the Raleigh criterion, i.e. $\delta_d = 0.61\lambda/\mu \sin\alpha$. In vacuum, $\mu = 1$ and at small angles, $\sin\alpha = \alpha$. By inserting the expression given previously for the wavelength in terms of the accelerating voltage, we obtain $\delta_d = 0.61\lambda/\alpha = 0.75/[\alpha\sqrt{V}(1 + 10^{-6}V)]$. It follows that, for a given divergence angle, it should be possible to improve the resolution by increasing the accelerating voltage. Experimental electron microscopes have been constructed with accelerating voltages of up to 3 MV, but commercial instruments have been limited to *ca* 1 MV. At these voltages, most samples are liable to experience extensive *radiation damage*, especially after prolonged exposure. Most high-resolution electron microscopy (by which we commonly mean imaging of the crystal lattice) is performed at 300 or 400 kV, the threshold for the onset of radiation damage in most materials. At these voltages, a point-to-point resolution (limited by the aberrations of the lenses) of the order of 0.15–0.2 nm is readily and routinely attainable.

4.1.2.2 Spherical Aberration

An electron beam parallel to, but at a distance from the optic axis of an electromagnetic lens will be brought to a focus at a point on the axis which depends

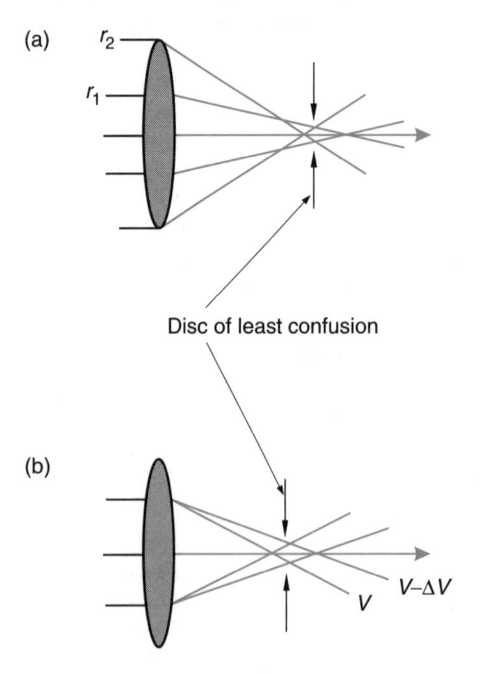

Figure 4.4 Spherical (a) and chromatic aberration (b) prevent a parallel beam from being brought to a point focus. Instead, a disc of least confusion is formed in the focal plane

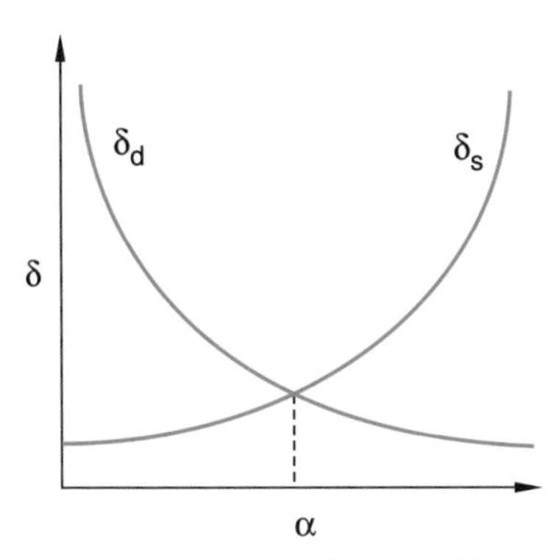

Figure 4.5 The diffraction and the spherical aberration limits on resolution have an opposite dependence on the angular aperture of the objective, so that an optimum value of α exists

on the distance of the beam from the axis (Fig. 4.4(a)), while a beam *further* from the optic axis will be focused *closer* to the lens, so that the plane of 'best' focus will correspond to a *disc of least confusion* whose size will depend on the spread of the beam. This phenomenon is referred to as *spherical aberration*, and the radius of the disc of least confusion constitutes an aberration-dependent limit on the resolution, which is given approximately by $\delta_s \approx C_s\alpha^3$, where C_s is the *spherical aberration coefficient* of the electromagnetic lens. While the *diffraction* limit on the resolution, δ_d, is *inversely* proportional to the angular aperture of the objective, α, the *spherical aberration* limit, δ_s, is proportional to the *third power* of the angular aperture (Fig. 4.5).

It follows that for any given lens of fixed spherical aberration coefficient, there is an *optimum* angular aperture for which δ_d equals δ_s, corresponding to $C_s\alpha^3 = 0.61\lambda/\alpha$, or $\alpha^4 = 0.61\lambda/C_s$. The optimum angular aperture is therefore a sensitive function of both the accelerating voltage (the wavelength) and the spherical aberration coefficient of the lens. Typical values for this coefficient for an electromagnetic lens are somewhat less than 1 mm. By inserting a reasonable value of 0.6 mm for C_s and the wavelength appropriate to 100-keV electrons, 0.0037 nm, we obtain a value of *ca* 8×10^{-3} for this optimum value of α.

4.1.2.3 Chromatic Aberration

Chromatic aberration arises because higher-energy electrons are *less* deflected by a magnetic field than those of lower energy, so that they are brought to a focus at a point further from the position of the lens, once again giving rise to a disc of least confusion, this time determined by the energy spread in the electron beam and the *chromatic aberration coefficient* of the lens.

There is more than one source of chromatic aberration, although that due to variations in beam energy is generally the most important. If the electrons are *thermally emitted* (as is usually the case), then the relative energy spread will be given by $\Delta E/E_o = kT/eV$, where k is the Boltzmann constant and e is the electronic charge. For $T = 2000\,\text{K}$ (a reasonable temperature for a tungsten filament source) and 100-keV electrons, the energy spread, $\Delta E/E_o$, is *ca* 1.5×10^{-6}. Electrons will also lose energy due to *inelastic scattering* in the thin sample, adding to the effect of the thermal-energy spread of the beam. This may be an appreciable fraction of the incident electrons if the specimen is thick or of high atomic number (for example a heavy metal). Variations in the electromagnetic lens currents also contribute to chromatic aberration, since they also result in changes in the focal length of the lenses. The equation relating the chromatic aberration limit on resolution to these variations in beam energy and lens current is given by the following relation:

$$\delta_c = C_c \frac{\Delta E}{E_0} \alpha \tag{4.4}$$

where C_c is the chromatic aberration coefficient of the lens, and ΔE represents the instabilities of both the accelerating voltage and the objective lens current. Once again, the resolution limit increases with α, but this time only linearly. Providing that the *voltage stability* of the electron gun and the *current stability* of the electromagnetic lenses are adequate, chromatic aberration should not be a problem for the *microscope*, although it may well be a problem, associated with *inelastic interactions*, in the *specimen*.

4.1.2.4 Lens Astigmatism

The *axial symmetry* of the electro-optical system is an extremely important factor in limiting the performance of the electron microscope, and the alignment of the lens components within the column of the microscope is a critical factor in ensuring that the performance is optimized. Of these components, the *objective lens* is the most sensitive to misalignment, while the axial asymmetry of this lens is highly sensitive to minor changes within the microscope (size, position and dielectric properties of

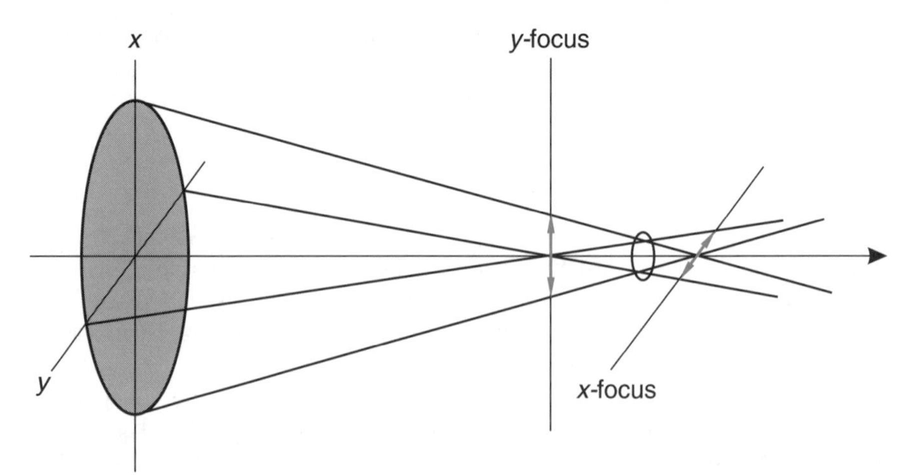

Figure 4.6 Astigmatism is the result of axial asymmetry and leads to variations in focal length about the optic axis, giving two positions of the line foci at right angles

the sample, or small amounts of contamination on the sample or the objective aperture).

A basic feature of the loss of axial asymmetry is a variation in the focal length about the optic axis, which results in *two* principal focal positions that will give two *line foci* at right angles (Fig. 4.6). This *astigmatism* cannot be prevented, both because of the inherent residual asymmetry in the lens and pole-piece construction, and because of the extreme sensitivity of the astigmatism to minor misalignment and contamination. However, it can be corrected. Twofold astigmatism correction is achieved by introducing sets of correction coils whose magnetic fields are at right angles to both the optic axis and the magnetic field of the main coils. The correction-coil currents are periodically adjusted to exactly balance the magnetic asymmetry. A number of arrangements are possible, for example four pairs of coils to form an 'octet' or *octopole* astigmatism correction system.

4.1.3 Transmission and Scanning Electron Microscopy Compared

We will now compare the performance of the *transmission* and *scanning* electron microscopes in order to distinguish clearly between their principal features.

4.1.3.1 The Optics of Image Formation

As noted previously, the primary image in transmission is focused by the objective lens, while a set of other imaging lenses enlarge the final image. In the scanning microscope, the image is developed, point-by-point, by collecting a signal which is generated by the interaction of a focused electron probe as it is scanned across the surface of the sample.

4.1.3.2 Depth of Field and Depth of Focus

As in the light-optical microscope, the depth of field of the transmission electron microscope is limited by the numerical aperture of the objective lens and the resolution of the microscope, but since the angular aperture of the lens is so small, the depth of field in transmission electron microscopy *exceeds* the resolution by at least two orders of magnitude.

In scanning electron microscopy, the electron probe is focused by a probe lens, whose operation is analogous to that of the objective lens in transmission. However, the *inelastic scattering* processes which occur during interaction of the probe with the specimen, together with the requirement for an adequate signal current, normally limit the probe size, and hence the resolution, to the nanometre range. It follows that, with an angular aperture for the probe lens of the order of 10^{-3}, the depth of field in the scanning electron microscope (δ/α) is typically of the order of *microns*, considerably better than that in optical microscopy, and at a much improved resolution.

Thus, while both the light-optical and the transmission electron microscopes generate a two-dimensional image of a thin, *planar* slice taken from the bulk material, the image in the scanning electron microscope may contain considerable in-focus information on the *three-dimensional topography* of the surface. Furthermore, since data collection in the scanning image is point-by-point, the question of depth of focus does not arise, since there is now *no* focused image in the optical sense.

4.1.3.3 Specimen Shape and Dimensions

The electro-optical image requirements in transmission electron microscopy place the specimen close to the back-focal plane of the objective lens, and usually within the magnetic field of the lens. The space available for the specimen is therefore minimal, so that, in addition to the stringent limitations on sample thickness which are dictated by the restricted penetration and inelastic scattering of the electron beam, there are also limitations on the lateral dimensions of the sample. The specimen diameter is typically less than 3 mm, of which only the central region is actually available for examination.

In contrast, a sample selected for scanning electron microscopy sits beneath the probe lens, and is usually separated from the lens magnetic field by several millimetres. With a *long-working-distance*, probe-forming lens, reasonable resolution is available even when the lens–sample separation is over 50 mm. In addition, there are *no* limitations on lateral dimensions, other than those dictated by convenience. Most samples have lateral dimensions similar to those used for optical microscopy (20–30 mm), but there is no inherent reason why much larger assemblies should not be inserted into the specimen chamber. Large specimen chambers are available that will accept specimens which are 10 cm across or even more.

4.1.3.4 Vacuum Requirements

The vacuum requirements of any electron microscope, transmission or scanning, are determined by the following three factors:

(a) the need to avoid scattering of the high-energy electrons by residual gas in the column;
(b) the requirement for thermal and chemical stability of the electron gun during operation;
(c) the reduction of beam-induced contamination of the sample.

The least significant factor is the first, since a vacuum of 10^{-5} torr is quite sufficient to ensure that the cross-section for scattering of the high-energy electrons in the beam by residual gas is negligible. The second factor is more important, and the traditional electron source, a heated *tungsten filament*, is steadily eroded during operation at 10^{-5} torr. Alternative sources, either a low-work-function lanthanum hexaboride crystal or *field-emission* source (both of which emit at lower temperatures, and hence generate a beam with a lower chromatic energy spread), require much better vacuum conditions, of the order of 10^{-7} torr for LaB_6 and 10^{-10} torr for field-emission guns.

However, the most important factor determining the vacuum requirements is usually the third, since specimen contamination is most frequently the result of an inelastic interaction between absorbed gas on the sample surface and the incident electron beam. *Hydrocarbons* arriving at the sample will be polymerized and pyrolysed by the incident beam to form a coherent, amorphous layer on the surface. After extended observation of a specific area, the layer of amorphous 'carbon' contamination will obscure all of the morphological detail.

The effect can be inhibited by *cryogenic cooling* of the specimen surroundings, which traps the condensable contaminant species, and this is the usual procedure adopted for transmission electron microscopy. However, the large specimens employed in the scanning electron microscope render a cryogenic trap much less effective, while the very high beam current concentrated in the electron probe exacerbates the rate of contamination. The only adequate solution is to take steps to ensure that contamination cannot originate from the specimen (plasma etching in an argon and oxygen gas mixture to oxidize any carbon on the surface), and to work with the best achievable chamber vacuum.

4.1.3.5 VOLTAGE AND CURRENT STABILITY

While chromatic aberration ought not to be a problem in either transmission or scanning microscopy, it is a mistake to assume that current and voltage instabilities only affect the performance through their influence on the objective or probe lens. In particular, the scanning electron microscope is susceptible to *image distortion* arising from electrical instabilities in the scanning system. These may be due to several causes, but it is the results that concern us here, namely differences in the effective *magnifications* for the x- and y-scan directions, *shear distortion* of the image in the x-direction, and *drift* of the image. All these imaging defects are a direct result of the point-by-point method of image collection and are a consequence of distortions of the *scanning raster* of the probe x–y coordinates with respect to that of the image. It is also important to note that charging of an electrically insulating specimen can be a major source of instability, but this can be eliminated by a conductive coating or minimized by working at optimized beam voltages and currents.

4.2 TRANSMISSION ELECTRON MICROSCOPY

It is not easy to prepare good specimens for examination in the transmission electron microscope, but techniques are available for doing so, and if a good specimen can be prepared then the information obtained is unique. However, there is nothing more frustrating than attempting to extract information from a poorly prepared specimen, and more than one graduate student has acquired a lifelong aversion to the transmission electron microscope for this sole reason.

In effect, successful transmission electron microscopy depends on three diverse skills, namely preparing a good *specimen*, acquiring good *data*, and adequate *interpretation*. In the following, we will try to provide a sufficient foundation in all of these three domains.

4.2.1 Specimen Preparation

A wide range of experimental methods and accompanying commercial aids are available to 'ease the pain' of specimen preparation. Providing that adequate facilities are available, there is no real excuse for failing to prepare a good thin-film specimen from any engineering material, and none whatsoever for wasting time over a poor specimen in the microscope.

In the following, we will comment on the methods available for obtaining a thin film from a *bulk sample* of an engineering material, ignoring those methods of sample preparation which have been developed for *soft tissues* (in the life sciences), and those which are concerned with the dispersion of *particulate matter* or which involve direct deposition from the gaseous or liquid phases. It should be recalled that *every* material is a 'special case', and every component has to be sectioned and a sample selected.

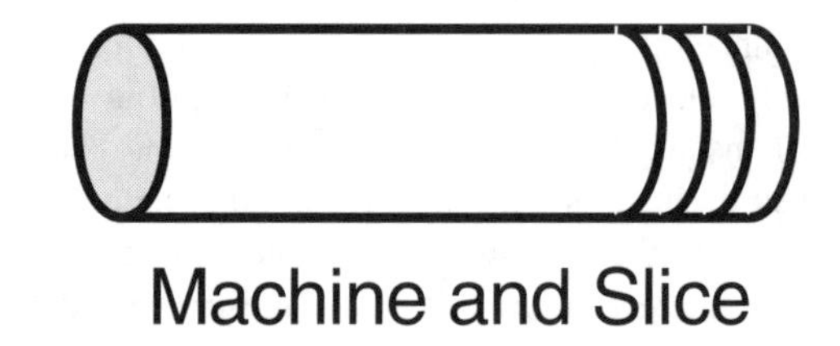

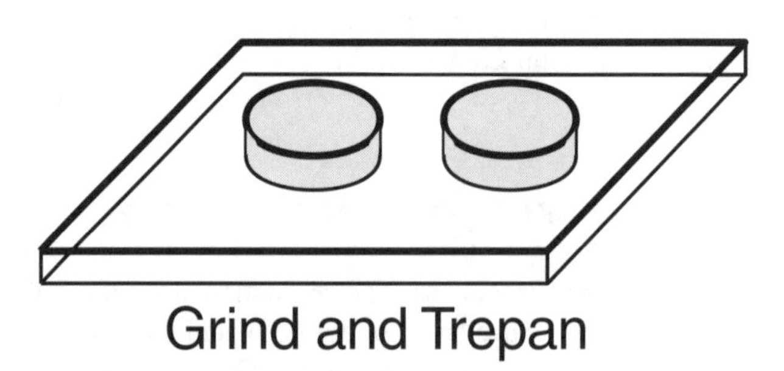

Figure 4.7 Examples of methods for sectioning a disc from a bulk component

4.2.1.1 MECHANICAL THINNING

The usual starting point for the preparation of a thin-film specimen for transmission electron microscopy is a sample taken from a bulk component. This sample is typically a 3 mm-diameter *disc*, several hundred microns in thickness. The disc may be punched out of a ductile metal sheet, trepanned from a brittle ceramic, cut from a bar, or machined from a larger section (Fig. 4.7). In all cases, it is necessary to minimize mechanical damage to the material and preserve a *flat* and *smooth* surface.

The next task is to reduce the thickness of the disc. This is accomplished by the same procedures of *grinding* and *polishing* as were described previously for the preparation of light-optical microscope specimens (Section 3.3). As in the previous discussion, the *stiffness* (elastic modulus), *hardness* and *toughness* of the material determine the optimum choice of grinding and polishing media, and *ductile* metals, *brittle* ceramics and *tough* alloys will all respond quite differently.

The following three mechanical thinning treatments are possible (see Fig. 4.8):

(a) A polished, parallel-sided disc is prepared and then thinned from one or both sides by using a rigid jig to maintain the geometry and a crystalline wax to fix the specimen to a polished base plate. (The wax is easily melted on a hotplate, both when attaching and when reversing the sample on the base plate.) As the thickness is reduced, so is the grit size of the grinding and polishing media, finishing with sub-micron diamond grit at sample thicknesses of 100 μm or less.

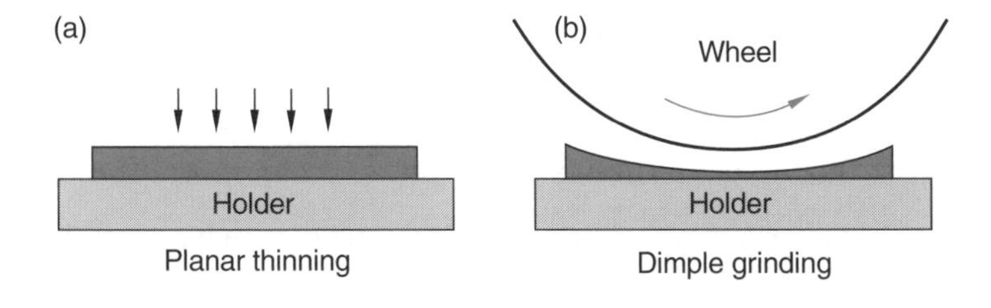

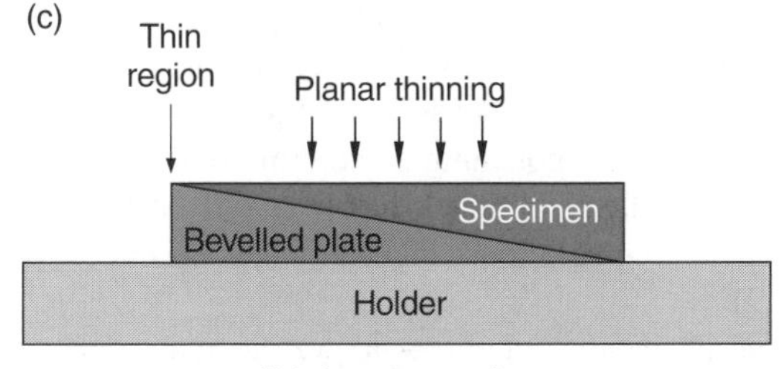

Figure 4.8 The mechanical thinning of a disc may be carried out by various methods: (a) using a parallel-sided geometry; (b) dimpling; (c) as a wedge

The greatest problem is associated with the relief of *residual stress* in the material, which leads to *curvature* of the sample when it is removed from the base plate (see below).

(b) Once the thickness of the sample is reduced to 100 μm or less, and at least one side has been polished, it can be secured on a polished base plate and '*dimpled*'. In this process, fine grinding media remove material from the contact area between the rotating sample and a rotating disc in what is, essentially, a *lapping* process, in which the particles of the grinding media are able to rotate in the region of contact shear, continuously exposing new cutting edges. Commercial *dimplers* are available, and the process has been successfully used to thin samples of hard and brittle materials to less than 20 μm in the central region of the 'dimple'. Dimpling is an excellent method for avoiding the complications of residual stress, since the thicker rim of the sample acts as a 'frame' to constrain the thin central region.

(c) For some samples, it is an advantage to prepare a '*wedge*' rather than a 'dimple'. This might be the case for an interface, where the microstructure of the latter is the subject of investigation, and the maximum thin area from the interface region will be obtained when the interface is *parallel* to the edge of the wedge. In the case of thin films and multilayer sandwich structures (semiconductor devices and composite materials), it is common practice to mount the sandwich *perpendicular* to the wedge, so that each layer is sectioned as a wedge and the morphology and interface microstructure of each layer can be studied in a single sample. The wedge is prepared by using a rigid jig, and the wedge angle can be pre-selected (typically less than 10°).

4.2.1.2 Electrochemical Thinning

No mechanical thinning process can avoid introducing *some* sub-surface mechanical damage, such as plastic shear or microcracking.

If the material is a metallic conductor, then it is frequently possible to thin the sample by *chemical dissolution*, most commonly carried out by electrochemical means. The techniques developed for achieving this were based on *electropolishing* solutions, but the conditions required for electropolishing a bulk sample differ markedly from those we are concerned with here, primarily because the area to be thinned is so small. In particular, it is possible to thin at far higher current densities than would be feasible in electropolishing. Furthermore, the problem of dissipating the heat generated during chemical attack has been solved by passing the current through a *jet* of the polishing solution which impinges on one or both sides of the disc sample (Fig. 4.9).

While thin films of all metals and alloys, as well as many other materials which posses the necessary conductivity, have been successfully prepared by *jet-polishing*, each of these materials requires its own polishing conditions (primarily the composition and temperature of the solution and the current density). These

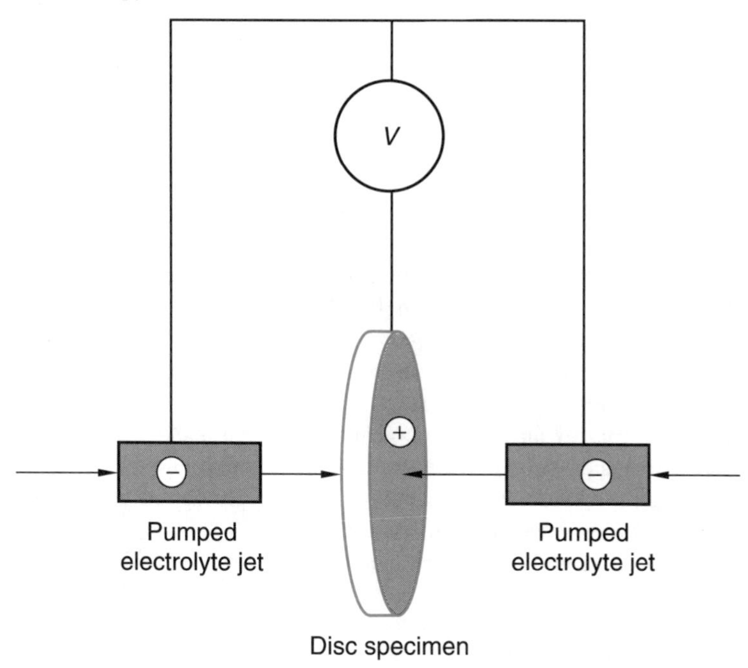

Figure 4.9 In jet polishing, a current is passed through a stream of the polishing solution impinging on the disc sample, with thinning is accomplished by electropolishing at a high current density

'recipes' are available in the literature or from the manufacturers of jet-polishing units.

Thinning is complete as soon as the first hole appears in the disc sample, and is usually detected by eye. Once the specimen has been rinsed and dried, it is ready for insertion in the microscope. The regions around the central hole in the sample should be transparent to the electron beam (typically 50–200 nm in thickness). If the areas available for investigation are too small, this is most commonly because the process was allowed to continue *after* formation of the first hole, thus leading to rapid attack at the hole and rounding of the rim. If these areas are found to have a roughened, etched appearance, then it is most probably because the polishing solution is exhausted, contaminated or over-heated, or because the current density is insufficient.

4.2.1.3 Ion Milling

The earliest successes in preparing thin-film specimens of ductile metals and alloys were achieved by Raymond Castaing in France, in the mid 1950s, who utilized a beam of energetic inert gas ions to sputter away the surface of aluminium alloy foil samples. While this technique was soon superseded by chemical methods, based on electrolytic polishing, the development of increasingly sophisticated commercial systems has lead to a resurgence of *ion milling* as the preferred method for removing the final surface layers from samples intended for transmission electron microscopy.

There are several reasons for this:

(a) Ion milling is a low-pressure, gas-phase process, so that contamination of the surface is, in principle, easier to control.
(b) Electrochemical methods are largely restricted to metallic conductors, while ion milling is much more generally applicable.
(c) Although some sub-surface *radiation damage* may be introduced during milling, this can be minimized by a suitable choice of the milling parameters, and there is no danger of *mechanical* damage.
(d) Ion milling *removes* surface-contaminant films, such as the residual anodic oxide layers associated with electropolishing. In addition, no changes in surface composition are expected, since milling is normally performed at temperatures well below those at which diffusion can occur.
(e) Multilayers composed of different materials on a substrate cannot be chemically thinned in cross-section.
(f) The sophisticated ion milling units now available are able to ensure that the region of the sample thinned for examination can be localized to *better* than 1 μm, a primary consideration in the study of thin film devices and quite beyond the range of chemical methods.

Of course, there are some disadvantages. *Sputtering* is a momentum-transfer process, and the rate of sputtering is a maximum when the ion beam is normal to the sample surface and the atomic weight of the sputtering ions matches that of the sample material. However, an angle of incidence of 90° maximizes both radiation damage to the surface and topographic irregularities associated with microstructural features, while the need to avoid ion-induced chemical reactions usually restricts the choice of sputtering ion to the inert gases (usually argon).

The sputtering rate can also be improved by increasing the incident-ion energy, but only at the cost of occluding the sputtering ions in the sub-surface region and creating a radiation-damaged layer. In practice, the ion energy is limited to a few kilovolts, at which energy the depth of ion injection is sufficiently limited to allow the injected ions to escape to the surface by diffusion, rather than nucleate sub-surface damage.

In general, the angle of incidence for ion milling is less than 15°, and at an energy of the order of 5 kV the rates of sputtering are often no more than 50 $\mu m\,h^{-1}$. It follows that the preferred specimen is one that has *already* been thinned mechanically (by dimpling or wedge-thinning) or electrochemically (by jet-polishing) to a thickness of the order of 20–50 μm before being inserted in the ion-miller. The specimen is rotated during the milling process, in order to ensure that milling is as uniform as possible, and the initial stages of milling, performed from both sides simultaneously, may be at an angle of up to 18° to improve the thinning rate. The angle (and hence the rate of milling) is then reduced. The *minimum* angle, resulting in a planar surface finish and large, uniformly thinned areas, is dictated by the ion-beam geometry. At glancing angles, a high proportion of the beam may strike the specimen mounting assembly, and material sputtered from this region may be deposited on the sample, thus seriously contaminating the specimen. Minimum sputtering angles for the final stages of milling are usually between 2 and 6°.

In general, milling is judged to be complete when the first hole is observed in the sample. In a *precision ion-milling* system, the area selected for thinning is observed *in situ* and milling is continued until the required region is ready.

4.2.1.4 Sputter Coating and Carbon Coating

Since the electron beam carries a charge, electrically insulating specimens will generally acquire an electrostatic charge during examination. If *charging* of the specimen proves to be a problem (and in many instances it *isn't*, since the small size of the sample and some residual surface conductivity often restrict the charge), then the specimen must be coated with an electrically conducting layer. The preferred material is carbon, since it has a low atomic number (6) and remains amorphous on deposition. Any sub-structure due to the carbon coating is therefore faint and on a nanometer scale.

The coating may be evaporated on to the surface by passing a high electric current through a point contact between two carbon rods, or *sputter-coated* by bombarding a carbon target with inert gas ions and depositing the sputtered material on the sample surface. 'Difficult' samples may require coating on both sides. The nanometre scale morphology of the thin (5–10 nm) coating is sometimes faintly visible in the microscope image.

4.2.1.5 Replica Methods

Instead of preparing a *thin slice* from a component for direct examination in the transmission electron microscope, it is possible to take a *replica* from a surface. Of course, this may not be necessary if the surface can be examined at sufficient resolution in the scanning electron microscope, but this is not always possible and there are many reasons why a replica may be desirable, for example:

(a) A *non-destructive* examination is necessary. In *failure analysis* the sectioning of the component is often the last resort, while taking a replica from the surface can be done in the field, far from the laboratory, and without destroying the component. Similar considerations apply to *forensic* investigations, where the evidence must be preserved for the court.
(b) In cases where evidence is sought for the presence of a specific phase on the surface. Collection on a replica will preserve the morphology of the phase with respect to the surface. Again, forensic examples are easy to imagine, such as gun-shot residues recovered from skin, or paint pigment particles at the scene of

a crash. However, *corrosion products* can also be isolated on a replica, when analysis of the chemical composition and phase content will not be obscured by the composition and structure of the bulk material.

(c) It is often possible to extract *one component* of a polyphase material, for example by suitable chemical etching, and isolate this component on a replica while still preserving its distribution in the section taken from the bulk material. Once more, the chemistry, crystallography and morphology of the extracted phase can now be studied in isolation, with no interference from any signal due to the microstructural features of the remaining constituents of the bulk material.

(d) Finally, it may be desirable to compare observations made on a *replica* of a surface with those made on the *same* surface in scanning electron microscopy. Surface markings associated with mechanical fatigue are one example of this. Clearly, *combining* the techniques may be an advantage. As an example, consider the possibility of observing the dimpled structure of a ductile failure in the scanning electron microscope, and then extracting the non-metallic inclusions associated with nucleation of the dimples on a replica and identifying their phase composition and distribution in the transmission electron microscope.

The usual procedure (Fig. 4.10), is to obtain a *negative* replica of the surface in a flexible, soluble plastic. The plastic may be cast in place and allowed to harden, or it may be a plastic sheet softened with a suitable solvent and then pressed on to the surface before allowing the solvent to evaporate. In some cases, it will be necessary

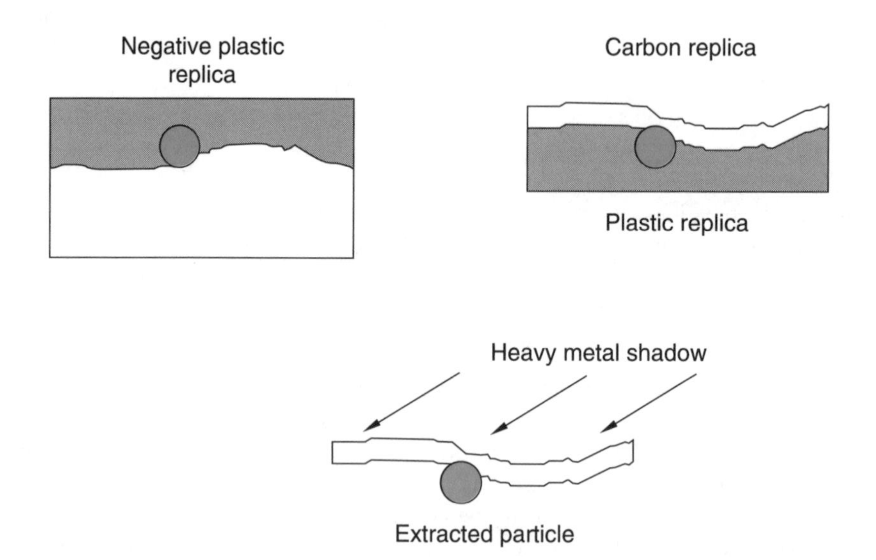

Figure 4.10 A negative plastic replica can be used to prepare a thin-film carbon replica of the original surface containing extracted particles; this is followed by shadowing to reflect the original morphology

to remove surface contamination by cleaning the surface ultrasonically or by using an initial 'cleaning' replica before making the final plastic replica. In other cases, it may actually be the 'contamination' which is the subject of interest.

Once a plastic replica has hardened, it must be peeled away from the surface and may be *shadowed* with a heavy metal (usually a gold–palladium alloy) in order to enhance the final contrast in the microscope. The metal shadowing is selected for minimum particle size and maximum scattering power. The cluster size is typically of the order of 3 nm. After shadowing, a carbon film is deposited on the plastic replica (100–200 nm) and the plastic is dissolved in a suitable organic solvent. The carbon film, together with any particles removed by the plastic from the replicated surface, as well as the heavy-metal shadow (reflecting the original surface topography), is then rinsed and collected on a fine-mesh copper grid for insertion into the microscope.

There may be no need for a 'negative' plastic replica. If the surface of an alloy is polished so that the particles of a second phase stand 'proud' of the surface, then a carbon film deposited *directly* on to the surface will adhere strongly to the particles, and further etching of the matrix will release a *carbon extraction replica* on which the particles of the second phase are distributed.

4.2.1.6 Preparing Cross-Sections

As noted previously, the development of a multitude of thin-film devices in the various branches of electronics has resulted in a major demand for specimen preparation techniques that will permit reliable and efficient transmission electron microscope examination of regions preselected perpendicular to a thin, *multilayer* assembly.

Several stages of preparation are required:

(a) The sample is *embedded* in a glass-stiffened epoxy resin and a slice sectioned perpendicular to the multilayer thin-film assembly on its semiconductor substrate (for example silicon).
(b) The slice is either *dimpled* or *wedge-polished* in a rigid mount with micrometre control of both the *thickness* removed, and (for wedge-polishing) the *angle* of the wedge. Several micro-grinding stages are necessary, finishing with 0.25 μm diamond powder.
(c) The mechanically polished sample is *ion-milled* (Fig. 4.11) until the thinned region is observed to be near the area of interest. *Precision milling* at an angle of less than 6° is then used to ensure a large area of interest having a sufficient thickness for transmission electron microscopy.

4.2.2 The Origin of Contrast

The electron beam interacts with the thin-film specimen both elastically and inelastically, but it is the *elastic* interactions which dominate the contrast observed in

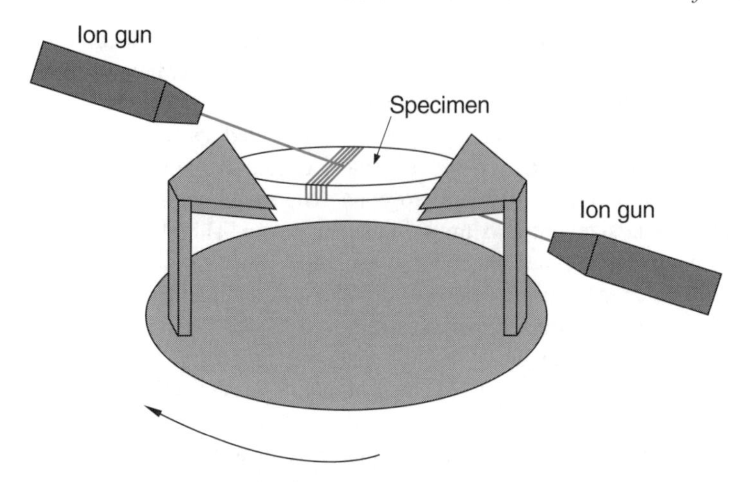

Figure 4.11 Precision ion-milling permits thin-film multilayer assemblies to be thinned perpendicular to the plane of the assembly

the transmitted electron image. On the other hand, inelastic scattering events contain information on the *chemical composition* of the sample—we will return to this further in Chapter 5.

Contrast arises from three mechanisms, termed *mass-thickness*, *diffraction* and *phase contrast*. The situation is illustrated schematically in Fig. 4.12. If the sample is *amorphous* (has a glassy microstructure with no long-range order), then the elastic scattering is *random* and the transmitted intensity varies with the scattering angle according to an approximate $\cos^2 \theta$ law. The intensity scattered out of the direct beam from a glassy specimen depends on the sample thickness and its density, and the resulting image contrast is said to be due to variations in *mass-thickness*. If an aperture is placed beneath the specimen, it will remove most of the scattered electrons from the imaging system and the image will be dominated by those direct transmitted electrons which have *not* been scattered.

In a *crystalline* sample, the electrons are scattered according to Bragg's law, generating diffracted beams at angles of $2\theta_{hkl}$ to the direct transmitted beam, which correspond to those crystal planes with Miller indices *hkl* which satisfy the Bragg condition, or, more exactly, whose reciprocal lattice vectors touch the reflection sphere (see Section 2.5). An aperture can now be placed so that it allows either the directly transmitted beam to pass into the imaging system, to form a *bright-field* image, or one of the diffracted beams to be accepted, forming a *dark-field* image. In both cases, the contrast is determined primarily by the presence of crystal lattice defects which affect the local diffracted intensity in the neighbourhood of the defect, with the imaging mode being termed *diffraction contrast*.

Finally, if the resolving power of the microscope is adequate, a larger aperture can be used to admit *several* diffracted beams to the imaging system, with or without the direct transmitted beam. These interfere in the image plane to yield a *periodic* image

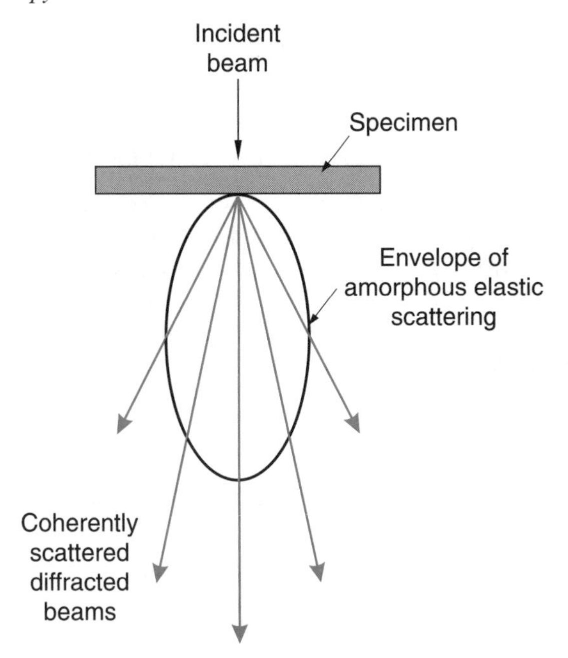

Figure 4.12 The incident beam is elastically scattered by the sample, either randomly (a glassy or amorphous specimen), or coherently (a crystalline phase). The image may be formed from the direct transmitted beam, by a diffracted beam, or by the interference of the diffracted beams with each other and/or the direct transmitted beam (see text for details)

which reflects the periodicity in the crystal in the plane *normal* to the incident beam, an effect which is termed *phase contrast.*

4.2.2.1 Mass–Thickness Contrast

The probability of an electron being elastically scattered out of the incident beam depends on the *atomic scatting factor*, which increases monotonically with the atomic number, and on the total number of atoms in the path of the beam, that is the thickness. *Mass–thickness* contrast thus reflects a combination of variations in specimen thickness and specimen density, and is roughly analogous to the *mass-absorption* effects discussed previously for X-rays (Section 2.3.1).

In the life sciences, mass–thickness contrast almost always dominates the image, and the contrast in the electron microscope is often enhanced by a variety of heavy-metal, *tissue-staining* procedures. In the natural sciences and in engineering studies, mass–thickness contrast predominates in *non-crystalline* materials (such as two-phase glasses), as well as in *replica* studies (where the contrast is commonly due to variations in the deposited mass of a *metal shadow* (see above), or to the presence of an extracted phase).

4.2.2.2 Diffraction Contrast and Crystal Lattice Defects

In a *perfect crystal*, the amplitude scattered from the incident beam into a diffracted beam which is associated with diffracting planes having a reciprocal lattice vector **g**, can be calculated by summing the diffracted amplitudes from all of the unit cells which lie along the path of the diffracted beam (Fig. 4.13). The phase difference, ϕ, in the amplitude scattered by a unit cell at a position **r** with respect to the origin is given by the rotation:

$$\phi = 2\pi \mathbf{g} \cdot \mathbf{r} \tag{4.5}$$

while the amplitude scattered by this unit cell can be written as:

$$A e^{i\phi} = A \exp[-2\pi i(\mathbf{g} \cdot \mathbf{r})] \tag{4.6}$$

The unit cells in the column being considered are each separated by the lattice parameter a and for *Bragg diffraction* (scattering in-phase) $\mathbf{g} \cdot \mathbf{r} = n$, an integer. Since the electron wavelength is much less than the lattice parameter, each unit cell in the crystal scatters *independently* of the others, so that the amplitude scattered by a *single* unit cell will be given by the structure factor for the cell, F (Section 2.4.2).

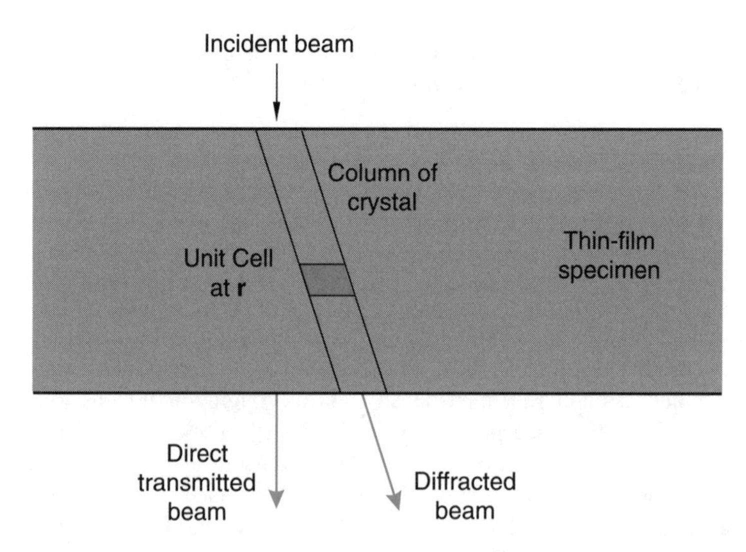

Figure 4.13 The unit cells in the column of the crystal corresponding to the Bragg scattering direction will each scatter a proportion of the incident beam into the diffracted beam

If for the time being we ignore any reduction in the amplitude of the direct transmitted beam, so that each unit cell scatters the *same* amplitude, then it follows that, at the Bragg position, the total amplitude scattered by a column of n unit cells is a *linear function* of n:

$$A_n + \sum_n F_n \exp[-2\pi i(\mathbf{g} \cdot \mathbf{r})] \quad (4.7)$$

If the number of unit cells in a column is sufficiently large, we can replace the sum by an integral, which for convenience we take over the thickness t of the thin-film sample, measured from the mid-thickness point, that is over $\pm t/2$, so that:

$$A_t = \frac{F}{a} \int_{-1/2}^{1/2} \exp[-2\pi i(\mathbf{g} \cdot \mathbf{r})] dr \quad (4.8)$$

If the angle of incidence of the incident beam deviates from the Bragg condition, then the vector $\mathbf{g}$ in reciprocal space is replaced by $\mathbf{g} + \mathbf{s}$. Similarly, if the structure of the 'perfect' lattice is distorted, due to the presence of a lattice defect which generates a displacement $\mathbf{R}$, then the position of the scattering element at $\mathbf{r}$ is shifted to $\mathbf{r} + \mathbf{R}$. The phase angle for the amplitude scattered from a unit cell at the position $\mathbf{r}$ in a misoriented and defective crystal is then given by the equation:

$$\phi = 2\pi(\mathbf{g} + \mathbf{s})(\mathbf{r} + \mathbf{R}) \quad (4.9a)$$

By expanding the brackets, we obtain:

$$\phi = 2\pi(\mathbf{g} \cdot \mathbf{r} + \mathbf{g} \cdot \mathbf{R} + \mathbf{s} \cdot \mathbf{r} + \mathbf{s} \cdot \mathbf{R}) \quad (4.9b)$$

Of these terms, $\mathbf{g} \cdot \mathbf{r} = n$ is an integer which has no effect on the phase of the amplitude scattered. The term $\mathbf{s} \cdot \mathbf{R}$ is obtained by multiplying two small vectors together, and so can be neglected. The two remaining terms, $\mathbf{g} \cdot \mathbf{R}$ and $\mathbf{s} \cdot \mathbf{r}$, are additive. They represent the *phase shift* in the amplitude scattered at the position, $\mathbf{r}$, into the diffracted beam, $\mathbf{g}$, due to *either* deviations from the exact Bragg condition ($\mathbf{s} \cdot \mathbf{r}$), *or* distortions of the crystal lattice (lattice strains) associated with the presence of lattice defects ($\mathbf{g} \cdot \mathbf{R}$). It follows that in diffraction contrast we are observing the *summed* contrast effects of *both* deviations in *reciprocal space* (changes in the specimen orientation with respect to the incident beam), *and* displacements in *real space* (displacements of the crystal lattice due to lattice strains).

Complete interpretation of diffraction contrast requires that these two effects be separated and their origin identified. This requires that the operating reflections (**g**-vectors) be known and the causes of the displacements in both real (**R**) and reciprocal (**s**) space be identified. This is often a difficult process, which requires a knowledge of the thickness of the sample, together with both bright-field and dark-field images taken using *different* **g** reflections. In practice, a complete analysis of diffraction contrast is often unnecessary, and image analysis may be limited to a *generic identification* of the types of lattice defects present (dislocations, stacking faults and defect clusters), rather than a *quantitative analysis* of the defect itself (the sign and magnitude of a dislocation Burgers' vector or stacking-fault vector, for example).

4.2.2.3 Phase Contrast and Lattice Imaging

The transmission electron microscope is commonly operated in two modes (see Fig. 4.14). When imaging mass–thickness or diffraction contrast, an aperture placed in the back focal plane of the objective lens limits the angle of acceptance for the beam to an angle α, which is less than the Bragg angle θ of any coherently scattered electron beam. The image is then a *shadow projection* of the electrons accepted into the imaging system in which the intensity variations reflect the variations in the electron current passed down the column of the microscope through the different regions of the sample. When the incident beam is tilted by a Bragg angle, so that one of the coherently diffracted electron beams is accepted into the imaging system, the objective aperture cuts out both the direct transmitted beam and all other diffracted beams, so that the *bright-field* image is replaced by a *dark-field* image.

If the objective aperture is removed or replaced by a larger aperture, which accepts both the direct transmitted beam and the diffracted beams into the imaging system, i.e. $\alpha > 2\theta$, then the differences in path length followed by the different beams will result in an *interference pattern* in the image plane (Fig. 4.15). However, if this interference pattern is to be observed in the microscope, then several conditions have to be met with respect to both the *chromatic* and *spherical aberration* coefficients of the microscope, the *focusing* of the image and the *coherence* and *energy spread* in the incident electron beam (the properties of the source).

The fact that *phase contrast* corresponds to an interference pattern is important. Imagine an object represented by a number of point sources having a total electron wave distribution described by $f(x, y)$. The *image function*, which represents the amplitude and phase of $f(x, y)$ after travelling down the microscope column, is $g(x, y)$. Each point in the image has contributions from *all* of the beams which have been transmitted by the objective aperture, so we can write the relation:

$$g(r) = \int f(r')h(r - r'n)\mathrm{d}r' = f(r)h(r - r') \tag{4.10}$$

where radial coordinates r are used instead of Cartesian coordinates (x, y), and $h(r)$ represents the contribution from each object point to any given point in the image. The function $h(r)$ is termed a *point-spread* function or *impulse-response* function, and $g(r)$ is the *convolution* of $f(r)$ with $h(r)$. Thus, the electron waves of the incident beam, modified by the electron-density distribution in the specimen, are *convoluted* with a function which describes the response of the microscope column (electron-beam source, electromagnetic lenses, apertures and lens aberrations).

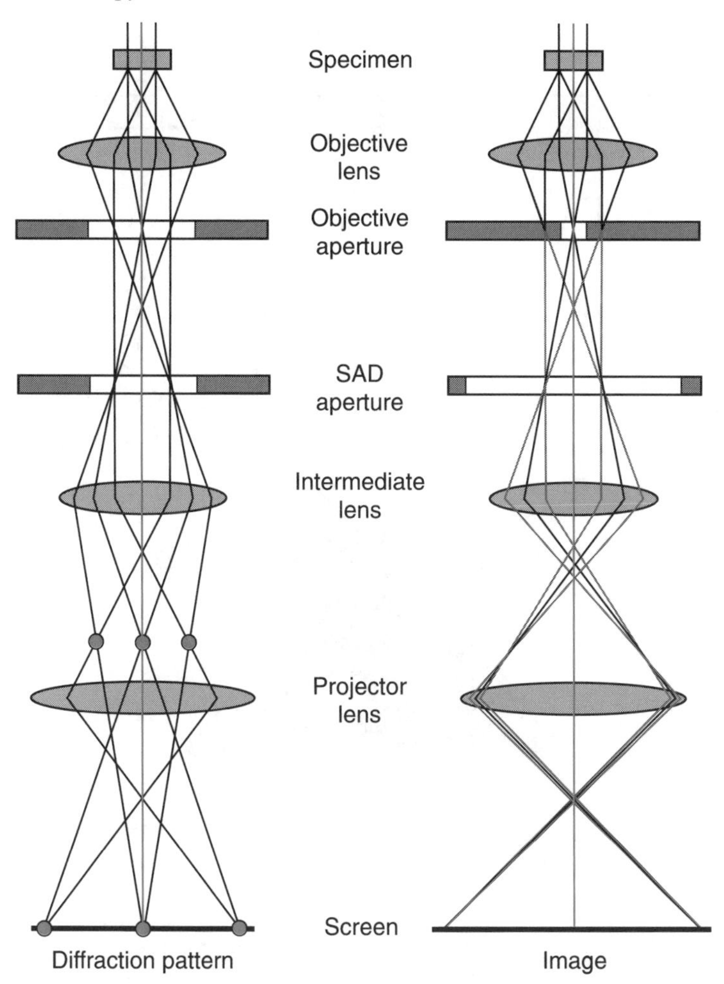

Figure 4.14 The transmission electron microscope can be used to image the specimen by focusing the final image in the plane of the fluorescent screen, or it can be used to image the diffraction pattern from the specimen. In the first case, to an excellent approximation, the image of the specimen is observed when the imaging system is focused on the front focal plane of the objective (the position of the specimen), while the diffraction pattern is observed when the imaging system is focused on the back focal plane of the objective (which corresponds to the first image plane for the diffracted and transmitted beams)

We represent $g(r)$ by a *Fourier transform* in reciprocal space where:

$$g(x, y) = \sum_{\mathbf{u}} G(\mathbf{u}) \exp(2\pi i \mathbf{u} \cdot \mathbf{r}) \tag{4.11}$$

and $\mathbf{u}$ is the reciprocal lattice vector, and we define the Fourier transforms of $f(r)$ $(F(\mathbf{u}))$ and $h(r)$ $(H(\mathbf{u}))$, where $H(\mathbf{u})$ is the *contrast transfer function* (CTF)

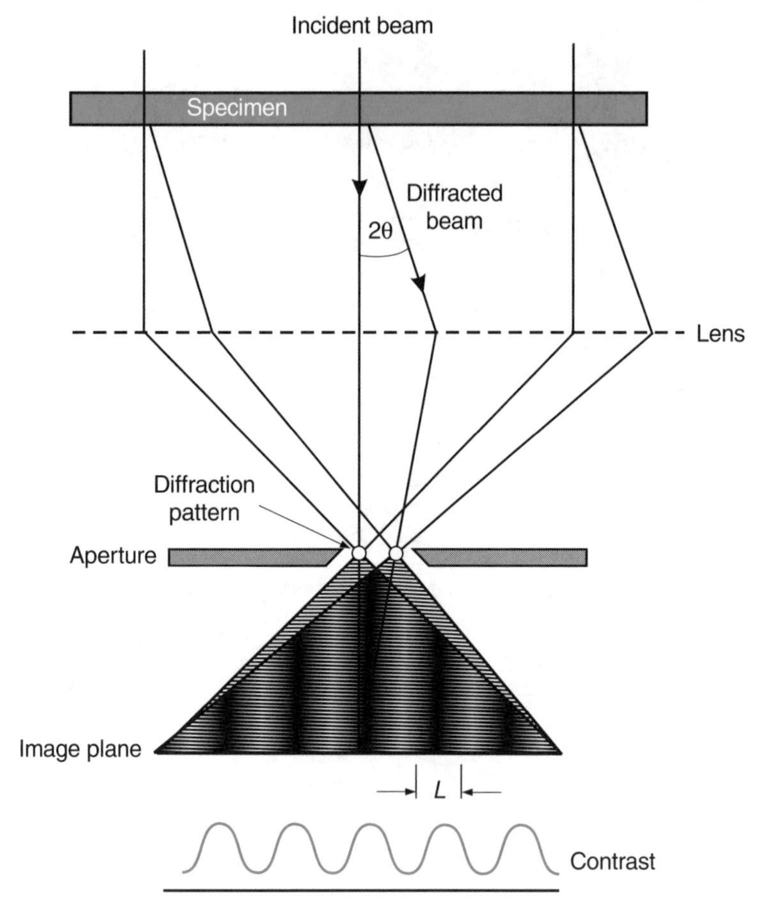

Figure 4.15 If the objective aperture accepts a Bragg-diffracted beam as well as the direct transmitted beam, $\alpha > 2\theta$, then an interference pattern will be formed in the image plane as a result of the difference in path lengths of the two beams

for the microscope. These Fourier transforms are related by the following relationship:

$$G(\mathbf{u}) = H(\mathbf{u})F(\mathbf{u}) \tag{4.12}$$

The CTF describes the influence of the various microscope parameters on the phase shift for an electron wave propagating down the microscope column. There are three major contributions to the CTF, as follows:

(a) an *aperture function* which defines the *cut-off* (removal) of spatial frequencies above a value determined by the aperture radius;

(b) an *envelope function* which describes the *damping* of the spatial frequencies, due to either chromatic aberration, instabilities in the objective lens current, or the limited coherency of the electron source;
(c) an *aberration function* which limits the available spatial frequencies and is due primarily to the spherical aberration of the electromagnetic objective lens.

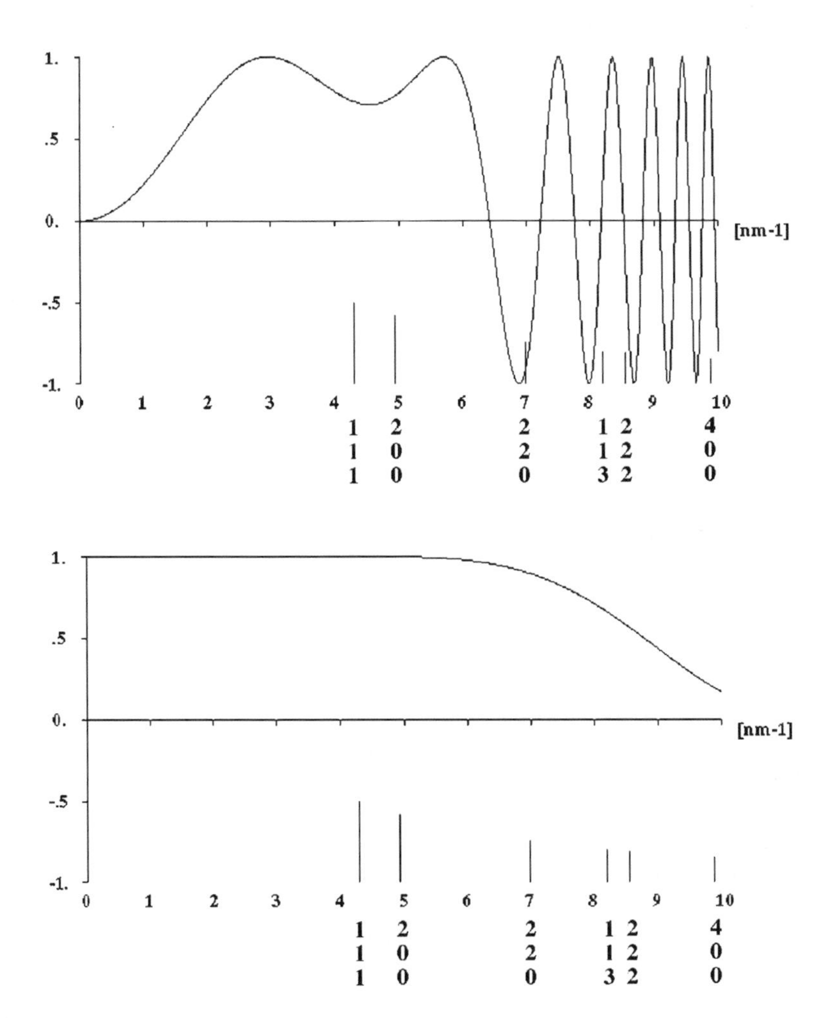

Figure 4.16 Contrast transfer functions for a high-resolution objective lens shown at Scherzer defocus and including the normalized structure factors for various crystallographic planes of aluminium: (a) the coherent contrast transfer function defined *only* by equation (4.14); (b) the spatial coherence envelope, defined by the aperture function; (c) the temporal coherence envelope defined by the damping of the spatial frequencies due to electromagnetic lens instabilities and the limited coherence of the electron source. Combining (a), (b) and (c) yields (d), the contrast transfer function (CTF) of the microscope

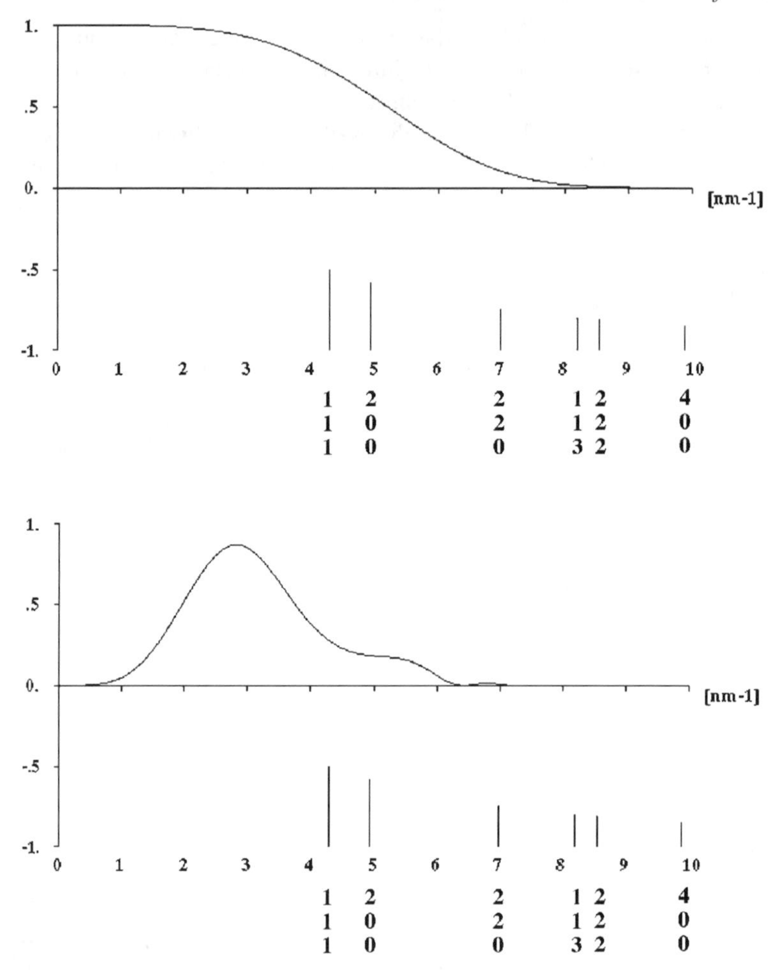

Figure 4.16 (*continued*)

The aberration function is usually given by the relation:

$$B(\mathbf{u}) = \exp[\mathrm{i}\chi(\mathbf{u})] \tag{4.13}$$

where

$$\chi(\mathbf{u}) = \pi \Delta f \lambda \mathbf{u}^2 + \tfrac{1}{2}\pi C_s \lambda^3 \mathbf{u}^4 \tag{4.14}$$

and Δf is the *underfocus* of the objective lens, λ is the wavelength, and C_s is the spherical aberration coefficient.

We will now examine the influence of the above three contributions on the CTF. Fig. 4.16 shows the CTF for a 300-kV transmission electron microscope with a *spherical aberration coefficient* of 0.46 mm at an objective lens defocus of −36 nm (an *underfocus*). Included in Fig. 4.16 are the *relative structure factors* for the lattice spacings in aluminium. Fig. 4.16(a) shows the *coherent contrast transfer function* defined *only* by equation (4.14), Fig. 4.16(b) is the *spatial coherence envelope*, defined by the aperture function, and Fig. 4.16(c) is the *temporal coherence envelope* defined by the damping of the spatial frequencies due to the lens instabilities and the coherency of the electron source. The *combination* of Fig. 4.16(a), (b) and (c) is given in Fig. 4.16(d), and corresponds to the *contrast transfer function* (CTF) of the microscope.

From Fig. 4.16(d) we see that the CTF of the microscope only transfers certain *spatial frequencies*, which are correlated with the lattice spacings of the specimen in the microscope. In regions where the CTF is zero, *no* information can be transferred to the image. Thus, for this example, *only* the $\{111\}$ and $\{002\}$ planes of aluminium will be visible in the image (assuming that the crystal is oriented in the correct zone axis!).

Note that the CTF depends on the *defocus* of the objective lens, so that the CTF can be altered by adjusting the objective lens current. This effect is shown in Fig. 4.17 for the *coherent* contrast transfer function only (both *positive* and *negative* values are transferred), and again in Fig. 4.18 for the *full* CTF, using the same wavelength and aberration constants which were used to produce Fig. 4.16.

By changing the objective lens defocus, we can shift the CTF and *enhance* the phase contrast for specific crystallographic planes. A calculation of the CTF and some careful thought about the information being sought should precede the microscope session!

One important piece of information is obtained directly from the CTF. Since *no* information is transferred when the CTF has a value of zero, the best CTF (or best microscope) will be that with the least zero values. *Underfocusing* the objective lens moves the first cross-over (first value of zero) to *larger* values of nm^{-1} (*smaller d*-spacings), partially compensating for the spherical aberration of the lens. The position of *optimum* compensation is termed the *Scherzer defocus* (Scherzer predicted this effect in 1949). The Scherzer defocus can be calculated from the equation:

$$\Delta f_{\mathrm{Sch}} = -1.2\sqrt{(C_s \lambda)} \tag{4.15}$$

The Scherzer defocus cross-over determines the *resolution limit* of the microscope (usually termed the *point resolution*) and corresponds to the *minimum* defocus value at which all beams below the first cross-over have approximately constant phase . Below the Scherzer defocus, crystallographic information is *still* available from spacings below the point resolution, but complete image simulation is necessary to interpret the information, and the information transferred by the microscope is severely damped, so that little contrast is visible from crystallographic planes with small *d*-spacings. This second resolution limit is termed the *information limit*. The distinction between the point resolution and the information limit is the reason why

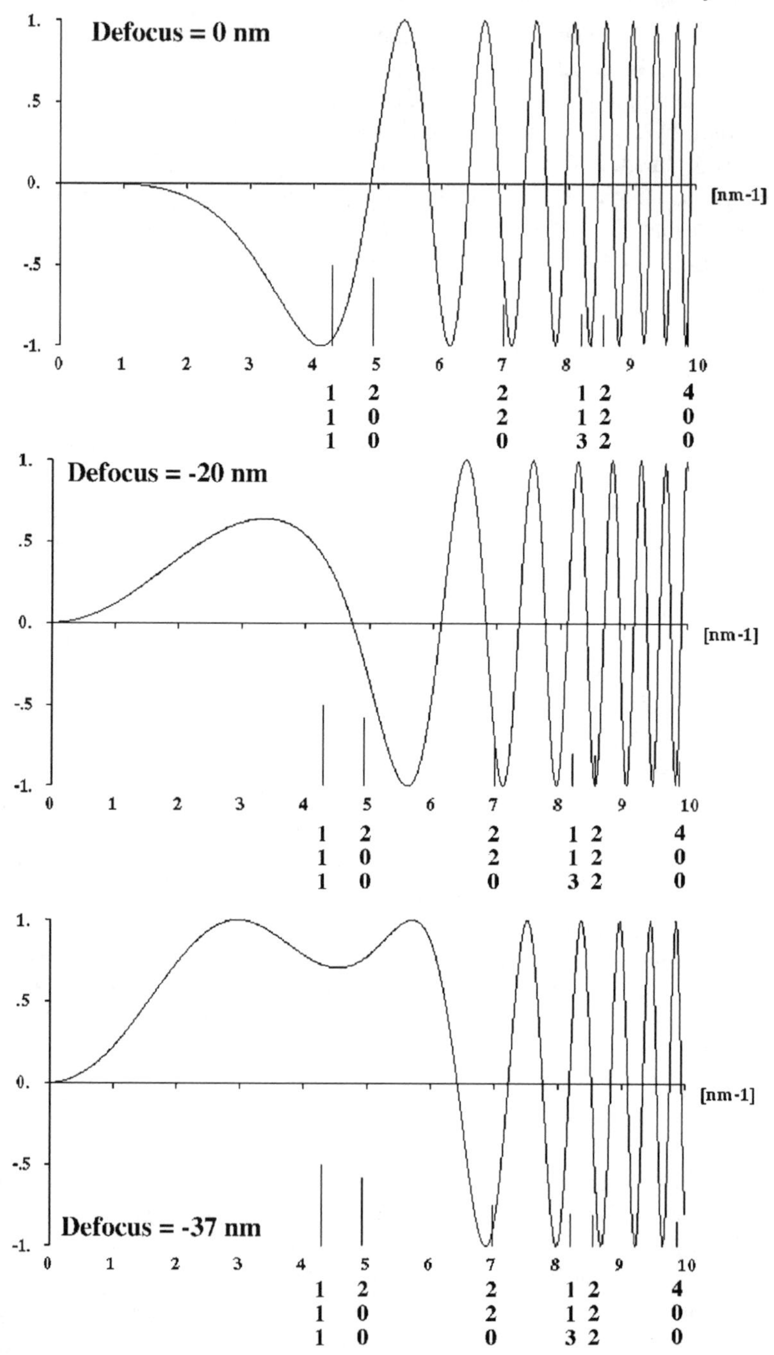

Figure 4.17 Examples of the influence of defocus on the coherent contrast transfer function (both positive and negative values are transferred)

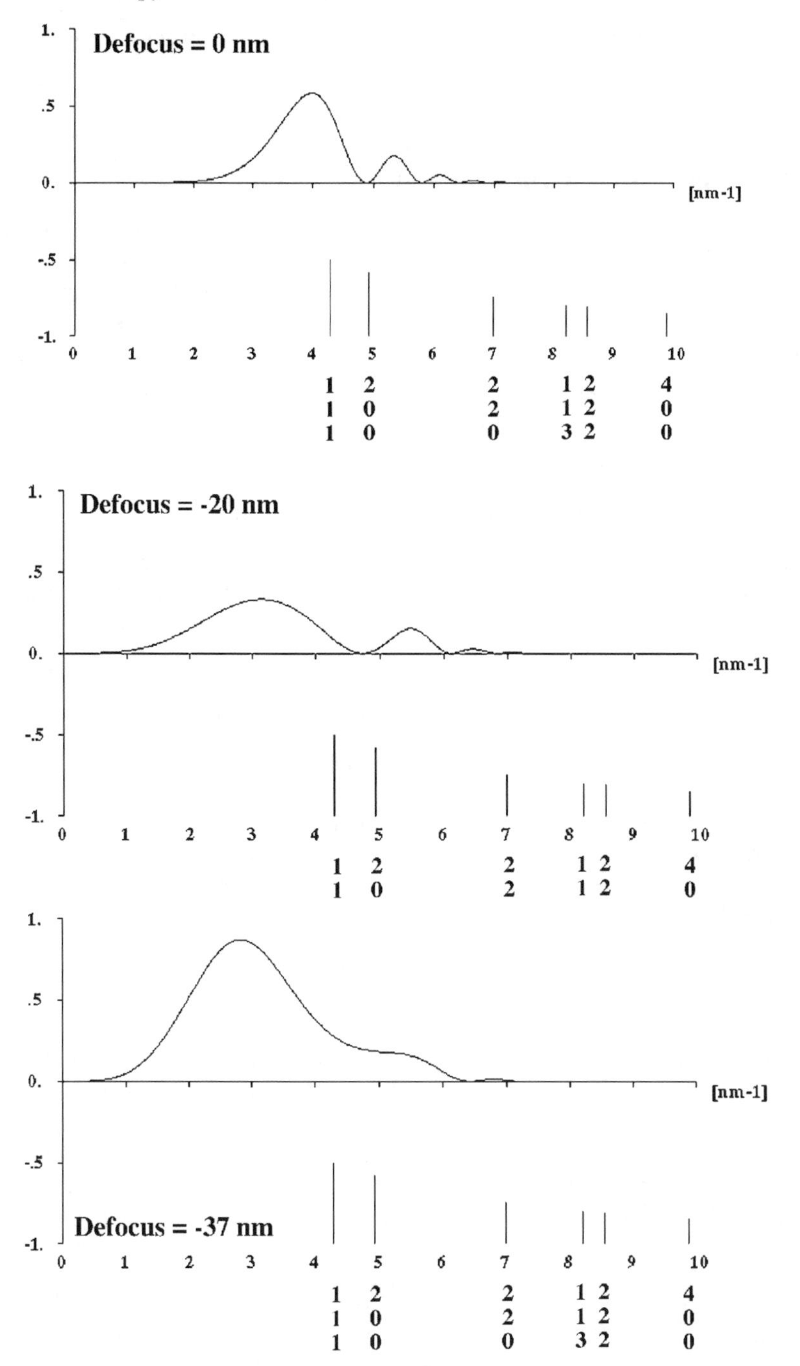

Figure 4.18 Examples of the influence of defocus on the full CTF, using the same wavelength and aberration constants which were used to produce the CTFs given in Fig. 4.16

lattice images are readily obtainable from crystallographic planes having d-spacings *less* than the point resolution of the microscope.

4.2.3 Kinematic Interpretation of Diffraction Contrast

The basic assumption of *kinematic diffraction theory*, that the amplitude diffracted out of the incident beam does not affect the intensity of this beam, is so patently false that it may seem surprising that we include this section in the text. However, the use of simple kinematic arguments, using *amplitude–phase diagrams*, drastically simplifies the discussion of diffraction contrast from lattice defects and allows the major qualitative features of the contrast to be demonstrated, so that the loss of quantitative rigour should be forgiven.

4.2.3.1 Kinematic Theory of Electron Diffraction

The basic equation for the kinematic theory of electron diffraction has previously been given in Section 4.2.2.2, and the generalized form that we will discuss here is as follows:

$$A_t = \frac{F}{a}\int_{-t/2}^{+t/2} \exp[-2\pi i(\mathbf{g}+\mathbf{s})(\mathbf{r}+\mathbf{R})] \qquad (4.16)$$

In order to justify the approximations of the kinematic theory, we will assume that the intensity of the diffracted beam is in all cases *much less* than that of the incident beam, so that the direct transmitted beam has a similar intensity to that in the incident beam.

4.2.3.2 The Amplitude-Phase Diagram

Returning to the *column approximation* (independent scattering by each unit cell in a stack viewed along the direction of the diffracted beam, Fig. 4.13), and assuming a *perfect crystal* ($\mathbf{R} = 0$), then the phase mismatch between each unit cell in the stack, $\Delta\phi$, is only determined by the deviation from the Bragg condition, $\mathbf{s}$, and is given by $\Delta\phi = 2\pi a$. Replacing the increment $\Delta\phi$ by $\mathrm{d}\phi$ and the sum by the integral (that is $\mathrm{d}\phi = 2\pi s \mathrm{d}r$), and ignoring the structure factor F (a constant for any given reflection), while noting that for $\mathbf{g}\cdot\mathbf{r} = 1$, $\exp(2\pi i\mathbf{g}\cdot\mathbf{r}) = 1$, gives the *total* amplitude diffracted by the column as follows:

$$A = \int_{-t/2}^{+t/2} \exp(-2\pi i\mathbf{s}\cdot\mathbf{r})\mathrm{d}r = \frac{\sin \pi ts}{\pi s} \qquad (4.17)$$

corresponding to a *relative diffracted intensity* of:

$$I/I_o = \frac{\sin^2(\pi ts)}{(\pi s)^2} \tag{4.18}$$

This relationship demonstrates the effect of both the specimen thickness and the deviation from the Bragg condition on the diffracted intensity and is best summarized by using an *amplitude phase diagram* (see Fig. 4.19). The radius is equal to $(2\pi s)^{-1}$ and decreases rapidly as the specimen is tilted out of the Bragg condition (increasing s). As the thickness *increases* at a fixed value of s, the diffracted amplitude *oscillates* sinusoidally between two maxima proportional to $\pm(\pi s)^{-1}$, while the relative diffracted intensity oscillates between zero and a maximum value proportional to $(\pi s)^{-2}$. A *wedge-shaped* sample at a *fixed orientation* will therefore give a series of *bright fringes* when viewed in the dark-field image from a particular reflection, with the fringes becoming closer together but fainter as the sample is tilted away from the Bragg position. On the other hand, a *curved* sample of *uniform thickness* will also show fringes whose intensity and

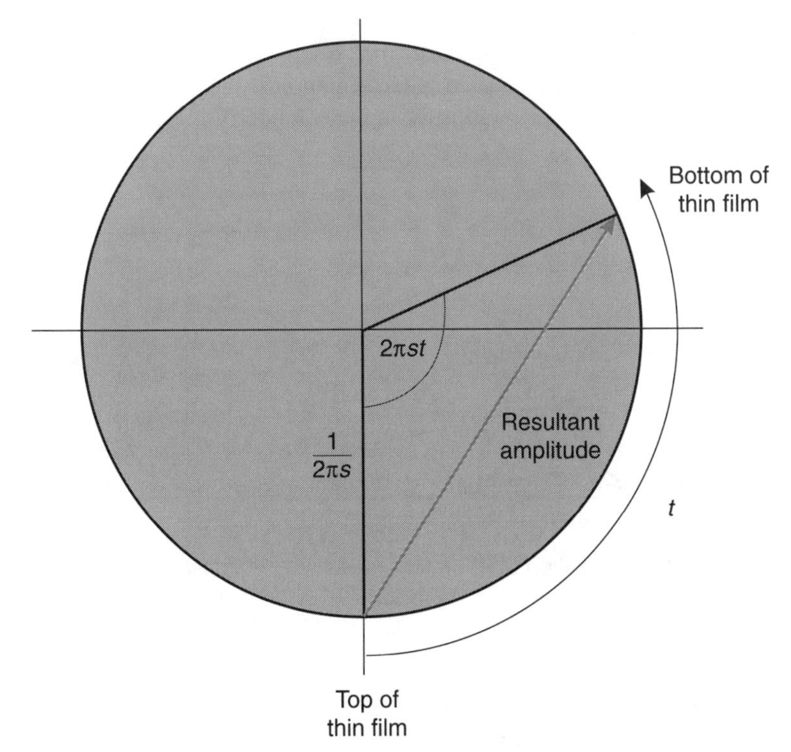

Figure 4.19 The amplitude phase diagram for the diffracted amplitude is a function of the specimen thickness t and the deviation from the Bragg condition **s**

separation *decrease* with *increasing* s, with the first extinction corresponding to the condition where $st = 1$.

Both *thickness fringes* and *bend contours* (so named because they follow contours of fixed orientation with respect to the incident beam) are prominent features of the contrast from thin crystalline films, although *dynamic diffraction* and *absorption* effects (Section 4.2.4) markedly modify the quantitative analysis of these features.

4.2.3.3 Contrast from Lattice Defects

The presence of a *lattice defect* modifies the amplitude phase diagram by introducing a second term into the *phase shift*, so that the integral determining the amplitude becomes:

$$A = \int_{-t/2}^{+t/2} \exp[-2\pi i(\mathbf{s} \cdot \mathbf{r} + \mathbf{g} \cdot \mathbf{R})]\mathrm{d}r \tag{4.19}$$

There are really only *two* possibilities here, either the additional phase shift due to the displacement field of the defect *increases* the curvature of the amplitude phase diagram, or it *decreases* this curvature (see Fig. 4.20). In the first case, the radius is *decreased* and the diagram collapses in the region near the defect, while in the second case the diagram is *expanded* in the region near the defect. For the sake of convenience, the zero for the coordinates in the column of the diffracting crystal is

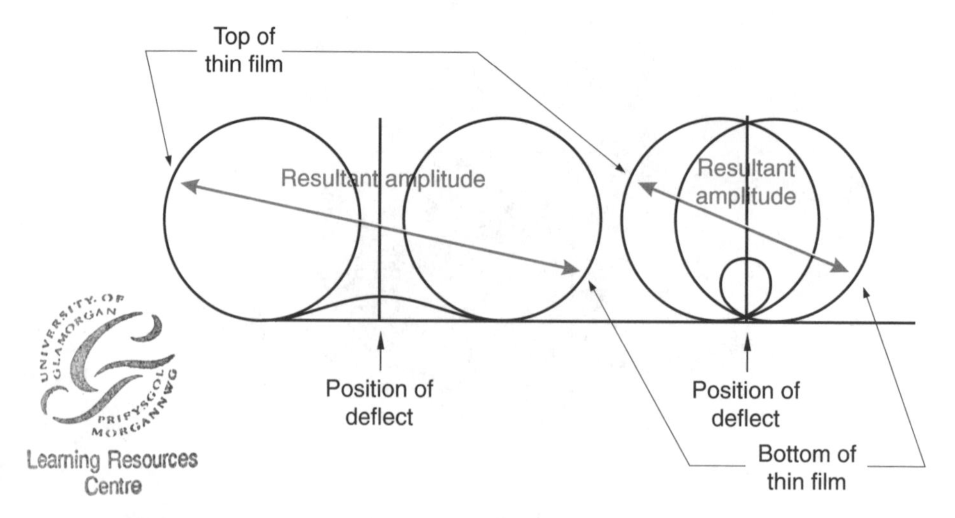

Figure 4.20 The presence of a lattice defect introduces a displacement field which either collapses the amplitude phase diagram (increases the curvature) or tends to open it out (decreases the curvature). (The origin of the amplitude phase diagram has been moved to correspond to the position of maximum displacement **R** in the column)

now chosen to coincide with the position of maximum lattice displacement (the 'centre' of the defect displacement field), rather than with the mid-thickness point of the thin film.

It follows that the displacement field **R** due to the defect has an effect equivalent to either *amplifying* or *suppressing* the effect of **s**, thus *increasing* or *reducing*, respectively, the effective deviation of the crystal from the Bragg condition. An important consequence of this conclusion is that the effect of the defect will be *reversed* if the sign of **s** is reversed, that is if the same defect were imaged with the incident beam on *opposite sides* of the exact Bragg orientation. The position of the maximum value of **R** in the column is also important, and in general the maximum diffraction contrast from defects is expected to occur when **R** and **s** are of opposite sign and when the position of maximum **R** is near the mid-point of the foil. In *thicker* regions, the intensity is expected to oscillate as the defect position moves from the top to the bottom of the foil, thus reflecting the oscillations in amplitude associated with the diameter of the amplitude phase diagram in the perfect crystal.

We have concentrated on the intensity in the diffracted beam, and hence on the dark-field image, but in practice samples are usually viewed in bright-field before any defect analysis is attempted in dark-field conditions. In general, the intensity observed in bright field is the inverse of that observed in dark-field. That this is not always the case is due to *absorption* effects, which become important for thicker films and will be discussed later (see Section 4.2.4).

Grain boundaries (and phase boundaries) constitute a special class of lattice defect whose contrast can be explained entirely in terms of *thickness extinction fringes*. Since the two crystals on either side of the boundary will not deviate from the nearest Bragg position in each crystal by the same amount, it follows that the crystal with the minimum value of **s** will dominate the contrast, showing a series of thickness extinction fringes at the boundary whose separation is dictated by the value of **s** and the tilt angle of the boundary with respect to the incident beam, and whose number depends on the thickness of the crystal and the structure factor for the reflection **g**. Diffraction from the neighbouring crystal will interfere with this contrast if the second crystal also approaches a Bragg diffraction condition, so that the appropriate value of **s** for the second crystal approaches that for the first.

4.2.3.4 Stacking Faults and Anti-Phase Boundaries

Stacking faults and anti-phase boundaries (APBs) constitute a special case in that the appropriate value of the displacement vector does not vary continuously throughout the column of the diffracting crystal, but rather changes *discontinuously* across the plane of the stacking fault or the APB. The amplitude phase diagram for the crystal *above* the point of intersection of the diffracting column with the fault plane is therefore undisturbed, while that *below* the fault plane (positioned for convenience at the zero of the diagram) is rotated by an angle equal to the phase angle associated with the fault vector, $2\pi\mathbf{g}\cdot\mathbf{R}$ (Fig. 4.21).

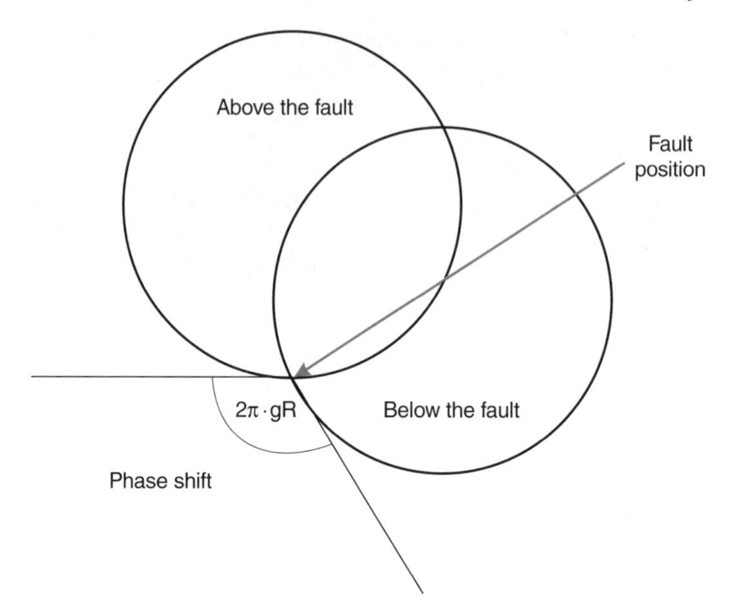

Figure 4.21 An amplitude phase diagram for a column of a crystal which contains a stacking fault or anti-phase boundary at the origin

For example, a stacking fault in the lattice of an FCC metal has a *fault vector* $1/6a\,\langle 112\rangle$ (which, it should be noted, is exactly equivalent to a fault vector of $1/3a\,\langle 111\rangle$ in the FCC lattice, since it can be combined with an appropriate *unit lattice vector* of $1/2a\,\langle 110\rangle$, e.g. $1/6a[112] + 1/2a[\bar{1}\bar{1}0] = 1/3a[\bar{1}\bar{1}1]$). The phase shift $2\pi\mathbf{g}\cdot\mathbf{R}$, where $\mathbf{R}$ is now the fault vector, will be zero if $\mathbf{g}$ is perpendicular to $\mathbf{R}$. If $|\mathbf{R}|$ is $1/3a\,\langle 111\rangle$ and $|\mathbf{g}|$ is an allowed $1/2a\,\langle 110\rangle$ reflection, then the phase shift can take one of three values, either 0 (the two vectors are mutually perpendicular) or $\pm 120^\circ$. The sign of $\mathbf{R}$ is determined by the type of fault, which may be either *intrinsic* or *extrinsic*, that is the fault may be due to a *missing* plane of atoms (ABC BCABC) or an *extra* plane (ABC**B**ABC).

In the case of an APB, the fault will only be visible if $\mathbf{g}\cdot\mathbf{R}$ is *non-integer*, so that it is not enough for the reciprocal lattice vector to have a resolved component parallel to the fault vector, but the reciprocal lattice vector must be a *superlattice* vector (a partial lattice vector of the parent lattice). Hence the APB will only be visible if the diffracted beam corresponds to a superlattice reflection.

If the sample is thick enough, then the amplitude phase diagram from a column of the crystal containing a fault plane will exhibit *thickness fringes* as the fault position moves from the top to the bottom of the crystal, similar to those observed in a wedge of crystal or at a grain boundary, but the origin of these fringes is very different, since diffraction is occurring in *both* regions of the crystal, above and below the fault.

4.2.3.5 Edge and Screw Dislocations

Diffraction contrast from dislocations is dominated by the Burger's vector of the dislocation line, **b**, since the strain field associated with the dislocation is proportional to the value of **b**. However, the *direction* of the dislocation line **l** with respect to the Burger's vector is also important. Dislocations which lie parallel to the Burger's vector are termed *screw dislocations*, and, in an isotropic crystal, have a cylindrically symmetric strain field, with no strain component perpendicular to the Burger's vector. It follows that a reciprocal lattice vector *perpendicular* to a screw dislocation should not result in any diffraction contrast, and the condition $\mathbf{g} \cdot \mathbf{R} = 0$ for no contrast can therefore be replaced by the condition $\mathbf{g} \cdot \mathbf{b} = 0$ for no contrast from a screw dislocation. By observing the contrast due to the presence of the dislocation in each dark-field image as a function of the diffracting vector **g**, and noting two values of **g** for which no contrast is observed, it is possible to identify the *direction* of the Burger's vector unambiguously; however, deriving the *magnitude* of the Burger's vector requires a more complete analysis. *Edge* dislocations have a residual component of the strain field *perpendicular* to **b**, which results from the *dilatation* (volume change) associated with the presence of an edge dislocation, so that some contrast is observed even if $\mathbf{g} \cdot \mathbf{b} = 0$, but the weakness of the contrast compared to that observed with other reflections, together with some knowledge of the Burger's vectors to be expected in the crystal lattice, usually permits some tentative conclusions to be drawn.

The vector product **bxl** gives the normal to the allowed *glide plane* for the dislocation and is undefined for screw dislocations, which are therefore free to *cross-slip*. In general, a preferred glide plane exists for screw dislocations, suggesting that they are *partially dissociated*, even when the separation of the partial dislocations is small and no stacking fault contrast is visible.

The strain field associated with a dislocation is *reversed* when the column of diffracting crystal being considered is moved across the projected position of the dislocation line, so that the sign of the phase shift is also reversed. It follows that on one side of the dislocation the amplitude phase diagram is *expanded*, thus enhancing contrast, while on the other side it *collapses*, so reducing contrast, depending on whether or not $\mathbf{g} \cdot \mathbf{R}$ has the same sign as $\mathbf{s} \cdot \mathbf{r}$ (see Fig. 4.20). Thus, as the projection of a dislocation line in the image crosses a bend-contour, the position of maximum contrast changes sides, and the true dislocation position will be determined by the *mid-point* between the two maxima (Fig. 4.22).

The *apparent width* of a dislocation depends on the numerical value of $\mathbf{g} \cdot \mathbf{b}$ and the value of **s**. Close to the Bragg position, the contrast will be a maximum and the width of the observed contrast maxima is typically of the order of 10 nm, which is a very poor resolution for transmission electron microscopy and a major barrier to the study of dislocation–dislocation interactions in sub-boundaries or during plastic shear. Much better resolution can be obtained in the dark-field image by moving *away* from the Bragg position (large values of **s**). The contrast is weak and may be difficult to see, but the resolution in this *weak-beam image* is appreciably better, down to of the order of 2 nm.

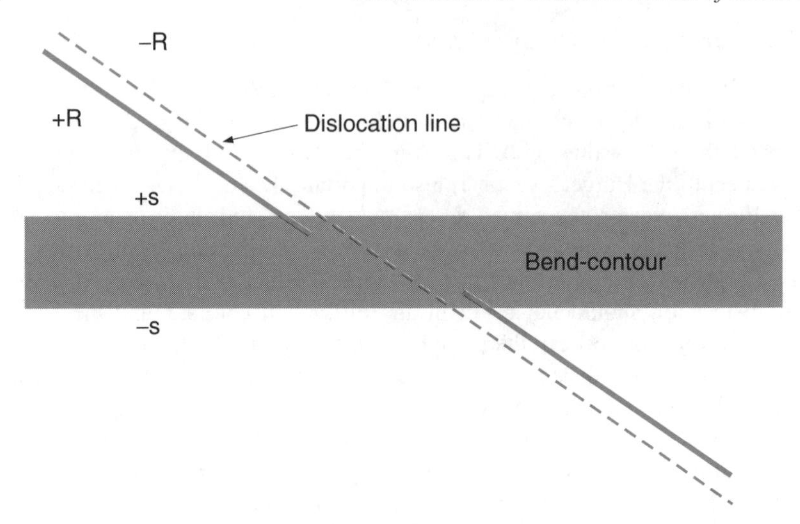

Figure 4.22 The contrast expected from a dislocation line as it crosses a bend-contour; when **s** and **R** are of the same sign, the amplitude phase diagram is expanded and strong contrast is expected

In samples of pure metals and ductile alloys, dislocations are often observed to *move* in the electron microscope, and such behaviour can be used to study the glide process. The phenomenon has been attributed to a combination of thermal stresses, associated with *beam heating* due to inelastic scattering processes, and, more generally, stresses induced by the build-up of a *contamination film* on the sample surface. When dislocation movement occurs, it is often accompanied by the appearance of *slip traces* in the image at the top and bottom of the film, where the dislocation has cut through any superficial surface deposit. These traces can be used to determine the *glide plane* or, if the glide plane is known, to derive an accurate value for the specimen thickness.

4.2.3.6 Point Dilatations and Coherency Strains

Point defects (vacant lattice sites and impurity or alloy atoms) cannot be resolved in transmission electron microscopy, but remarkably small *clusters* of such defects do give rise to detectable diffraction contrast. Such contrast has been observed for three specific types of defect, *radiation damage* (associated with lattice-atom displacement by an energetic particle), faulted or unfaulted *vacancy loops* and more complicated assemblies such as *stacking-fault tetrahedra* (all associated with the condensation of a supersaturation of lattice vacancies), and the early stages of *precipitation*, especially the formation of solute-rich *Guinier–Preston zones*.

In all cases, the dominant lattice displacement for most of these small defects is a *hydrostatic strain field* which can be approximated by a *point defect dilatation*. In the case of radiation damage, a wide range of effects has been observed, ranging from the condensation of *interstitial* atoms, generated by long-range 'knock-on' collisions, to *vacancy loops* associated with the annealing out of vacant lattice sites by diffusion, and the formation of small *helium bubbles* due to α-particle condensation.

The clustering of vacant lattice sites to form a collapsed dislocation loop may result in a planar *stacking fault* in the crystal lattice (for example in Ni, see Fig. 4.23), but if the stacking fault energy is high, subsequent *shear* in the plane of the loop will result in a vacancy loop bounded by a unit lattice dislocation whose Burger's vector is at an angle to the plane of the loop. These loops can be resolved in *weak-beam* contrast when they exceed 2–5 nm in diameter, and analysed by using the criteria described previously for dislocations, noting in particular, the change in contrast from *inside* to *outside* the loop as the sign of **s** is changed by tilting the specimen so that a bend contour sweeps across the image.

Smaller loops behave only as *point dilatations*, and typically show up as twin black/white crescents, as do the larger loops when viewed edge-on. Again, the sign of this contrast (black/white as opposed to white/black) changes depending on the sign of **s**. However, as we will see later, the situation is usually rather more complicated than can be accounted for by *kinematic theory* and the observed contrast also depends on the *position* of the defect within the thickness of the foil. In effect, point dilatations observed near the *top* surface of the foil show the *same* contrast in both bright- and dark-field, while those near the *bottom* surface show a *reversed* contrast. It is only the dark-field image that meets the criterion noted previously, namely that the contrast is determined by whether or not the phase shift associated with the defect reinforces that associated with deviation from the Bragg condition.

In addition to the care required to interpret the image contrast associated with *non-kinematic* effects, we should also note the potentially confusing effects associated with *rotation* or *inversion* of the image in the electromagnetic imaging system. This is particularly the case in recording diffraction patterns for comparison with lattice images in either bright- or dark-field, where inversion is equivalent to a sign reversal.

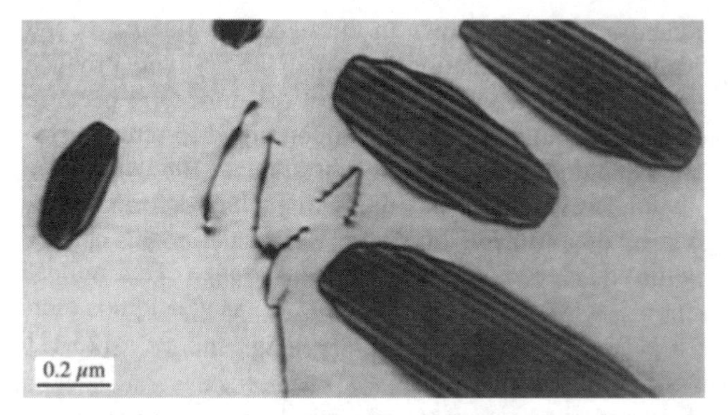

Figure 4.23 Faulted dislocation loops in Ni. Reproduced by permission of Plenum Press from Williams and Carter, *Electron Microscopy: A Textbook for Materials Science* (1996)

Distinguishing between *vacancy* and *interstitial* defects (or *tensile* as opposed to *compressive* dilatation) is straightforward in theory, but may be confusing in practice.

Small *precipitates* and *Guinier–Preston* zones behave qualitatively like *vacancy clusters* and *radiation damage*, but now the displacements are determined by lattice misfit associated with mismatch in the effective size of the solvent and solute atoms. As the precipitates grow, the strains associated with lattice mismatch accumulate, until these *coherency strains* are eventually relieved (or partially relieved) by the nucleation of *interfacial dislocations* at the phase boundary.

In effect, the precipitate and the matrix are in a unique orientation relationship (their orientations are *epitaxially related*), but since the lattice constants of the second phase do not match those of the matrix, they will give rise to *interference* effects in the image which are associated with *double diffraction*. The doubly diffracted beam will result in a *non-localized* set of interference fringes (termed moiré fringes) whose spacing depends on the mismatch in the two **g** vectors from the matrix and the precipitate:

$$d_{\text{moiré}} = \frac{d_1 d_2}{\left((d_1 - d_2)^2 + d_1 d_2 \beta^2\right)^{0.5}} \tag{4.20}$$

where β is the angle of rotation between the two **g** vectors and $|\mathbf{g}| = 1/d$.

It is not difficult to distinguish between the moiré fringes and the interface (van der Merwe) dislocations, since the latter are fully *localized* and have Burger's vectors which can be analysed by comparing dark-field images from different reflections, while the latter are *non-localized* in the image plane and have a separation that depends on the operating reflection.

In passing, it should be noted that moiré fringes may be associated with either a difference in lattice spacing or a rotation between the two diffracting regions. It follows that moiré effects may also be associated with dislocation tilt or twist sub-boundaries.

4.2.4 Dynamic Diffraction and Absorption Effects

The kinematic theory of contrast is a poor approximation for all but the thinnest of specimens. We will not give the full theory of *dynamic diffraction* here, but rather a physical explanation of its significance and an outline of some of the more important conclusions which affect diffraction-image contrast. One consequence of dynamic effects has already been noted, namely that the position of a small defect (point dilatation) with respect to the foil thickness determines whether or not bright-field image contrast is inverted with respect to dark-field contrast.

The phenomenon of *dynamic electron diffraction* is to some extent analogous to the effects discussed earlier concerning optical wave propagation in anisotropic materials (Section 3.4.3), namely the splitting of an incident beam into two propagating beams on entering the crystal which differ in their wavelength. In the case of electrons, it is the *electrical potential* inside the crystal that affects the wave vector (the wavelength), and an electron beam incident at the Bragg angle propagates through the crystal as *two* beams, one with its probability maxima *at* the atomic positions and one with its maxima *between* the atomic positions. The small difference in electrical potential experienced by these two beams results in a slightly different wavelength, and hence in a difference in *phase angle* which increases with the path length through the crystal. For a given reciprocal lattice vector, this phase difference will lead to extinction of the *direct transmitted beam* at a critical thickness, termed the *extinction thickness* for that reflection, which corresponds to a phase difference of π between the two beams exiting the crystal. At the same time, the energy scattered into the diffracted electron beam from the direct transmitted beam also behaves as two beams of slightly different wave vector propagating through the crystal, and will lead to a diffracted *maximum* at the same extinction thickness, thus conserving the total energy in the diffracted and direct transmitted beams leaving the foil.

Within the crystal, solutions to the wave equation for the incident beam behave as *four* waves, termed the *Bloch* waves, with *two* propagating as the direct transmitted beam and *two* propagating as the diffracted beam. These two *pairs* of waves are nearly $\pi/2$ out-of-phase, and within each pair there is a small wavelength difference, corresponding to the oscillating periodic potential in the crystal. The small difference in wavelength leads to a '*beat pattern*' of standing waves inside the crystal, which gives rise to a *thickness-dependent* series of complementary maxima and minima in the amplitudes of the direct and diffracted beams. On exciting the crystal, the two pairs of waves *recombine* into two beams, one corresponding to the *direct transmitted beam* and the other to the *diffracted beam* (Fig. 4.24).

Typical values for the *extinction thickness* at 100 kV vary from of the order of 20 to 100 nm, depending on the *atomic number* of the material (the electron density) and the *reciprocal lattice vector* (the most densely packed planes have the *strongest* scattering power and therefore the *shortest* extinction distance). When the crystal is oriented with a direction of high symmetry parallel to the incident beam (so that the crystal is accurately aligned parallel to a low-order Laue zone), then several reflections will be active simultaneously (multibeam diffraction conditions) and an anomalously *low* extinction distance will result, which leads to a particularly strong sensitivity to lattice defects. Such conditions are exceptionally favourable to the analysis of small defect clusters and the early stages of precipitation (Fig. 4.25).

Inelastic scattering is associated with those Block waves which have their maxima at the atomic positions, thus *maximizing* the probability of an inelastic scattering event, while those waves whose minima fall at these positions are unlikely to interact inelastically with the atomic nuclei. It follows that *inelastic absorption* is effectively confined to one member of each pair of Bloch waves, so that, in a sufficiently thick crystal, the contrast oscillations associated with thickness fade out (only the wave travelling 'between' the atoms is transmitted), and the intensity scattered by a defect near the centre of the foil into the diffracted wave depends only on the deviation

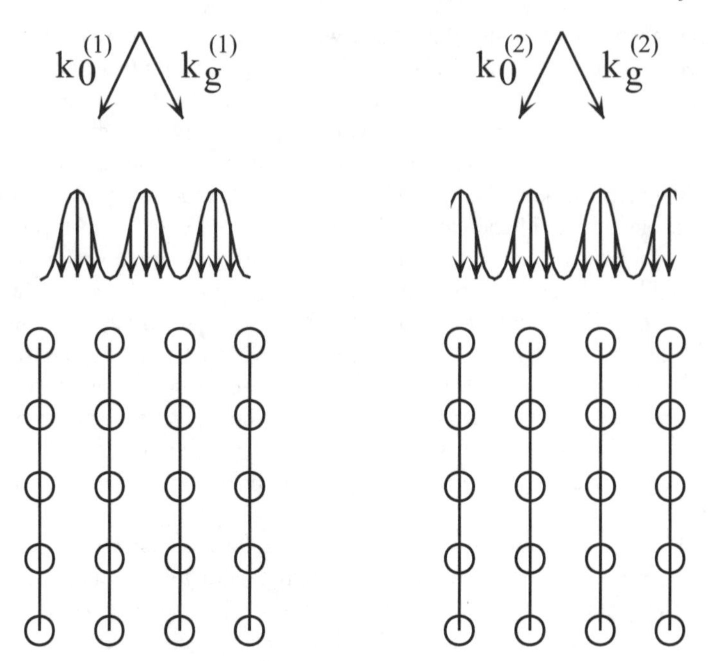

Figure 4.24 The Bloch wave model for dynamic diffraction. Electron energy is transmitted through the crystal as two pairs of waves with a small wavelength difference within each pair; the two wave pairs have their maxima at and between the atomic positions, respectively. Amplitude is transferred from one pair of Bloch waves to the other as the electrons propagate through the crystal. Upon exciting the crystal, each pair of Bloch waves recombines to form the direct beam and the diffracted beam

from the Bragg condition and *not* on the foil thickness. Hence in thick crystals the oscillations in contrast along a dislocation running at an angle through the foil are restricted to the regions at the top and bottom of the specimen.

4.2.4.1 Stacking Faults Revisited

We noted previously, when discussing the contrast from a point dilatation, that this depended on the *position* in the foil and whether the contrast was observed in *bright-* or *dark-field*. The case of a stacking fault observed at an inclined angle in a thick foil provides a clear and instructive example of the way in which *dynamic effects* (the extinction thickness) and *absorption* (removal of one of the Bloch waves) determines the observed diffraction contrast (Fig. 4.26).

When the stacking fault is near the *bottom* of a thick foil, the *second* Bloch wave (2) in the diffracted beam from the region of crystal above the fault is *absorbed*, so that only the *first* Bloch wave (1) is available to be scattered into the diffracted beam below the fault (2′). The two beams exiting the crystal are then the *first* Bloch wave from the region above the fault (1) and the *second* Bloch wave from the region below

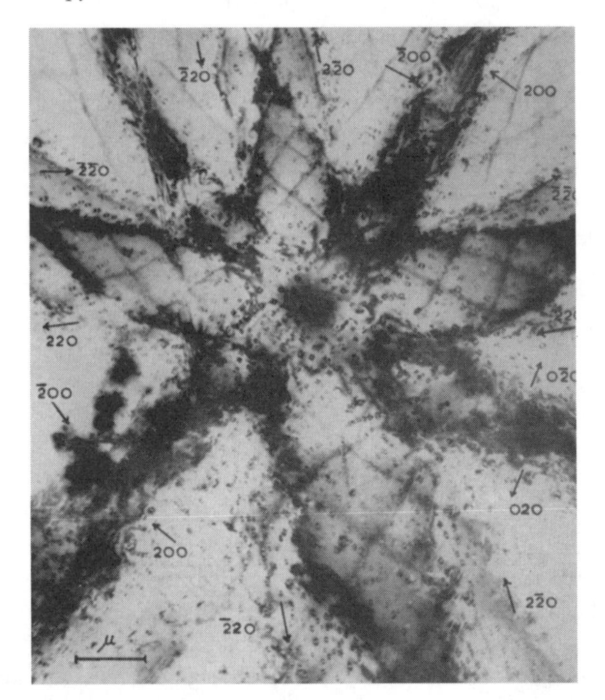

Figure 4.25 Contrast from precipitate nuclei observed in a specimen of uniform curvature in the region of a low-order Laue zone (a symmetry axis). Reproduced from Hirsch *et al.*, *Electron Microscopy of Thin Crystals*, published by Butterworth

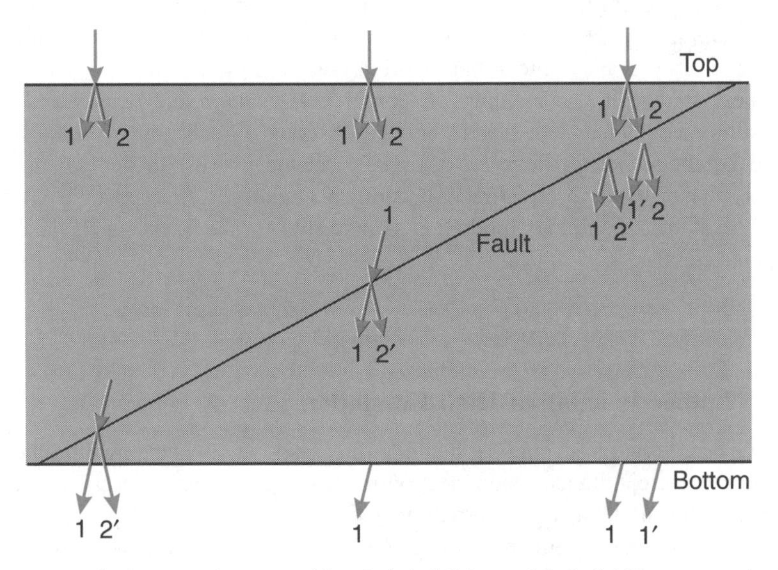

Figure 4.26 The contrast from a stacking fault in bright- and dark-field images results from a combination of extinction interference and absorption

the fault (2′). When the stacking fault is near the *top* of the foil, the *first* Bloch wave above the fault is scattered, (1) and (2), and both beams are then transmitted, with the appropriate phase shift, through the crystal below the fault, (1′) and (2′). However, *both* beams (2) and (2′) are now *absorbed*, so that the two beams exiting the crystal are (1) and (1′). It follows that the *bright-field* and *dark-field* images of a stacking fault will be *complimentary* when the fault is near the *bottom* of the foil (black stripes in the dark-field image corresponding to white stripes in the bright-field image), but *identical* when the fault is at the *top* of the foil. Similar arguments can be applied to strain contrast from small *precipitates* or *dislocations* and used to rationalize the contrast observed and identify their position in the specimen in three dimensions.

4.2.4.2 Quantitative Analysis of Contrast

Enough has already been said to make it clear that the *qualitative* analysis of diffraction contrast requires some care to ensure that the sign of the displacement field associated with a defect is correctly identified. It is important to image the same defect in both bright-field and dark-field, to obtain images from *more than one* reflection, and to check the effect of reversing the sign of **s**, the deviation from the Bragg condition. It is also important to calibrate the microscope, in particular any inversion or rotation of the image with respect to the diffraction pattern.

Quantitative analysis requires considerable additional knowledge, well beyond the scope of this present text. However, it should be possible to analyse displacements *semi-quantitatively*. This includes the nature of the *coherency strains* around point defect clusters (vacancy, interstitial or solute), as well as any anisotropy associated with these strains. It also includes the complex *dislocation* and *partial dislocation* interactions associated with plastic flow or observed in sub-grain *boundaries* and semi-coherent *interfaces* between epitaxially related phases. In general, too little effort is made to interpret diffraction contrast *quantitatively*, especially when the power of modern computer modelling is available to check any analysis through computer simulation.

4.2.5 Lattice Imaging at High Resolution

We have noted that phase contrast imaging of crystal lattices is well within the reach of good commercial transmission electron microscopes and have described the main features of this mode of image formation, especially its sensitive dependence on the physical parameters of the microscope, as summarized in the *contrast transfer function* (CTF). In the following, we concentrate on the relationship between the lattice image and the crystal structure.

4.2.5.1 The Image and the Contrast Transfer Function

The *contrast transfer function* (CTF) summarizes the *phase shift* introduced by the imaging system of the microscope into the wave function of the electrons forming the image at the screen, the photographic recording emulsion or the charge-coupled device. The phase shift associated with this imaging system, x, is expressed as $\sin x$, and the CTF gives $\sin x$ as a function of the *wave number* or *frequency*, equivalent to the scattering angle subtended by that frequency at the objective aperture and commonly defined in reciprocal space as $1/d$ in nanometres (Fig. 4.16). The final phase and amplitude for each frequency is then obtained by *convoluting* the CTF with the *calculated* phase and amplitude of the electrons transmitted through the sample. The phase shifts and the amplitudes are calculated from the *dynamical theory of electron diffraction* and then 'transmitted' through the imaging system of the microscope, using the CTF, to interfere in the image plane. The intensity for any image *pixel point* is obtained by multiplying the integrated image *amplitude* by its *complex conjugate*.

The factors that enter into the CTF include parameters associated with both the *source* and the *objective lens*. The source parameters are its *coherency* and *energy spread*. The large dimensions of a LaB_6 crystal result in poor coherency for this source of electrons. A *field-emission gun* has a very small emitting diameter with excellent coherency, behaving approximately as a *point source*, as well as having a small energy spread. The traditional tungsten filament is inferior on both counts. An additional parameter is the *angular spread* (divergence) of the incident beam, determined by the condenser system and the beam focusing conditions.

The *objective-lens* parameters include the *chromatic* and *spherical aberration* constants, as well as the *current stability* in the lens. Additional minor corrections are concerned with second- or third-order lens defects. *Astigmatism* (a second-order defect) should be eliminated by a suitable correction, and modern correction systems can also remove *coma* (a third-order defect). The single most important parameter limiting the performance of the transmission electron microscope is the *spherical aberration constant* of the objective lens. All other performance-limiting factors can be corrected to better than this limit.

4.2.5.2 Computer Simulation of Lattice Images

The lattice image in phase contrast is commonly recorded at or close to the *Scherzer focus* (the underfocus that *maximizes* the frequency at which the imaging system introduces oscillations in the sign of the phase shift (see Section 4.2.2.3). However, the *periodicity* in the image is preserved when a series of images are recorded while the focus is incrementally changed, and it is only the *intensity* at each point that changes. These intensity changes include *contrast reversal* at defocus values below the Scherzer focus, the *doubling* of characteristic periodicities, and apparent changes in the prominence of specific crystallographic directions (rows of bright maxima in one direction being replaced by similar rows in another direction).

It is difficult to resist the temptation to assign specific features of the *crystallography*, deduced from prior knowledge of the crystal structure or from the electron diffraction patterns, directly to the prominent features of the *lattice image*. However, minimal practical experience of the sensitivity of the lattice image contrast to both the *defocus* of the objective lens and the *thickness* of the sample (distance from the edge of a tapered foil specimen) should quickly convince any scientist of the dangers of subjectively 'seeing' those features that are compatible with previous conceptions and ascribing any disparities to imaging artifacts.

The only acceptable solution is *computer simulation* of the image, and several software packages are now available for this purpose. For successful computer simulation three conditions should be satisfied. These are as follows:

(a) *Accurate calibration* of the microscope parameters, in particular the spherical and chromatic aberration constants and the beam divergence.
(b) A *contamination-free*, uniformly thinned specimen, preferably no thicker than the extinction thickness for the imaging reflections.
(c) *Accurate alignment* of the optic axis along a prominent zone axis, with an angular aperture accepting as many reflections into the imaging system as is compatible with the critical spatial frequency at the *Scherzer focus* (Section 4.2.2.3).

If this information can be supplemented by an estimate of the *film thickness*, then so much the better, but in many cases film thickness is a variable in the image simulation needed to estimate errors associated with an imperfect knowledge of this parameter.

A model for the unit cell of the *crystal lattice* must be inserted into the program, including the positions of all atoms and their expected occupancy. *Lattice defects*, such as faults and dislocations, can be simulated by suitable displacement fields. *Boundaries* are commonly simulated *parallel* to the optic axis, but can also be included by using a transformation matrix for the region of the crystal below the boundary plane.

Although *visual matching* is most commonly employed to decide the 'best fit' simulation, significant progress has been made in deriving a 'difference' image, in which the simulated intensities for each pixel are compared with recorded intensities for a through-focus series inputted to the computer using a CCD camera. This technique is a two-dimensional extension of the *Rietveld method* for comparing an experimental X-ray spectrum with a calculated spectrum derived from a hypothetical crystal lattice unit cell.

4.2.5.3 Lattice Image Interpretation

The common supposition that the lattice image 'shows you where the atoms are' is actually incorrect.

In the first place, as in any magnified image, the lattice image is recorded in *two dimensions*, as a periodic pattern of varying intensity. The *amplitudes* leading to these intensity variations can be deduced, but the *phase information* has been lost. The image interference pattern is *not* localized in space and, as we noted previously, is a sensitive function of both *defocus* and *sample thickness*, in addition to other, less critical parameters. Providing that the imaging conditions are well known and that the simulated image is a good match to the observed image, then there is adequate justification for the conclusion that the model lattice used in the simulation (atomic positions, occupancies and periodicities) is a good description of the microstructural morphology.

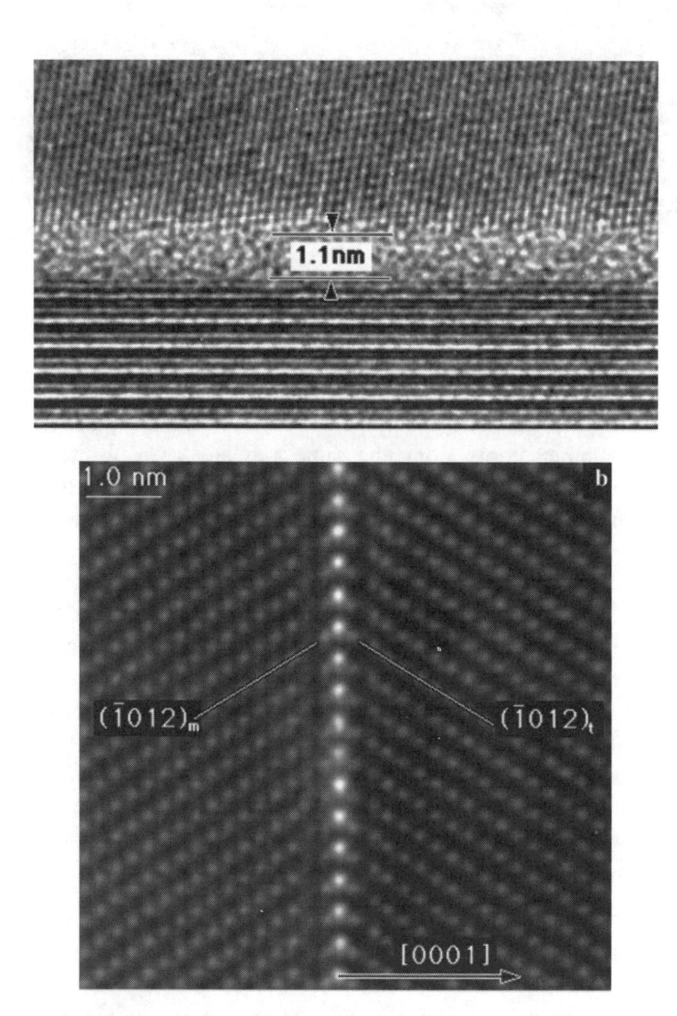

Figure 4.27 Some examples of lattice images from grain and phase boundaries: (a) an amorphous glassy layer in equilibrium at a grain boundary in silicon nitride (permission of X. Pan); (b) a layer of calcium cation segregation at a twin boundary in alumina; (c) A 1 nm-thick Si–Ca glass at an interface between alumina and copper

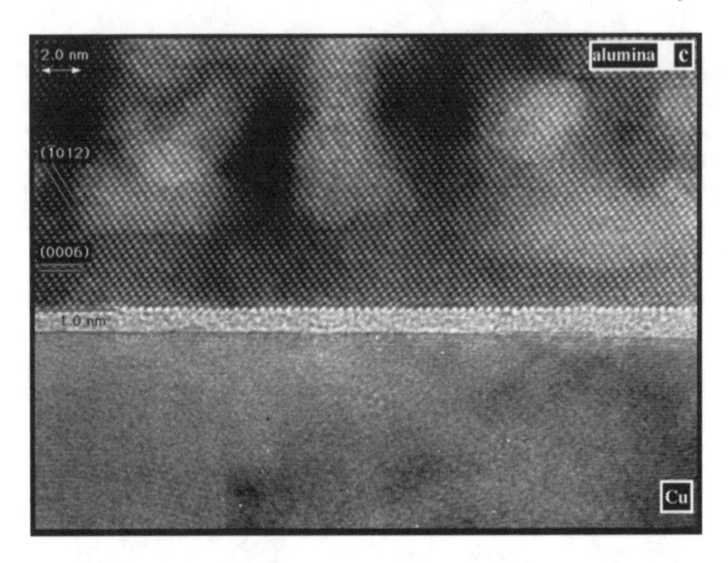

Figure 4.27 (*continued*)

There are still serious problems, however, primarily due to the non-localized nature of the lattice image. Superimposing a unit cell of the model lattice on the image and identifying 'channels' along the optic axis with black patches, while rows of high-atomic-number atoms along the same axis are 'seen' as white patches is unlikely to commend itself to the expert. An excellent example is the apparent structure of a *phase boundary*, which is complicated by the difference in the lattice potential between the two phases. This is a discontinuity in 'refractive index' which can result in a shift in the interference pattern of one phase with respect to the other *perpendicular* to the boundary, thus making it extremely difficult to deduce the atomic displacements at the boundary, even though they have been clearly 'observed'!

Nevertheless, the morphological information which can now be obtained from lattice images of both phase and grain boundaries is extraordinary (see Fig. 4.27), and there is little doubt that the need to make this information *quantitative* will lead to a continued improvement in the methods for computer simulation and evaluation of digitally recorded lattice images to optimize a model for the various structures.

4.3 SCANNING ELECTRON MICROSCOPY

The scanning electron microscope provides the closest approximation to what the eye expects to see, since the *depth of field* for resolved detail is very much greater than the *spatial resolution*, as it is for the world as we perceive it. The curious 'flatness' of the topological and morphological detail observed in the light-optical or

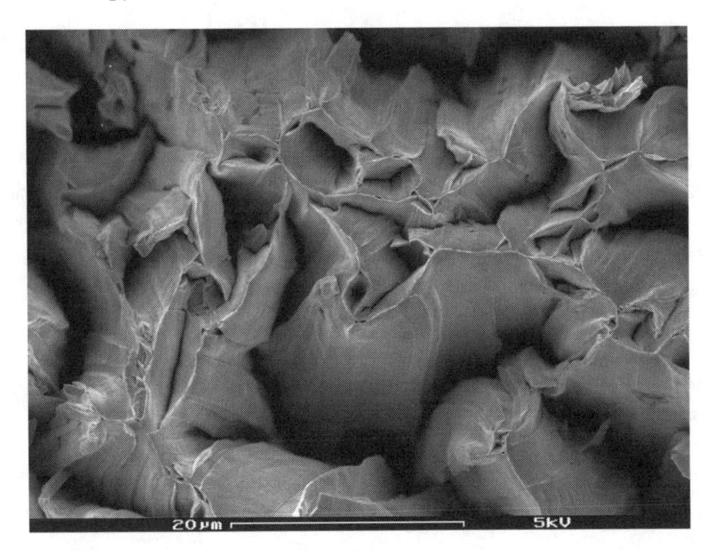

Figure 4.28 A secondary electron image of a ductile fracture surface

transmission electron microscope is replaced in the scanning electron microscope by what looks very like a play of light and shade over the hills and valleys, hollows and protrusions of a three-dimensional object (see Fig. 4.28). Only two features are absent from a complete 'optical illusion'. First, a true evaluation of image *depth* (which can be provided by recording two images from slightly different viewpoints, the technique of *stereoscopic imaging*, see Section 4.3.6.3). Secondly, the presence of *colour* (which is sometimes used to enhance contrast or compare images during data processing).

This visual impact of scanning electron microscope images and the ability to observe details displaced along the optic axis has led to the rapid penetration of *scanning electron microscopy* into all branches of the sciences and engineering from the time it was first developed, in the late 1960s. We will describe the various *signals* that are detectable in the scanning electron microscope due to the interaction of a focused beam of electrons with a solid sample, the *information* that may be derived from these different signals, and some of the special *techniques* which are available to enhance and assist image interpretation.

4.3.1 Electron Beam–Specimen Interactions

An energetic electron penetrating into a solid sample undergoes both *elastic* and *inelastic* scattering, but it is the *inelastic scattering* that will eventually predominate, reducing the energy of the electrons in the beam to the kinetic energy kT. The various processes which occur are complex, but they are generally well understood, so that there is little ambiguity about the interpretation of image contrast in the scanning microscope.

4.3.1.1 Beam-Focusing Conditions

The *probe lens*, used to focus the electron beam on to the specimen surface in the scanning electron microscope, has similar characteristics to the *objective lens* in transmission electron microscopy, with the best '*resolution*' that can be obtained being determined by the focused probe size. A simple consideration of the geometry of the source, condenser system and probe lens demonstrates that the electron path is in effect *reversed* in the scanning electron microscope, and the probe-forming system is just the *inverse* of the imaging system in transmission. In other words, the electron *source* (the 'gun') is where the *image* would be observed in *transmission* geometry. The *condenser* system *reduces* the apparent size of the source (in place of magnifying the image), and the *probe lens* forms the beam probe where the *source* would be in the transmission microscope. In short, the *probe* is a *minified* image of the *source*.

In practice, there are three limitations on the *minimum* size of the probe in the plane of the specimen, as follows:

(a) the *spherical* and *chromatic aberrations* of the probe lens, with the spherical aberration being the dominant factor;
(b) the maximum *beam current* that can be focused into a probe of a given diameter;
(c) the need to allow *sufficient space* beneath the probe lens pole-pieces to accommodate large specimens (typically of the order of 20 mm in diameter, but up to 10 cm in some cases).

In practice, it is the *beam current* limitation which has proved the most serious problem, since the beam current varies approximately as the third power of the *beam diameter*, and hence drops dramatically for a finer probe. However, the *working distance* between the probe and the specimen is also an important factor, and typically some compromise is reached whereby the minimum *required* probe diameter (resolution) determines the *allowable* spherical aberration, which then sets the *maximum* working distance.

The commercial introduction of the *field-emission gun* has drastically reduced the size of the primary electron source and increased the available current density in the probe (by some four orders of magnitude!), thus allowing for either a much reduced *probe size* (better ultimate resolution) or, often more important, the use of lower incident *beam energies* (longer-wavelength electrons). It is now possible to obtain resolutions of the order of 2 nm at beam energies down to 200 V. This may be compared to previously available scanning electron microscopes, which were seldom capable of generating data at below 5 kV or achieving a resolution better than 20 nm.

The *distribution* and *stability* of the electron current within the beam is also important. While the beam diameter is usually defined as the width of the current distribution at half the measured maximum beam current (*full-width at half-*

maximum (FWHM) diameter), this may be a poor guide to the *total* current, which commonly includes a long tail, extending up to several microns from the central spot. For many imaging purposes this is unimportant, but can significantly affect X-ray data collection if quantitative analysis across a phase boundary is being attempted.

A few words should be added here concerning the *scanning system*. Since data is acquired by scanning the electron probe across the surface of the specimen and collecting one of the signals generated, the rate of data collection is not limited only by the intensity of the probe and the efficiency of signal collection, but also by the *scanning speed*. A signal which is *weak* will require a *slower* scanning speed, and in the collection of *characteristic X-ray data* (where not only the inelastic scattering cross-section, but also the collection efficiency is low) the *statistics* of data collection determine the resolution available, while the beam current *stability* often affects the accuracy of analysis. We will return to these statistical considerations later (Section 4.3.2).

4.3.1.2 Inelastic Scattering and Energy Losses

The calculation of inelastic scattering paths can be accurately simulated by *Monte Carlo* methods, providing that some specifically *crystallographic* effects are ignored (for example lattice anisotropy and channelling processes along preferred directions). The electrons in the beam entering the crystal follow an irregular scattering path, losing energy as the path length in the crystal increases (Fig. 4.29). It is not possible to calculate any average trajectory for multiply scattered electrons, but it *is* possible to define (and measure) two critical depths, as well as to estimate the envelope which defines the boundaries of the electron trajectories for any given average energy. Thus, the *diffusion depth* x_D is defined as that depth beyond which the electrons are randomly scattered in all directions, so that an electron is equally likely to be scattered in any direction. At depths below x_D, there is a net drift of electrons towards increasing depths. The *penetration depth* or range, x_R, is defined as the depth at which the electron energy is reduced to the thermal energy kT. Both x_D and x_R *decrease* with *increasing* atomic number Z and *decreasing* incident beam energy E_0, but whereas the change with beam energy is more or less self-similar, that with atomic number is not. The shape of the envelope of scattered paths changes markedly with the atomic number, primarily because the lateral spread of the beam is roughly proportional to the difference (x_D–x_R), while x_R is less sensitive to Z than x_D. These effects are summarized schematically in Fig. 4.30.

4.3.2 Electron Excitation of X-rays

If the incident electron energy *exceeds* the energy required to eject an electron from an atom in the specimen, then there exists a finite probability of such an *ionization event* occurring. *Ionization* of the atom occurs by an inelastic scattering event which raises the energy of the atom above the *ground state* by an amount equal to the

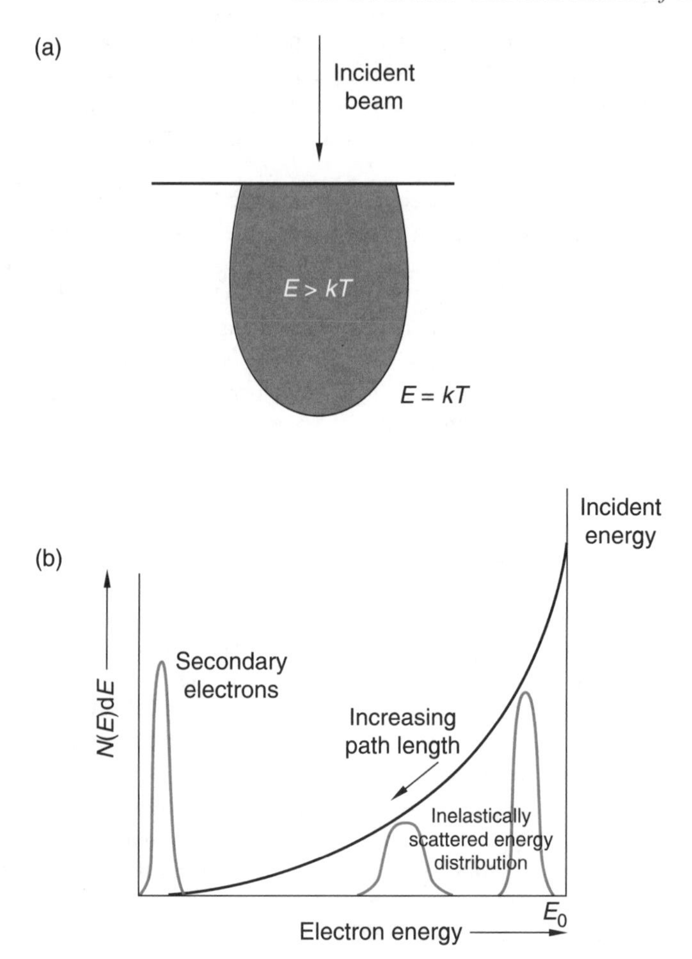

Figure 4.29 (a) The electron beam is inelastically scattered within an envelope bounded by the condition that the average energy has reached the thermal kinetic value kT. (b) The energy spread increases and the average energy of the electron falls as the path length within the solid increases (channelling effects and lattice anisotropy being ignored). (c) Random scattering models for individual electrons (Monte Carlo simulation) provide a vivid image of both the energy distribution and the spatial distribution of the electrons in the volume of the material beneath the beam, as well as the origin of the back-scattered electron signal

ionization energy. The energy of the atom can then decay by an *electron transition* into the now vacant state. All such transitions are accompanied by the emission of a *photon*, and if the excited state of the atom corresponds to the ejection of an electron from one of the *inner shells* of the atom, then this photon will have an energy in the *X-ray* region of the spectrum.

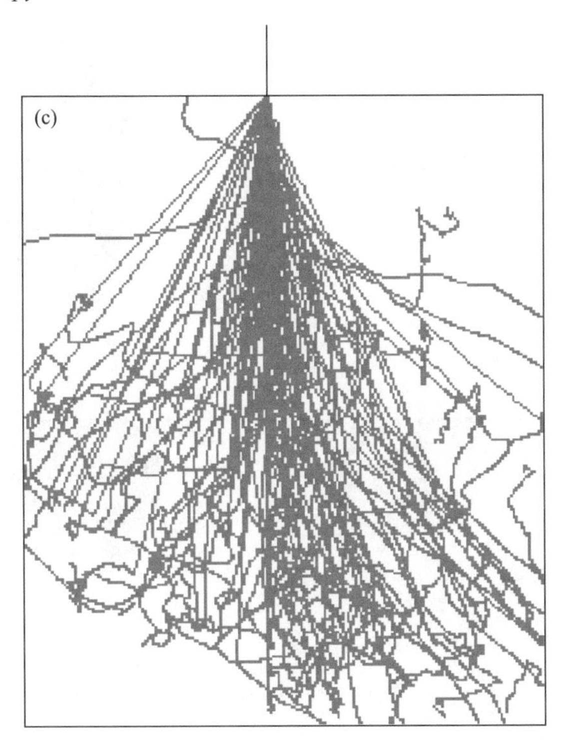

Figure 4.29 (*continued*)

In general, decay of an atom from the excited state takes place in more than one stage, with the emission of several photons of different wavelengths, each corresponding to a transition of the excited atom back towards the ground state. It follows that the energy lost by the incident electron *must* exceed the threshold energy for the ionization state if that particular ionization state is to be achieved, while the energy of the *most energetic* photon which can be emitted will always be *less* than this threshold for excitation. Furthermore, if we consider a particular inner shell of electrons (for example the innermost K-shell), then, as the atomic number increases, the *ionization energy* for electrons occupying this shell must also increase (since the inner-shell electrons are more deeply embedded in the atom for the higher atomic numbers). An approximate representation of the situation is shown in Fig. 4.31.

The X-ray spectrum generated when an electron beam hits a solid target contains *all* wavelengths, starting from the *minimum* wavelength derived from the relationship $\lambda_0 = hc/eV$, where h is the Planck constant, c is the speed of light, e is the charge on the electron and V is the accelerating voltage applied to the electron beam (see Section 2.3.1).

The wavelengths of the *characteristic lines* which are emitted constitute a *fingerprint* for the elements present in the solid and provide a powerful method of

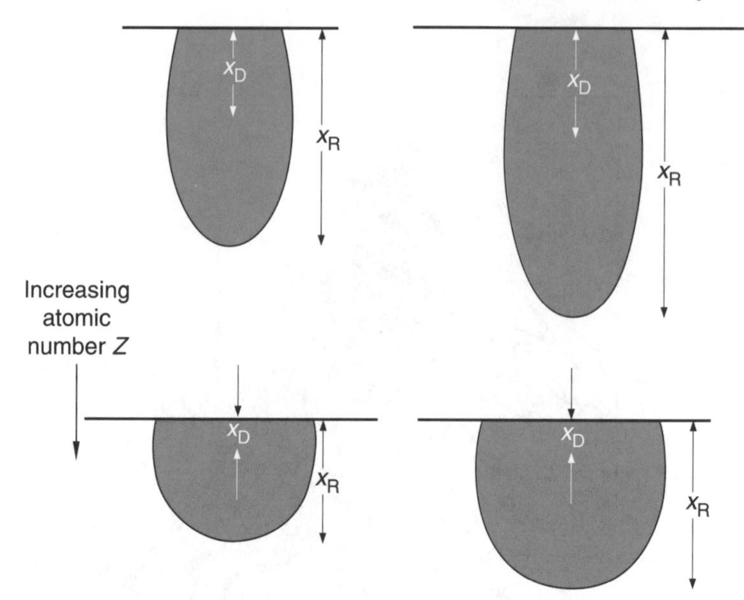

Figure 4.30 The inelastic scattering envelope for an incident beam of energetic electrons depends on both the incident energy and the atomic number of the target, and is qualitatively characterized by the two parameters, diffusion depth, and penetration depth (or range)

identifying the chemical constituents and their distribution. The atomic-number dependence of these wavelengths was illustrated approximately in Fig. 2.13, but it should be recognized that, rather than a single line, *groups* of lines are emitted, while there may be *overlap* between the lines emitted by one chemical constituent and those emitted by a second component.

In parallel to the *emission spectrum*, X-rays generated by the incident electron beam also have an *absorption spectrum*, since X-ray photons themselves have a finite probability of exciting atoms to higher-energy states. Therefore, a K photon from a *higher*-atomic-number element will have sufficient energy to excite an atom of *lower* atomic number to the K state, thus resulting in *absorption* of the photon. The excited atom will now decay back towards the ground state, generating a new, lower-energy photon characteristic of this atom, a process which is termed *X-ray fluorescence*. X-ray *absorption*, as noted previously, is characterized by an *absorption coefficient* μ which depends on the wavelength of the X-rays and the atomic number of the solid. Critical *absorption edges* in the X-ray absorption spectrum (see Section 2.3.1), correspond to those photon energies which are at the threshold of excitation for fluorescent radiation of a constituent of the solid.

Both the *excitation* and the *absorption* spectra of both *electrons* and *X-rays* can be used to derive information on the *chemical composition* of the sample, and we will return to these topics in more detail in Chapter 5.

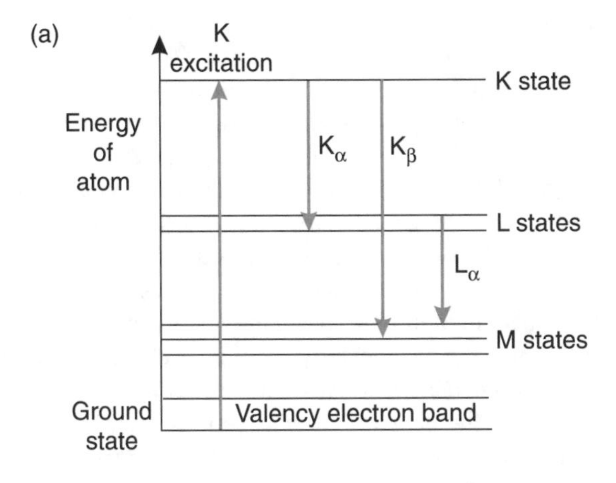

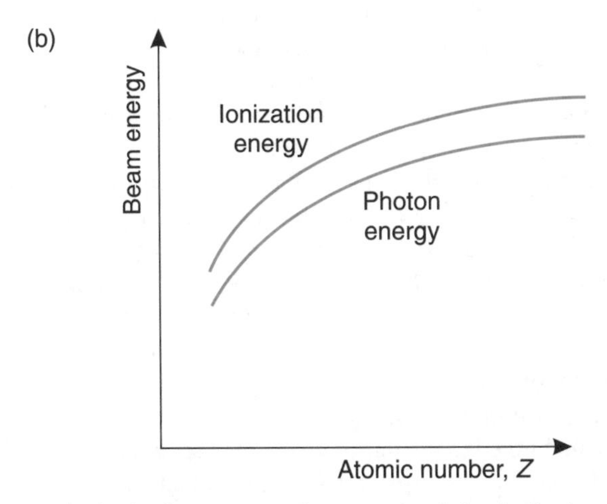

Figure 4.31 (a) An inelastic electron scattering event involving ionization of an inner-shell electron raises the energy of the atom to the appropriate ionization state, with subsequent decay to a lower energy being accompanied by photon emission. The photon emitted is characteristic of the energy difference between the two energy states, but must always be less than that required for ionization. (b) The ionization energy required to eject an electron from a particular inner shell increases with the atomic number

4.3.2.1 Characteristic X-ray Images

The X-ray signal generated from a focused electron probe comes from a *volume element* defined by the envelope of electron energies which exceed the energy required to excite the characteristic radiation of interest. As the beam voltage is *reduced*, so this volume element *shrinks*, thus improving the potential spatial resolution but at the same time reducing the emitted X-ray signal. A compromise has to be made in order to ensure that the signal remains *statistically significant* but still *spatially localized*. The intensity of the emitted signal first *increases* as the electron beam energy is *increased* above the minimum required to excite the characteristic signal, but then, when the incident energy exceeds the critical excitation energy by about a factor of four, starts to *decrease*. This decrease is associated with an increased absorption of the primary signal within the sample, since the signal now originates deeper below the surface due to the increased diffusion depth of the electrons. It follows that the 'optimum' beam energy is of the order of four times the excitation energy, but that the beam energy should be selected for the *highest* energy of interest (the *shortest* characteristic X-ray wavelength to be detected).

The *characteristic X-ray generation efficiency* is very low, and the X-rays generated are emitted at all angles, so that a high proportion are absorbed in the sample. *X-ray collection* is also inefficient. Collection is usually accomplished with a *solid-state detector*. The alternative is a *crystal spectrometer*, which employs a different bent crystal for each range of wavelengths of interest. The advantage of the crystal spectrometer (a *wavelength-dispersive spectrometer* (WDS)) is that the *spectral resolution* is excellent, considerably reducing the chances of any ambiguity in ascribing a characteristic wavelength to a specific constituent.

Solid-state detectors are based on the energy discrimination capability of a cryogenically cooled semiconductor crystal. The *charge* generated in the detector by each photon absorbed is proportional to the photon energy, thus leading to a detectable *current pulse* proportional to the photon energy (an *energy-dispersive spectrometer* (EDS)). While the X-ray collection efficiency in the EDS is far higher than can be achieved in a WDS system, *signal overlap* is much more frequent, although the problem can usually be avoided by selecting *alternative* characteristic emission lines. The detector crystal is commonly shielded from contamination in the microscope by a thin window, transparent to most X-rays, which is capable of detecting photons of wavelengths exceeding 5 nm, corresponding to emission associated with *light elements* (B, C, N, and O).

All detectors are limited in the rate at which they are able to accept X-ray 'counts', corresponding to the time required for each individual charge pulse to decay. This *dead-time* is typically less than 1 μs, during which time no further counts are recorded. It follows that count rates should not exceed 10^6 s^{-1}. Some counts are inevitably lost, since the photons are randomly generated in time, and the proportion of dead-time between counts is registered by the counting system. Acceptable dead times are of the order of 20 %, with lower values corresponding to lower rates of data accumulation. The effect of an increasing dead-time on the collection efficiency is shown in Fig. 4.32.

The X-ray signal may be displayed in three distinct modes, as follows:

(a) As a *spectrum*, in order to identify the elements present from their characteristic 'fingerprints' (Fig. 4.33). Such a spectrum may be collected with the beam *stationary* at a specific location on the sample surface (*point analysis*), or the

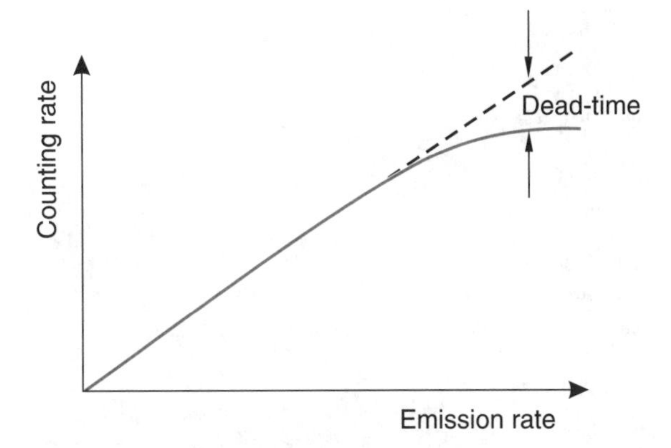

Figure 4.32 The counting rate as a function of the signal incident on the detector, illustrating the effect of dead-time on the counting efficiency

spectrum may be collected while the beam is *scanned* over a *selected area*. Typical *collection times* required to ensure detection of all elements present in concentrations of a few percent are of the order of 100 s.

(b) As a *line-scan*, in which the beam is traversed across a selected region of the sample (usually in discrete steps) and the *rate of detection* for one or more selected characteristic X-ray wavelengths is displayed as a function of the position of the beam (Fig. 4.34). The *detection windows* are set to exclude all wavelengths other than those selected and the number of counts for each characteristic wavelength of interest is displayed for each position of the beam traverse.

(c) As an X-ray *dot image* or *elemental map* in which each characteristic photon detected is recorded as a white dot in a position on the screen corresponding to the coordinates of the beam as it is scanned over a *selected area* of the specimen surface. The *range of wavelengths* is preselected in a 'window' corresponding to the *characteristic wavelength* of the element of interest, and more than one element may be detected simultaneously and displayed in *separate* windows, which are usually colour-coded for the different elements (Fig. 4.35).

The overriding consideration in recording X-ray data in the scanning electron microscope is the *statistical significance* of the signal. The simplest assumption is that the counting statistics for each pixel point obey a *Poisson distribution*, so that the counting error associated with each element's characteristic line is just $1/N^{0.5}$, where N is the number of counts recorded for the element in the area defined by the pixel point.

Point analysis, corresponding to collection of a complete spectrum, can provide data corresponding either to a fixed point or averaged over a selected scanned area, but the statistical significance is usually high and can be determined after correcting for the *background counts* measured outside the selected window (it is common

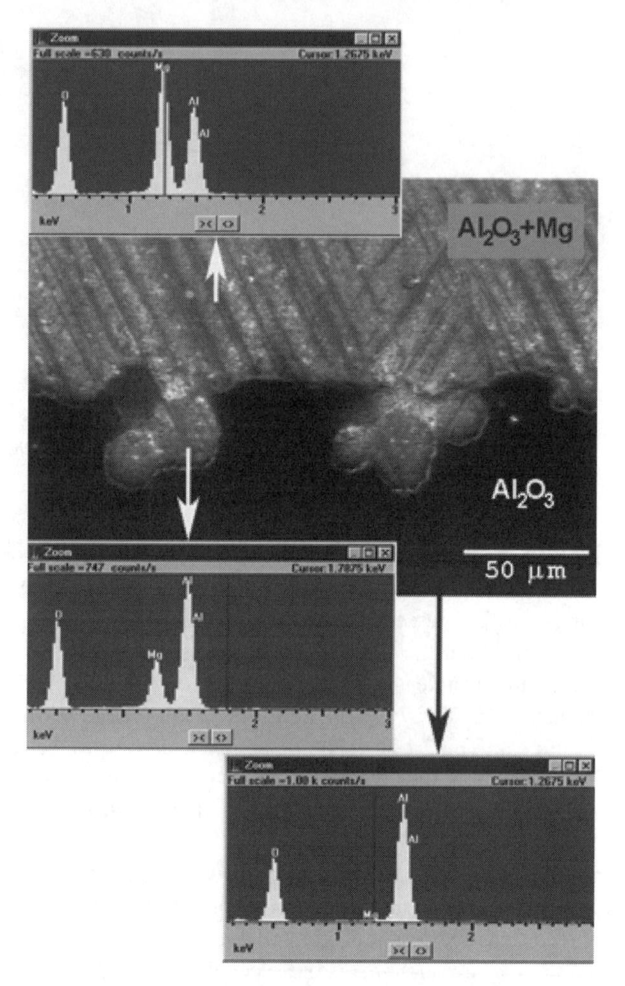

Figure 4.33 X-ray spectra obtained from a porous alumina specimen partially infiltrated with magnesium, showing the presence of oxygen, magnesium and aluminium in different regions of the microstructure via characteristic lines superimposed on a background of white radiation

practice to define the collection window using the FWHM of the characteristic peak, $N = N_T - N_B$, where N_T and N_B are the *total* counts within the selected window and the *background correction*, respectively.

Line analysis generally results in *reduced* statistical significance, in proportion to the square root of the number of pixel points along the line selected, but is nonetheless extremely useful in determining changes in concentration near surfaces and grain boundaries (segregation effects) or associated with second phases.

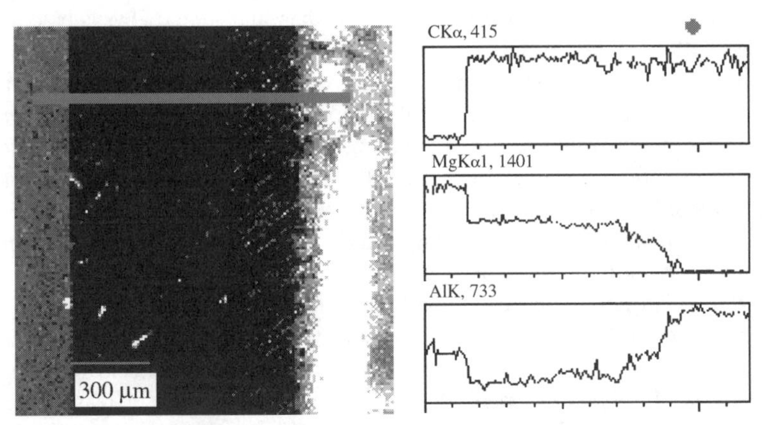

Figure 4.34 An X-ray line-scan across a porous alumina specimen partially infiltrated with magnesium, showing the intensity of selected characteristic lines (oxygen, magnesium, and aluminium) as a function of beam position

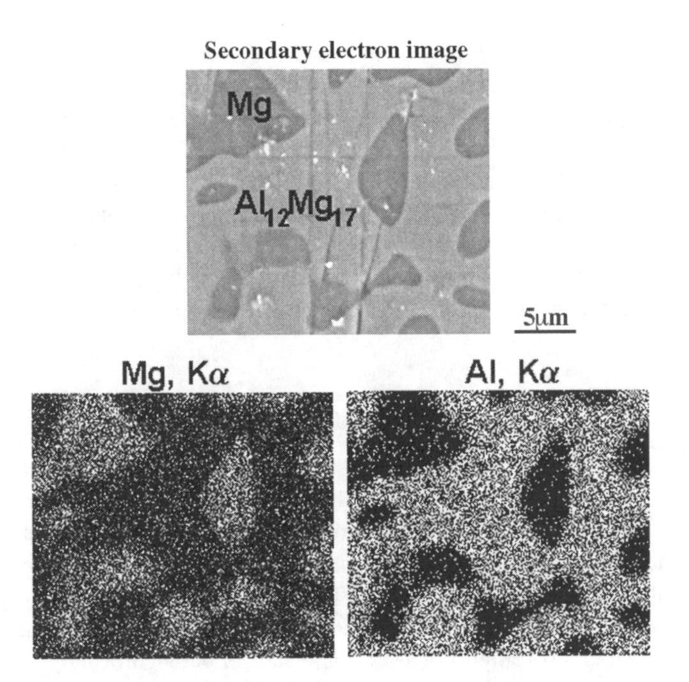

Figure 4.35 X-ray dot images for aluminium and magnesium reveal the elemental distribution within a selected area of a two-phase Mg–$Al_{12}Mg_{17}$ alloy. The dot image can be correlated with the contrast in the secondary electron image to identify the grains of the two phases

Elemental maps are in a different class altogether, and should always be collected together with the corresponding high-resolution *secondary* or *back-scattered electron images* (see below). Since a high-resolution secondary electron image may well contain up to 10^6 or more resolved pixel points, an X-ray *elemental map* of the *same* resolution would require a counting time of the order of 10^6 times *longer* than that required for a single point count. Clearly this is quite impractical, not only in terms of the limited lifetime of the operator but also due to the limited *stability* of the probe, *contamination* of the sample and the need to maximize information collection in the scanning microscope.

What do we mean by the expression 'practical'? If we assume a counting time of 100 s, as for the point analysis spectrum, and we require a statistical significance corresponding to a 95 % confidence level, then the *minimum* number of counts for a resolved *X-ray pixel point* is *ca* 1000. At a counting rate of 10^4 s^{-1} (reasonable for a constituent present at a concentration level of a few percent) this gives a total count of 10^6, or *ca* 10^4 resolved X-ray image points, corresponding to an X-ray elemental map with a statistically limited resolution of *an order of magnitude worse* than that of the corresponding secondary electron image (with of the order of 10^6 resolved pixels). Apart from the *counting statistics*, the *dimensions* of the emitting region, dictated primarily by the electron diffusion distance in the sample, will also limit the spatial resolution. The size of this zone (the diameter of the envelope of energetic electrons) depends on the *beam voltage* and the *atomic number* (density) of the sample, but is typically between 0.5 and 2 μm for a standard scanning instrument. By inserting the typical value of 1 μm, and using the previously estimated maximum number of X-ray image points, 10^4, we conclude that useful elemental maps can be obtained from regions of the order of $100 \times 100\ \mu m^2$. Attempts to obtain elemental maps from a *larger* region (magnifications less than 100) will only be successful if the counting times are significantly increased, which is usually impractical, while imaging *smaller* areas (higher magnifications) will simply result in fewer effective image points, since the resolution is then dictated by the dimensions of the excitation envelope.

4.3.3 Back-Scattered Electrons

A proportion of the incident electrons will be scattered by angles greater than π and may escape from the surface. The proportion of these *back-scattered* electrons, **R**, will depend on the *average atomic number* of the specimen but is almost independent of the incident beam energy. The back-scattered electrons originate in a *surface layer* corresponding to the diffusion distance and come from an area beneath the beam which is also proportional to this distance, but significantly less than the diameter of the envelope of inelastically scattered electrons.

The *average energy* of the back-scattered electrons is of course *less* than that of the primary incident beam, but nevertheless of the same order of magnitude. These electrons are usually detected in an annular region close to the probe lens pole-pieces. The collection efficiency is high, but the back-scattered electron current is only a fraction of the incident beam current.

4.3.3.1 Image Contrast in the Back-Scattered Image

Contrast in a *back-scattered electron image* may arise from either of the following two sources;

(a) Any region of the specimen surface which is tilted *towards* a back-scattered electron detector will give rise to an *enhanced* signal, while the signal will be *reduced* if the surface is tilted *away* from the detector. A segmented annular detector can therefore be used to obtain a *topographic image* of the surface in which the signals collected from *diametrically opposite detectors* are first subtracted and then amplified, thus enhancing differences in contrast from regions tilted in opposite directions.
(b) On the other hand, collecting a backscattered image from a *conical detector* surrounding the probe lens pole-pieces (or equivalently, *summing* the signals detected from all of the segments) will effectively *decrease* contrast associated with changes in surface topography. Most features detected in the image are then due to *atomic-number* contrast, and reflect variations in the density (usually the composition) of the sample.

The resolution in the *back-scattered electron* (BSE) *image* is typically an order of magnitude better than can be obtained from an *X-ray elemental map*, but not nearly as good as that available in the *secondary electron image* (discussed below). The direct relationship between the BSE image and the *diffusion distance* in the material typically results in a resolution of the order of 10–20 nm when working at 10–20 kV. The BSE image can give very useful information on the *distribution* of the phases present, providing that they differ sufficiently in density (see Fig. 4.36), if the secondary electron image is lacking in contrast.

4.3.4 Secondary Electron Emission

Most of the electron current generated in a sample due to the impact of a high-energy incident beam is due to the release of *secondary electrons* from the surface. In fact, the *secondary electron emission coefficient*, the number of secondaries released per incident high energy electron, is *always* greater than one and may reach values of several hundred. All of these secondaries have rather similar energies, up to 100 or 200 eV, but typically in the range 10–50 eV, and they are therefore readily deflected by a low-bias voltage and collected with very high efficiency (close to 100 %). Moreover, their low kinetic energy severely restricts their mean free path in the sample, so that the secondaries escaping from the surface are generated very close to the latter, typically within 1–2 nm, and are almost unaffected by beam spreading beneath the surface.

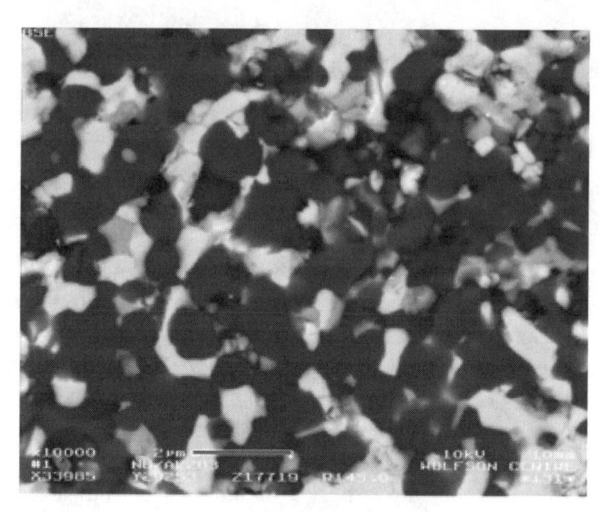

Figure 4.36 Back-scattered electron–atomic number contrast image showing a niobium-rich intermetallic phase (bright contrast) dispersed in an alumina matrix (dark contrast)

4.3.4.1 Factors Affecting Secondary Emission

Four factors directly affect *secondary emission* from the surface:

(a) The *work function* of the surface, that is the energy which has to be supplied to an electron in the solid which is at the *Fermi level*, in order to permit it to escape from the surface. Typical work functions are a few eV in magnitude, with the work function depending on both the *composition* and the *atomic packing* (crystal structure) at the surface. The work function is sensitive to both *surface adsorption* and films of *contamination*.
(b) The *beam energy* and *intensity*. As the beam energy is increased, more secondaries might be expected to be created, but a high-energy beam is inelastically scattered further beneath the surface, so that the *proportion* of secondaries escaping from the surface is reduced. On the other hand, the secondary electron current is *directly proportional* to the current in the incident beam.
(c) The *density of the sample* has a relatively limited influence, and is usually masked by the effect of surface films or surface contamination. Since higher-atomic-number materials have a smaller diffusion distance, the number of high-energy electrons is higher in the surface region for any given beam intensity, thus *increasing* the number of secondaries. This effect is most pronounced at *low*-beam energies, when the diffusion distance is comparable to the mean free path of the secondary electrons.
(d) The most pronounced effect is that of *surface topography*, or more precisely, the local *curvature* of the surface. Any region protruding from the surface (*positive* radius of curvature) improves the chances of secondaries escaping, while any

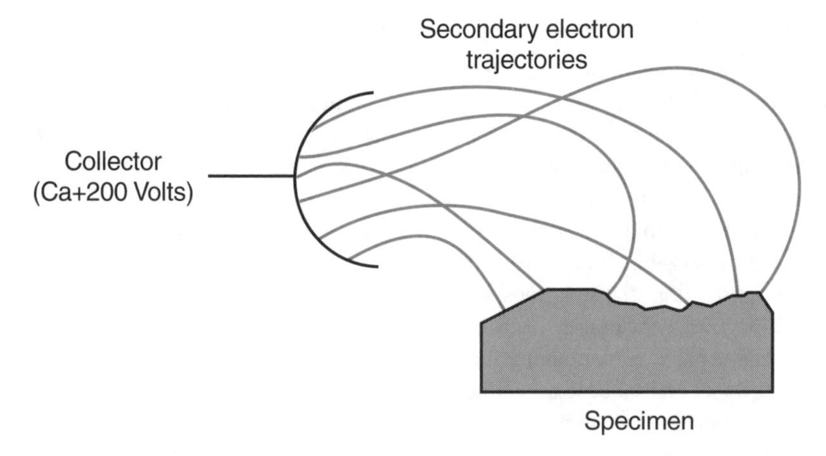

Figure 4.37 Secondary electrons can be collected from regions of the surface which are not in the 'line of sight' of the collector

recessed region (*negative* radius of curvature) will generate a reduced secondary electron current (associated with local trapping of the secondaries). Since the secondaries are collected by a bias voltage applied to the collector, even though some regions of the sample are out of the 'line of sight' (Fig. 4.37), we conclude that the *secondary electron image* should excel for studies of rough surfaces at high resolution and in good contrast (Fig. 4.38).

Figure 4.38 A high-resolution secondary electron image of nano-sized TiCN particles

4.3.4.2 Secondary Electron Image Contrast

Some complications in interpreting *secondary electron image contrast* arise from the two characteristic features of signal generation, namely the *production* of secondaries by the primary electrons and their *escape* from the surface. As long as the sample is reasonably planar, the *escape* of secondaries is associated with variations in work function, while the *generation* of secondaries is associated with the number of primaries in the surface region. There is no way of distinguishing these two effects, but it is an advantage to reduce the *beam voltage* to enhance the contrast, providing that this does not sacrifice resolution.

High-resolution scanning instruments which are equipped with a *field-emission source* are capable of nanometre resolutions at beam energies as low as 200 V, and give excellent contrast based on either *atomic number* or *work function*. At these low beam voltages, there is often no need to conductively coat a non-conducting sample, so that contrast features are only associated with the sample material, not with any conducting layer deposited to avoid electrostatic charging.

Topographic contrast may also present some problems. Protruding parts of the specimen may trap secondary electrons and screen recessed regions from the detector, despite the high collector efficiency. More commonly, *enhanced* emission from a region of high *positive* curvature may extend some distance *beyond* the high-curvature region (typically by a distance corresponding to the electron range, since inelastically scattered high-energy electrons will generate free secondaries when they approach the surface). Similarly, a *darkened* area associated with a region of *negative* curvature may also extend for some distance. It follows that there is a 'shadowing' of *hollows* and a 'lighting-up' of *protrusions* which can be misleading, particularly when operating at high magnifications close to the resolution limit.

4.3.5 Alternative Imaging Modes

In addition to the *three imaging signals* which we have discussed so far (X-rays, back-scattered electrons and secondary electrons), there are some other modes of operation which have quite specific and useful applications. We shall briefly discuss just two of these, namely *cathodoluminescence* and *electron beam image current*.

4.3.5.1 Cathodoluminescence

Many optically active materials will emit electromagnetic radiation in the visible range when suitably excited. Under an energetic electron beam, optically active regions of the sample will glow. The sample may be observed with an *optical microscope*, in which case the resolution is limited to that which is characteristic of the optical microscope, or the light emitted may be picked up by a suitable *detector*, and then amplified and displayed by using the same time base as the beam-scanning coils. In this case, the resolution may be appreciably better than that associated with the *wavelength* of the emitted radiation.

4.3.5.2 Electron Beam Image Current

If the specimen is electrically isolated from the surroundings and the current flowing through the specimen is monitored, then it is possible to display this *specimen-current* signal synchronously with the scanning of the beam over the sample.

Several variants of this mode of operation have been used to study *semiconductors* and *semiconductor devices*. For example, the conductivity of a semiconductor sample is often a sensitive function of its defect structure and dopant concentration, and the conductivity can be monitored as the beam is scanned across the surface (with or without an appropriate bias voltage applied to the sample). This mode of operation of the scanning electron microscope is referred to as *electron beam induced current* (EBIC). Alternatively, a complete device assembly may be inserted in the specimen chamber and operated *in situ* while under observation. The electric field developed in the various regions of the assembly may then modify both the secondary electron signal and the specimen currents flowing in the various components of the device. These effects can also be imaged at high resolution, but once again, this topic is beyond the scope of the present text.

4.3.6 Specimen Preparation and Topology

The most important specimen requirement in scanning electron microscopy is that *electrostatic charging* of the surface should be avoided, since the associated charge instability will lead to unstable secondary emission and destroy both the resolution and the image stability. Charging of non-conducting samples can be prevented or inhibited by operating at *low beam voltages* or by coating the sample with a thin layer of an *electrically conducting film*. The specimen must fit into the sample chamber without seriously constraining the geometrical freedom to select any area of interest for examination or to tilt the specimen to bring any required region under the beam at the appropriate angle. As noted previously, most specimen chambers readily accept quite large specimens (up to 10 cm).

Apart from these requirements, specimens must be stable in the vacuum system and under the electron beam. They should be free of any organic residues, such as oil and grease, that might lead to the build-up of contamination either on the specimen, or in the electro-optical system, the spectrometer, and the sample chamber. Loose particles should be removed from the surface, usually by *ultrasonic cleaning* in a suitable solvent, followed by rinsing and drying in warm air. These precautions are particularly important for *low-voltage*, high-resolution scanning electron microscopy. At voltages below 1 kV, nearly all of the secondary electrons come from a region very close to the surface. Electron beam induced carbon contamination then becomes the sole source of secondaries!

4.3.6.1 Sputter-Coating and Contrast Enhancement

Coating of the samples to enhance contrast and improve electrical conductivity is usually performed in a sputtering unit, as discussed previously for non-conducting transmission electron microscope specimens. Two types of coating are commonly used, either a heavy metal or carbon. *Heavy-metal* coating (usually a gold–palladium alloy, which gives a 5 nm particle size and only interferes with the resolution at the highest magnifications) improves the contrast appreciably. However, it may well interfere with chemical microanalysis and is not suitable if the best resolution is to be achieved. *Carbon coatings* have a much finer particle size (*ca* 2 nm), which is generally below the resolution limit of the instrument, but they do nothing to enhance the contrast. Nevertheless, a carbon coating is mandatory for non-conducting samples intended for microanalysis.

The best solution to *electrostatic charging* of the specimen is always to reduce the beam voltage in so far as this is feasible.

4.3.6.2 Fractography and Failure Analysis

A major area of application of the scanning electron microscope is in *fractography* and *failure analysis*, not only for engineering metals and alloys, but also for *plastics* and *composites* (polymer, ceramic or metal matrices), as well as *semiconductor devices* and *engineering ceramics* (see Fig. 4.39). Many other classes of materials have also been studied by using the same techniques (see Fig. 4.40).

There are a few simple rules to be followed if the maximum information is to be extracted from *scanning electron fractography*. Although these have been covered in previous sections, it is worth repeating them here, as follows:

(a) The sample selected for insertion in the microscope must bear a known *spatial* and *orientational* relationship to the component from which it was taken and the system in which it was used. *Without* this information, it may be extremely difficult to evaluate the significance of the observations made.
(b) The surface should *not* be damaged or altered in any way by any prior specimen preparation procedure. It is difficult to believe how often the two halves of a failed component are 'fitted together' prior to examination, thus resulting in superficial damage and often destroying important evidence of the cause of failure.
(c) The specimen should be inserted in the microscope in a simple relationship to the available specimen stage directions of *traverse* (the x–y coordinates) and axes of *tilt* (parallel and perpendicular to a direction of crack propagation, for example). A little forethought can save a *lot* of microscope time and simplify the process of investigation.
(d) Images should be recorded over the *full* range of magnifications which show any significant features. It is particularly important to be able to relate the *topography* of the observed features to the results of any other observations. If these are purely *visual*, then an *initial* magnification in the scanning microscope of ×20 is not too little. A good procedure to adopt is to locate the features of interest by first scanning rapidly over the sample, experimenting with the magnification, and only then to record a *series* of magnified images, identifying

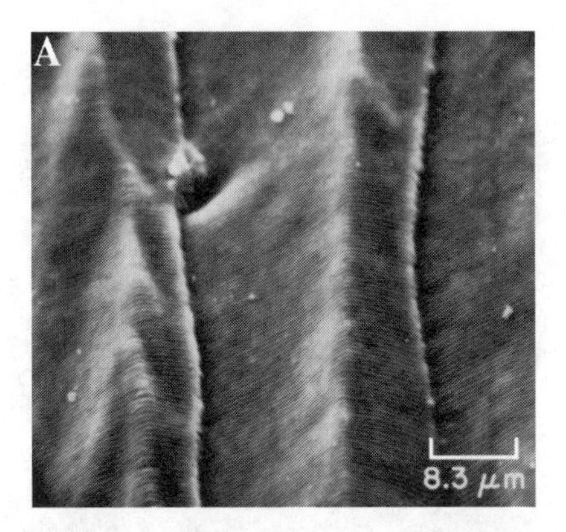

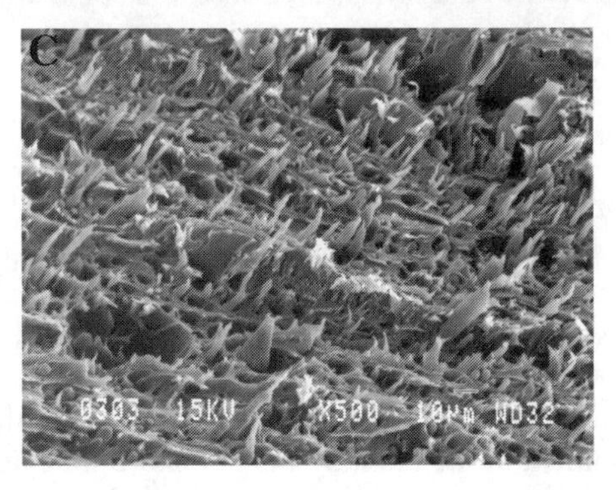

Figure 4.39 Some examples of failures in engineering materials as imaged by scanning electron microscopy: (a) mechanical fatigue failure in steel (*Metals Handbook, Eighth Edition*, Vol. 9, *Fractography and Atlas of Tractograph* (1987). ASM International, Materials Park, OH 44073-0002, p. 69 (Fig. 16)); (b) brittle failure in porous TiCN; (c) failure of a fibre-reinforced polymer matrix composite (reproduced by permission of A. Siegmann)

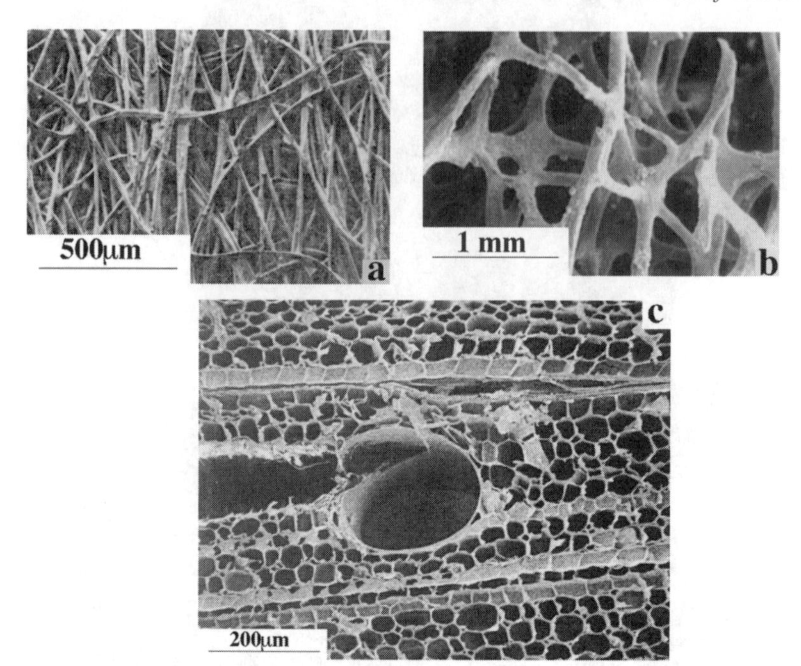

Figure 4.40 Some examples of various materials studied by scanning electron microscopy: (a) paper; (b) bone; (c) wood. From *Cellular Solids; Structure and Properties*, L. J. Gibson and M. F. Ashby, Pergamon Press

the highly magnified features by their *geometrical relationship* to those features which can be observed with the naked eye. A factor of $\times 3$ between images ensures that some 10 % of the surface area observed at the lower magnification is visible in the next highest magnification image (Fig. 4.41).

4.3.6.3 Stereoscopic Imaging

As noted at the beginning of Section 4.3, the scanning electron microscope is unique in its ability to focus detail over a range of *depth*, along the axis parallel to the incident beam. This information can be extracted by recording a pair of images of the same area. The two images, a *stereo-pair*, correspond to two observations of the specimen surface taken from different points of view. The situation is equivalent to the *stereoscopic*, three-dimensional visual image observed by the superposition of the retinal data transmitted to the brain by the left and right eyes (Fig. 4.42).

Several commercial systems have been developed for directly viewing *stereoscopic images* in the scanning electron microscope, for example by a lateral displacement of the axis of scan of the probe, about either the x- or y-axes, for two sequential scan sequences or for alternate lines of the x-sweep in a single scan. These systems have never been particularly popular, since it is quite easy to record a pair of

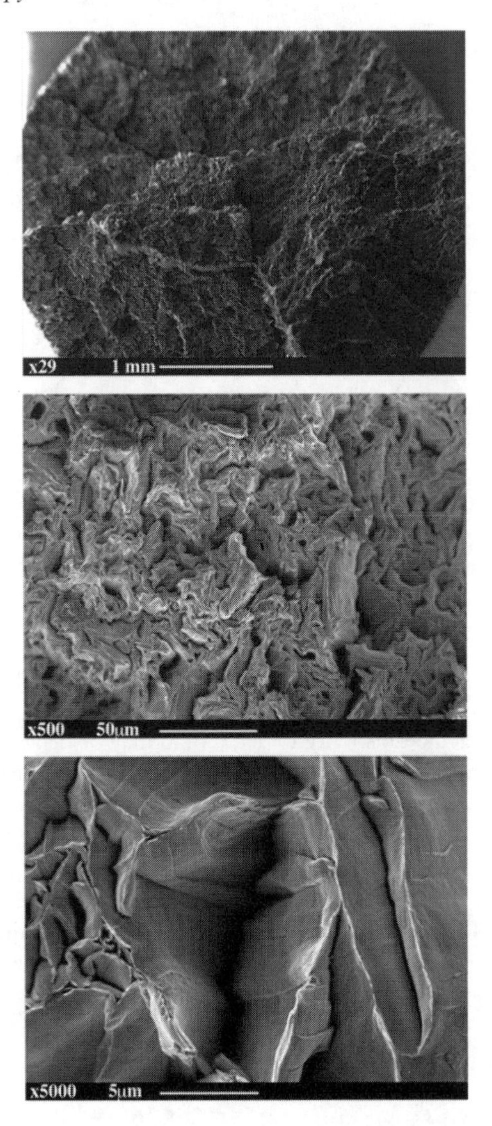

Figure 4.41 Micrographs of ductile failure in molybdenum, obtained at increasing magnification in order to reveal both the general features and the fine details of the fracture surface

images without disturbing the scanning-coil settings, but rather by tilting the specimen itself.

Given that the microscope screen is commonly viewed at a distance of some 30 cm (a comfortable reading distance) and that the eyes are set some 5 or 6 cm apart, the required *angle of tilt* needed to give an impression of depth identical to

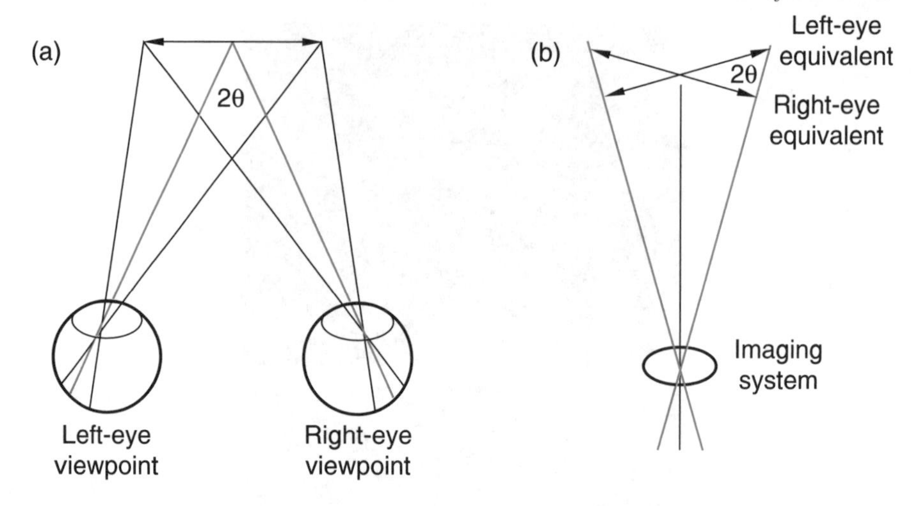

Figure 4.42 The twin images observed by the left and right eyes (a) are equivalent to the image pair recorded before and after tilting a specimen about a known angle (b)

that received by the eye is $\pm 12°$. In effect, this is equivalent to setting the *depth* 'magnification' equal to the *lateral* magnification in the image. Tilt angles *less* than $\pm 12°$ will then *reduce* the sensitivity to depth (and may be necessary for very rough surfaces), while tilt angles *greater* than $\pm 12°$ will *amplify* the impression of depth (and may be useful for recording shallow features). An example of a *stereo-pair* is given in Fig. 4.43.

Stereo-images, with the axis of tilt vertical and placed with corresponding points at the separation of the eyes, may, with some practice, be viewed at the normal reading distance by focusing the eyes on infinity, without any additional optical aids. More commonly, commercial *stereo-viewing* systems are available and can provide striking in-depth information which is unobservable in the two-dimensional image.

4.3.6.4 Parallax Measurements

In addition to the *visual* impact of a stereo-image, it is also possible to obtain *quantitative information* on the vertical displacement of features along the axis parallel to the incident beam. This is accomplished by measuring the *horizontal displacement* of features projected perpendicular to the tilt axis in the two stereo-images (Fig. 4.44). This displacement of the two features is termed *parallax*, and can be represented by the following expression:

$$x_L - x_R = 2h \cos\theta \qquad (4.21)$$

Figure 4.43 A stereo-pair showing a brittle fracture. Hold the pair at a comfortable reading distance and then focus the eyes at a distant object above the page. It should then be possible to fuse the two images into a single three-dimensional view of the surface topography, although most observers fail at the first attempt and some are unable to view a stereo-pair without a suitable viewing system

where x_L and x_R are the projected distances in the two images from any given position perpendicular to the axis of tilt (y), h is the 'height' difference between the two features measured parallel to the incident beam, and the images are obtained at a tilt angle of $\pm\theta$ with respect to the beam normal.

Parallax measurements are particularly useful for checking the *thickness* of deposited films or measuring the *height* of slip steps or other features. They have also been used to estimate the *roughness* of ground and machined surfaces, and the *fractal dimensions* of fracture surfaces.

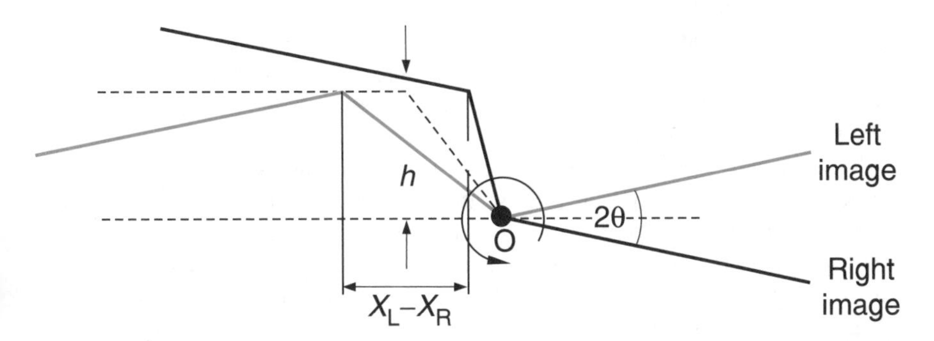

Figure 4.44 Schematic representation of parallax geometry: the difference in separation of two features seen in a stereo-pair and projected perpendicular to the tilt axis is directly related to the difference in height of the two features, measured along an axis parallel to the incident beam

Summary

In the *transmission* electron microscope, high-energy electrons are *elastically* scattered as they penetrate a thin specimen. The transmitted electrons are then focused by *electromagnetic* lenses to form a well-resolved image which can be viewed on a fluorescent screen or recorded on a photographic emulsion.

In the *scanning* electron microscope, an electron beam is focused into a fine probe which is *inelastically* scattered when it strikes the surface of a solid sample. The inelastically scattered electrons generate a signal from the sample which can be collected and amplified. An image is formed by *scanning* the beam probe across the surface of the sample in a television raster and displaying the collected signal on a monitor which has the same time-base as the probe scan. The commonest signal is obtained from *secondary electrons*, but *characteristic X-rays*, high-energy *back-scattered electrons*, visible *cathodoluminescence* and the net *specimen current* have all been used to obtain microstructural information in the scanning electron microscope.

It is important to distinguish clearly between the *optical* image formed by focusing elastically scattered electrons from a given *area* of the specimen, and a *scanning* image, generated by recording information *point-by-point* as the electron probe is scanned over the surface.

The *wavelength* of the energetic electrons used in electron microscopy is well below the interatomic spacing in solids, so that *atomic resolution* is possible, although not easy to achieve. The electromagnetic lenses used to focus the beam and the image in electron microscopy operate over an appreciable proportion of the optical path in the microscope and suffer from very severe lens defects which limit the divergence angle for a sharp focus to a fraction of a degree. The most important lens defects are *spherical* and *chromatic aberration*, and *astigmatism*. The diffraction limit on the resolution improves with increasing objective-aperture angle, while the aberration limit on the resolution deteriorates. The optimum objective aperture is typically between 10^{-2} and 10^{-3} rad. The electron microscope requires a vacuum in order to prevent scattering of the beam in the microscope column, a high voltage stability for the electron source (in order to ensure a monochromatic beam), and an equally good current stability for the electromagnetic lenses (in order to maintain their focusing characteristics).

Since transmission electron microscope specimens must be less than 100 nm in thickness in order to avoid inelastic scattering of the transmitted beam, specimen preparation is critical. Combinations of mechanical, chemical and electrochemical methods are used, and the final stage is often by *ion milling*, in which the atomic layers are sputtered away from the surface by an incident ion beam of inert gas. In order to avoid surface charging of insulator materials, it may be necessary to deposit a conductive coating. The morphology of some specimens may be most conveniently studied by taking a *replica* of the sample.

Contrast in the transmission electron image may be associated with the *mass–thickness* of the constituents, *diffraction* of the incident beam out of the objective aperture, or *phase contrast* associated with interference of the diffracted beams, with or without the directly transmitted beam, in the image plane. In biological samples, which are usually amorphous, mass–thickness contrast is the usual source of image information, and a variety of heavy-metal staining agents have been developed to enhance this source of contrast. Crystal lattice defects are most commonly imaged in diffraction contrast, since lattice defects sensitively affect the amplitude diffracted out of the incident beam. Periodic lattices, on the other hand, can be resolved by phase contrast, providing that sufficient resolution is available in the microscope. Changes in lattice periodicity at interfaces can also be detected and interpreted by using suitable modelling procedures.

A simple qualitative treatment of diffraction contrast is possible by using *kinematic diffraction theory*, in which dynamic effects are ignored. The diffracted amplitude in the kinematic theory depends on the sum of two terms. The first describes the effect of deviations from the Bragg condition on the amplitude scattered by a unit cell, while the second describes the effect of displacements from the ideal lattice position due to the presence of the defect. These effects are most conveniently summarized by an *amplitude–phase* diagram which can be used to explain qualitatively the contrast from stacking faults and anti-phase boundaries, edge and screw dislocations, and the strain fields associated with small precipitates and damage clusters.

Dynamic diffraction theory and the inclusion of affects associated with absorption allow diffraction contrast to be interpreted *quantitatively*, as well as qualitatively, but requires considerable expertise in both the operation of the microscope and the interpretation of the image data.

The *lattice* or *phase-contrast* image is an interference image in which the coherently diffracted beams are recombined in the image plane. Since these beams have followed different paths through the electromagnetic lens fields, they experience phase shifts which depend on the electro-optical properties of the microscope, and which are described by the *contrast transfer function*. The *periodicity* in a lattice image is essentially independent of the specimen thickness and the focal plane, but gross variations in the *amplitude* occur, so that the observed image cannot be directly interpreted as a projection of the periodic lattice. Nevertheless a slightly underfocused image (the *Scherzer* focus) can be shown to compensate for phase shifts associated with the lens system, and images taken from very thin specimens at the Scherzer focus do correspond roughly to the contrast to be expected from a direct projection of the crystal lattice. Much better agreement can be achieved by simulating the contrast to be expected from a model of the periodicity, and *image simulation* has been used to construct models of localized lattice structure to an accuracy better than a fraction of the interatomic spacing.

In the *scanning electron microscope*, the beam is focused by a probe lens to a diameter of as little as 2 nm. However, the probe current decreases rapidly as the probe diameter is reduced, and some signals require quite large probe diameters (up to 1 μm). This is particularly the case for the characteristic X-rays which are excited by the incident beam, since the excitation cross-section is small and the resolution in an X-ray image is usually limited by the counting statistics as much as by the size of the excitation volume beneath the probe.

The characteristic X-ray signal is an important source of information for the microchemistry of the sample (see Chapter 5), but can also be used simply to demonstrate the presence of one or other of the constituents in the sample. The X-ray signal is typically generated from a volume beneath the probe of the order of 1 μm in diameter, corresponding to the depth of penetration of the energetic electrons into the sample and their lateral dispersion associated with inelastic scattering. The X-ray signal can be displayed in three distinct formats, a *spectrum* in which the intensity of the signal from a given region is displayed as a function of the X-ray energy or wavelength, a *line-scan* in which the intensity of the characteristic signal from a specific element is displayed as a function of position (for example a scan across an interface or second-phase particle), or an *elemental image map* in which all photons arriving within a given energy window are displayed as white dots on the screen, with their position on the screen correlating with the position of the beam at the time of detection.

High-energy, *back-scattered electrons* are useful because the intensity of the back-scattered signal reflects the density of the sample and not just the surface topology. The brighter regions are therefore a clear indication of denser material of higher atomic number. Nevertheless, the *secondary electron* signal is undoubtedly the most useful, for two reasons. First, the number of secondary electrons which are emitted greatly exceeds the current in the primary beam and they can be collected with close to 100 % efficiency, and secondly, the signal originates in the surface layers of the specimen and the resolution is not degraded by inelastic scattering of the primary beam. It follows that there are few statistical limitations on the resolution of the secondary electron image, with this resolution only being limited by the ability to focus the probe beam.

Other types of signal are also available for the scanning electron microscope, for example the beam current to the sample or cathodoluminescence, but these are of secondary importance.

Specimen preparation for the scanning electron microscope is very straightforward, although it must be remembered that some specimens (especially polymers and biological tissue samples) may be degraded by the beam and give rise to contamination. A conductive coating is commonly required to prevent charging and can often usefully enhance the contrast. In general, contrast is associated with both surface topology and variations in density (atomic number), and it is not always easy to separate the two.

Insufficient use is made of *stereoscopic* analysis. The tilting of a sample by a fixed angle about a known axis and the subsequent recording of the two images allows a stereo-image to be observed in which the depth distribution of the microstructural features is clearly evident. In addition, accurate measurements of lateral displacements (*parallax*) permit the displacements along the optic axis to be determined, so that the scanning electron microscope can be used to form a *three-dimensional* image of the sample.

Bibliography

1. D.B. Williams and C.B. Carter, *Transmission Electron Microscopy: A Textbook for Materials Science*, Plenum, New York (1996).
2. P. Buseck, J. Cowley and L. Eyring (eds), *High-Resolution Transmission Electron Microscopy and Associated Techniques*, Oxford University Press, Oxford (1998).
3. J.C.H. Spence, *Experimental High-Resolution Transmission Electron Microscopy*, Clarendon Press, Oxford (1981).
4. J.I. Goldstein and H. Yakowitz (eds), *Practical Scanning Electron Microscopy*, Plenum, New York (1975).
5. O.C. Wells, *Scanning Electron Microscopy*, McGraw-Hill, New York (1974).
6. J.J. Hren, J.I. Goldstein and D.C. Joy (eds), *Introduction to Analytical Electron Microscopy*, Plenum, New York (1979)

Worked Examples

Once again, we will demonstrate the techniques that we have just discussed, this time using transmission electron microscopy (TEM) and scanning electron microscopy (SEM) for two quite different material systems, namely polycrystalline alumina and a thin film of aluminium deposited by chemical vapour deposition (CVD) on a $TiN/Ti/SiO_2$-coated silicon substrate.

The first example is polycrystalline alumina. As always, it is extremely important to define the questions being asked before preparing specimens or specifying the characterization techniques. For our alumina sample, we wish to check for residual porosity, determine the grain size, and establish that the grain boundaries are free from secondary phases.

The presence of residual porosity and grain size measurements can best be determined by SEM. However, it is unlikely that small secondary phase particles or glass at grain boundaries would be detected by SEM, so we will therefore use TEM for this investigation.

For SEM, we can prepare specimens in two ways, either by mechanical polishing, down to a sub-micron diamond grit polish, followed by thermal etching to form grain boundary grooves, or by breaking a specimen, so that we can examine the fracture surface. Alumina is an electrical insulator, so we would usually coat the surface of the specimen with a conducting layer prior to the SEM work. However, we have access to a low-voltage SEM with a field-emission gun, so we will use a low accelerating voltage to minimize charging.

Fig. 4.45 shows SEM micrographs of a polished and thermally etched alumina sample; two micrographs of the same region are shown. The first was recorded by using secondary electrons at an accelerating voltage of 5 kV, while the second was recorded at 20 kV, also using secondary electrons. The higher accelerating voltage leads to unstable surface charging of the specimen which affects both the secondary electron emission coefficient and the trajectory of the secondary electrons. The image is blurred, and unsatisfactory for morphological analysis. When using 5 kV

Figure 4.45 Scanning electron micrographs of thermally etched polycrystalline alumina, recorded by using (a) 5 kV electrons, to minimize charging, and (b) 20 kV electrons. No conductive coating was applied to the specimen

electrons, the incident electron current is compensated by the exiting secondary and back-scattered electrons, so charging is reduced, and the image is sharper.

Secondary electrons are generated throughout a thin surface region of interaction of the incident beam with the sample, but lowering the accelerating voltage can ensure that the secondary electrons are only generated from regions very close to the surface. These have been termed 'type-1' secondary electrons, or SE1. With a field-emission gun and a specialized secondary electron detector (discussed below), it is possible to detect the very few SE1 electrons generated at low accelerating voltages (Fig. 4.46). The resolution in this figure is sufficient to detect fine surface facets formed during thermal etching, because the signal contains fewer electrons generated from *below* the surface.

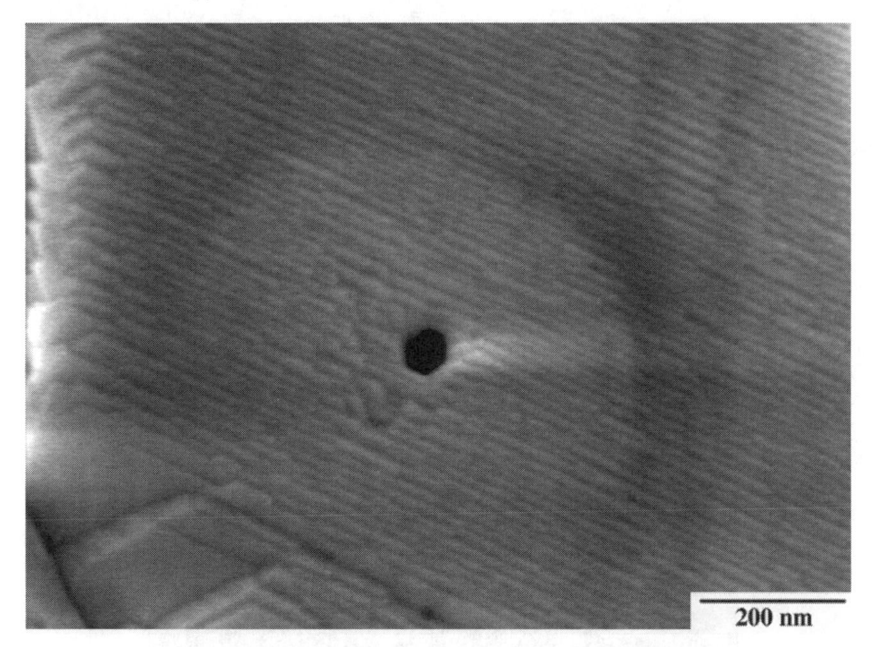

Figure 4.46 Micrograph showing the surface facets of alumina recorded by using a field-emission gun and a specialized secondary electron detector; most of the contrast comes from SE1 electrons

Secondary electrons are generated from the surface (SE1 electrons), but they can also escape from the sub-surface if the voltage is more than a few hundred volts ('type-2', SE2 electrons). In addition, a secondary electron current is generated by back-scattered electrons and stray electrons striking the microscope chamber, column or specimen holder ('type-3', SE3 electrons). The secondary electron detector is usually located within the chamber, above and to the side of the specimen, so the SE3 electrons contribute to signal noise. However, an annular secondary electron detector can be placed above the specimen and in the microscope column to screen out SE3 electrons, and improve both the contrast and the resolution of the image. However, since the detector is *above* the specimen, rather than to the side, we will lose some of the three-dimensional shadow effect associated with the surface topology and which are strikingly visible when using the conventional secondary electron detector (Fig. 4.47). Some scanning microscopes allow on-line mixing of signals from both types of detectors in order to optimize resolution and contrast.

Returning now to our alumina specimen: Fig. 4.48 shows a 5 kV SE1 micrograph of the thermally etched specimen. Grain boundaries are easily identified, since the grain boundary grooves give off fewer secondary electrons than the groove shoulders. By using the methods discussed in Chapter 1, it is easy to determine the average grain size to any required accuracy (see Chapter 7).

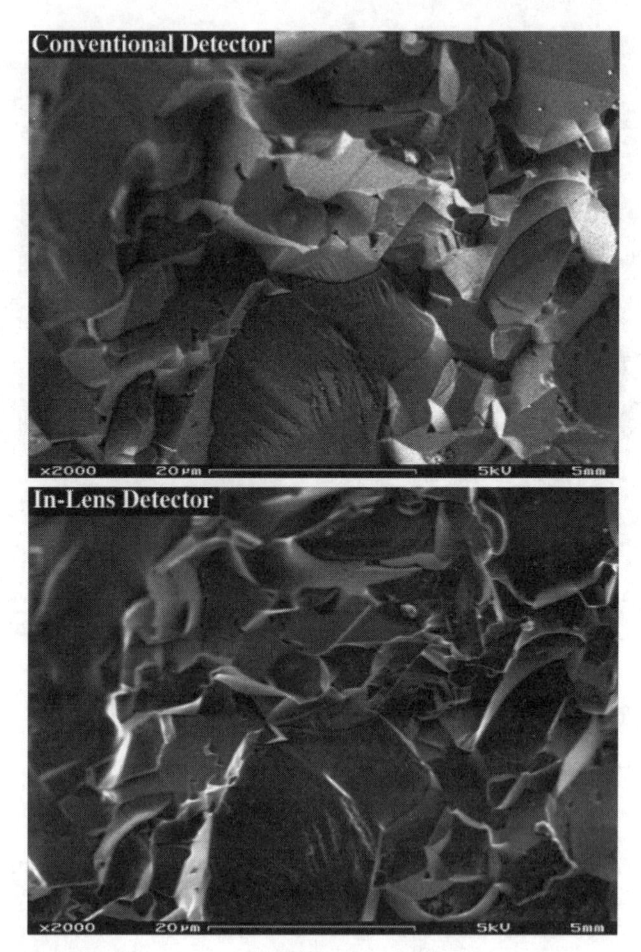

Figure 4.47 Micrographs of the fracture surface of alumina recorded by using (a) a conventional secondary electron detector and (b) an in-lens secondary electron detector

Fig. 4.49 shows a fracture surface of the same alumina. Two fracture modes are visible, namely *intergranular* fracture and *transgranular* fracture. Transgranular fracture is characterized by the appearance of planar, crystallographic *cleavage* planes, which give sharp contrast. Residual porosity is readily visible, both at the grain boundaries, and within the grains.

In order to study the grain boundaries we use TEM. For a bulk polycrystalline ceramic sample, the easiest way to prepare a specimen is by first cutting a thin (600 µm) slice from the bulk specimen with a diamond saw. The TEM specimen is limited in diameter to 3.0 mm, so a 3.0 mm disc must be trepanned from the 600 µm slab with an annular ultrasonic tool. The disc is then mechanically thinned to *ca* 80 µm by using diamond polishing media. At this point the *centre* of the disc should

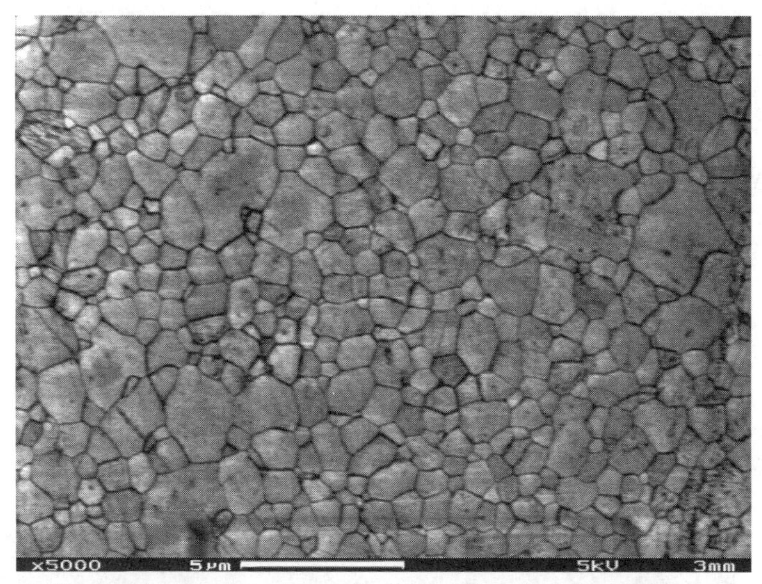

Figure 4.48 Scanning electron micrograph of thermally etched alumina

Figure 4.49 Scanning electron micrograph of the fracture surface of alumina, showing transgranular fracture (cleavage), intergranular fracture, and pores at grain boundaries and within the alumina grains

be thinned to perforation by ion milling. In order to ensure that the perforation occurs at the centre of the disc, it is first mechanically 'dimpled' to a thickness of *ca* 30 μm. Finally, the specimen is ion milled for *ca* 60 min at 5 kV using argon ions at an incident angle of 6° on both the top and bottom surfaces of the specimen (Fig. 4.50). For TEM, the non-conducting specimens must be coated on one side with a ~10 nm layer of carbon.

Fig. 4.51 shows a bright-field, 200 kV transmission electron micrograph of an alumina specimen observed in diffraction contrast, obtained by using a small

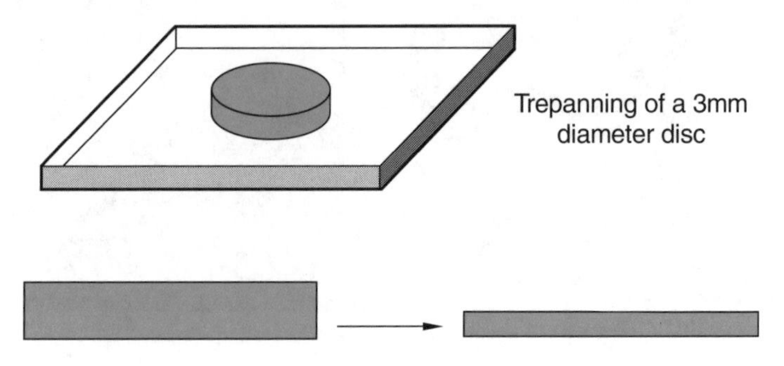

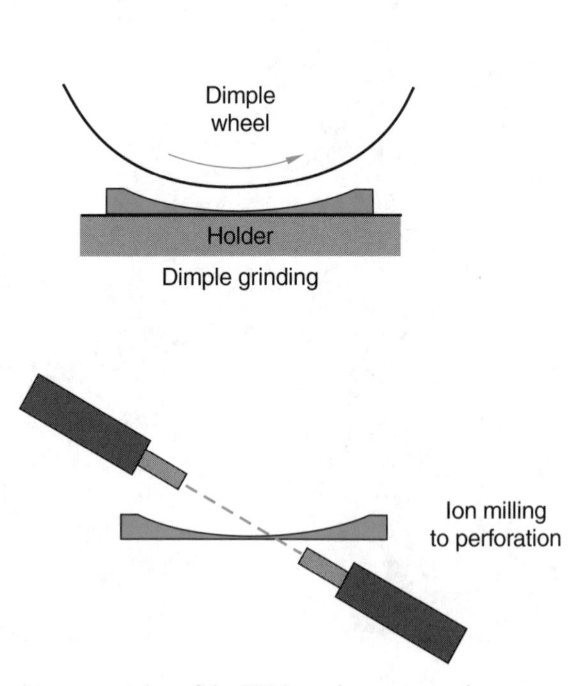

Figure 4.50 Schematic representation of the TEM specimen preparation process for a plan-view bulk specimen

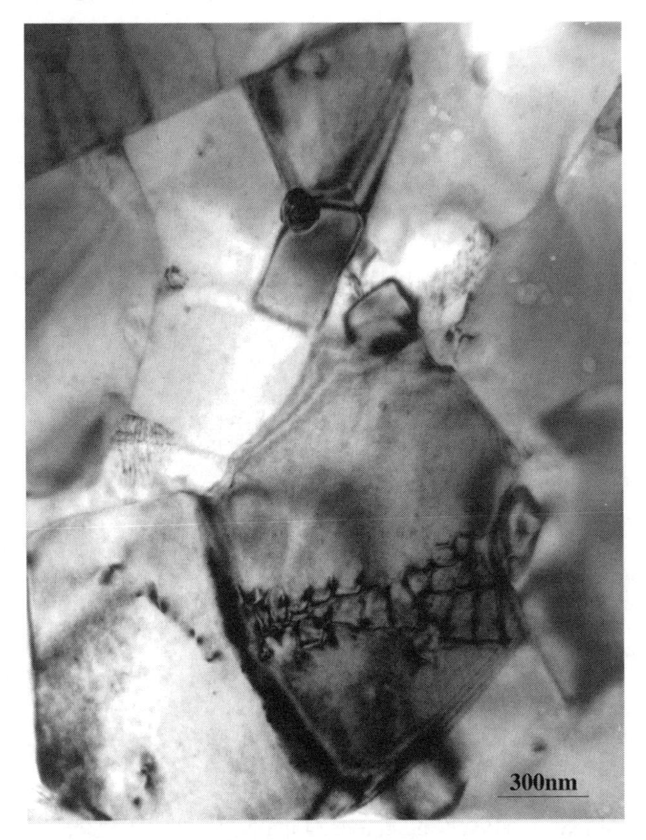

Figure 4.51 Bright-field diffraction contrast transmission electron micrograph of alumina reinforced with SiC particles

objective aperture to cut out the diffracted beams. The polycrystalline alumina contains sub-micron SiC particles, and the contrast variations are due to the different crystallographic orientations of each individual grain, as well as a dislocation network in one alumina grain. If we align a particular grain along a low-index-zone axis, by using *Kikuchi* diffraction, the *bright-field* diffraction contrast image will show this grain in dark contrast. An example is given in Fig. 4.52, which shows an $AlNb_2$ particle located within an alumina grain. We can record an image by using a selected diffracted beam from the aligned grain (a *dark-field* image), in which the diffracting grain will appear light when compared to the neighbouring grains.

The detection of secondary phases at grain boundaries can be achieved by several methods. In the first of these, high-magnification bright-field images are used. Secondary phases with a different chemical composition form the matrix will give *mass–thickness* contrast and appear lighter or darker than the neighbouring regions. Fig. 4.53 shows an example for NbO particles at grain boundaries in polycrystalline

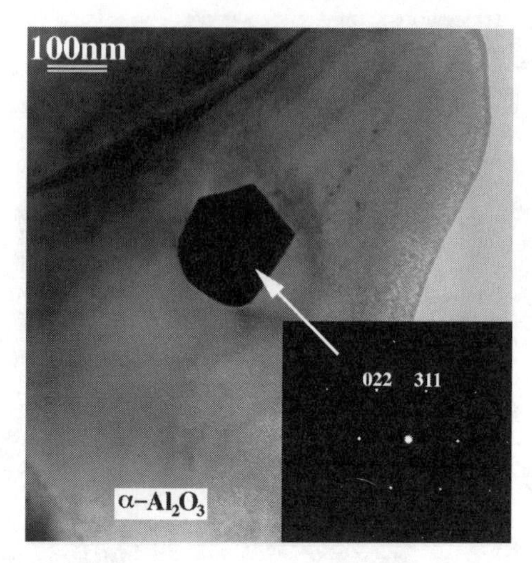

Figure 4.52 Bright-field diffraction contrast transmission electron micrograph of an $AlNb_2$ particle within an alumina grain. The $AlNb_2$ particle lies along a low-index-zone axis, and thus has a dark contrast in the bright-field image

alumina. An amorphous phase at grain boundaries or grain-boundary triple junctions can be highlighted in a dark-field image. This image is recorded by using diffuse scattering of electrons from the amorphous material. When the glass-containing grain boundary is *parallel* to the incident beam, the glassy region will appear lighter than the neighbouring crystalline grains. Finally, if the TEM resolution limit is sufficient to detect the lattice planes of alumina, we can use *phase contrast* to study the grain-boundary regions. To be able to record a *lattice image* of a grain, the latter should be oriented to diffract exactly along a low-index-zone axis. For recording a lattice image of a grain-boundary region or interface, *both* of the boundary-forming grains should lie in low-indexed zones, and the grain boundary *must* be parallel to the incident beam. This is a very rigid condition, and is exceedingly difficult to obtain in a polycrystalline sample unless a special orientation relationship exists between the grains. Fig. 4.54 shows a lattice image of a special grain boundary in alumina (a rhombohedral twin boundary), where *both* the boundary-forming grains are along a low-index-zone axis and the boundary plane $\bar{1}012$ is parallel to the incident beam. In some cases, information on the faceting and size of small secondary phases at grain boundaries can be obtained when the interface is parallel to the incident beam and at least *one* boundary-forming grain lies on a zone axis. Fig. 4.55 shows a lattice image of a SiC particle located within an alumina grain. A lattice image is visible for the alumina, while only *moiré fringes* are visible within the SiC particle, since it is not oriented along a low-index-zone axis.

Let us now turn our attention to the CVD Al system. For this example, we wish to characterize the initial deposition conditions, when the first aluminium nuclei form

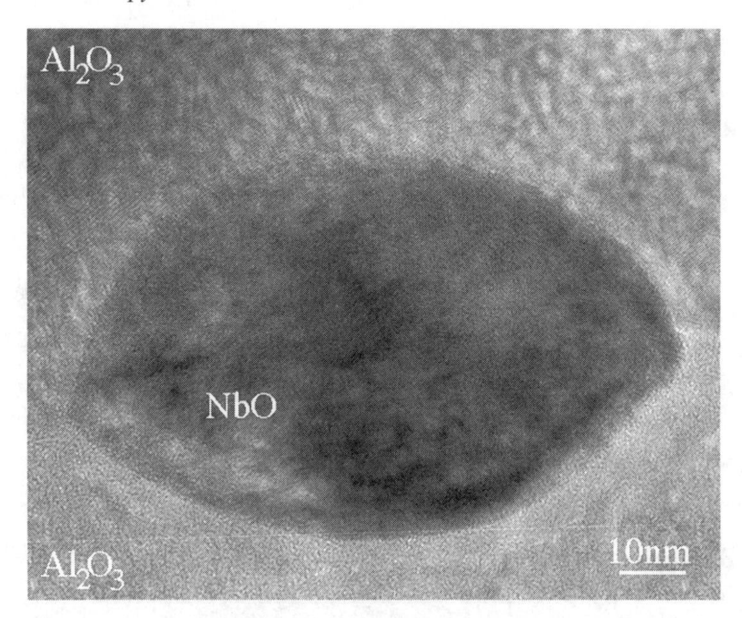

Figure 4.53 Transmission electron micrograph of a NbO particle located at a grain boundary in polycrystalline alumina. Phase contrast (lattice fringes) and mass–thickness contrast vary from the alumina grain to the NbO grain

Figure 4.54 Lattice image of a rhombohedral twin in alumina

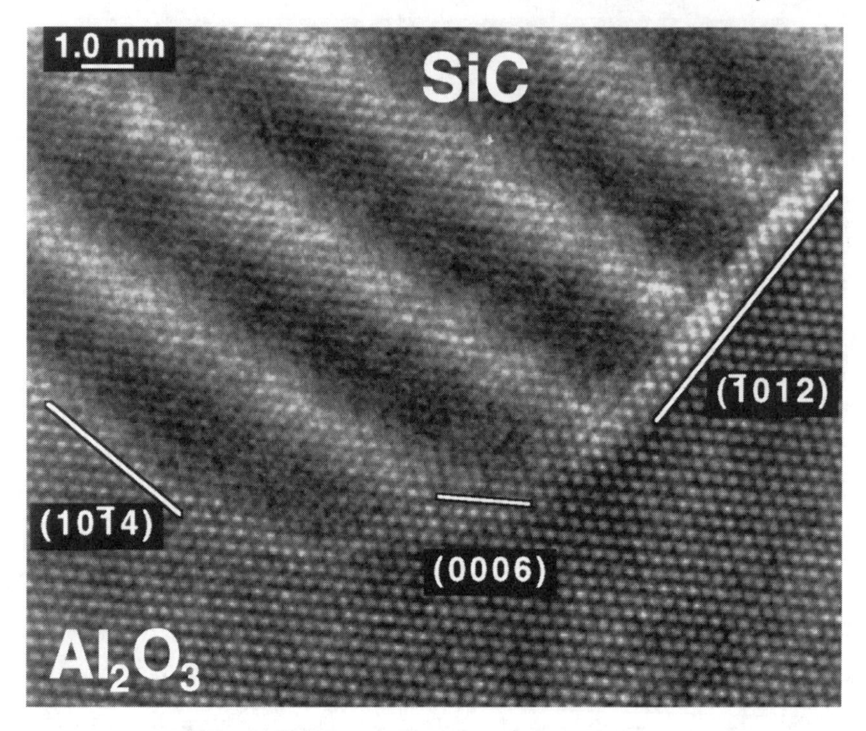

Figure 4.55 Lattice image of a SiC particle located within an alumina grain. The alumina lies along a low-index-zone axis, and is the source of the lattice image. A moiré pattern appears within the SiC particle due to overlap between the alumina and SiC (in the direction of the electron beam)

on the surface of the TiN, the morphology of the final aluminium film, and the morphologies of the underlying TiN and titanium films.

Preparation of these samples for SEM is straightforward, since the sample is a metallic conductor and we wish to investigate the surface in plan-view. However, the aluminium nuclei are very small, requiring the highest possible resolution *low-voltage* SEM, which generates secondary electrons only from the surface layer. Contamination of the surface is a critical factor limiting the image quality. Fig. 4.56 shows a secondary electron SEM image of the surface (using an in-lens secondary electron detector). The central region of this figure was exposed to the electron beam for about 30 s, and then the magnification was reduced and the micrograph was recorded. Thus the outer regions of Fig. 4.56 were recorded immediately after focusing on the area of interest, thus reducing the contamination to a minimum. The buildup of contamination is evident, and the reason for a clean specimen is clear. There are three options available for cleaning the specimen, (a) plasma etching prior to placing the specimen in the microscope, (b) swabbing with ethanol to remove

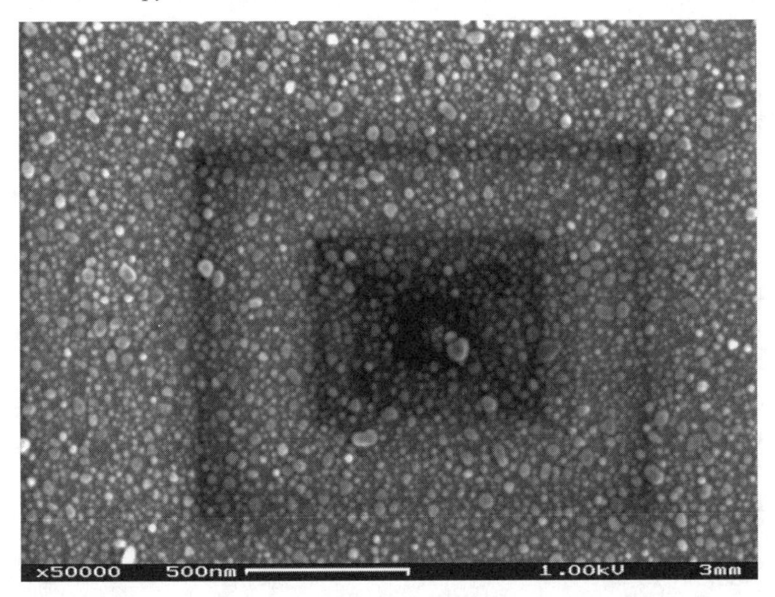

Figure 4.56 Scanning electron micrograph of Al on TiN. The beam was first focused on the central region of the micrograph for *ca* 30 s, and then the magnification was reduced and the micrograph recorded immediately. The buildup of contamination in the central region strongly affects the contrast of the image

grease and oil, and (c) heating the specimen above 100°C, to drive off absorbed water. Since heating of the specimen surface by the electron beam could cause changes in the aluminium morphology, we need to obtain the images as quickly as possible.

The final, thick aluminium film has a different morphology, and SEM shows that open voids have formed during the deposition, Fig. 4.57. These voids are process defects which change both the electrical and optical properties of the aluminium.

Scanning electron microscopy can also be used for cross-sectional specimens. Fig. 4.58 shows a cross-sectional secondary electron micrograph of a cleaved specimen. The aluminium film is clearly visible on the $TiN/Ti/SiO_2/Si$ stack. Open voids are just visible, as well as possible closed voids which could not have been detected from the plan-view specimens.

The thickness of the various films in this sample is at the limit of detection for SEM, so we need TEM to characterize the material more completely. TEM cross-sections are not easy to prepare, so we will use this opportunity to review the specimen preparation process. For TEM, we need very thin specimens, with as much as possible of the interface region being thin enough for this form of electron microscopy. There are a number of methods available to produce thin specimens, but for a multilayered material chemical thinning is *not* a good idea, since the different layers will react differently to a chemical or electrochemical etch. *Ion milling* is really the only option.

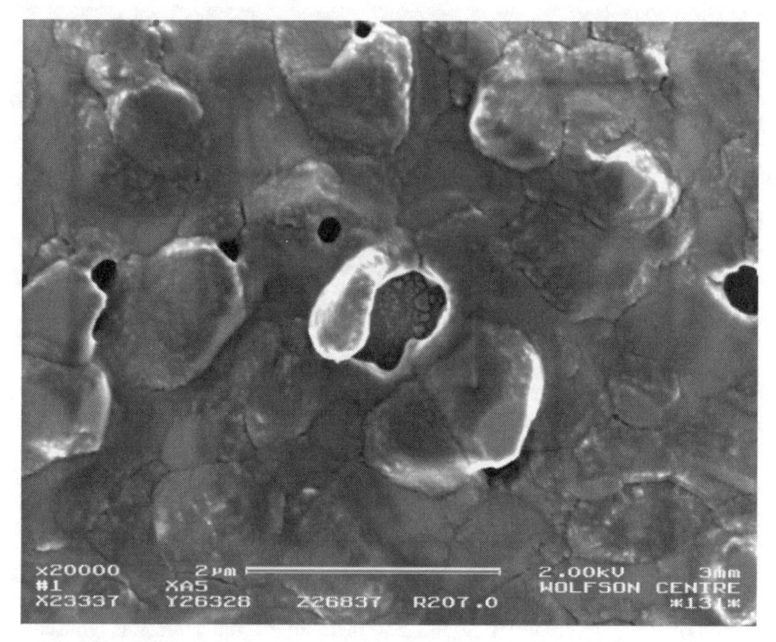

Figure 4.57 Scanning electron micrograph of a thick Al film, showing open voids and a relatively rough surface

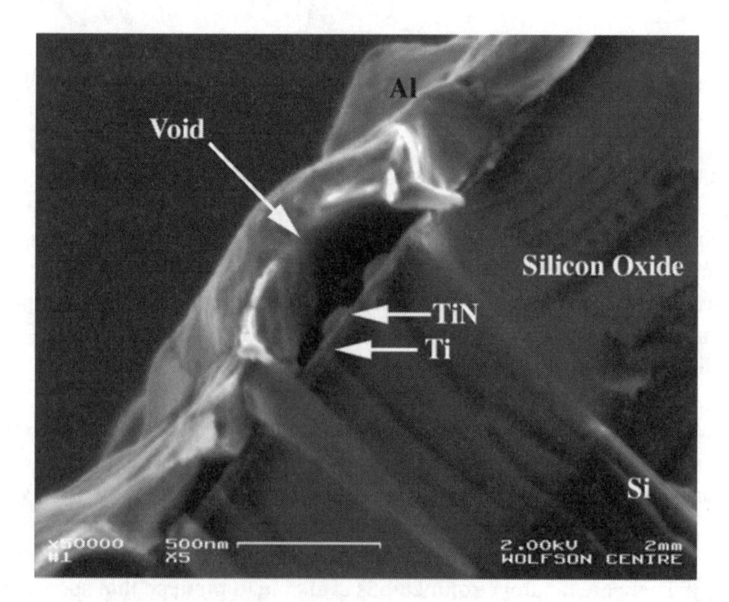

Figure 4.58 Scanning electron micrograph of a cross-sectional specimen prepared by cleaving the silicon wafer upon which an Al/Tin/Ti stack of films was deposited. The Al shows a considerable amount of plastic deformation compared to the underlying layers, and a void in the Al film is clearly visible

In order to ensure that the largest possible area of the interface region is at or near the perforation produced by ion milling, we need to glue four specimens together—one pair 'face-to-face', and the second pair 'face-to-back', as shown schematically in Fig. 4.59. We require a cross-section which is at least 3 mm in width, so two more silicon wafers (with a thickness of 300 μm) are then glued to the specimen. We now drill ultrasonically, using an annular drill bit, down the length of the cross-section, and then glue the resulting 'rod' inside a copper tube of outside diameter 3 mm. Once the glue has set, slices are cut from the end of the rod with a diamond wafering saw. These slices are now mechanically thinned to less than 100 μm, dimpled at their centre to less than 20 μm, and then ion milled to perforation. By gluing the sample wafers face-to-face and face-to-back, we therefore are able to locate the interfaces of interest near to the centre of the 3-mm diameter specimen. Ion milling, combined with dimpling, improves the probability that perforation will occur in the central region, thus producing a specimen which is thin enough for TEM in the region of

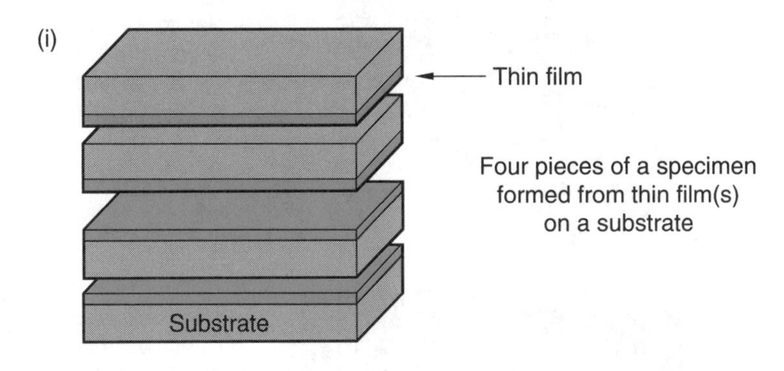

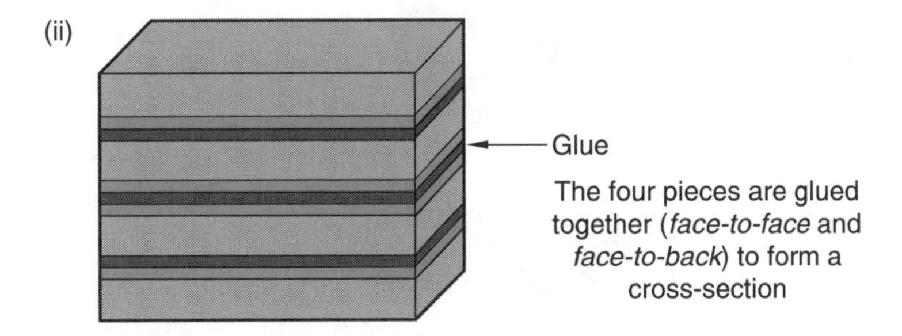

Figure 4.59 Schematic representation of the process used to prepare a cross-sectional TEM specimen

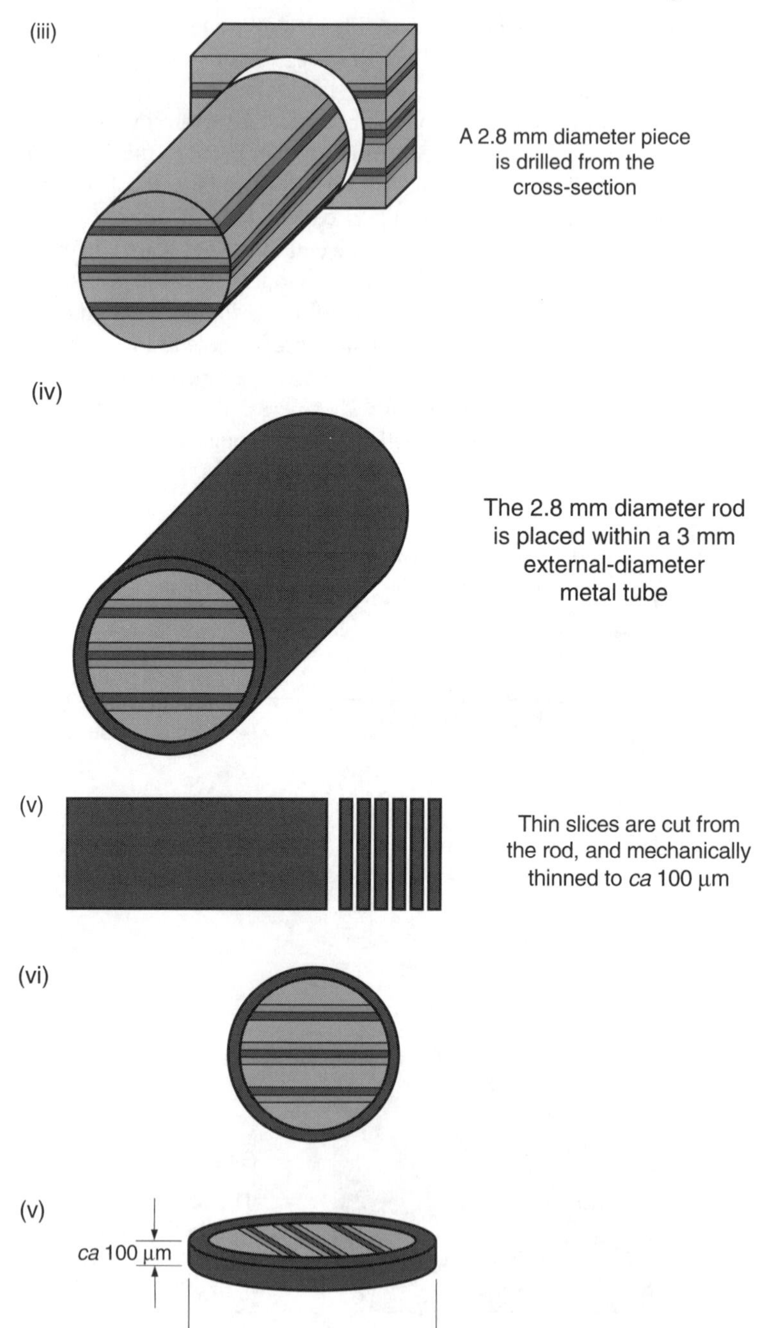

Figure 4.59 (*continued*)

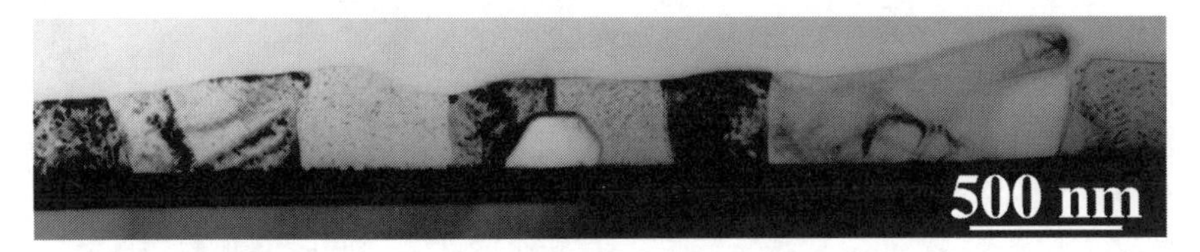

Figure 4.60 Low-magnification, bright-field transmission electron micrograph of a cross-sectional specimen of Al on $TiN/Ti/SiO_2/Si$. The individual Al grains, as well as open and closed voids in the Al film, are clearly visible

interest. This seems to be a lot of work for just one specimen, but with practice the time required to prepare a sample becomes more reasonable.

Fig. 4.60 shows a bright-field diffraction contrast transmission electron micrograph of the specimen that we have just prepared. The thick aluminium film is clearly visible in cross-section, as are the separate grains within the film. The open and closed voids are now easy to detect, and statistical analysis of both the grain size

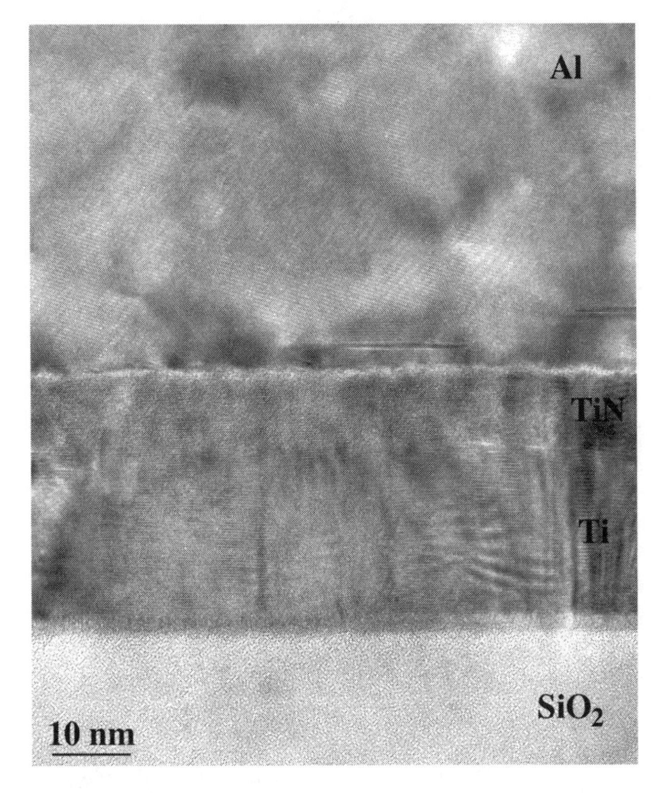

Figure 4.61 Higher-magnification transmission electron micrograph (showing some phase contrast) of the interface region, displaying the morphology of the TiN and Ti layers

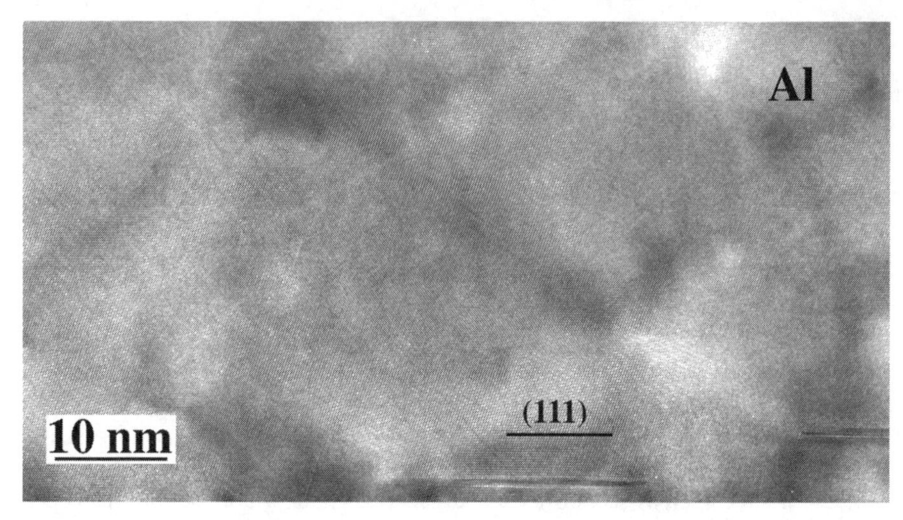

Figure 4.62 HRTEM lattice image of an aluminium grain oriented along a [110] zone axis

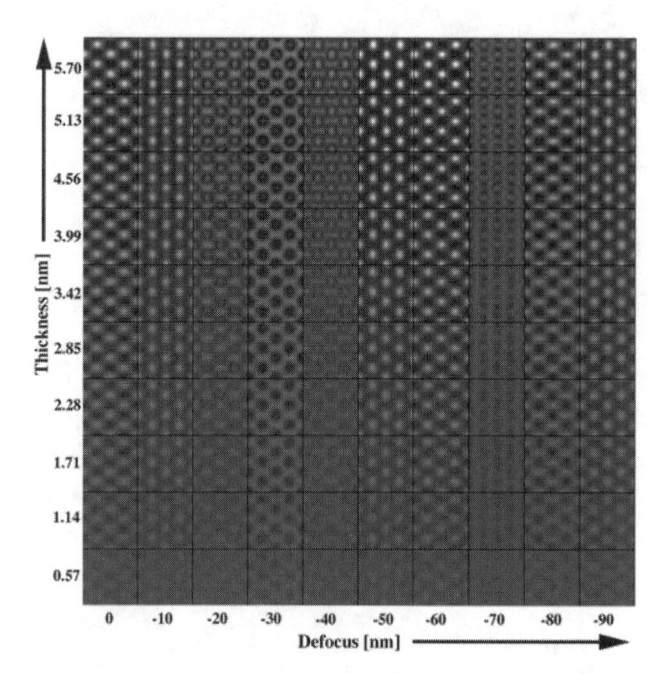

Figure 4.63 Simulated defocus–thickness map of Al in a [110] zone axis. The map is 'contrast invariant', and the contrast of each individual simulation has not been changed when the images were 'pasted' together, so the thinner regions have a lower contrast than the thicker regions; note that this feature is not due to an absorption effect

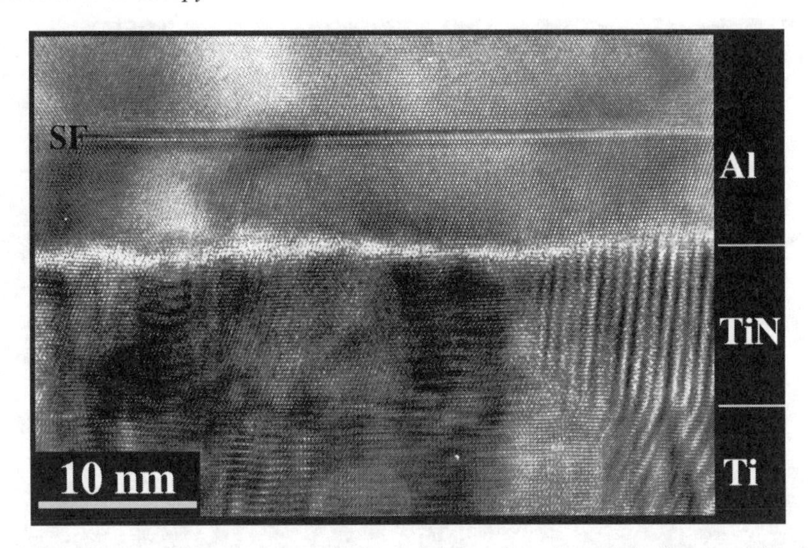

Figure 4.64 Lattice image of a stacking fault in an Al grain, with the latter running parallel to the interface with the TiN layer

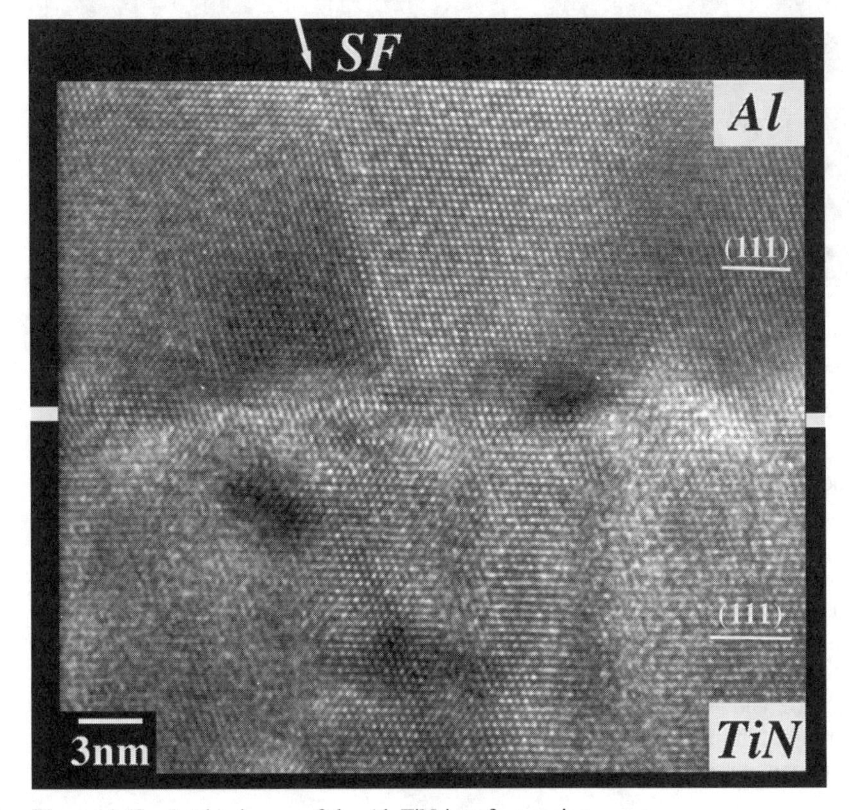

Figure 4.65 Lattice image of the Al–TiN interface region

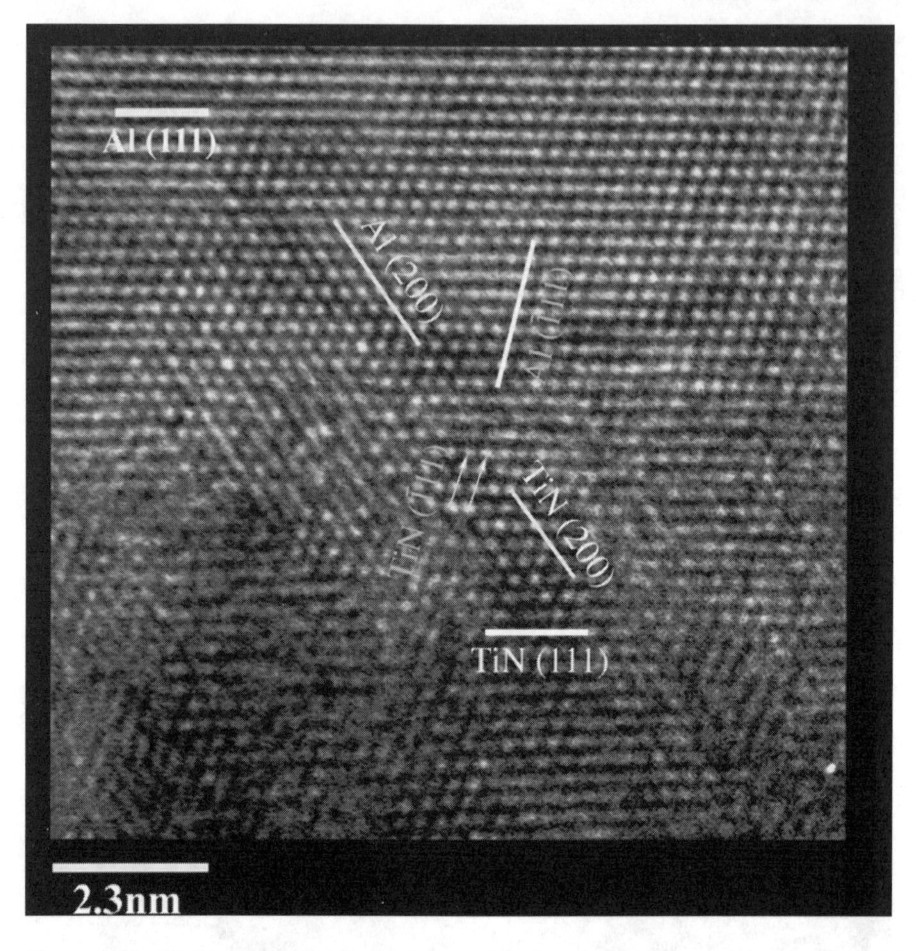

Figure 4.66 High-magnification lattice image of the interface between an Al grain (along a [110] zone axis) and the TiN layer. An orientation relationship can be found between the Al grain and one of the TiN grains; this is the topotaxy which influences the preferred orientation of the Al film

and the void density is feasible. Higher-magnification images (Fig. 4.61) show the morphology of the TiN/Ti layers, which have very small grains. The primary crystal defect observable in Fig. 4.61 are grain boundaries, and this image contains information on the average grain size of the TiN and Ti constituents. However, Fig. 4.61 is actually a projection of a three-dimensional thick slice on to a two-dimensional image, and the boundaries visible in the image come from different grains within the film. We will discuss the problem of image *overlap* in Chapter 7.

Fig. 4.62 shows a high-resolution (HR) transmission electron micrograph of an aluminium grain in the same specimen taken after the grain had been oriented along a [110] zone axis. The aluminium film has a strong $\langle 110 \rangle$ texture, the $\{111\}$ planes are predominately parallel to the interface, and tilting the aluminium grain into a [110] zone axis has brought the interface parallel to the optic axis (the incident-beam direction). There are several reasons for the contrast variations in the aluminium grain shown in Fig. 4.62. First, changes in thickness from the interface region to the edge of the specimen result in changes in contrast (thickness extinction effects), and we can determine the specimen thickness for any region by comparing the experimental image to the simulated image, providing that we know the defocus value of the objective lens and have a defocus–thickness 'map' in order to determine both independent variables (as shown in Fig. 4.63).

Additional sources of contrast variations in the lattice image shown in Fig. 4.62 are local bending, due to *residual stresses*, or the presence of *dislocations*. Stacking faults are also visible in the aluminium grain near the interface (see Fig. 4.64). These faults lie on the {111} planes, some of which are parallel to the interface. The faults were most likely formed during deposition of the film, and are expected to influence the electrical properties.

Fig. 4.65 shows a higher-magnification HRTEM image of the same interface region. The lattice image from a [110] zone axis projection of the aluminium lattice is visible together with individual grains from the TiN layer. Fig. 4.66 displays a lattice image from an Al/TiN/Ti interface which shows *topotaxy* between the aluminium grain and one of the underlying TiN grains. This is the most likely nucleation site for the Al grains, which will then grow both laterally and in thickness during the deposition process.

Problems

4.1 Distinguish between an optically focused image and a scanned image. What do you consider to be the major advantages and disadvantages of each of these two approaches to forming an image?

4.2 Appreciable inelastic scattering is incompatible with the formation of an optically focused image. Why is this?

4.3 Electromagnetic lens systems employ very small angular apertures, primarily because of the large spherical aberrations which are encountered. Explain how the spherical aberration and other lens defects limit the resolving power of a transmission electron microscope.

4.4 Illustrate, by means of a graph, the dependence of electron wavelength on accelerating voltage. Which factors prevent the development of ultra-high voltage electron microscopes for ultra-high resolution work?

4.5 Estimate the minimum voltage stability required to ensure a resolution of 0.2 nm at 100 kV.

4.6 Outline the steps required to prepare a transmission electron microscope specimen from the following samples: (a) a tungsten light bulb filament; (b) a pinch of talcum powder; (c) a steel bolt; (d) a brazed vacuum seal.

4.7 Distinguish between *mass–thickness* contrast and *diffraction* contrast.

4.8 List the experimental and specimen parameters which affect diffraction contrast and define precisely each parameter in your list.

4.9 Draw the kinematic amplitude phase diagrams for a *stacking fault* and for a *dislocation line* for the *two* cases in which traversing the defect either *increases* or *decreases* the phase shifts between neighbouring atomic columns.

4.10 Explain the diffraction conditions under which lattice defects will fail to give diffraction contrast and give examples for the following: (a) a stacking fault in a FCC lattice; (b) a lattice screw dislocation in a BCC metal.

4.11 To what extent does a lattice image mirror the atomic positions in the unit cell of a crystal?

4.12 How does the working distance of the probe lens from the sample surface affect the minimum probe size in a scanning electron microscope?

4.13 Given that the signal collected in a scanning system is determined by inelastic scattering and secondary excitation processes, discuss the effect that these probe-specimen interactions have on the scanning resolution. Compare, in particular, *characteristic X-ray* and *secondary electron* excitation.

4.14 Both *surface topology* and local *mass density* can influence the scanning electron image contrast. How would you expect the *beam voltage* to affect the atomic number contrast in a back-scattered electron image?

4.15 The secondary electron image is the one that is generally used for routine examination in the scanning electron microscope. Why is this?

4.16 Discuss some ways in which samples sensitive to degradation in a vacuum or under an electron beam could nevertheless be imaged in the scanning electron microscope.

4.17 What angle of tilt would be needed to distinguish, using stereoscopic imaging, between two features which are separated by a vertical distance h, given that the lateral resolution is δ?

Chapter 5

Microanalysis in Electron Microscopy

We have already noted (Section 4.3.2) that a proportion of the X-rays emitted under electron excitation are characteristic of the chemical constituents of a solid sample. These *characteristic X-rays* may be selected from the spectrum of emitted electromagnetic radiation and their distribution displayed in a line-scan or dot image to provide *qualitative* information on the morphological relationship between the microstructure and the chemical composition.

In this and in the following chapter, we will explore several ways in which chemical information about the sample can be made *quantitative*. In all of the methods to be discussed, *inelastic scattering* of the probe (either electrons or X-rays) excites a signal which depends on the *chemical composition* of the material beneath the probe, and the challenge is to interpret this signal as quantitatively as possible. The *sensitivity* of each method (the minimum concentration of a constituent which can be detected) will be different, as will the *accuracy* (the errors involved in quantitative analysis). We will also be concerned with the *spatial resolution*, both in the plane of the sample surface (or more accurately, the image plane) and in the depth which is sampled (the localization of the signal to the surface).

In this present chapter, we will only consider those methods which are commonly available as *microanalytical facilities*, attached to either the *transmission* or the *scanning electron microscope*, while in the following chapter we will describe some additional methods commonly used to characterize the composition of a *surface* or *surface layer*.

5.1 X-RAY MICROANALYSIS

In X-ray microanalysis, the *characteristic X-rays* emitted from a sample viewed in the electron microscope (most often *scanning*, but also in *transmission*) are analysed both qualitatively and quantitatively in order to determine the relationship between the local chemical composition of the sample and the morphological features.

The physical basis for X-ray emission after excitation by a high-energy electron beam which is incident on a solid target has already been discussed (Section 4.3.2), and all that remains here is to explain those factors which influence the *sensitivity* and *accuracy* of microanalysis. These include both the detection system and the computer software used to convert the raw data to a quantitative estimate of composition, as well as the specimen preparation procedures and the sample geometry in relationship to the detector, together with the composition and microstructure of the sample.

In order to simplify the discussion, we assume that the specimen surface is *smooth* and *planar*, and that the only geometrical parameters of interest are the angle at which the electron beam strikes the specimen surface and the angle subtended by the detector with respect to this surface. If the surface roughness is on a scale which is small compared to the diffusion depth for the incident electrons (typically a few tenths of a micron), there are no significant errors in assuming that the surface is planar. Fracture surfaces, on the other hand, and surfaces which have been heavily corroded are certainly *not* planar, and the existing computer software for quantitative X-ray microanalysis is not intended for such specimens. The same applies to powders, fibres and grits, and the results of analysis on these materials must be regarded as only being *qualitative*. In fact, *all* of the available computer correction procedures for quantitative microanalysis assume that the composition is *homogeneous*, so that quantitative microanalysis is only reliable for regions far from phase boundaries and in the absence of strong concentration gradients.

5.1.1 Excitation of Characteristic X-Rays

As was noted in the previous chapter, the characteristic X-ray signal generated by the incident beam constitutes a *fingerprint* of the local chemistry. To carry this fingerprint analogy further, *qualitative* analysis consists in identifying the print as from the index finger or the thumb, while *quantitative* analysis has as its goal the identification of the perpetrator, which is a rather more difficult task.

The characteristic X-rays are generated in a region of the envelope of scattered electrons for which the energy of the electrons exceeds the threshold for excitation, with the size of this region depending on the *incident beam energy*, the *atomic number* of the element being detected and the *mass density* of the sample. The mass density and the beam energy determine the diffusion distance and the range of the electrons, respectively. To take an example (Fig. 5.1), copper at high beam energies will give rise to both K and L characteristic radiation, with the L radiation from a region just within the thermal energy envelope and the K radiation from a much smaller zone. At beam energies *below* the threshold for K excitation, only the L radiation can be excited, but since the range of the electrons is now much reduced, the *source* of the L radiation will have a smaller volume, in principle improving the spatial resolution for the identification of copper in the sample.

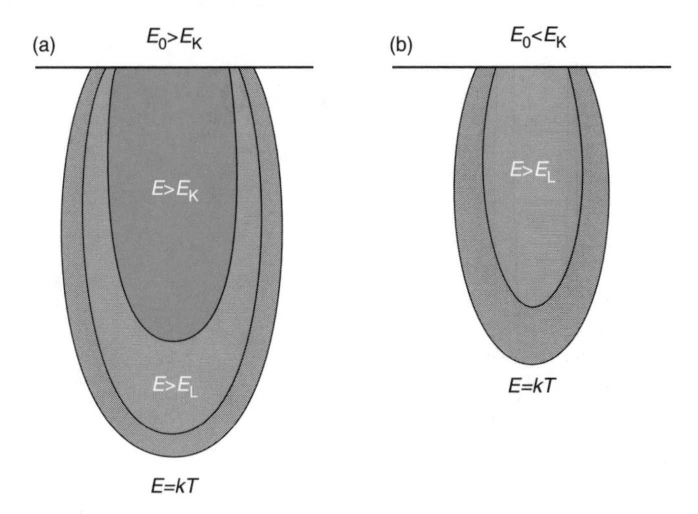

Figure 5.1 Schematic representation of the volume elements for characteristic X-ray excitation for copper (a) above and (b) below the energy threshold for K excitation

Increasing the energy of the incident beam increases the total X-ray signal and ensures that most, if not all, characteristic lines are excited. However, the total amount of *white background radiation* is also increased, and, when the beam energy exceeds about four times the threshold energy for excitation, the ratio of the characteristic line intensity to that of the background intensity (the *signal-to-noise ratio*) starts to decrease. In addition, the size of the volume element generating the signal is also increasing, thus reducing the spatial discrimination (that is the *resolution*). There are therefore good reasons for limiting the energy of the incident beam.

Although *reducing* the beam energy improves the spatial resolution and limits the background level of white radiation, as the beam energy approaches the critical excitation energy the intensity of the characteristic line decreases dramatically. As a general rule, there is no advantage to be gained by reducing the beam energy to *less* than four times the excitation energy for the *shortest-wavelength* (highest energy) characteristic lines of interest. In fact, *light element analysis* (for which only characteristic lines of long wavelength are available) can often be combined with analysis of the heavier constituents by using the characteristic wavelength from the L- or even M-shells of these high-atomic-number elements.

5.1.1.1 Signal-to-Noise Ratio

The background 'noise' must be subtracted from the intensity in a characteristic 'line' by fitting a function to the 'white' background signal (Fig. 5.2) and calculating the difference. The usual function is a straight line which is based on *background counts* summed on either side of the intensity peak and at a sufficient distance to ensure that the main peak does not interfere with the background measurements.

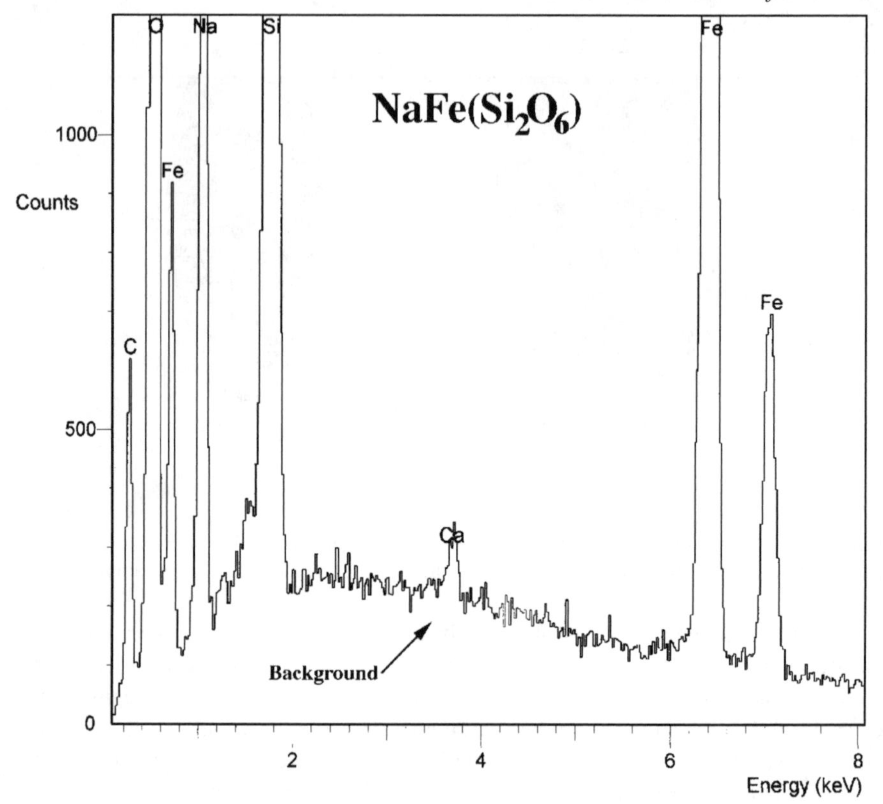

Figure 5.2 If a background correction can be calculated, then the intensity in a characteristic signal can be obtained by subtracting this background from the total number of counts in the full range of the detected intensity peak. Here the background is seen for a spectrum recorded from $NaFe(Si_2O_6)$

In practice, and especially in *energy-dispersive spectrometry* (EDS), the limited energy resolution of the spectrometer, together with the potential for overlap between characteristic intensity peaks, makes it impractical to set the acceptance channel for a characteristic line to the full width of the peak. A satisfactory compromise is to set the width of the channel to coincide with the intensities on either side of the maximum characteristic intensity which correspond to *half* the observed peak height, a condition which is termed the *full width at half maximum* (FWHM).

The *counting error* associated with the background can always be improved by counting for longer times, but there are limits to both the patience of the operator and the stability of the instrument. In general, the counting error (s) is given by the following expression:

$$s = \frac{\sqrt{(N + N_B)}}{(N - N_B)} = \frac{\sqrt{(N/N_B + 1)}}{(N/N_B - 1)} \tag{5.1}$$

where N is the number of counts associated with the characteristic emission, $(N + N_B)$ is the total number of counts registered in the channel selected and N_B is the background count correction estimated for this channel. The counting error is a very sensitive function of $(N - N_B)$, and peaks that are at the limit of detection require very long counting times if they are to be statistically significant. While it is common to refer to the *signal-to-noise ratio*, N/N_B, the significant parameter is actually s, as given in equation (5.1).

5.1.1.2 Resolution and Detection Limit

The spatial resolution for microanalysis in *scanning* electron microscopy is generally not much better than 1 μm, limited primarily by the diffusion depth and range of the electrons. In *transmission* electron microscopy (see Section 5.1.3.3), the limited thickness of the thin-film sample limits spreading of the beam by inelastic scattering, and resolutions of the order of 0.5–10 nm have been obtained, more easily when a *field-emission source* is available, but with a very much weaker characteristic X-ray signal, and a considerable accompanying loss of sensitivity.

In effect, the *spatial resolution* and the *concentration sensitivity* are in competition, as we noted when the size of the envelope which defines the volume element generating the characteristic X-ray signal was discussed. If detection sensitivity could be improved for longer-wavelength X-rays, then the better spatial resolution at lower beam energies could be exploited. This is possible with the *field-emission gun*, which increases the beam current by two orders of magnitude and can be used at incident beam energies of the order of 1 kV, thus localizing the excitation volume to the initial probe size multiplied by the film thickness. The analysis of *long-wavelength* radiation (1 kV $\equiv$ 1.24 nm) should also provide quantitative information on the nature of the chemical bonding, but this has yet to be demonstrated.

The *detection limit* in X-ray microanalysis at the beam energies commonly used (5–20 kV) is of the order of 0.5 atom %, while the *error* in analysis for a well-calibrated system is about ±2 % of the measured concentration when the latter exceeds a few percent.

5.1.2 Detection of Characteristic X-Rays

The detection of characteristic X-rays presents twin problems of *discrimination* and *detection efficiency*.

To achieve 100 % *efficiency*, every photon emitted by the sample would have to be recorded. This is quite impossible for two reasons. First, the detector always subtends a limited solid angle at the sample, and only those photons that reach the detector have any chance of detection. Secondly, the detector itself has a limited detection efficiency which also depends on the energy (wavelength) of the incident photon.

Similarly, perfect *discrimination* (energy resolution) would require that the wavelength of every photon counted should be accurately known and distinguishable from any activation events taking place at the detector which are associated with photons of different wavelength. This also is impossible, both because of overlapping background white radiation (which implies an uncertainty in the source of the photon) and because the detector can only identify the photon energy to of the order of 50–100 eV.

5.1.2.1 Wavelength-Dispersive Spectrometry

The best discrimination of the emitted X-ray signal is achieved by using a wavelength-dispersive spectrometer (WDS). This system (Fig. 5.3), employs a series of bent single crystals to cover the range of wavelengths which are of interest, and the wavelength within each range is scanned, adjusting the angle 2θ by rotating the crystal and moving the detector, while keeping the position of the crystal fixed. In this geometry, the angle at which the X-rays are collected from the sample is *fixed*, but the angle subtended at the collecting crystal will vary with 2θ, while the diameter of the focusing circle will change. (Thus, the system is designated '*semi-focusing*',

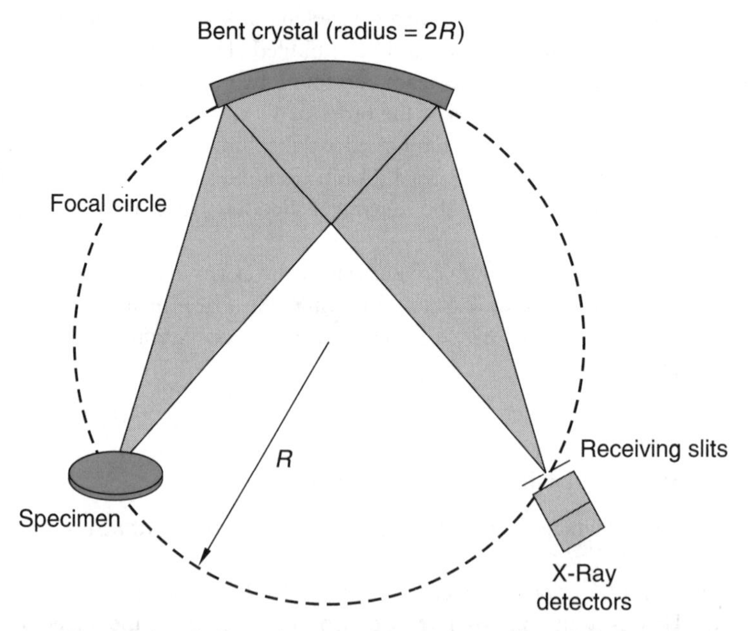

Figure 5.3 The wavelength-dispersive spectrometer is usually a fully-focusing system in which a curved crystal accepts the radiation emitted by the specimen surface over a specific solid angle. The distance D from the source to the crystal, the average take-off angle for the X-rays and the diffraction angle 2θ for the wavelength of interest are adjustable. The diagram shown here is that of a fully focused system

since the radius of the collecting crystal is constant and only 'fully-focusing' for one specific focusing circle.)

There is one additional reason why the system can only be 'semi-focusing', and that is because the X-rays are generated over a finite depth of the sample, which depends on both the beam energy and the sample density. However, this effect is not usually significant.

An important consequence of the geometrical discrimination provided by the WDS system is that all of the peaks must be scanned sequentially and there is no way of recording more than one characteristic line at once (unless multiple spectrometers are available). In practice, a WDS unit can be programmed to scan through a *series* of peaks, including background measurement settings, in order to maximize the efficiency of data acquisition. Nevertheless, the improved discrimination available when using wavelength- rather than energy-dispersion spectroscopy carries a heavy penalty in this *sequential*, as opposed to a *parallel*, counting mode.

5.1.2.2 Energy-Dispersive Spectrometry

In the *energy-dispersive spectrometer* (EDS), the pulse height recorded by a detector is directly related to the energy of the photon responsible for the pulse. The original detectors used for this purpose, namely gas proportional counters, have been replaced by *solid-state detectors*, but the principle remains the same. An incident photon is absorbed and creates ionization events in the active layer of the detector. The charge developed is then detected as a current pulse. A number of solid-state materials are potential candidates for EDS counters, but silicon is still the detector material of preference and good silicon crystals, suitably doped and cryogenically cooled, can have an energy discrimination of only 130 eV.

There are two problems associated with EDS, as opposed to WDS systems. The first concerns their poor discrimination or energy resolution. Good WDS systems can still do much better, especially in the detection of long-wavelength radiation or in cases where characteristic lines from different elements overlap. WDS systems can also resolve the lines of L- and M-spectra unambiguously, thus improving the accuracy of quantitative analysis (Fig. 5.4).

The second problem is connected with low-energy (long-wavelength) photon detection (Fig. 5.5), which requires either a windowless detector, or a detector protected from the rest of the system by only a thin, fragile window. Care is then needed to ensure that the detector retains its long-wavelength capability, and is neither damaged nor contaminated. All solid-state detectors are operated at liquid-nitrogen temperatures, and cryogenic condensation on the surface of the detector can be a serious life-limiting concern.

5.1.2.3 Detection of Long Wavelengths

The wavelengths of X-radiation commonly used for crystallographic structure analysis are usually less than 0.2 nm, and it was not until electron probe

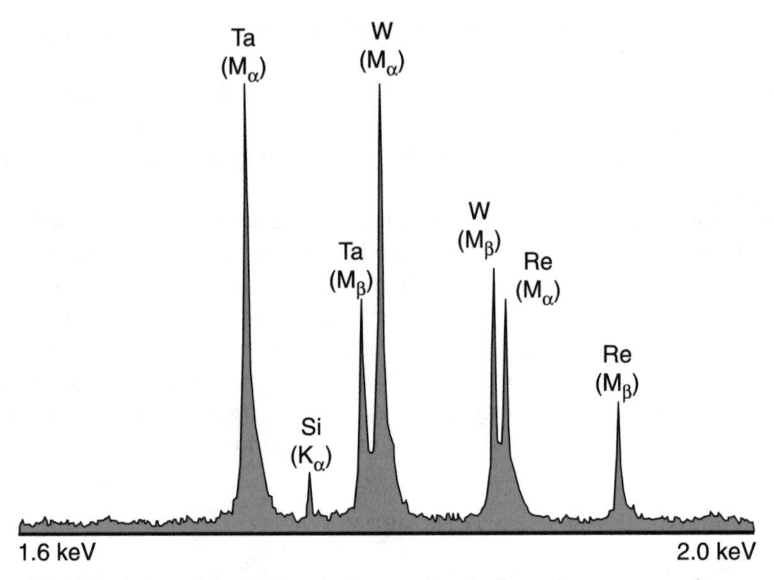

Figure 5.4 Resolution of the M lines in the wavelength-dispersive spectrum of a superalloy. (Oxford Instruments)

microanalysis was developed, in the late 1950s, that any commercial need existed for the detection of longer wavelengths. The past two decades have seen a steady improvement in the reliability, the sensitivity and the discrimination available for the detection of wavelengths longer than *1 nm*, corresponding to excitation energies of *1 kV* or less.

There are two major problems involved when considering these longer wavelengths. The first concerns the absorption of *soft X-rays* in the sample itself. Even in low density specimens, the absorption coefficient for soft X-rays is large. Signal detection is significantly improved if the specimen is inclined towards the detector so as to reduce the absorption length in the sample. The same is true of the detector itself, and *windowless* detectors are commonly used for the detection of very long wavelengths.

The second problem is the deterioration in *wavelength discrimination* at long wavelengths. In WDS systems, the crystals used for these applications have long periodicities, such as those developed by depositing layers of an organic molecule from the surface of a Langmuir trough. (The molecules are segregated to the surface of the liquid in a self-ordered array, which can be collected on a suitable curved substrate. Subsequent layers are then deposited in the same way, one upon another.) These WDS crystals are very fragile and easily damaged, even assuming that they do not contain defects introduced during deposition, and their ability to discriminate the wavelengths of the incident radiation is generally less satisfactory than for the more perfect, ionic crystals used for shorter wavelengths.

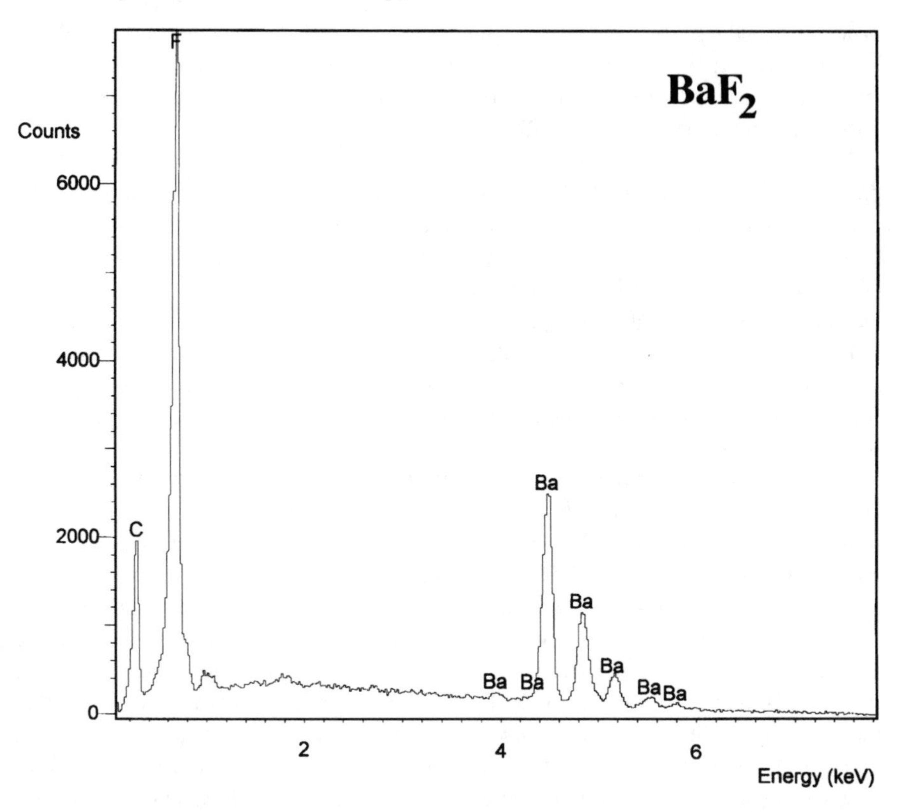

Figure 5.5 An energy-dispersive spectrum of BaF_2 showing the resolution of light-element characteristic lines; the carbon signal results from surface contamination

In spite of the problems, both EDS and WDS systems are easily capable of detecting light elements down to boron ($Z = 5$) for *quantitative* analysis, providing that the specimen is kept free of any contamination films which would affect the results. This ability should be compared with the situation that existed as recently as 10 years ago, when quantitative analysis of elements below magnesium ($Z = 12$) presented problems, and detection of light elements below carbon ($Z = 6$) was not realistic. Today, it is more often *specimen contamination*, originating from the sample itself under the electron beam, that restricts quantitative light element analysis, rather than any lack of capability in the equipment.

5.1.3 Quantitative Analysis of Composition

From the point of view of the operator, the composition of the sample is to be derived from the spectrum, and this calculation of the composition is an *iterative process* in which the observed characteristic intensities are first corrected for background and then compared with expected (calculated) or measured intensities from standard specimens, after corrections have been made for fluorescence, absorption and atomic number effects.

In this calculation process, the *relative characteristic intensities* (minus the background) are used as a 'first guess' for the *relative concentrations* of the corresponding elements in order to calculate the magnitude of the *corrections*. This first calculation of the corrections then allows a 'second guess' for the relative concentrations to be made, and hence a *revised* (better) estimate of the corrections needed to convert relative characteristic intensities to relative concentrations. In practice, remarkably few iterations are needed before the variation between successive iterations reaches a more or less constant value, thus leading to a 'best' estimate of the *relative concentrations* and a *residual error* which is typically better than 2 % of the calculated corrected concentrations.

The available computer software for performing these corrections varies considerably, but all such codes follow similar logic. The *geometrical parameters* must be entered, including the angle of the specimen to the incident beam and the angle subtended by the detector at the specimen (the *take-off* angle for the X-ray signal). The *accelerating voltage* (the beam energy) must also be noted (and the calculated background and intensity standards need to be adjusted whenever the voltage is changed). The program will then ask the operator if the results are to be *normalized*, that is whether the calculated values should be further adjusted so that they add up to 100 %. This is usually *not* a good idea. If the sum of the calculated concentrations does *not* add up to close to 100 %, then this is an excellent indicator of a serious problem in applying the correction procedures. Either one element has not been detected, the sample differs significantly in density from the expected value (porosity?), or some other feature of the microstructure is affecting the data. Alternatively, the wrong take-off angle or operating voltage, or some other operating parameter, has been entered into the program.

The software may also allow you to *omit* data from one element (which must nonetheless be identified), and the concentration of this element will then be determined by assuming that the total of the *estimated* concentrations is always 100 %. This may be useful for the light elements, which are hard to detect accurately, particularly in *non-stoichiometric materials*, for example for lithium in an Al–Li casting alloy.

A final option is also commonly available, and that is to assume *stoichiometry* during the calculation. This can be used to improve the accuracy when working with known *stoichiometric* compounds, such as ceramics, but it may be misleading when two phases with more than one valency are possible, such as TiN (Ti^{3+}) and TiO_2 (Ti^{4+}), or where the compound is known *not* to be stoichiometric, for example $Fe_{1-x}O$.

In the present text, we will describe the various corrections for quantitative analysis, not as they would be performed in an iterative quantitative analysis correction program, but rather as these corrections affect the *X-ray signal* when it is first generated by the incident beam and subsequently emitted from the sample. In other words, we will first consider the *volume element* of the material in which the X-rays are generated, then the *absorption losses* as the X-rays travel through the sample, and finally the possibility of *fluorescent excitation* of one characteristic line by X-rays of higher energy (shorter wavelength).

For the sake of completeness, we shall also include a short discussion of the application of X-ray microanalysis in *transmission electron microscopy*, that is the case of thin films.

5.1.3.1 Atomic Number and Absorption Corrections

For the most part we restrict ourselves to K-excitation in a two-component alloy, and ignore more complicated events (which do not differ in principle). We first discuss the factors affecting the *excitation efficiency*, and which depend primarily on *atomic number*, and then treat the subsequent process of *X-ray absorption* within the specimen.

Atomic Number Correction

In the simplest approximation, the number of *K ionization events* for a component A which occur over an element of the *electron path*, d*x*, is given by the relation:

$$\mathrm{d}n_{\mathrm{K}} = \frac{Q_{\mathrm{K}} N_0 \rho C_{\mathrm{A}} \mathrm{d}x}{A} \tag{5.2}$$

where Q_{K} is the ionization cross-section for K radiation from A, N_0 is Avogadro's number, ρ is the density, C_{A} is the mass concentration of A, and A is the atomic weight.

Excitation can occur as long as the electron energy exceeds E_{K}, and, assuming that the electron is not lost by *back-scattering*, the total number of ionization events *per electron* will be given by:

$$n_{\mathrm{K}} = \frac{N_0 C_{\mathrm{A}}}{A} \cdot \int_{E_{\mathrm{K}}}^{E_0} \frac{\rho Q_{\mathrm{K}}}{\mathrm{d}E/\mathrm{d}x} \mathrm{d}E \tag{5.3}$$

The *stopping power* for the electrons is defined as $S = -(1/\rho)\mathrm{d}E/\mathrm{d}x$, and the ratio of the number of ionization events in the sample to the number that would be obtained in a pure standard is given by the equation:

$$k_{\mathrm{A}} = \frac{(n_{\mathrm{k}})_{\mathrm{AB}}}{(n_{\mathrm{K}})_{\mathrm{A}}} = C_{\mathrm{A}} \frac{R_{\mathrm{AB}}}{R_{\mathrm{A}}} \frac{\int_{E_{\mathrm{K}}}^{E_0} Q_{\mathrm{K}}/S_{\mathrm{AB}} \mathrm{d}E}{\int_{E_{\mathrm{K}}}^{E_0} Q_{\mathrm{K}}/S_{\mathrm{A}} \mathrm{d}E} \tag{5.4}$$

where A refers to the pure standard and AB to the alloy. The parameter R, the *back-scatter coefficient*, is introduced to allow for that fraction of the high-energy electrons which is lost by *back-scattering* before their energy has fallen to E_K. R and S are expected to be *correlated*, since a high stopping power should lead to a high back-scatter coefficient, so that to a first approximation, $k_A \approx C_A$.

The *stopping power* is a function of the *concentration* and increases with the *average atomic number* as follows:

$$S = D(\bar{Z}/A)\frac{1}{E}\ln\left(\frac{2E}{11.5\bar{Z}}\right) \tag{5.5}$$

where

$$\overline{Z/A} = \sum_i (C_i Z_i/A_i) \tag{5.6a}$$

$$\bar{Z} = \sum_i (C_i Z_i) \tag{5.6b}$$

and D is a constant.

The *scattering cross-section* for K excitation, Q_K, depends on the overvoltage, $U = E/E_k$, and can also be calculated.

The *back-scatter* coefficients have been derived from experimental measurements of the energy distribution in the back-scatter signal for targets of different atomic number. The important points to note about the back-scatter process are:

(a) as the incident beam energy, E_0, is reduced by *inelastic scattering* to the excitation energy, E_k, the *back-scatter coefficient* approaches unity;
(b) as E_0 is *increased*, R falls *more* rapidly for a *high*-average-Z material, in accordance with the dependence of the ratio of the diffusion depth to the penetration depth, that is $x_D/x_R \approx 12/(Z+8)$;
(c) *providing* that E_0 exceeds the excitation threshold for *all* of the components by a sufficient factor ($\sim$3), the ratio R_{AB}/R_A is almost independent of the excitation voltage.

Absorption Correction

The total intensity generated by a specimen of pure A *in the direction of the spectrometer* is given by the relation:

$$I = \int_0^\infty \phi_A(\rho Z)\mathrm{d}(\rho Z) \tag{5.7}$$

where ρZ is the *mass thickness* normal to the surface and $\phi_A(\rho Z)$ is a function of the mass thickness *only*. If the *collection angle* is θ and the mass absorption coefficient for the characteristic wavelength is μ/ρ, then the intensity generated, I, is *reduced* by absorption to the intensity emitted, I', as follows:

$$I' = \int_0^{\infty} \phi_A(\rho Z) \exp[-(\mu/\rho)\rho Z\theta] \mathrm{d}(\rho Z) \tag{5.8}$$

These functions can be rewritten as $I = F(0)$ and $I' = F(\chi)$, where χ is the Laplace transform of the function $\phi(\rho Z)$, and the normalized *absorption correction* function, $f(\chi) = F(\chi)/F(0)$, is determined as a function of E_0–E_K (Fig. 5.6).

An example of the expected effect of the *absorption correction* on the relationship between the K-intensity ratio and the concentration for Cu–Au binary alloys is shown in Fig. 5.7. As expected, the correction is *greater* than unity for the lower-atomic-number copper and *less* than unity for the higher-atomic-number gold, while the *magnitude* of the correction varies inversely as the *concentration* of the component.

In different *correction software* packages, both the 'standard' data employed in calculating the corrections, as well as the algorithm for the calculations, vary appreciably. In most cases, the supplier of the package is more than willing to discuss the assumptions and approximations involved. All of the packages should be able to convert *relative characteristic intensity* measurements to *quantitative* estimates of the concentration at an accuracy of better than 2% of the true concentration, providing that the concentration levels are at least a few percent and all elements are detected. However, this accuracy will *only* be achieved if the

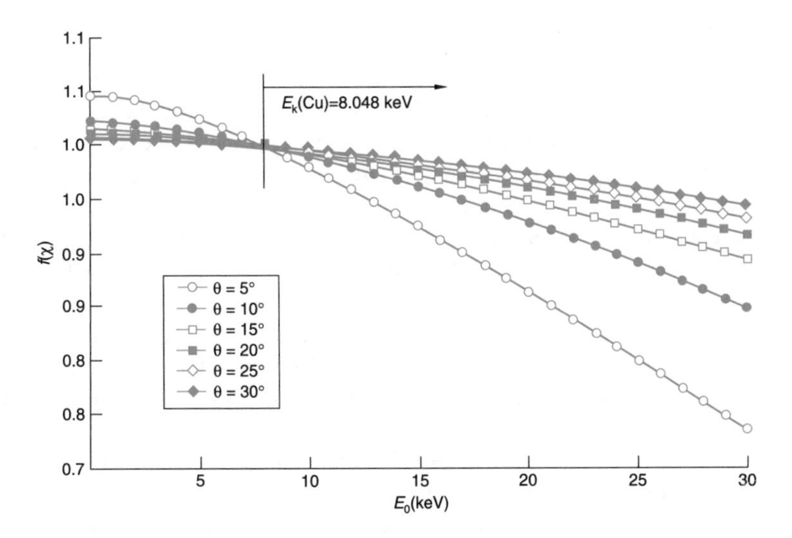

Figure 5.6 The absorption correction function, $f(\chi)$, as a function of the overvoltage, E_0–E_K, and the absorption parameter, $\chi = (\mu/\rho)$ cosec θ

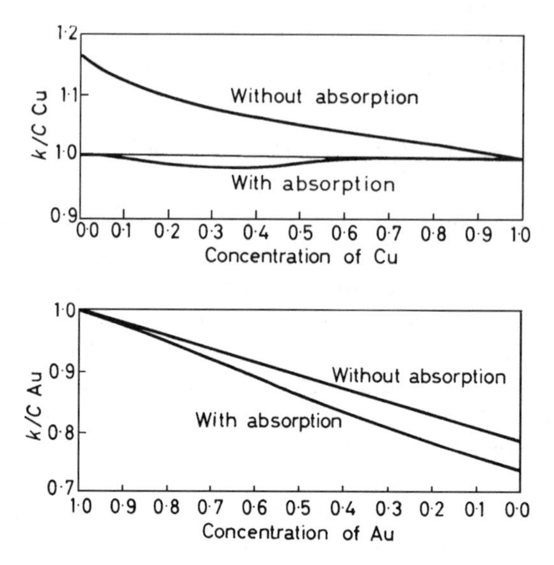

Figure 5.7 Calculated ratios of k/c for a Cu/Au alloy with and without an absorption correction: (a) copper constituent; (b) gold constituent

specimen preparation is satisfactory, the settings of the *microanalysis system* are optimized and the *counting statistics* are adequate.

5.1.3.2 The Fluorescence Correction

The *fluorescence correction* is, in a sense, the inverse of the *absorption correction*, since strong *absorption* in the sample of the X-rays generated by the incident beam implies that some sort of *secondary excitation* is taking place. The case of nickel and iron is instructive. Fig. 5.8 shows the *absorption coefficient* for both elements as a function of photon energy. Since nickel has the higher atomic number, the absorption coefficient is generally higher at any given wavelength, but the excitation threshold for displacing a K-shell electron occurs at a shorter wavelength (higher energy), so that the K *absorption edge* for nickel appears at a lower wavelength, and for a band of critical wavelengths between the two absorption edges, iron has a very much larger absorption coefficient than nickel. The characteristic K-lines of any element must lie to the *longer*-wavelength (lower-energy) side of the excitation threshold (the *absorption edge*), and in the region of low absorption by the same element. However, the K-lines for nickel lie in the high-absorption region for *iron*, so that nickel radiation will be strongly *absorbed* in an iron alloy, and thus result in *fluorescent excitation* of the iron matrix, so enhancing the characteristic signal for iron. More seriously, in a nickel alloy the strong excitation of iron will result in a fluorescence correction for iron of up to 30 % of the total signal.

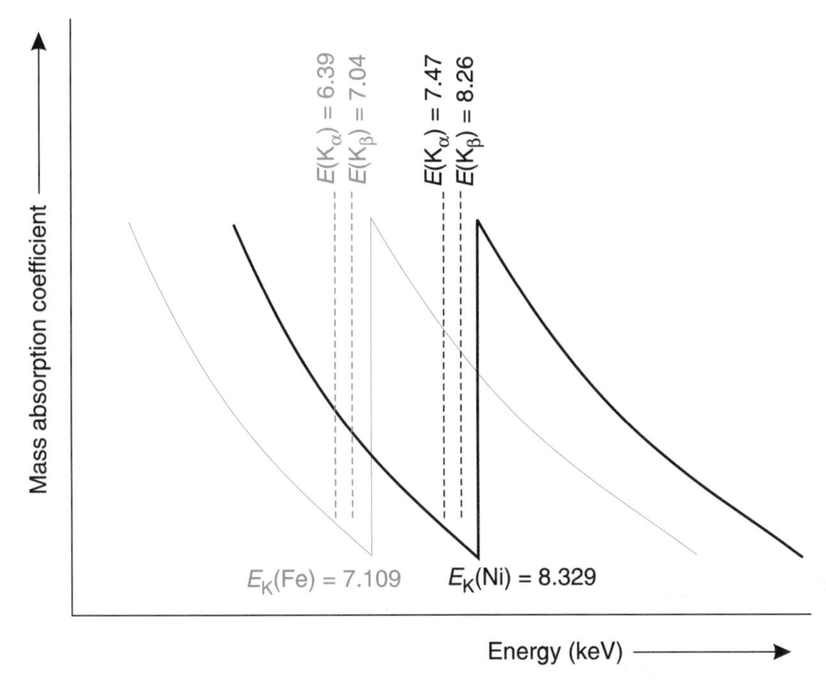

Figure 5.8 The energy dependence of the mass absorption coefficients of Fe and Ni, showing the positions of the characteristic K_α and K_β lines; the absorption edges correspond to the critical K-excitation thresholds of the two elements

The major problem in the microanalysis of samples where strong *fluorescence* is to be expected arises from the comparatively large volume over which fluorescent excitation may occur. The penetration depth for X-rays in the sample decreases with increasing X-ray wavelength and sample density, but is typically *at least* an order of magnitude greater than the penetration depth for the incident beam. It follows that, whereas the *primary* excitation events occur in a volume of the order of 1 μm in diameter, *secondary fluorescent excitation* will occur in a volume of the order of 10 μm or more in diameter (and so larger by a factor of 10^3). The situation is illustrated schematically in Fig. 5.9, where it can be seen that the *fluorescent radiation* is much less localized than the *primary radiation*.

Consider a phase boundary between a nickel-rich and an iron-rich phase in which the X-ray detection system is placed perpendicular to the boundary, either on the side of the nickel-rich phase, or on the side of the iron rich phase. Two quite different excitation spectra will be obtained from the boundary region (Fig. 5.10):

(a) When the boundary is *perpendicular* to the X-ray take-off direction and the detector is positioned on the *nickel-rich* side of the boundary, so as to maximize absorption by the *nickel*, then neither the iron nor the nickel signals are strongly absorbed. However, when the *probe* is on the nickel-rich side of the boundary,

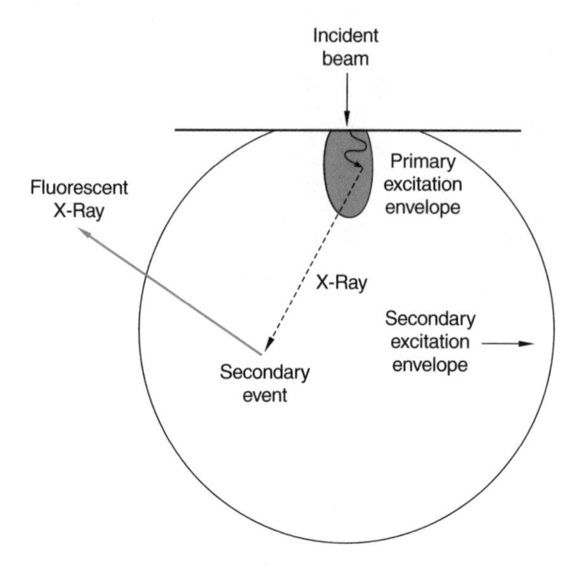

Figure 5.9 Schematic representation of secondary fluorescent excitation occurring well outside the envelope of excitation events for the primary X-radiation

the *nickel* radiation penetrating into the *iron-rich* region will generate a strong fluorescent iron signal, which cannot be corrected for, even when the probe is some way away (up to 10 μm or more) from the boundary.

(b) Rotating the specimen by π so that absorption is for the most part through the *iron-rich* region, will generate roughly the *same* fluorescent excitation of iron by nickel when the probe is on the nickel-rich side of the boundary. Now, however, strong absorption of the *primary nickel radiation* will also occur, since the nickel radiation must pass through the *iron-rich* region in order to reach the detector.

Clearly, the only sensible course of action in this case is to position the detector on either side of the boundary (by rotating the sample) to check for these artifacts, and to note the strong dependence of these effects on the *X-ray take-off angle*. Only semiquantitative analysis is possible in such a case, for example by determining apparent concentrations as a function of take-off angle (tilting the specimen towards the detector) and extrapolating the results to a $\pi/2$ take-off angle. Bearing in mind that spurious fluorescence may be significant at distances of 10 μm from a phase boundary, concentration measurements near such interfaces should be treated with caution.

5.1.3.3 Microanalysis of Thin Films

The limitation on specimen thickness in transmission electron microscopy is set by the onset of inelastic scattering processes, and some inelastic scattering is always to be expected, in addition to the (dominant) elastically scattered signal. It follows that

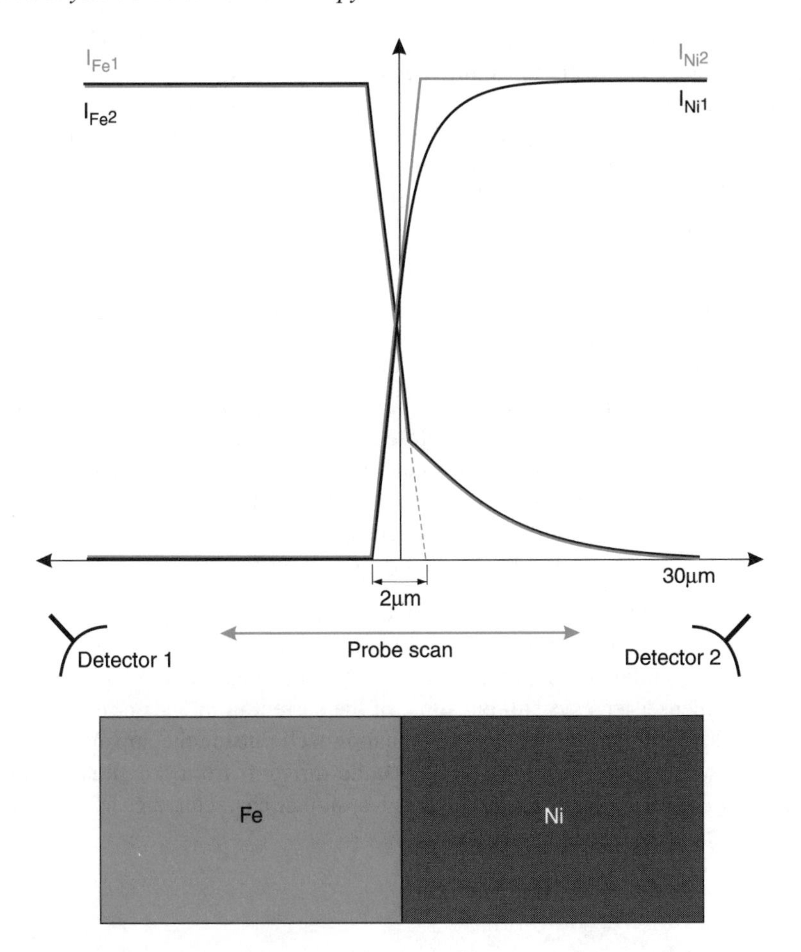

Figure 5.10 The Fe and Ni intensities perpendicular to a Fe–Ni interface. Two different detectors are used, i.e. detector 1 on the left side, and detector 2 on the right side of the sample. Using a 1 μm diameter probe and ignoring fluorescence, the Fe signal will drop in a linear fashion at the interface over a distance of 2 μm. Fluorescence will account for a Fe signal from the Ni side of the couple (no matter which detector is used) of an intensity of *ca* 6 % of the pure Fe signal, which decays to zero at *ca* 30 μm from the interface. The Ni signal, when collected from detector 1, shows a sharp decrease due to absorption through the Fe, while the Ni intensity, when collected through detector 2, will not be affected by absorption

a *characteristic X-ray signal* is available in thin-film transmission electron microscopy. Since the specimen is thin and the electron energy is high, very few characteristic X-ray photons are generated and it is not generally possible to detect constituents which are present in concentrations of less than 5 % unless a highly focused electron probe is used, for example that generated by a field-emission source.

Nevertheless, the very high *spatial resolution* of the transmission electron microscope, together with the limited *lateral spread* of the electron beam, has provided motivation for the development of a number of X-ray detection systems and software packages for *quantitative microanalysis*, specifically geared to the requirements of transmission electron microscopy.

The factors which have to be taken into account when considering the special requirements of *thin-film microanalysis* when using characteristic X-rays are the following:

(a) *background noise*, associated with stray electrons in the column, which may generate both white radiation and spurious characteristic X-rays (particularly from copper, a major constituent of the specimen stage);
(b) the need to optimize the *collection efficiency* by ensuring a wide solid angle for detection (placing the detector as close as possible to the sample);
(c) the *thickness dependence* of the signal and the possibility of using thickness-dependent intensity measurements to replace standard correction procedures;
(d) the desirability of *reducing* the accelerating voltage to improve the X-ray excitation probability and hence the counting statistics.

The problem of excitation from components of the column is exacerbated by the optical limitations on beam-probe focusing in *transmission microscopy*. It is not difficult to focus a fine probe (of the order of 10 nm diameter at FWHM) by using a standard condenser lens system, but some of the current in the beam remains in the tail of the distribution, thus leading to excitation well outside the apparent diameter of the focused probe (Fig. 5.11). This is quite different from the situation in the *scanning microscope*, where the *probe* lens has similar characteristics to an *objective* lens in the transmission microscope.

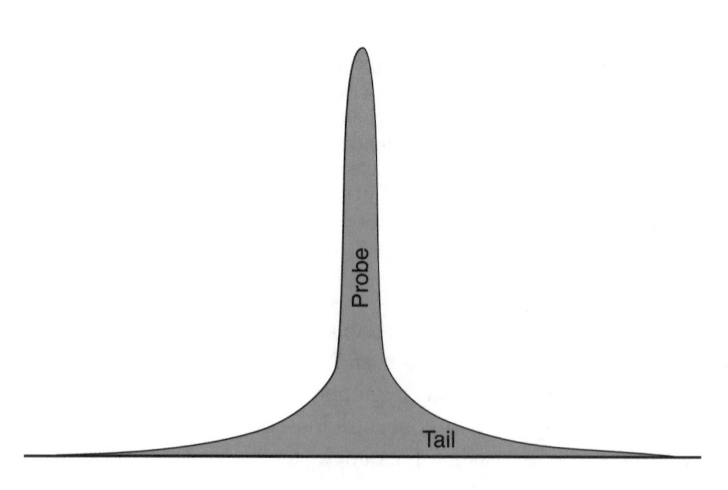

Figure 5.11 The distribution of the current in a focused probe includes a significant component in the tail of the distribution

Some improvement is possible by constructing the components of the specimen holder from low-atomic-number elements and mounting the specimen using low-atomic-number grid materials. Graphite and beryllium suggest themselves as suitable materials, since both are electrically conducting and will not cause electrostatic charging. Specimens which do not require a supporting grid are certainly preferred.

Solid-state X-ray detectors are used for microanalysis in transmission electron microscopy, and, in an inclined geometry, can be placed as close as 10 cm from the specimen. The *take-off angle* is high and variable, due to the wide solid angle used for data collection, but should not affect the measured relative intensities providing that the geometry is held constant.

Specimen thickness variations can be a positive *advantage*. Assuming a wedge-shaped specimen, the relative intensities measured as a function of distance from the specimen edge can be extrapolated to zero thickness, for which *no* corrections are necessary since the absorption of the generated X-rays is then zero. In any event, all X-ray microanalysis measurements made in the transmission electron microscope are free of fluorescence effects as a result of the thin section of the sample.

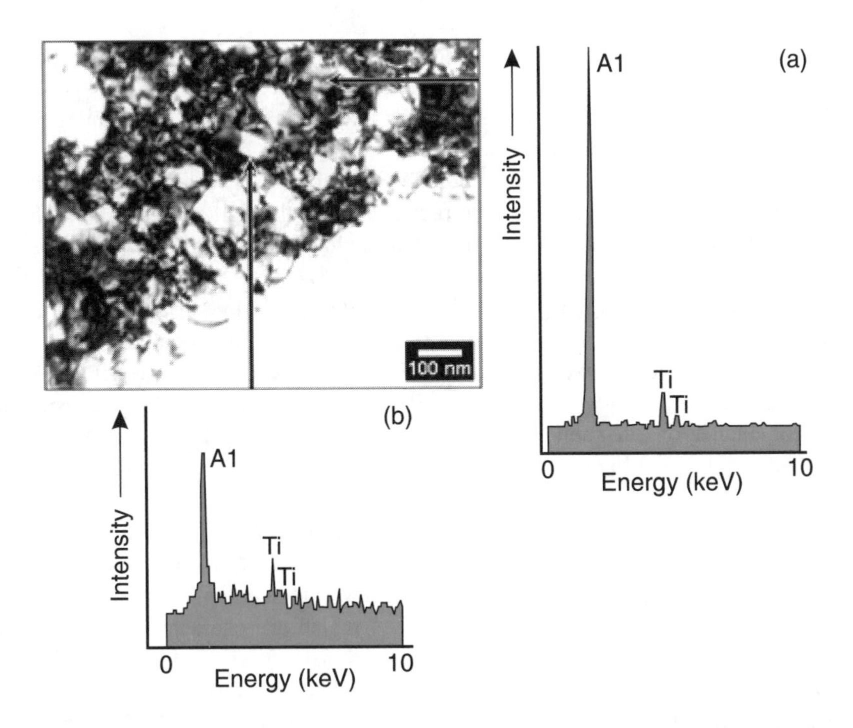

Figure 5.12 An EDS X-ray spectrum obtained from a small titanium carbonitride particle in aluminium, illustrating the power of the transmission electron microscope for overcoming the resolution limitations for X-ray detection in scanning electron microscopy. Two spectra are shown, one from the aluminium metal (a) and one from the TiCN particle (b). Note the change in relative intensities (Al/Ti) between the two spectra

As the accelerating voltage decreases, the probability of *inelastic* interaction (and the generation of characteristic X-rays) increases, so it should help to *decrease* the accelerating voltage of the transmission electron microscope. However, as we decrease the accelerating voltage the *brightness* (current density) of the electron beam probe *decreases*, and in all practical cases the best spectra are recorded at the *highest* accelerating voltage available. This also corresponds to the highest *spatial resolution*, since beam spreading *decreases* with accelerating voltage, thus limiting the diameter of the region from which the X-rays are generated.

Finally, it is worth repeating that the *primary* motivation for energy dispersive X-ray microanalysis in thin-film transmission microscopy is the improvement in *spatial resolution*. Fig. 5.12 illustrates this point. Nevertheless, there is no reason why thin-film samples should not also be analysed in the scanning electron microscope, with a dramatic improvement in spatial resolution for microanalysis in this instrument. It is not clear why the latter is not used more often, particularly since prior examination of a transmission electron microscope specimen in the scanning electron microscope can be an enormous time-saver in deciding which samples are most suitable for transmission microscopy, and which areas of the sample ought to be given priority.

5.2 ELECTRON ENERGY-LOSS SPECTROSCOPY

Microanalysis based on characteristic X-ray excitation has two disadvantages. The first is the very low *collection efficiency* for X-rays (no better than 10^{-3} for WDS systems and of the order of 10^{-2} for EDS detectors). The second concerns the inefficient excitation of characteristic radiation from the lighter elements. Although the characteristic spectra from the elements below magnesium ($Z = 12$) are now readily detectable (at least down to boron, $Z = 5$), these elements are difficult to analyse quantitatively, for two reasons. First, *absorption* by surface contamination and surface films tends to falsify quantitative results, and secondly, only a small fraction of the primary excitation events lead to characteristic X-ray emission (the remainder generate *Auger electrons*, see Section 6.2). For example, it is estimated that the chance of a K-shell excitation of carbon ($Z = 6$) yielding a photon is only 1 : 400, and this yield increases only slowly with atomic number, rising to 1 : 40 for sodium, at $Z = 11$.

Answers to both these limitations are provided by analysing the *energy spectrum* of the inelastically scattered electrons, since very high electron *collection efficiencies* are then possible at the detector (better than 50 %), while the maximum *sensitivity* is achieved precisely for the low energy losses which are characteristic of the low-atomic-number elements. Moreover, the *energy loss spectrum* is measured in the thin-film transmission geometry, precisely the case for which X-ray microanalysis is most difficult due to the poor counting statistics (Section 5.1.3.3). Two modes of operation are possible. In normal transmission electron microscopy, the region of interest can be selected either by placing an aperture in the first image plane (as if for selected-area diffraction) or by focusing the beam on the area using the condenser lens system. Alternatively, a much finer probe can be focused in a *scanning transmission electron microscope* (STEM) mode, in which the image is first formed in a *scanning raster* (as in scanning electron microscopy, but now in transmission), and the beam is then positioned on the area of interest. In general, areas several microns in diameter can be analysed at low magnifications, while areas down to less than 1.0 nm diameter can be subjected to a '*point analysis*'.

Instrumental developments in *electron energy-loss spectroscopy* (EELS) have improved the energy resolution to better than 1 eV (compared to 130 eV for the energy resolution of the best EDS systems), while the provision of parallel recording systems (PEELS, see Section 5.2.3.1) and *energy-loss imaging filters* show promise of providing *quantitative* chemical information with a spatial resolution of the order of 0.5 nm and a detection limit of less than 100 atoms.

One *caveat*, however, is in order. In both crystalline and amorphous thin-film specimens, *elastic* scattering events are far more likely than *inelastic* events, and much of the inelastic energy absorption spectrum is generated by electrons which have first been elastically scattered, so that *inelastic* scattering in a *diffracted* beam is a common occurrence. It is usually the energy spectrum from the direct transmitted beam which is collected, while the electrons which have been inelastically scattered in the diffracted beams are lost (that is, prevented from entering the spectrometer by an aperture). Relatively little work seems to have been done to compare the energy-loss spectrum from the direct transmitted beam with that from a diffracted beam (the energy-loss spectrum corresponding to a dark-field image), but the introduction and steady development of imaging filters promises to correct this omission, thus improving current capabilities to detect, for example, segregation around lattice defects imaged by diffraction contrast.

5.2.1 The Electron Energy-Loss Spectrum

Four processes contribute to the energy losses of the electron beam, although only two of these can be detected and resolved by the *magnetic spectrometers* commonly available. These are as follows:

- *Phonon excitations* result in very small energy losses, typically less than the kT energy spread in the incident beam. These peaks are within the zero-loss peak of the energy-loss spectrum, and are not resolvable.
- *Electron transitions*, both within and between the different shells of the atom, commonly lie in the range 1–50 eV. These peaks can be detected in the spectrum, and have been used to identify the phase in which a specific element is present by a comparison with known, standard spectra. Analysis of the very-low-loss region (in the 10 eV range) can be used to measure the dielectric properties of a material.
- *Plasmon excitation* is associated with quantized oscillations in the conduction band of a metallic conductor, typically resulting in energy-loss peaks in the range 5–50 eV. Plasmon peaks have been interpreted in chemical terms (they are

concentration-sensitive), but the interpretation is controversial. However, if the mean free path of the plasmons is known, then we can use the intensity of the plasmon peaks relative to the zero-loss peak to determine *specimen thickness*. *Relative* values of specimen thickness are in any case always proportional to the plasmon losses, which can therefore be used to check any *thickness dependence* of the analysis.

- The high-energy region of the spectrum ($\Delta E > 50$ eV) contains the *absorption edges* which are associated with atomic *ionization* and are accessible for chemical analysis in EELS. The remainder of the account given here will be restricted to these signals alone.

5.2.2 Limits of Detection and Resolution

The energy resolution of the magnetic spectrometers now available for EELS is certainly better than 1 eV, and resolutions of the order of 0.5 V have been reported when using a field-emission gun. This is of the order of the thermal-energy spread in an incident electron beam from a thermal emitter (*ca* 0.1 eV), but excludes other chromatic factors, associated with the electromagnetic optics and the beam/specimen interaction, which will also degrade the energy resolution. In most cases, it is the effective energy spread in the beam, not the characteristics of the EELS detector, which limits the energy resolution. It is therefore common practice to record the *zero-loss peak*, corresponding to the primary Gaussian peak for the beam exiting the specimen when observed at a nominal zero-energy loss. This primary peak is then used to calibrate both the absolute zero for the energy-loss spectrum and to estimate the available energy resolution, usually from the FWHM of the zero-loss peak.

The intensity of the *first plasmon peak* relative to the *zero-loss peak* is also a good benchmark for judging the suitability of a thin-film specimen for EELS analysis. In practice, if the intensity in the first plasmon peak is less than *one-tenth* of the zero-loss peak, then the specimen is thin enough for EELS. The *inelastic signal* from very thin samples is extremely weak, but increases as the thickness increases. However, in thick samples *multiply scattered electrons* obscure the edge structure of the energy-loss spectrum, while multiple scattering (between 1 and 20 scattering events per electron) requires a rather unreliable deconvolution correction if quantitative EELS analysis is to be attempted. In practice, there exists an *optimum* specimen thickness for EELS at which the *counting statistics* are adequate, while the probability of *multiple scattering* is small.

The sensitivity of the EELS system can be discussed in terms of either the *minimum detectable mass* (the minimum signal that can be identified from a given constituent) or the *minimum mass fraction* (the minimum detectable concentration of a given element). The EELS technique is remarkably sensitive and can pick up a signal from less than a few thousand atoms, with the detection limit varying with atomic number.

In many cases, the best *spatial resolution* is not required and it is possible to work with much higher incident beam currents, either by increasing the probe size or by working with the focused transmission image and a selected-area aperture. The sensitivity is then not limited by the *signal statistics* but rather by the *errors of extrapolation* in determining the relative contributions to the signal from the different absorption edges and from the background. Unfortunately, these errors are typically at least 10 % for K-excitation. Data determined from L-edges are markedly less accurate (the edge is much broader than the K-edge), while the results obtained from an M-edge are really only useful for confirming the *presence* of a constituent, and not for estimating *concentration*. The spectra can be displayed for any *energy range* and at any *gain* (Fig. 5.13). The change in gain is particularly important, given the exponential decay of the signal with increasing energy loss.

Identification of the constituents responsible for the edges observed in the energy-loss spectrum depends on accurate *calibration* of the energy scale of the magnetic spectrometer. The *zero-loss peak* is used to define the origin as the *mass centre* of this Gaussian peak, while, as noted previously, the *width* of the zero-loss peak (FWHM) is used to define the *energy resolution*. The carbon K-edge (Fig. 5.14) is usually selected to define the linear-energy scale, since carbon contamination makes this signal an ubiquitous feature of nearly all spectra. The position of any edge is defined as the position of maximum slope.

Excitation events for the *same* excited state of the atom will also result in *higher* energy losses, beyond the initial edge, which continue to the maximum possible

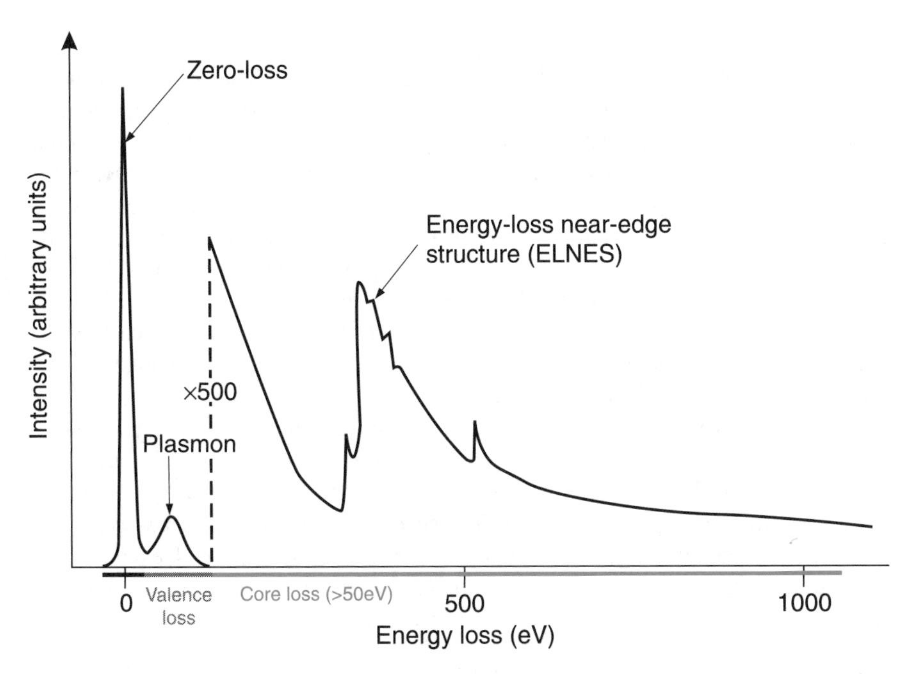

Figure 5.13 A schematic EELS spectrum showing the zero-loss peak and low-energy-loss region at reduced gain and an excitation absorption edge

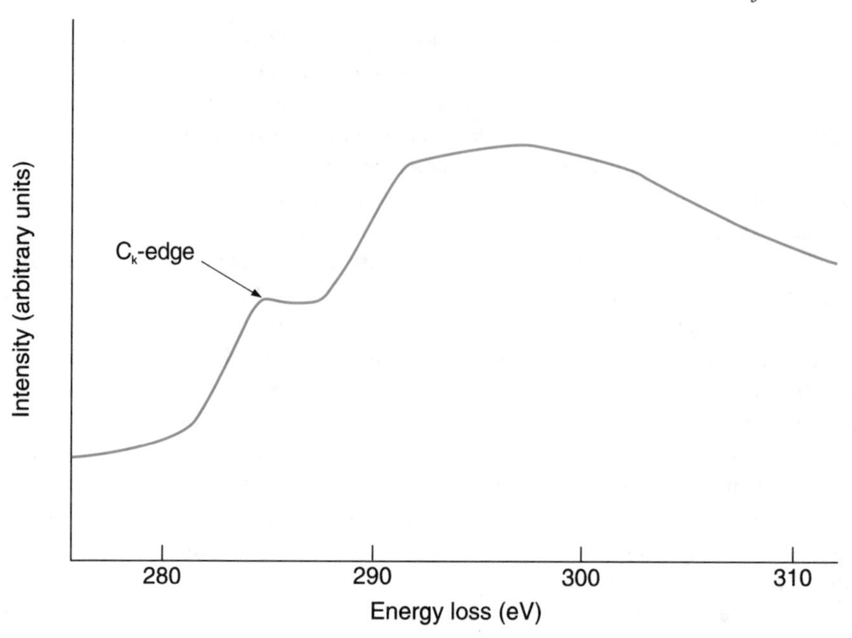

Figure 5.14 The carbon K-edge at 284 eV showing the fine structure

incident beam energy with rapidly decreasing probability. The number of events associated with any particular absorption edge is estimated by fitting an empirical function to the pre-edge data, extrapolating this *background* curve to the high-energy-loss region, and then subtracting it from the measured data (Fig. 5.15). Typical empirical functions are of the form $I = AE^{-r}$, where A and r are empirical constants, determined separately for each pre-edge energy-loss region. Since the signal from each edge decreases rapidly at higher energy-loss values it is not necessary to sum the counts derived for a particular edge over *all* energies above the edge, particularly since the extrapolation errors will increase with distance from the edge. However, it *is* important that all data for all edges should be summed over the *same range of energy loss*. This 'energy loss window', Δ, may be typically 50–100 eV, and the same window can be used to derive an *absolute mass data* scale by applying it to the *zero-loss* and low-energy, *plasmon-loss* region. This is carried out by assuming that the data are collected over a *fixed solid angle* (usually determined by the spectrometer aperture, β) and that the *inelastic scattering cross-section* for this angle and energy range, $\sigma(\beta\Delta)$, is known (estimated values for K-excitation are available). In this case, the *absolute number of atoms excited* is given by the equation:

$$N_K = \frac{I_K(\beta\Delta)}{I_0(\beta\Delta)} \frac{1}{\sigma_K(\beta\Delta)} \tag{5.9}$$

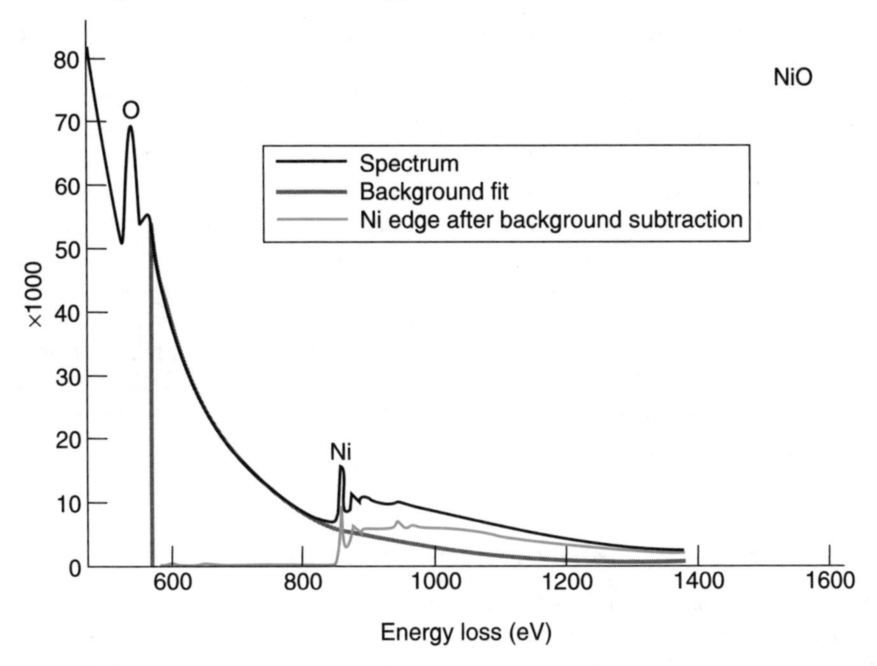

Figure 5.15 Illustration of the method of determining the number of excitation events associated with a specific edge for Ni in NiO. The background is fitted before the Ni edge, and then extrapolated to the edge region and subtracted (see text for details)

where I_K refers to the *K-edge signal* for a particular element and I_0 is the *zero-loss* and *low-energy-loss signal* collected for the same energy interval and collection angle.

5.2.3 Quantitative Electron Energy-Loss Analysis

Quantitative estimates of the *mass of an element present* in the excited region are generally less useful than determinations of *relative concentrations*. The simplest method of obtaining quantitative estimates of relative concentrations is by taking the ratios of the data derived from each edge. In this case, the zero-loss intensity data cancel out, and the zero-loss scan is only required to define the origin of the energy-loss coordinate:

$$N_i = \frac{I_i(\beta\Delta)}{I_0(\beta\Delta)} \frac{1}{\sigma_i(\beta\Delta)} \tag{5.10}$$

and

$$C_A = \frac{N_A}{\sum_i N_i} = \frac{[I_A(\beta\Delta)]/\sigma_A(\beta\Delta)}{\sum_i [I_i(\beta\Delta)/\sigma_i(\beta\Delta)]} \tag{5.11}$$

Of course, this also requires a good estimate of *all* of the relevant values of the *partial inelastic scattering cross-sections*, σ.

If two *separate* edges come within the range Δ, then it is clearly important to subtract the extrapolated energy-loss curve for the *higher* energy-loss edge *before* subtracting the former from the lower loss edge in order to obtain the excitation intensity for the lower-atomic-number constituent.

The major pitfall in applying EELS to microanalytical problems is in selecting an appropriate sampling thickness from a wedge specimen. Fig. 5.16 shows the calculated intensity ratios (normalized) for three different binary material combinations, B/N, Al/O and Al/Ni, as a function of the relative intensity of the first plasmon (low-energy-loss peak) peak to the zero-loss peak. The relative height of the first plasmon peak is proportional to the absorption thickness. Clearly, it is an advantage to choose a region as thin as possible, while still commensurate with obtaining adequate counting statistics.

5.2.3.1 Parallel Electron Energy-Loss Spectroscopy

In place of *scanning* the energy-loss spectrum and recording the number of counts in a given time interval for each energy channel *sequentially*, it is now possible to record many channels in *parallel* over a selected range of energy loss. Clearly, a multichannel analyser with a thousand channels can speed up data collection by a factor of one thousand, and the *parallel electron energy-loss spectrometer* (PEELS)

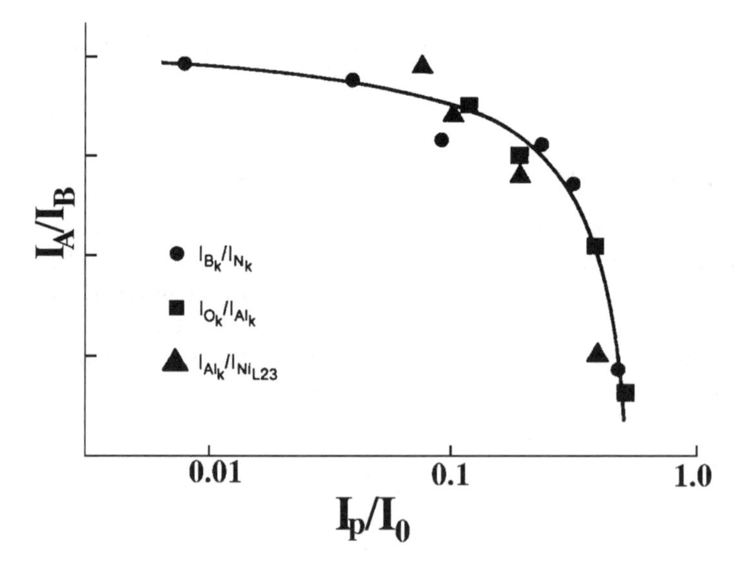

Figure 5.16 Intensity ratios for two ionization edges in three different materials as a function of the thickness, plotted as the ratio of the intensities of the first plasmon, low-energy-loss peak to the zero-loss peak (Philips Electronic Instruments)

has revolutionized the efficiency of energy-loss data collection, not only improving the *counting statistics* but also significantly reducing the effects of *radiation damage* and *contamination*, so that on the one hand it is now possible to use energy-loss spectrometry for *damage-sensitive* organic or polymeric materials, while on the other hand the improvement in counting statistics has made EELS practicable for high-atomic-number K-edge excitation, despite the very weak signals involved.

While EDS methods of microanalysis still have the advantage in accuracy over EELS methods at *high atomic number*, and EELS retains its superiority at *low atomic number*, both techniques effectively overlap over the full range of atomic number in the periodic table. EDS possesses greater analytical *accuracy*, while EELS possesses far better *spatial resolution* and *detection limits*.

5.2.4 Near-Edge Fine Structure Information

The 1 eV resolution of EELS detectors has revealed a wealth of structure in the region of the absorption edge which is still only partially understood, even though the general physical principles are clear. This *energy-loss near-edge structure* (ELNES) reflects both the *chemical state* of the atom and its *coordination* with respect to its neighbours. Changes in the *bonding state* correspond to *shifts* of the primary edge and the positions of the fine-structure peaks, while *peak splitting* is associated with a reduction in *coordination symmetry*. In alumina, cation disorder, present in the (higher-symmetry) transient structures gives rise to a few broad, comparatively simple peaks at the edge, while in the stable (less symmetric) corundum phase several sharp peaks are visible (Fig. 5.17). Calculations of the energy shifts associated with energy exchange in the nearest-neighbour coordination sphere are consistent with differences in coordination symmetry, as indicated for the aluminium $L_{2,3}$-edge in alumina (Fig. 5.18).

This additional EELS information is not limited to a determination of *chemical composition*, but is related to the *atomic bonding* in the solid. There is little doubt that the ability to distinguish bonding states and coordination symmetry is now a major motivation for expanding EELS applications.

5.2.5 Far-Edge Fine-Structure Information

Oscillations in the energy-loss signal are also observed at some distance *above* the position of an edge and are referred to as *extended energy-loss fine structure* (EXELFS). The wavelength of these oscillations in the energy spectrum is of the order of 20–50 eV, while the amplitude amounts to some 5 % of the edge signal. The effect is analogous to that observed in X-ray absorption, which is referred to as *extended absorption fine structure* (EXAFS). EXELFS is due to *elastic scattering* of the *inelastically scattered* electrons by the surrounding (periodic) atomic structure and may extend over several hundred electron volts above the edge.

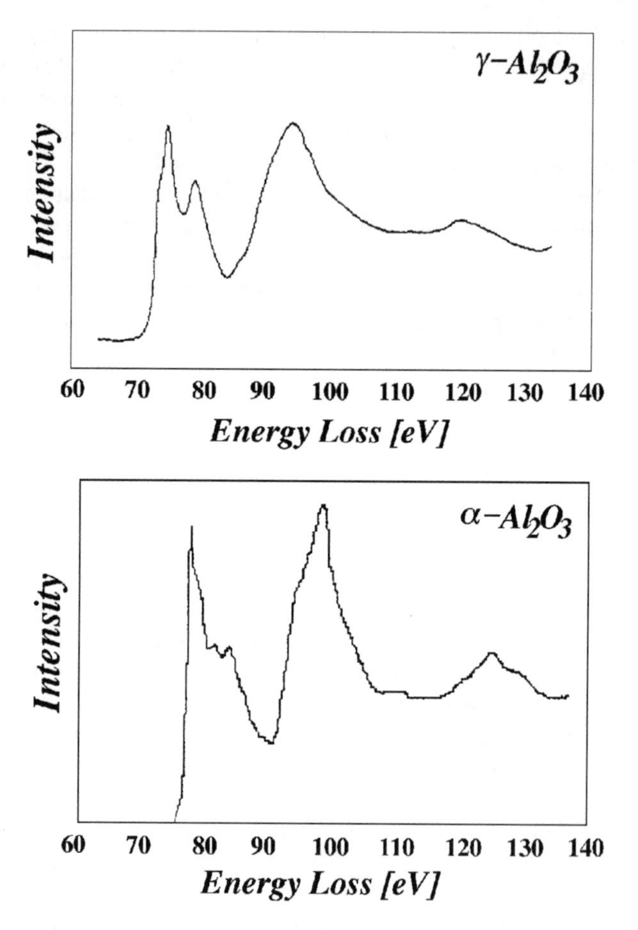

Figure 5.17 The aluminium $L_{2,3}$-edge in (a) γ- (cubic, metastable) and (b) α (rhombohedral, stable) Al_2O_3 (I. Levin)

The information contained in this EXELFS signal is related to the local atomic density and can sometimes be extracted to give a *radial distribution function* (RDF) for the atomic environment of the component responsible for the edge. The RDF records the probability that an atom exists (occupies a specific coordination sphere) at a given distance from the source atom. Extracting this information first requires subtraction of the background to reveal the oscillations, followed by a Fourier transform of these oscillations to convert them into a radial distribution function; in principle, this technique could supply information on the *near-neighbour atomic environment*, thus dramatically extending the usual concept of microanalytical information.

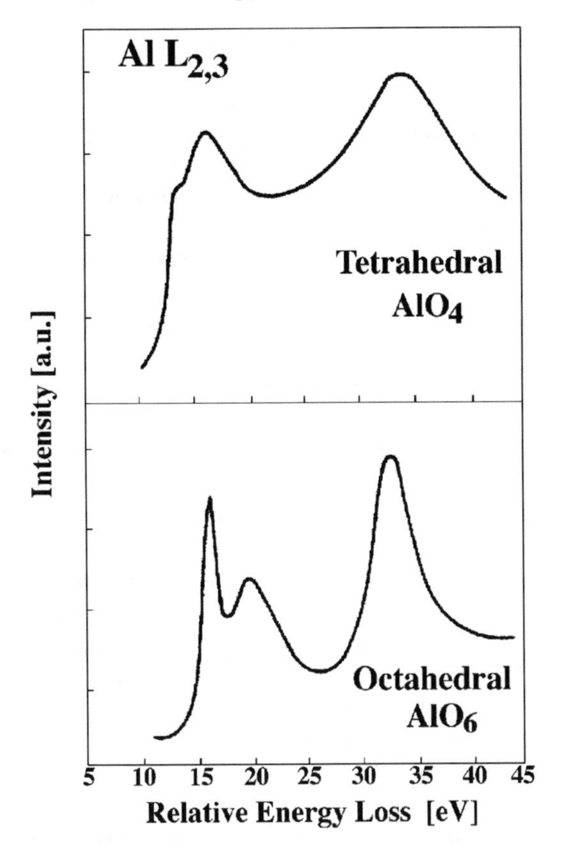

Figure 5.18 Calculated ELNES spectra ($AlL_{2,3}$-edge) for (a) tetrahedrally and (b) octahedrally coordinated Al cations (R. Brydson)

Summary

Information on the chemical composition of microstructural features is often of crucial importance, and considerable effort has been given to developing *microanalysis* in order to complement morphological and crystallographic data. Beyond the primary *qualitative* requirement of microanalysis, namely to identify the chemical elements present in a specific region, there are two *quantitative* aspects to be considered. The first is the *spatial resolution* of the analysis, while the second is the *spectral sensitivity*. The characteristic X-ray signal in the scanning electron microscope is the most important of the microanalytical tools which are available. This signal emanates from a volume element of material near the surface of the sample, of the order of 1 μm^3, with the corresponding spatial resolution for microanalysis being of the order of 1 μm. Better spatial resolution is available in the *energy-loss spectra* of the transmitted beam in the transmission electron microscope. X-ray microanalysis also has a limited spectral sensitivity, which is associated with the 'white' background of X-radiation which obscures the characteristic signal. Errors in *quantitative X-ray microanalysis* are typically of the order of 1 or 2 %, and the limit of detection for a constituent element is usually no better than 0.5 %.

Characteristic X-ray microanalysis may use K-, L- or M-spectra, depending on the atomic number of the constituents and the electron energy of the incident beam. In general, the incident beam energy should be selected to optimize the *spatial resolution* (which is best at *lower* beam energies) with respect to the *counting statistics* (which generally improve at *higher* beam energies). Two methods are commonly used to collect the characteristic X-ray spectra, namely *wavelength dispersion* and *energy dispersion*. In wavelength dispersion, a proportion of the X-rays emitted from the sample falls on a curved crystal and those photons that have a wavelength which fulfils the Bragg condition for diffraction by the crystal are focused on to a detector. By rotating the crystal and detector to change the Bragg angle, it is possible to scan across a range of wavelengths and thus record the excited X-ray intensity as a function of wavelength.

In energy dispersion, a solid-state detector is used which absorbs the photon energy and gives rise to current pulses whose intensity reflects the absorbed energy. These pulses are counted and a histogram developed of the number of pulses as a function of the energy of the photons. While wavelength dispersion has better *spectral resolution*, energy-dispersive systems have the advantage of recording *all* of the photons admitted to the detector simultaneously, and therefore has much better *counting statistics*. *Both* systems are important, depending on the application requirements, and *both* systems are limited in their detection efficiency for the long-wavelength radiation which is characteristic of *light elements*. Even so, it is still possible to identify all elements down to boron ($Z = 5$) and it is problems of *surface contamination* of the sample, rather than limitations of the spectrometer, which restrict the accuracy of light-element microanalysis.

In order to convert the recorded characteristic intensities into quantitative analyses of *chemical composition*, it is necessary to subtract the background count from the integrated characteristic intensity peaks, and reliable methods for optimizing *background subtraction* have been developed. In many cases, it is not possible to analyse for *all* of the elements present, and many software programs allow for additional chemical information, for example the assumption of *stoichiometry*, to be introduced into the quantitative analysis procedure. The size of the volume element of the sample in which an electron beam of given energy will generate characteristic X-rays depends on the *atomic number* of the sample, while the number of electrons available for X-ray excitation in the sample will depend on the proportion which are lost to *back-scattering* of the primary beam. The *atomic number*, or *Z-correction*, and the *back-scatter correction* are combined. A *high-Z* component in a region of *low* average Z will have a *higher* probability of exciting characteristic X-rays than would be expected just from the chemical composition, while the reverse is true for a *low-Z* constituent in a region of *high* average Z.

The X-rays detected by the spectrometer are generated beneath the surface of the sample, and may themselves undergo inelastic *absorption* processes before they escape from the surface. If microanalysis is to be made *quantitative*, then these absorption processes have to be corrected for, based on a knowledge of the *X-ray absorption coefficients* of *all* of the elements present for the *characteristic radiation* generated by *each* element. This has to be an iterative process, in which the composition of the sample is first assumed to be given approximately by the *relative intensities* of the characteristic lines of the X-ray spectra (after subtracting the *background radiation*), and the corrections for *atomic number*, *back-scattering* and *X-ray absorption* are estimated on this basis. The new corrected values for the relative concentrations are then used to re-evaluate the various *correction factors*, which in turn serves to provide a still more accurate estimate of the *composition*. In general, several iterations are required in order to yield values for corrected compositions which cease to converge. The extent of the *microanalytical errors* can be estimated by summing the calculated concentrations and determining the deviation from 100 %. In many cases, the *calculated compositions* are then further *normalized* so that they sum to 100 %, and these values are presented as the 'best estimate' of the composition.

One further correction is important for quantitative microanalysis, namely that due to *fluorescence*. If X-rays excited by the incident electron beam are absorbed before exiting the sample, then they must give rise to other, additional excitation processes. The most important of these is *fluorescent excitation* of characteristic radiation from other constituents present in the sample. While it is in principle possible to correct for fluorescence when necessary, it is not always possible to know the origin of the fluorescent radiation. The *electron excitation* of X-rays occurs in a limited volume beneath the electron probe, but once generated, the X-rays then have a mean free path in the sample of at least 10 μm, and subsequent fluorescent excitation may emanate from material some three orders of magnitude larger in volume than that corresponding to the original, primary X-ray signal.

While the most important application of X-ray microanalysis is to provide chemical information in *scanning electron microscopy*, it is also possible to detect characteristic X-rays generated in the thin-film specimens examined in the *transmission electron microscope*. In this case, the signal is very much weaker, both because of the small volume of material available in the thin film and also because the probability of inelastic scattering of the electrons is much reduced at the high beam energies which are used. *Fluorescence* effects are negligible in thin films, while *absorption* effects are much reduced and are inversely proportional to the film thickness. The latter also limits spreading of the incident beam, so that the *spatial resolution* for X-ray microanalysis in transmission electron microscopy is greatly improved, being limited only by the *counting statistics*.

Far better counting statistics for microanalysis in *transmission electron microscopy* can be obtained from the *energy-loss spectra*. In this case, the *lighter elements* give the most readily detectable signals, since the absorption edges are at low energy losses adjacent to the zero-loss beam in the spectrum. Nevertheless, the edges can also be detected in the tail of the energy loss spectra, and L-, M- and N-edges can be identified with specific *high-atomic-number* constituents. In addition to the characteristic edges, inelastic processes responsible for the energy-loss spectra include *photon excitations* (at the kT level), *electron transitions* within the atom (1–150 eV) and *plasmon excitations* associated with resonance in the conduction band of a metallic conductor (5–50 eV).

There has been a steady improvement in the performance of microanalytical systems, both in the 'hardware' and in the computer 'software' available for data processing and quantitative analysis. This has been particularly true for *energy-loss spectroscopy*, where the energy resolution has reached the point where quantitative composition analysis is approximately equivalent in accuracy to X-ray microanalysis in the scanning electron microscope, but with a spatial resolution of the order of 2 nm. A major improvement is achieved by *parallel electron energy-loss spectroscopy* (PEELS) in which the energy spectrum is collected simultaneously over a given energy band, instead of being limited to a step-by-step scan over the energy range required.

The improvement in energy resolution for electron energy-loss spectroscopy (EELS) has revealed considerable fine structure in the neighbourhood of the absorption edges. This *energy-loss near-edge structure* (ELNES) is associated with the chemical state of the atom (the nature of the chemical bonding) and its coordination in the solid (local symmetry and number of near neighbours). The *extended energy-loss fine structure* (EXELFS) observed at energies above the edge is associated with local composition changes in the atomic shells surrounding the constituent responsible for the absorption edge, and hence reflects changes in the degree of order. Microanalysis at this level is no longer a question of simply determining a local chemical composition, but is beginning to provide information on the chemical state and the local atomic symmetry and order.

Bibliography

1. D.B. Williams and C.B. Carter, *Transmission Electron Microscopy: A Textbook for Materials Science*, Plenum, New York (1996).
2. P. Buseck, J. Cowley and L. Eyring (eds), *High-Resolution Transmission Electron Microscopy and Associated Techniques*, Oxford University Press, Oxford (1998).
3. J.I. Goldstein and H. Yakowitz (eds), *Practical Scanning Electron Microscopy*, Plenum, New York (1975).
4. J.J. Hren, J.I. Goldstein and D.C. Joy (eds), *Introduction to Analytical Electron Microscopy*, Plenum, New York (1979).
5. R.F. Egerton, *Electron Energy-Loss Spectroscopy in the Electron Microscope*, Plenum, New York (1986)

Worked Examples

In this section, we will demonstrate some microanalytical techniques by using EDS and EELS on samples of polycrystalline alumina and a 1040 constructional steel.

We first focus on the use of EDS in the scanning electron microscope. The first specimen is polycrystalline alumina. Alumina is usually sintered (densified) with

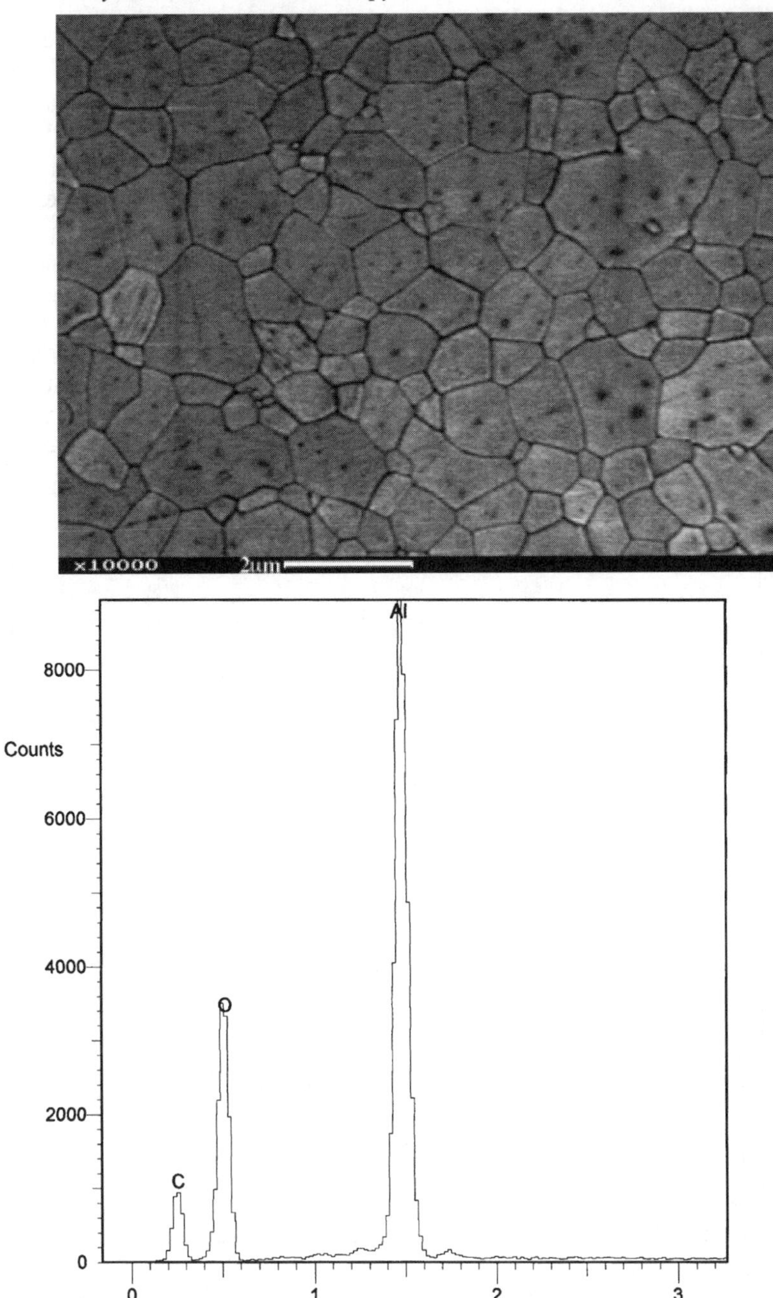

Figure 5.19 (a) Secondary-electron scanning electron micrograph obtained from thermally etched alumina. (b) The energy-dispersive spectrum from the entire region in (a) shows only aluminium and oxygen, as well as carbon due to surface contamination

additives (dopants) in order to prevent excessive grain growth or improve the sintering rate. In contrast, impurities are commonly found in alumina (Ca, Fe or Si for example) which may increase the rate of grain growth during sintering, and it is important to be able to detect these impurities when they are present. Fig. 5.19 shows a scanning electron micrograph from a polished and thermally etched alumina specimen, together with an energy-dispersive spectrum acquired from the entire region shown in the micrograph. The specimen was prepared using pure alumina, and intentionally doped with Mg, together with a small amount of Si and Ca. Energy calibration and peak identification was performed automatically by a computer program, and only the principle Al and O peaks were detected. A thin-window EDS detector allows light elements, such as O, to be detected. The failure to detect Mg, Si, and Ca in the energy-dispersive spectrum taken from the entire region shown in Fig. 5.19 was expected, since the total amount of these dopants over this rather large volume is below the detection limit. However, a spectrum taken with the probe positioned at a single point located at a grain boundary clearly shows the presence of

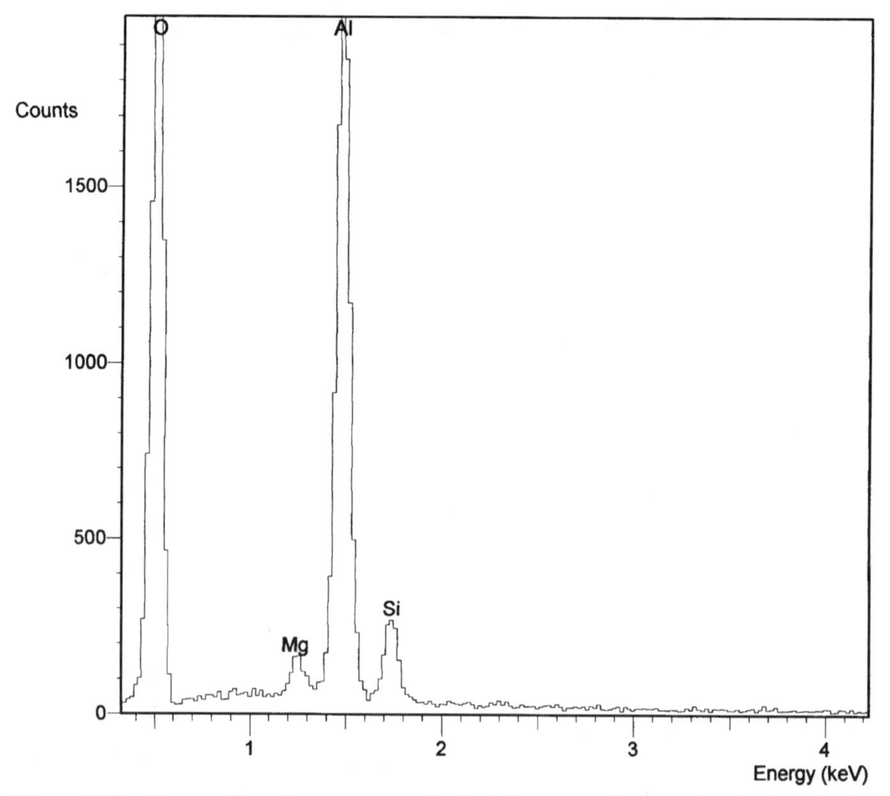

Figure 5.20 Energy-dispersive spectrum obtained from a grain boundary in the alumina specimen shown in Fig. 5.19. Al and O are of course found, as well as Mg and Si which have segregated (been enriched) at the boundary. In this particular case, Ca (which was present in the specimen) is either below the detection limit or is not present at this specific grain boundary

Figure 5.21 Transmission electron micrograph of the alumina shown in Fig. 5.19; an amorphous phase can be detected at the triple junction (labelled glass)

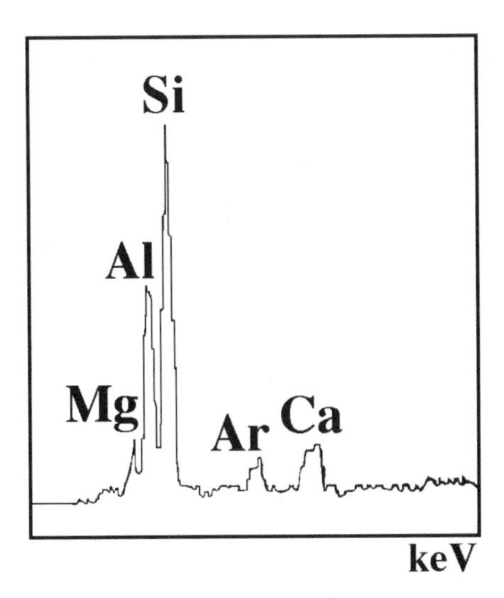

Figure 5.22 Energy-dispersive spectrum taken from the triple junction shown in Fig. 5.21; Mg, Si, and Ca can all be detected. Al is probably from the alumina grain, while argon is an artifact from ion milling; oxygen, however, could not be detected with the EDS system used to collect this spectrum

both Mg and Si, although the amount of Ca at this grain boundary is still below the detection limit (Fig. 5.20).

It is clear from this result that both Mg and Si have segregated at the alumina grain boundaries. We cannot detect any of the dopants within the grains because the solubility limit of the cations is well below the detection limit of EDS. However, the presence of Si and Mg at the grain boundaries is an important finding, which implies that the alumina grains are saturated with these impurities.

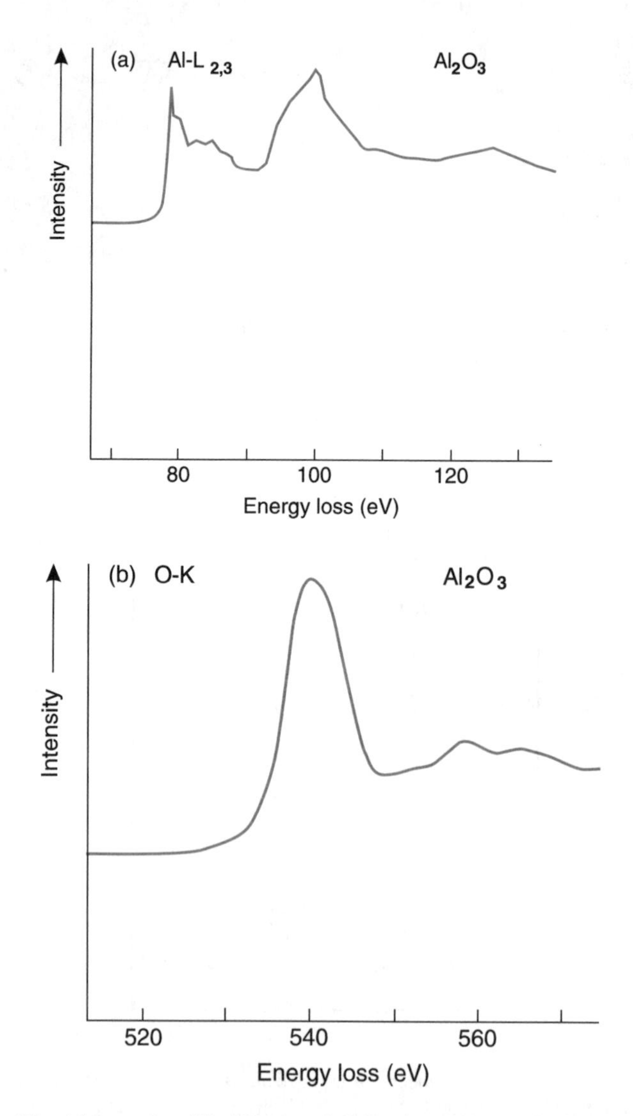

Figure 5.23 The Al $L_{2,3}$-edge (73 eV) (a) and O K-edge (532 eV) (b) in alumina

Fig. 5.21 shows a transmission electron micrograph of a region taken from the *same* alumina specimen. Grain boundaries are indicated by arrows. EDS in the electron microscope can also be used to determine if Si and Ca are present at specific boundaries or triple junctions. Fig. 5.22 shows an energy-dispersive spectrum taken from the triple junction which is indicated as containing glass in Fig. 5.21. Significant amounts of both Si and Ca are detected, as well as Mg. The Al signal is most likely from one of the alumina grains adjacent to the triple junction, while the argon signal is from argon ions which have been occluded in the thin TEM specimen during ion milling.

We can also use PEELS to microanalyse our alumina TEM specimen. Fig. 5.23 shows portions of the PEELS spectra acquired from an alumina grain, in the energy-loss region of the OK-edge (532 eV), and the $AlL_{2,3}$-edge (73 eV). Quantitative analysis of the PEELS spectra from the alumina grain confirms the stoichiometric concentration of Al_2O_3, and we can compare the ELNES of the Al edge from alumina to that from metallic Al (Fig. 5.24). The observed differences in the ELNES spectra reflect the differences in the nature of the chemical bonding of aluminium in the ionic solid and the metal.

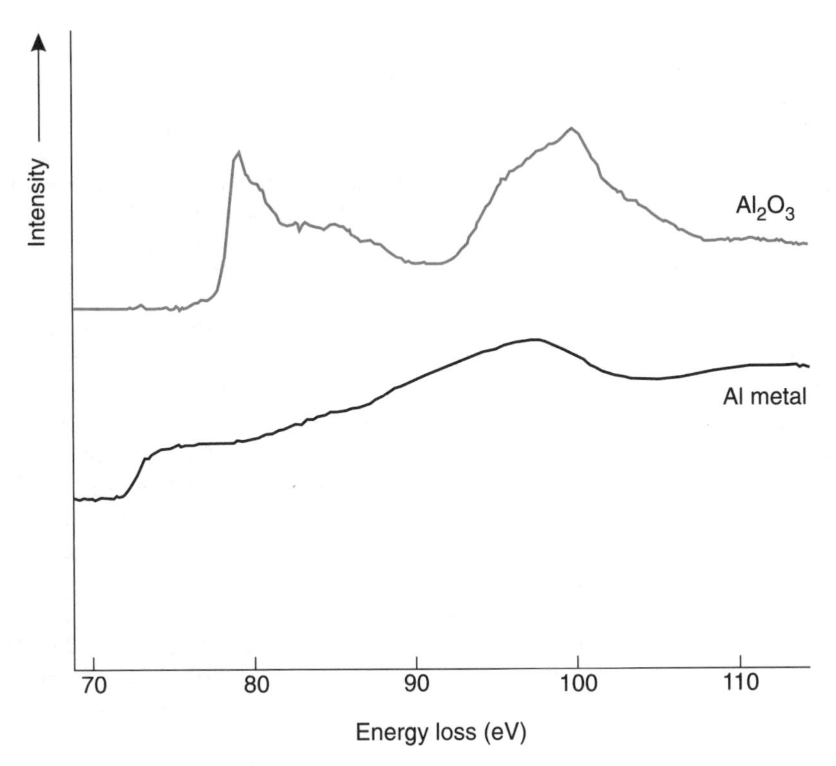

Figure 5.24 The Al $L_{2,3}$-edges in alumina and in aluminium metal; note the distinct differences in the positions of the peaks and in the peak shapes

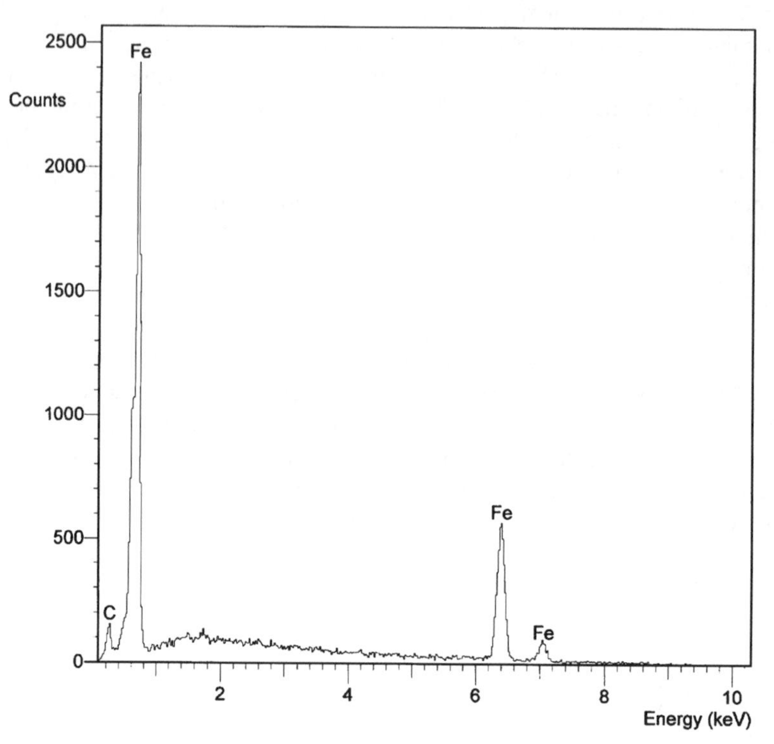

Figure 5.25 Energy-dispersive spectrum obtained for 1040 steel (from an α-Fe grain). The carbon signal is too high to result only from its 0.4 wt % content in the steel, arising in part from surface contamination

We now examine a 1040 steel sample by using EDS in the scanning electron microscope. EDS analysis of the carbon content of any steel is practically impossible, since the carbon content is at the limit of detection for the technique. Fig. 5.25 shows an energy-dispersive spectrum taken from a large region of the specimen. A carbon peak is visible, suggesting that the carbon content is much higher than expected for 1040 steel. However, *carbon contamination* of the surface due to breakdown of volatile hydrocarbons is a common problem in electron microscopy, and, in fact, the carbon signal detected has nothing to do with the carbon content of the alloy. However, Fe_3C is present in the specimen and has a carbon content well above the detection limit for EDS. As a consequence, carbide platelets can be clearly seen in the pearlitic regions of the microstructure (Fig. 5.26). The presence of a higher carbon content in the pearlite can be confirmed from an energy-dispersive spectrum taken from a point in the pearlite (Fig. 5.27—compare with Fig. 5.25).

In addition to pearlite, small dark areas are visible in the α-Fe grains and in the pearlite microstructure. Some of these dark regions may be larger Fe_3C regions which have formed during heat treatment. This can be confirmed by EDS line-scans

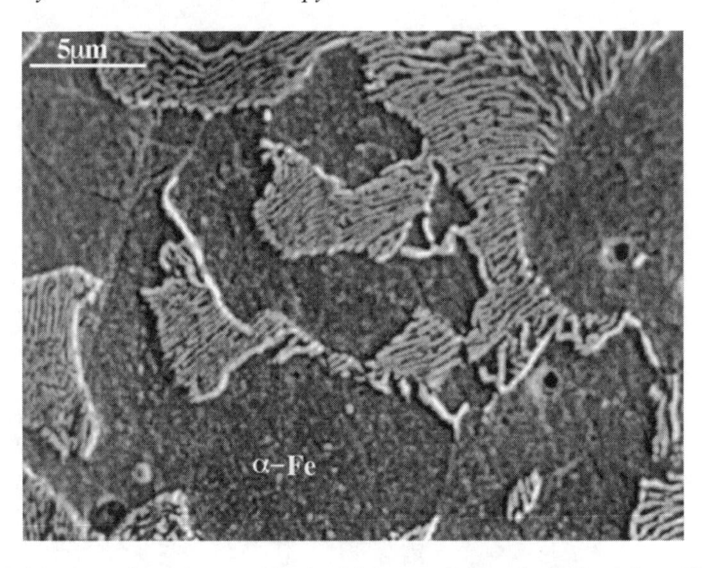

Figure 5.26 Secondary-electron scanning electron micrograph obtained from 1040 steel

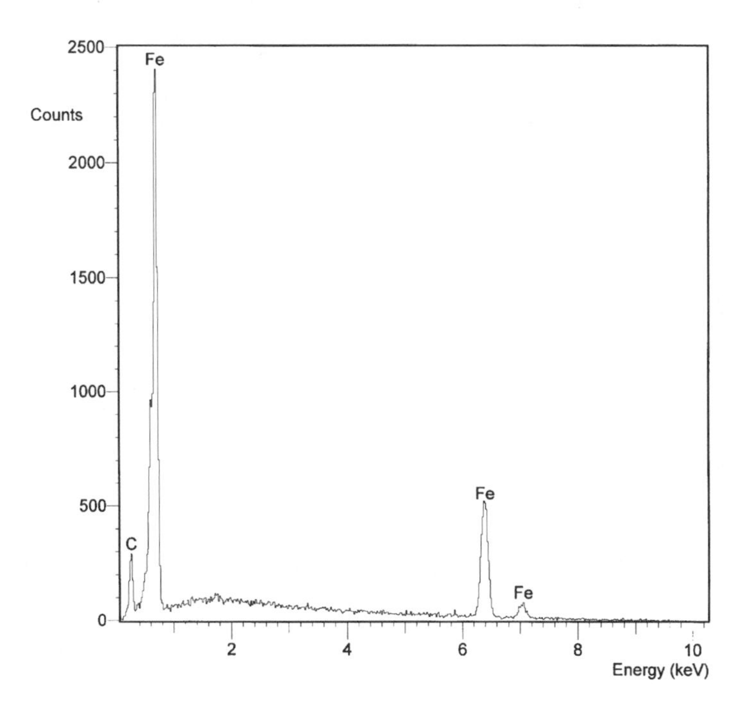

Figure 5.27 Energy-dispersive spectrum taken from the pearlite microstructure shown in Fig. 5.26. Note the increase in the carbon signal (relative to the Fe signal) when compared to that shown in Fig. 5.25

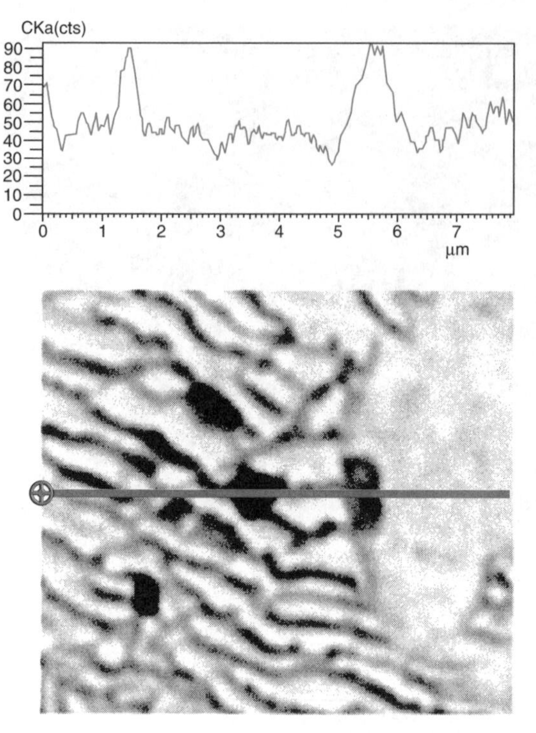

Figure 5.28 EDS line-scan (carbon signal) and corresponding secondary electron scanning electron micrograph of the interface between pearlite and α-Fe

across the dark regions (Fig. 5.28). In this figure, an X-ray line-scan for the carbon peak was run across two dark regions, with one being located within the pearlite area, and the other at the boundary between the pearlite microstructure and an α-Fe grain. The dark region at the interface between the pearlite and α-Fe has a much higher carbon content, and we conclude that this is a large carbide particle. The dark region within the pearlite does *not* show a higher EDS carbon count than the background, and this is most likely to be a surface pore. The EDS line-scan also shows the presence of another carbide particle towards the left side of the micrograph, which is not immediately evident from the micrograph alone. The presence of Fe_3C can therefore be detected by EDS line-scans, but a comparison of these line-scans to the secondary electron images of the same area is necessary to interpret the results with any confidence. Quantitative analysis of the carbide phase is strongly affected by the presence of carbon contamination.

Problems

5.1 Why is it important to know the chemical composition associated with microstructural features? Give three examples.

5.2 Why would you expect a fracture surface to be a difficult sample for microanalysis in the scanning electron microscope? What purpose might be served by comparing a microanalysis of a fracture surface with that of a polished and etched specimen of the same material?

5.3 How does the optimum electron beam energy for microanalysis depend on the density of the sample? What are the disadvantages of working with either too high or too low a beam energy?

5.4 Which factors determine the width of the characteristic line selected for quantitative analysis and how would you optimize the counting statistics to obtain a maximum signal-to-noise ratio?

5.5 Distinguish between sensitivity and accuracy with respect to microanalysis.

5.6 What are the statistical limitations on spatial resolution in microanalysis?

5.7 There is little point in attempting to analyse regions much less than 1 μm in diameter in the scanning electron microscope. Why is this?

5.8 Given that a characteristic X-ray spectrum should be corrected for background, atomic-number effects, back-scatter losses, X-ray absorption and X-ray fluorescence, in which order should these corrections be made?

5.9 An aluminium–copper alloy contains 3.5 % of copper. Would you expect the intensity of the Cu K_α radiation to be increased or decreased by the atomic-number effect and the back-scatter losses? Give reasons for your answer.

5.10 A series of iron–nickel alloys are to be subjected to microanalysis. Discuss the relative importance of the absorption and fluorescence corrections for iron and nickel as a function of the composition of the alloy.

5.11 A brazed joint is to be analysed. Would you recommend that the line of intersection of the joint with the sample surface be aligned parallel or normal to the X-ray collection system? (However, perhaps it doesnt matter?)

5.12 Why is electron energy-loss spectroscopy slowly displacing thin-film microanalysis in transmission electron microscopy?

5.13 How would you expect the limit of detection in energy-loss spectroscopy to vary with the atomic number of the constituent?

5.14 What are the advantages of parallel electron energy-loss spectroscopy as an analytical tool in transmission electron microscopy?

Chapter 6

Chemical Analysis of Surface Composition

The *chemical analysis* of composition is an important part of the characterization of a material at all stages of the materials cycle (extraction, manufacture, service and, ultimately, disposal or recycling). Nevertheless, it is not *per se* a topic for this text, which we have restricted to *microstructural* aspects of the characterization process. In Chapter 5, we have already explored *microanalytical methods* of chemical analysis associated with electron microscopy, both for bulk samples, using the *scanning electron microscope*, and for thin-film specimens in the *transmission electron microscope*, and this discussion covered both *X-ray microanalysis* and *electron energy-loss spectroscopy.*

In this present chapter, we will consider three further analytical methods which are especially important in surface engineering and science, predominantly in the study of solid-state device materials, including both opto-electronic and superconducting applications, particularly for thin-film devices and detectors. However, first we should mention some of the many available methods of analysis which will *not* be covered here. These include the following:

- *Atomic absorption (AA) spectroscopy* depends on the detection of characteristic absorption spectra for individual atomic species. The beam of individual atoms is commonly generated by *laser vaporization* of a sample, and the *absorption spectra* recorded by using a white light source.
- *Optical emission (OE) spectroscopy* is the inverse of atomic absorption spectroscopy, using *characteristic emission* lines of the different atomic species. A *spark* source is often used in this technique.
- *Infrared (IR) spectroscopy* relies on absorption spectra detected in the *infrared* range when the material is placed in the path of a suitable infrared source. There are a number of possibilities for extracting information, for the most part related to the chemical bonding in the system.
- *Raman spectroscopy* is a specific form of optical excitation (usually in the infrared region) in which *absorption* at one wavelength is accompanied by

emission at another wavelength. This is an *inelastic phonon scattering* process, and the emitted *Raman emission lines* can reflect the decay of a wide range of excitation states, primarily associated with variations in the chemical bonding.
- *Electron spin resonance (ESR) spectroscopy* is carried out on *paramagnetic* materials, which are subjected to *microwave* frequencies. The observed resonances correspond to specific electronic states in the material, such as those associated with specific *valencies*.
- *Nuclear magnetic resonance (NMR) spectroscopy* is carried out at *radio-frequencies* and corresponds to resonance of the magnetic moment of the *nucleus* for certain specific isotopes. Imaging by using NMR has revolutionized non-invasive diagnostic medicine. In particular, it has vastly extended the resolution and sensitivity of *X-ray radiography* and *ultrasound imaging* methods.
- *Fluorescence spectroscopy* is based on *characteristic X-ray excitation* by a beam of high-energy white X-rays. In contrast to the X-ray spectrum generated by an electron beam, there is *no* background radiation generated and the characteristic lines from trace elements can often be detected.
- *Rutherford back-scattering* makes use of the simple laws of *momentum transfer* which apply when a beam of MeV-energy ions (usually helium) is back-scattered after colliding with the individual atoms in a target. The *energy spectrum* of the back-scattered ions reflects their momentum distribution and can be quantified with excellent depth resolution.

Many other techniques for *bulk chemical analysis* exist. With the partial exceptions of high-resolution scanning secondary-ion mass spectrometry (SIMS), microbeam excitation of Raman spectra and infrared absorption, *none* of these techniques has any microanalytical capability and we will not discuss them further here. Table 6.1 summarizes the different *probes* used to excite the samples and the corresponding *signals* for the various analytical techniques which are either mentioned or discussed in this present text.

6.1 X-RAY PHOTOELECTRON SPECTROSCOPY

The photons in a beam of *monochromatic characteristic* X-rays which is incident on a solid target will be *absorbed* by the atoms and *secondary electrons* will be ejected whose kinetic energy (E_s) is equal to the difference in energy between that of the incident photon ($h\nu$) and the energy required to remove the electron from the target. The latter energy consists of two components, namely the *binding energy*, E_b, required to raise the electron to the Fermi level and the (orientation-dependent) *work function*, ϕ, required to bring the electron from the Fermi level into the vacuum. In other words, $E_s = h\nu - (E_b + \phi)$, so that the binding energy of the photoelectron (E_p) is given by $E_b = h\nu - (E_p + \phi)$, and *increases* as the energy of the photoelectron *decreases*. A schematic representation of secondary emission of a

Table 6.1 Common analytical methods used in chemical analysis

Method	Probe used	Signal detected
Optical spectroscopy	Visible light	Absorption and emission in the visible range
Infrared spectroscopy	Thermal excitation	Infrared radiation
Raman spectroscopy	Thermal excitation	Infrared emission
Electron spin resonance spectroscopy	Microwave excitation	Chemical bonding states
Nuclear magnetic resonance spectroscopy	Radiowave excitation	Selected isotope resonance
X-ray fluorescence spectroscopy	'White' X-rays	Characteristic X-rays
X-ray photoelectron spectroscopy[a]	X-rays	Secondary electron emission
Auger electron spectroscopy[a]	Energetic electrons[b]	Secondary electrons from Auger transitions
X-ray microanalysis[a]	Energetic electrons	Characteristic X-rays
Electron energy-loss spectroscopy[a]	Energetic electrons	Absorption edges
Secondary-ion mass spectrometry[a]	Inert (keV) ion beam	Sputtered target atoms and ions
Rutherford back-scattering	Helium (MeV) ion beam	

[a] Methods treated in the present text
[b] Most commonly used

photoelectron from copper is shown in Fig. 6.1, while Fig. 6.2 shows the corresponding copper photoemission spectrum. In addition to the lines corresponding to the emission of electrons from the various energy levels of the inner electron shells, the spectrum also includes *Auger electrons* which are associated with energy transitions *within* an atom *after* excitation by an incident X-ray photon. The Auger peaks are readily separated from the photoelectron spectrum by recording *two* spectra using two characteristic wavelengths (photon energies) for the excitation. The Auger peaks will remain in the *same position* in the two spectra, while the photoemission lines will be *shifted* by an energy corresponding to the energy difference between the two incident X-ray lines.

The electron binding energies of interest may exceed 1 keV. The relationship between the X-ray energy and its wavelength is $\lambda = hc/E = 1.24/V$ nm, where c is the speed of light, E is the energy of the quantum and V is in keV, so we conclude that suitable *excitation wavelengths* are of the order of 0.1 to 1 nm.

6.1.1 Depth Discrimination

The *background signal* in the photoelectron spectrum arises from *multiply scattered* secondary electrons which are generated in the deeper layers of the sample. If these secondaries are *inelastically scattered* before they can escape from the surface, then

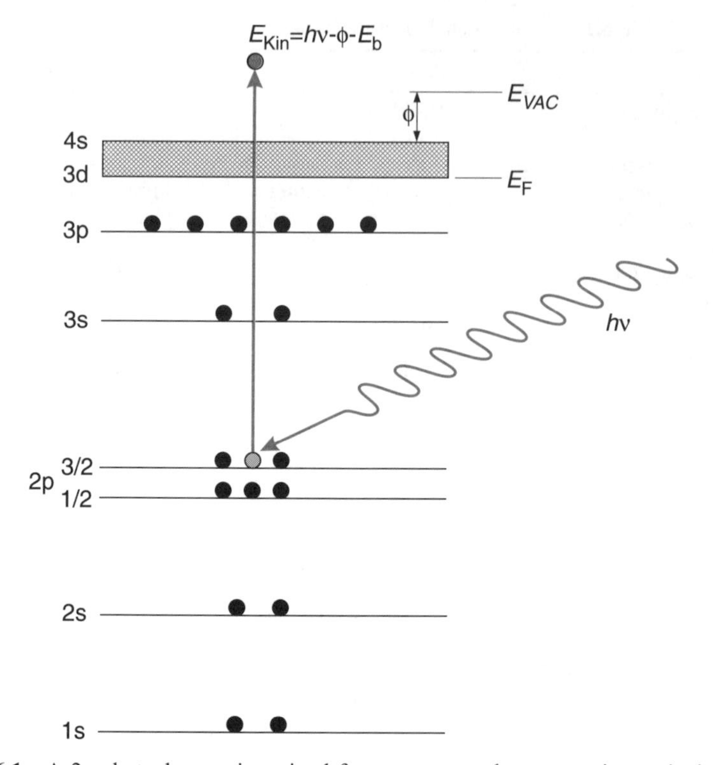

Figure 6.1 A 2p photoelectron is emitted from copper, and accompanies excitation of the atom by absorption of an X-ray photon

they will contribute to the *background* rather than to the *characteristic* peaks. It follows that the observed peaks in the spectrum are due to photoelectrons which are generated in the *surface layers only*, at a depth which is *less* than the mean free path of the secondaries in the material, while the higher background levels, which lie to the *high*-binding-energy side of each peak, are due to those inelastically scattered electrons which originate in the deeper layers and correspond to a *lower-energy* collected signal.

In practice, both *elastic* and *inelastic* scattering can occur, and the *attenuation length* is a measure of the loss of intensity in a signal detected normal to the surface as a function of depth (Fig. 6.3). The broad minimum for photon energies between 10 and 500 eV corresponds to emission depths of 2 to 5 atomic layers and accounts for the power of this technique for analysing *surface composition* and *chemical binding*.

X-ray photoelectron spectroscopy (XPS)† is extremely sensitive to chemical changes occurring at the surface, with a *depth resolution* close to atomic dimensions.

† XPS is sometimes called electron spectroscopy for chemical analysis (ESCA). The authors prefer XPS since this name is more related to the physics of the technique than ESCA.

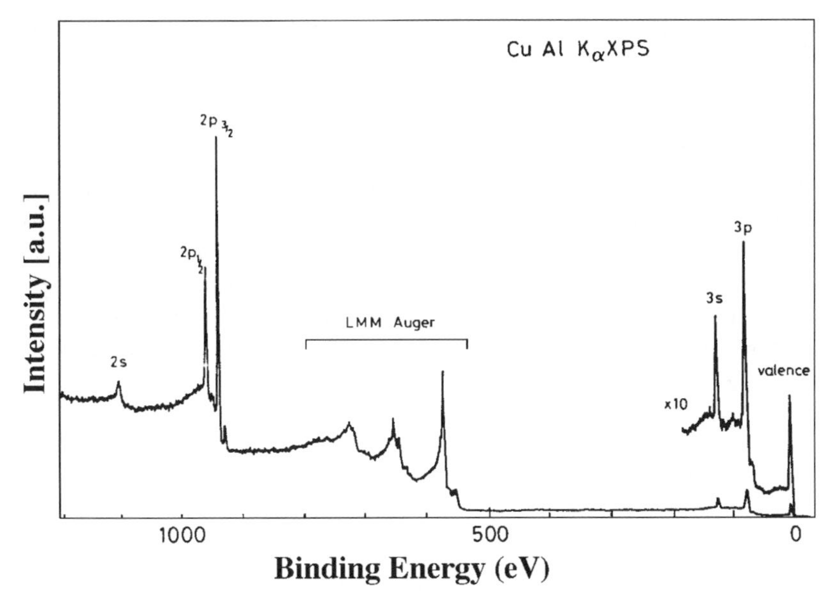

Figure 6.2 The photoelectron spectrum of pure copper includes both the photoelectrons and Auger electrons associated with energy transitions within the atom after excitation (see Section 6.2) (Reprinted from *Materials Characterization*, Vol. 25, G. C. Smith, Qualitative surface analysis by Auger and X-ray photoelectron spectroscopy, pp. 37–71, Copyright 1990, with permission from Elsevier Science)

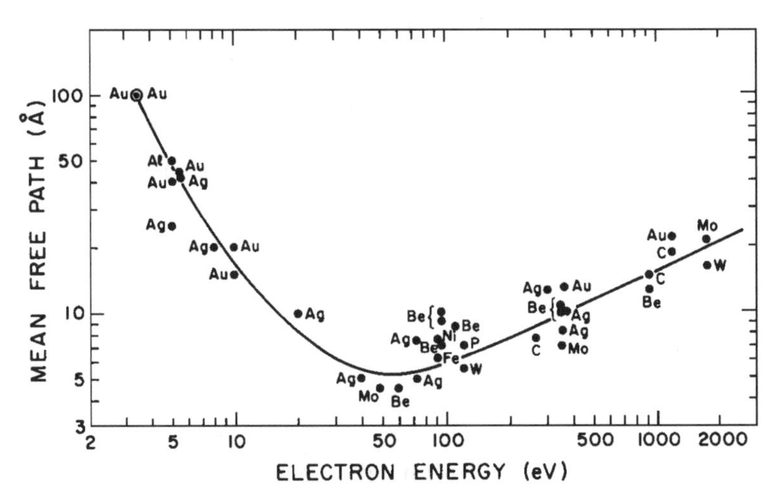

Figure 6.3 The attenuation lengths for electrons detected normal to the surface of a number of different solids as a function of their energy; a broad minimum is shown in the range 10–500 eV, corresponding to a few atomic layers. (Reprinted from Feldman and Mayer, *Fundamentals of Surface and Thin Film Analysis*, Copyright 1986, with permission from Elsevier Science)

However, unlike the other techniques of microstructural characterization which we have discussed, it possesses no imaging capability.

6.1.2 Chemical Binding States

The *binding energies* for electrons in the outermost shell of an atom are strongly affected by the *chemical state*, so that photoemission from the *same* atomic species in a different *coordination* or *binding state* will give rise to *multiple* peaks in the photoelectron signal. Fig. 6.4 displays a particularly elegant example of this behaviour, in which the 1s binding energies for carbon in an organic molecule are clearly distinguished by *peak splitting*, corresponding to energy differences of a few electronvolts.

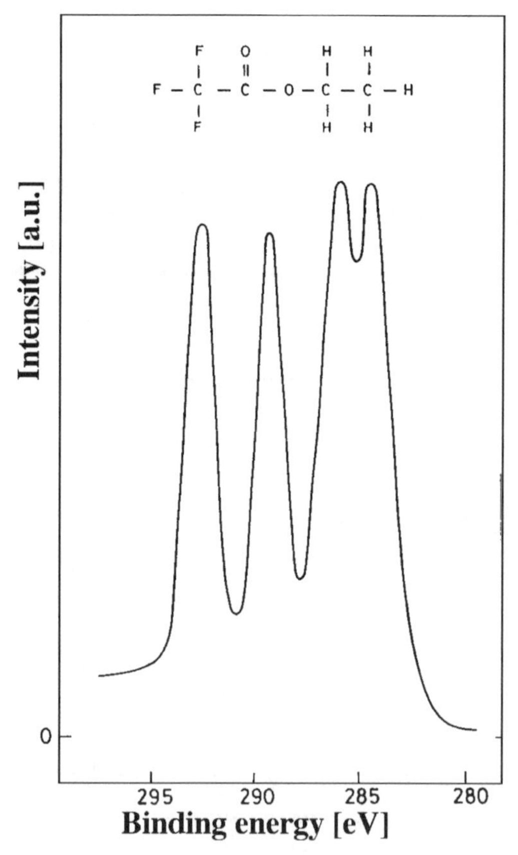

Figure 6.4 The binding energy of a 1s electron in carbon is a clear indication of its chemical binding state, with well-resolved peaks for each of the four carbon atoms in an ethyltrifluoracetate molecule (Reprinted from *Materials Characterization*, Vol. 25, G. C. Smith, Qualitative surface analysis by Auger and X-ray photoelectron spectroscopy, pp. 37–71, Copyright 1990, with permission from Elsevier Science)

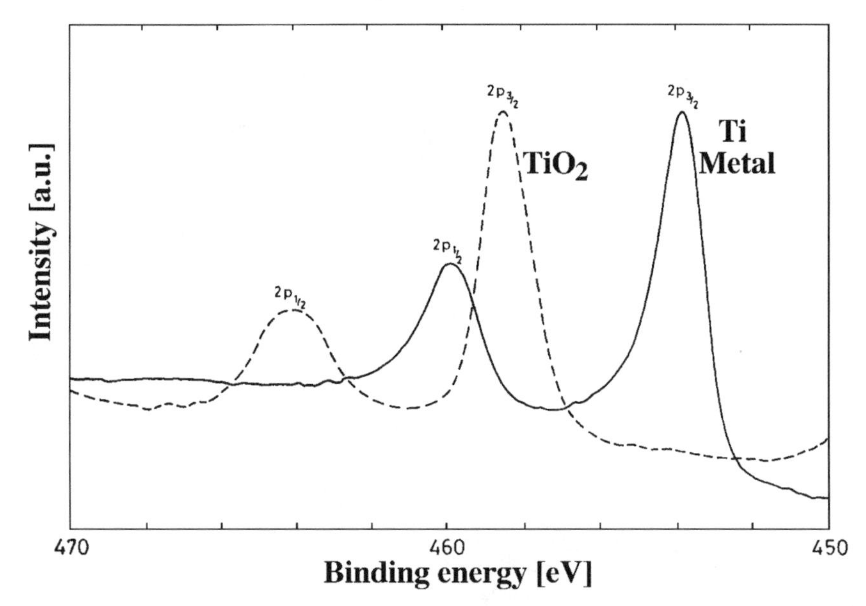

Figure 6.5 Oxidation states on metal surfaces can be resolved by using XPS; the growth of TiO_2 on titanium metal is accompanied by 2p peak shifts of 4–5 eV. (Reprinted from *Materials Characterization*, Vol. 25, G. C. Smith, Qualitative surface analysis by Auger and X-ray photoelectron spectroscopy, pp. 37–71, Copyright 1990, with permission from Elsevier Science)

Similar effects are seen in the *oxidation states* of metals, and an example (Fig. 6.5) is given for the 2p states in titanium, for which the higher binding energy in the *oxide* gives a peak shift of 4–5 eV.

6.1.3 Instrumental Requirements

Since the signal comes from the first few atomic layers, it is important to ensure that the surface of the sample remains *uncontaminated* during the analysis. An *ultra-high vacuum chamber* is required, and the removal of surface layers of contamination is usually achieved by argon ion *sputtering* at ion-beam energies of a few kilovolts. The arrival rate of gaseous species from the environment depends on their molecular weight, the temperature of the gas and the pressure. For air at room temperature, a monolayer of the gas will arrive (but not necessarily 'stick') in *one second* at a pressure of *ca* 10^{-6} *torr*. The time required for monolayer coverage depends linearly on the pressure, so we may conclude that pressures of 10^{-8} torr are mandatory if reasonable time is to be available to collect data without environmental contamination of the clean surface. This is usually the lower limit for the vacuum required in an XPS chamber.

In principle, there is no reason why any system of *electron spectroscopy* should be limited to only one type of excitation probe, and it is possible to purchase ultra-high

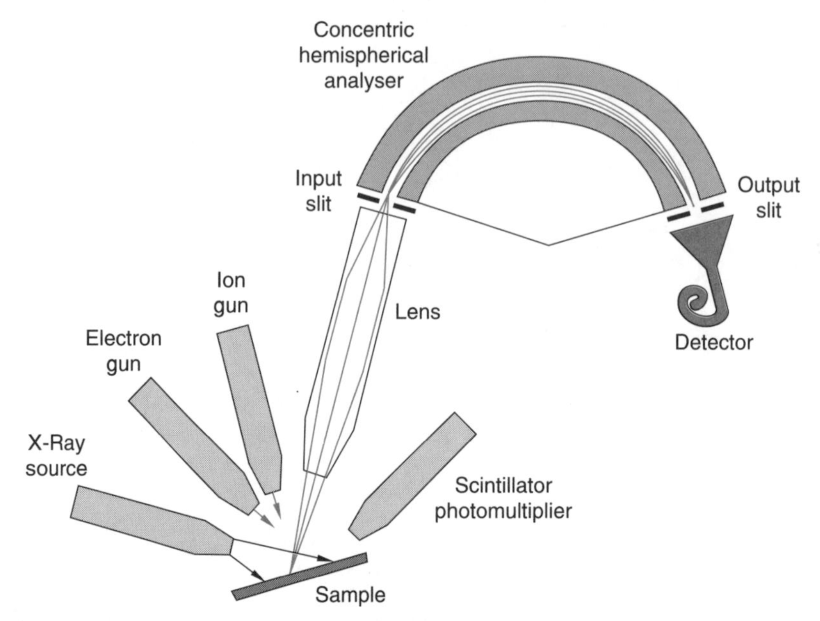

Figure 6.6 Schematic representation of the layout for an ultra-high vacuum electron spectrometer which can be tailored to incorporate various probes and specimen stages, thus providing facilities for both Auger analysis and X-ray photoelectron spectroscopy. (After Smith, *Materials Characterization*, Vol. 25, Qualitative surface analysis by Auger and X-ray photoelectron spectroscopy, pp. 37–71, Copyright 1990, with permission from Elsevier Science)

vacuum chambers which combine an analyser system with a *range* of probe sources, as well as specimen stages and carousels. Fig. 6.6 indicates schematically some of these possibilities. The *specimen* is cleaned by *sputtering* under an incident ion beam (usually argon). The *signal* is generated either by incident X-rays or by an electron beam. The latter can be scanned over the surface in a fine-focus raster and a *secondary electron image* of the surface observed, using a suitable detection system, in order to position the sample.

The electron signal is usually analysed *electrostatically*, in a focusing analyser with an energy resolution of the order of the inherent peak width (typically 1 eV). The *secondary electrons* are focused on a defining *input slit* and the electrostatic field in the *analyser* brings the electrons of a *specific energy* to a corresponding focus at an *output slit* where they are detected and the signal amplified. Scanning the analyser voltage then scans the focused energy across the output slit, whose width will limit the energy resolution.

The provision of a *specimen carousel*, capable of accommodating a number of samples, drastically reduces the average time required to change specimens, while high-vacuum *airlocks*, which include pre-bake facilities for *degassing* the sample, cut sample exchange times to no more than a few hours, despite the stringent vacuum requirements. Several degrees of freedom are provided with the *specimen stage*, which allow *tilting* of the specimen with respect to both the *probe source* and the *spectrometer*. This is particularly important, since both the take-off angle of the signal and the incident-beam angle can affect the recorded spectrum, either as a result of the crystalline anisotropy of the specimen or the increased secondary-electron path length at shallow take-off angles at a given emission depth.

6.1.4 Applications

The information available from X-ray photoelectron spectroscopy fills an important gap in our ability to characterize *surface structure* and *surface chemistry*. Reactions in solids for the most part take place at surfaces and interfaces, and the sensitivity of XPS analysis makes it a first choice in studies of *gas-phase absorption* and *catalysis* involving partial coverage of less than a monolayer. Given the sensitivity of the technique, it should be quite possible to pick up energy differences associated with surface atoms of the *same* species which have *different* coordination numbers (atoms on low index surfaces, close-packed ledge atoms and atoms at kink sites).

With the rapid development of *electronic device technology*, particularly thin-film detectors and optical systems, XPS is proving a powerful method for quantifying chemical changes at surfaces. What it *cannot* easily do is analyse the surface *composition* of complex samples or provide any useful lateral resolution. For this purpose, we need *Auger electron spectroscopy* (AES).

6.2 AUGER ELECTRON ANALYSIS

A typical energy scheme for Auger *non-radiative emission* (that is when no photon emission is involved) is shown in Fig. 6.7. *Excitation* of the atom may be either by an incident energetic *electron* beam or by *photon excitation* (hence the presence of Auger peaks in XPS spectra), but electron excitation is generally preferred. By using an electron beam scanned over the surface, a *secondary electron image* can be recorded, as in the scanning electron microscope, and the region for *Auger analysis* selected. Focusing and scanning of the incident beam can be accomplished with good lateral resolution of the Auger information so that, providing the signal intensity is sufficient, *Auger imaging* is feasible. The characteristic energy of the Auger electron is determined by *three* energies characteristic of the atom, namely the energy of the *excited state*, E_{L3} (in the case illustrated in Fig. 6.7), the energy *released* by the M-electron which fills the vacant L-hole, E_{M1}, and the energy *absorbed* in allowing the *Auger electron* to escape to the vacuum from its original M-level, E_{M23}, with an energy ${}^{A}E_{L3M1M23}$. From this, we obtain ${}^{A}E_{L3M1M23} = E_{L3} - E_{M1} - E_{M23}$.

In this discussion, we have simplified considerably the potential complexity of all *absorption* and *emission spectra*, primarily because most methods of analysis do not really distinguish the *fine energy structure* in either absorption or emission spectra. This is *not* the case in *Auger spectroscopy*, and the resolvable *fine structure* constitutes an additional tool for analysing *chemical bonding* and distinguishing neighbouring Auger peaks. The energies of the K-, L- and M-states are well separated, and while not all quantum excitations are observed, all of the *possible*

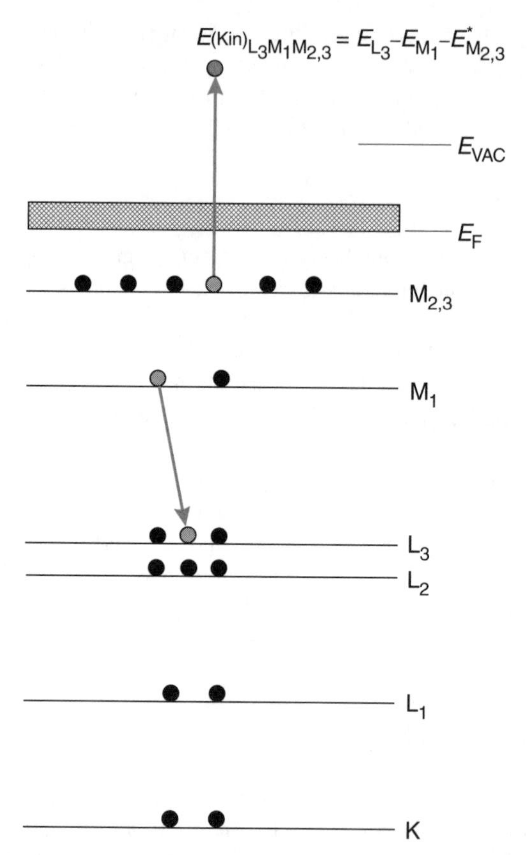

Figure 6.7 An Auger electron originating from the M-level is emitted from copper as a result of an atom which is excited to the L-state then decaying to the M-state. The energy of the Auger electron is determined by the energy of the excited state, the energy released by the M-electron which fills the vacant L-hole, and the energy absorbed in allowing the Auger electron to escape to the vacuum from its original M-level

transitions (and, for the most part, their relative *transition probabilities*) are both well known and well understood. There is therefore little chance of confusion, always providing that the operator remains aware of the need to refer to the relevant literature and the importance of meticulous *calibration* procedures.

Auger excitation is particularly complex, since three *separate* energy states contribute to the energy of the Auger electron, thus giving a far wider range of possible Auger energies than is possible in *X-ray photoelectron spectra* or the characteristic X-ray spectra in X-ray microanalysis. Auger electron excitation is *in competition* with characteristic X-ray excitation, and the energy absorption edges, which are the basis of EELS analysis when the 'probe' is an electron beam, are associated with *both* Auger excitation *and* characteristic X-ray excitation. As we have noted previously, the *low*-atomic-number constituents give *weak* characteristic

X-ray signals, while their corresponding *absorption edges* are strongly defined. It should not be a surprise, therefore, that the Auger electron emission signal is *strongest* for these same low atomic number elements, and the greatest strength of *Auger spectroscopy* is its excellent sensitivity to changes in chemical concentration and binding of these *low-atomic-number* constituents.

6.2.1 Spatial Resolution and Depth Discrimination

The range of *Auger electron energies* is the same as the range of energies observed for *photoelectrons*, and the signal therefore originates from the *same* surface layers of the sample, which are typically two to five atomic layers from the surface. However, in Auger excitation by *electrons* it is possible to focus the incident electron beam to a fine *probe* and limit the region for analysis to the small surface area beneath the probe. As in the scanning electron microscope, the total electron current in the probe depends on the *third power* of the probe diameter and falls rapidly as the size of the probe is reduced, while the ultimate limit on probe size is set by the electro-optical parameters of *electron wavelength* (energy) and *lens aberrations*.

Since the signal is derived from the first few atomic layers at the surface, it is to a large extent determined by the vacuum conditions in the instrument and the *rate of contamination* under the electron beam is a far more important factor limiting performance in *Auger electron spectroscopy* (AES) than it was for any of the previous methods of microstructural characterization which we have discussed. (This even includes *high-resolution lattice imaging* in transmission electron microscopy, where the thin-film specimen is also contamination-susceptible.) As in XPS, *ultra-high-vacuum* chambers are essential, with residual pressures of the order of 10^{-9} torr. These conditions require periodic *bake-out* (heating to de-gas the specimen and surrounding surfaces) of the assembly. The optimum electron probe size is a compromise between the limitations on the *mechanical tolerances* imposed by a bakeable, ultra-high-vacuum system, the *statistical limitations* determined by the probability of observing a given Auger electron signal, and the *electron optics* of probe formation.

With the exception of the vacuum conditions, these requirements are all rather similar to those which we have discussed before with respect to both electron energy-loss spectroscopy and thin-film X-ray microanalysis. In the case of AES systems, beam energies of several keV are generally used, with probe diameters of the order of 50–100 nm, although resolutions of better than 10 nm have been reported for high-intensity LaB_6 or field-emission electron sources and high-performance electron optics.

As for scanning electron microscopy, the total secondary electron signal (which contains the Auger signal) typically exceeds the incident beam current by at least two orders of magnitude, so that there is no problem whatsoever in recording a secondary electron *scanning image* of the surface at a better spatial resolution than that achievable for Auger analysis (typically limited to *ca* 20 nm).

6.2.2 Recording and Presentation of Spectra

The accepted method for presenting Auger data is as the *differentiated-signal* rather than the *collected-energy* spectrum. Fig. 6.8 compares these two methods of presentation. In the *differentiated* form, the *background signal* is effectively removed and each *Auger peak* is defined by a maximum and minimum in the differentiated signal, corresponding to the maximum *slope* of the leading and trailing edges, respectively, of the Auger peak. In the differentiated signal, the *position* of a peak is defined by the zero-*point* between the positions of maximum and minimum slope, while the distance between the maximum and minimum in the differentiated curve defines the *peak width*. These two parameters, *peak position* and *peak width*, are usually quite sufficient to identify the *atomic species* responsible for the signal, often by a simple visual comparison with published spectra.

Nevertheless, *quantification* of the Auger signal, as opposed to simple element *identification*, requires the same attention to *correction procedures* as X-ray microanalysis, for which the *direct* signal is necessary in order to derive the total integrated characteristic X-ray intensities (Section 6.2.4).

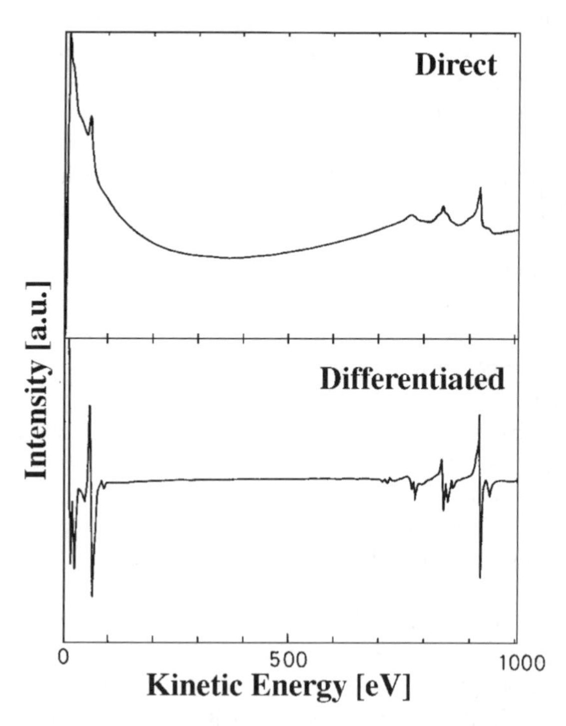

Figure 6.8 The Auger spectrum obtained from copper can be presented as either the energy dependence of the collected intensity (the direct signal) or as its differential; the latter form is that which is usually published in the literature. (Reprinted from *Materials Characterization*, Vol. 25, G. C. Smith, Qualitative surface analysis by Auger and X-ray photoelectron spectroscopy, pp. 37–71, Copyright 1990, with permission from Elsevier Science)

6.2.3 Identification of Chemical Binding States

Since the *Auger electrons*, like the *photoelectrons*, originate from the *outer* shells of the atoms, they are sensitive to the *chemical binding states*. In general, two effects are possible, i.e. either a simple *energy shift* of a main Auger peak, which directly reflects a change in the binding energy of the atom, or a change in the *energy loss structure* on the low-energy side of a peak. An example is given in Fig. 6.9 for an aluminium Auger peak taken from both the metal and its oxide. The *low-energy-loss structure* is analogous to ELNES (Section 5.2.4).

One difficulty in Auger fine-structure analysis is to separate *crystallographic* effects from *chemical* effects. For example, the *channelling* of the electrons in the incident beam down a low-index direction may greatly affect the *intensity* and *structure* of the Auger signal, independent of any chemical effects, and lead to an *orientation dependence* of the recorded spectra.

6.2.4 Quantitative Auger Analysis

As in the case of *quantitative* X-ray microanalysis, the role of *quantitative analysis* in both XPS and AES is to derive data on the *chemical composition* from the measured intensities of specific spectral lines from an unknown specimen compared to the intensities from a *known standard* determined under the *same* experimental conditions. For a binary system, and assuming a linear dependence of composition

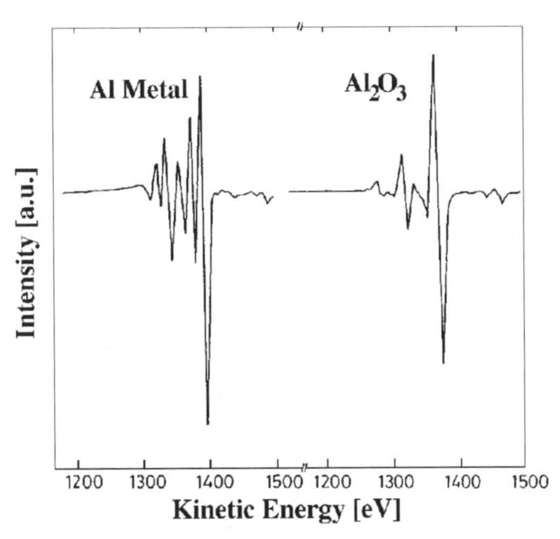

Figure 6.9 A differentiated Auger peak for metallic aluminium compared with that recorded for aluminium oxide, showing an 18 eV shift in the peak position and extensive differences in the energy-loss structure. (Reprinted from *Materials Characterization*, Vol. 25, G. C. Smith, Qualitative surface analysis by Auger and X-ray photoelectron spectroscopy, pp. 37–71, Copyright 1990, with permission from Elsevier Science)

on the measured intensity, this implies an atom fraction of A(X_A), which will be given by the relation:

$$X_A = \frac{I_A/I_A^0}{(I_A/I_A^0 + I_B/I_B^0)} \tag{6.1}$$

In practice, the standard intensity values are available as calibrated *sensitivity factors*, S, and the atom fraction of A for a multicomponent system may be approximated as follows:

$$X_A = \frac{I_A/S_A}{\sum_{i=1}^{n}(I_i/S_i)} \tag{6.2}$$

Although this equation may be satisfactory as a rough approximation, it omits corrections arising from differences in *emission probability* which are associated with *composition* and which are due, among other things, to changes in the *back-scatter coefficient* with composition (compare X-ray microanalysis, Section 5.1.3.1).

For the binary-alloy case, a *matrix factor* F_{AB} can be included so that

$$X_A = F_{AB}\frac{I_A/S_A}{I_B/S_B} \tag{6.3}$$

where F_{AB} is given by the equation:

$$F_{AB}^A(X_A \to 0) = \left(\frac{1 + R_A E_A}{1 + R_B E_A}\right)\left(\frac{a_B}{a_A}\right)^{3/2} \tag{6.3a}$$

or

$$F_{AB}^A(X_A \to 1) = \left(\frac{1 + R_A E_B}{1 + R_B E_B}\right)\left(\frac{a_B}{a_A}\right)^{3/2} \tag{6.3b}$$

where a_A and a_B are the *atomic diameters* and R_A and R_B are *back-scatter coefficients*. As in quantitative X-ray microanalysis, the problem is how to handle the *iteration procedure* required for a *multicomponent* system. In such cases, the matrix factors have to be estimated from initial intensity ratios (by assuming that the intensity ratio gives an approximate composition ratio) and then recalculated. As in the case of X-ray microanalysis, concentration data presented in the literature very often omit details of the correction procedures which have been used, leaving the reader very much in the dark as to the true experimental significance of the published concentrations.

6.2.5 Depth Profiling

The excellent *depth discrimination* (resolution *perpendicular* to the surface) of both XPS and AES makes them ideal tools for the investigation of thin-film devices, multilayered structures and distributed interfaces, and not just for the analysis of features which are associated with the free surface.

Controlled sputtering of the surface by a *chemically inert* beam of energetic incident ions is the preferred method for obtaining a *concentration profile.* Sputtering rates depend both on *relative mass* and on *incident angle*, as well as on the *ion energy* and *ion beam intensity.* A Newtonian, billiard-ball model for the sputtering process is quite successful in predicting the main features of the process. In particular, the *maximum energy*, E_{max}, which can be transferred to the sputtered atom at the surface by an incident ion with kinetic energy E_0 is given by the relation:

$$\frac{E_{max}}{E_0} = \frac{4m_1m_2}{(m_1 + m_2)^2} \tag{6.4}$$

where m_1 and m_2 are the atomic masses of the *struck* atom and the *incident ion*. The primary assumption is that the struck atom reacts *independently* of its surroundings, so that the *reaction time* for the collision with the sputtered atom is small compared to the reaction times concerning any neighbouring atoms. This will be true for high values of E_0. For the special case $m_1 = m_2$, and $E_{max} = E_0$. The larger the mismatch in the atomic masses of the incident and the sputtered particle, then the smaller the value of E_{max}. The choice of argon as the preferred sputtering ion is dictated by its chemical inertness and its atomic mass, which is not too different from that of the major constituents in many engineering materials. The maximum value of energy transferred corresponds to a direct, *knock-on* collision in the forward direction, and the sputtering process is complicated by *multiple collisions* in which the knock-on collision ejects a neighbouring atom. Careful calibration of the sputtering rate is necessary, for example with a multilayer sample of known layer compositions and dimensions, if the true concentration profile is required. Samples can be prepared by several standard methods, such as *sputter deposition*. In this case, a *taper section* taken through any interlayer will migrate steadily across the field of view as sputtering proceeds, and provides a reliable method of thickness calibration.

Auger spectra are typically recorded at set sputtering intervals, which may correspond to the partial removal of *single atomic layers*. This method has also proved exceptionally successful in studies of grain-boundary embrittlement in steels and other alloys, for which the segregant responsible for embrittlement is localized within a few atomic layers of the surface. The embrittled material is first fractured in a jig mounted inside the ultra-high-vacuum system and then transferred directly to the Auger analysis chamber for AES. (Emission of *occluded* gas at the moment of fracture makes it undesirable to break the sample in the Auger chamber itself.)

Even if the alloy is not embrittled by a segregant, it is possible to study grain-boundary segregation phenomena in some detail, by first cathodically loading the sample with hydrogen to induce hydrogen embrittlement, and then fracturing the sample in the vacuum chamber. An example is shown in Fig. 6.10 for a *nickel alloy.* The Auger peaks for *phosphorus*, *boron* and *carbon* are clearly visible, but decrease rapidly as the boundary layer is sputtered away, thus yielding good quantitative data on the *degree of segregation* of these elements, even in the absence of reliable morphological evidence as to their spatial distribution.

6.2.6 Auger Imaging

As in the case of an *X-ray* '*dot map*' (Section 4.3.2.1), the primary factor limiting acquisition of a resolved *Auger* '*image*' is the poor counting statistics, and the same relationships apply between the number of counts needed per *pixel point* and number of pixel points required to develop a useful image of any given microstructural feature. The *human factor* also plays the same role—100 s is close to the limit of acquisition times which the (non-dedicated) user is prepared to accept for routine applications!

As in the case of *characteristic X-ray microanalysis*, there are some further possibilities for improving the spatial 'resolution' in AES, in the sense that a spectrum can be derived from a highly localized area. Thus, X-ray microanalysis in thin films can provide localized information from the region of a focused probe less

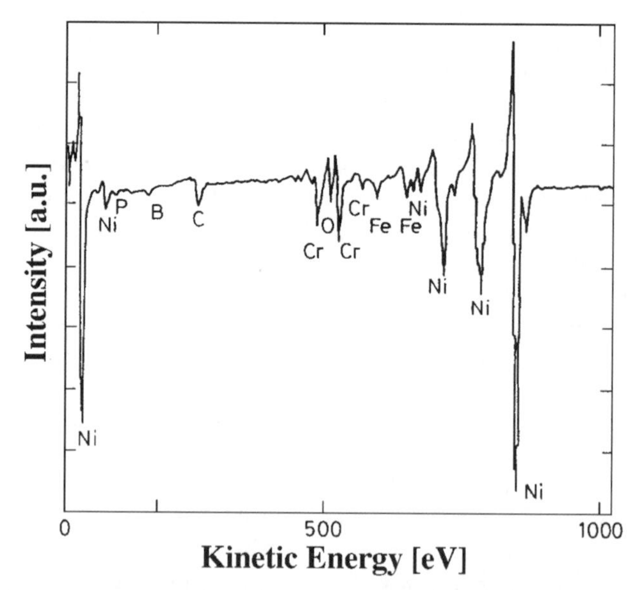

Figure 6.10 Auger spectrum obtained from a brittle grain-boundary failure in a nickel alloy. Carbon, boron and phosphorous are all enhanced at the boundary, thus indicating segregation of these elements. (Reprinted from *Materials Characterization*, Vol. 25, G. C. Smith, Qualitative surface analysis by Auger and X-ray photoelectron spectroscopy, pp. 37–71, Copyright 1990, with permission from Elsevier Science)

than 20 nm in diameter, because there is very little spreading of the incident electron beam by scattering in the thin film.

In Auger spectroscopy, the signal is obtained from the first few atomic layers of the sample, and may be generated either by primary inelastic collisions of the incident beam, or by secondary collisions of back-scattered electrons. For incident beam energies of the order of 10–20 kV, the area of emission for *back-scattered electrons* is of the order of 50–100 nm in diameter, while the *back-scatter coefficient*, *R*, increases rapidly with atomic number (Section 5.1.3.1). It follows that most of the Auger signal from a *low*-atomic-number matrix will be localized within the diameter of the focused incident-beam probe, while that from a *high*-atomic-number material will be distributed over the diameter of the *back-scattered electron distribution*. The expected behaviour is illustrated schematically in Fig. 6.11.

Providing that the incident beam can be focused to a fine nanometre probe and the source intensity improved by using *field-emission* or a LaB_6 source, good resolution should be no more difficult to obtain from an Auger signal than from a characteristic X-ray signal, and considerably easier in the case of low-atomic-number matrices. If we set the upper limit at $Z = 15$, this will include silicon technology, the aluminium alloys and all polymers! An example of a reasonably resolved Auger image is shown in Fig. 6.12. Some care is required to avoid errors of interpretation which are associated with the *topography* of the sample surface. In Fig. 6.13, the edge of a *step* may shadow the signal from the underlying matrix at the detector, while electrons

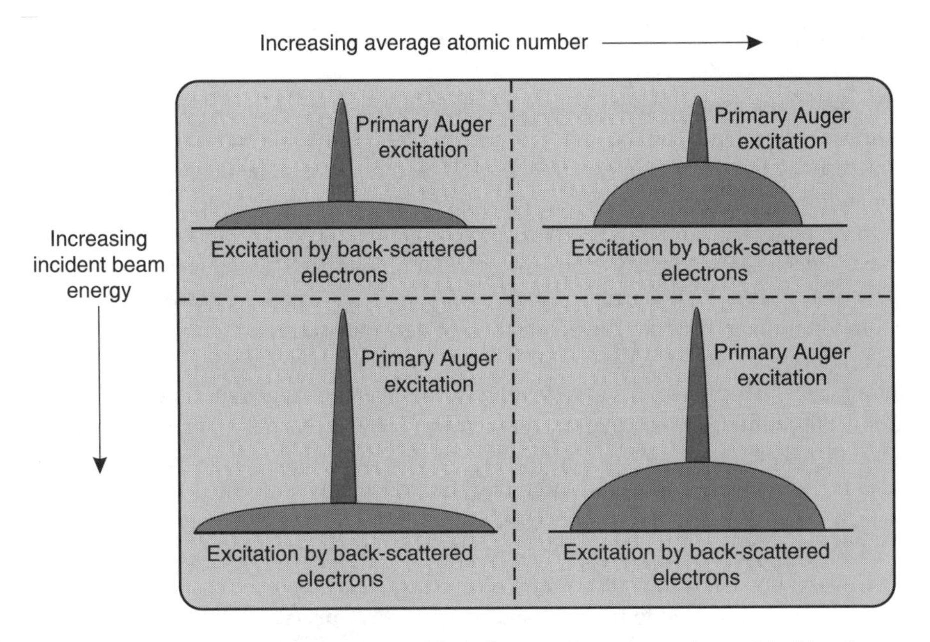

Figure 6.11 Schematic representation of the influence of atomic number and incident-beam energy on primary Auger excitation and excitation by backscattered electrons

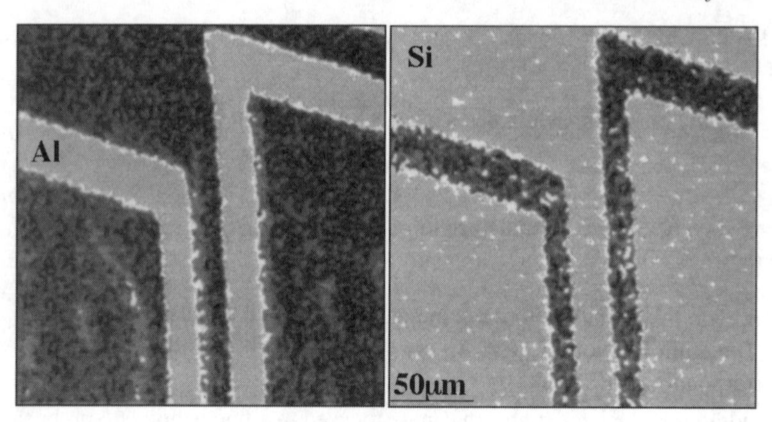

Figure 6.12 An Auger image taken from aluminium conduction lines on silicon (from Wachtman, *Characterization of Materials*, by permission of Butterworth Heinemann, a division of Reed Educational & Professional Publishing Ltd., 1993)

penetrating an *edge* may give rise to a spurious matrix signal. These effects should be compared with those discussed earlier in interpreting contrast in the scanning electron microscope (see Sections 4.3.3.1 and 4.3.4.2).

6.3 SECONDARY-ION MASS SPECTROMETRY

We have seen how a beam of *energetic* ions incident on a surface will *sputter* the surface layers and can be used to prepare thin films for transmission electron microscopy by *ion milling* (Section 4.2.1.3), and how the same method can be used in an ultra-high-vacuum *Auger spectrometer* chamber, either to clean the specimen surface or to sequentially remove successive atomic layers for *depth profiling*. The next logical step is to analyse the *sputtered ion signal* itself, and this is precisely the function of ultra-high-vacuum *secondary-ion mass spectrometry* (SIMS).

In ion milling, used for the preparation of thin-film electron microscope samples, the rate of removal may be several microns per hour, corresponding to perhaps 50 atom layers per minute. This is far too fast for cleaning Auger specimens *in situ* or for depth profiling, which may be at the rate of a few layers per minute, but no more than that. The *sputtering rates* of interest in SIMS are even lower, down to a *fraction of a layer* per minute, thus reflecting the extreme sensitivity of the detection system, which is capable of measuring *concentration* in the ppm range or even lower.

The sputtering ions in a SIMS system are usually, but not necessarily, positive, inert-gas ions. For some purposes, a *chemically active* sputtering ion is desirable. Cs^+ is commonly used to promote sputtering of *electronegative* elements, while O_2^+ promotes sputtering of *electropositive* constituents. In order to ensure maximum conversion of *neutral* sputtered atoms to ions, the region of the sample surface is

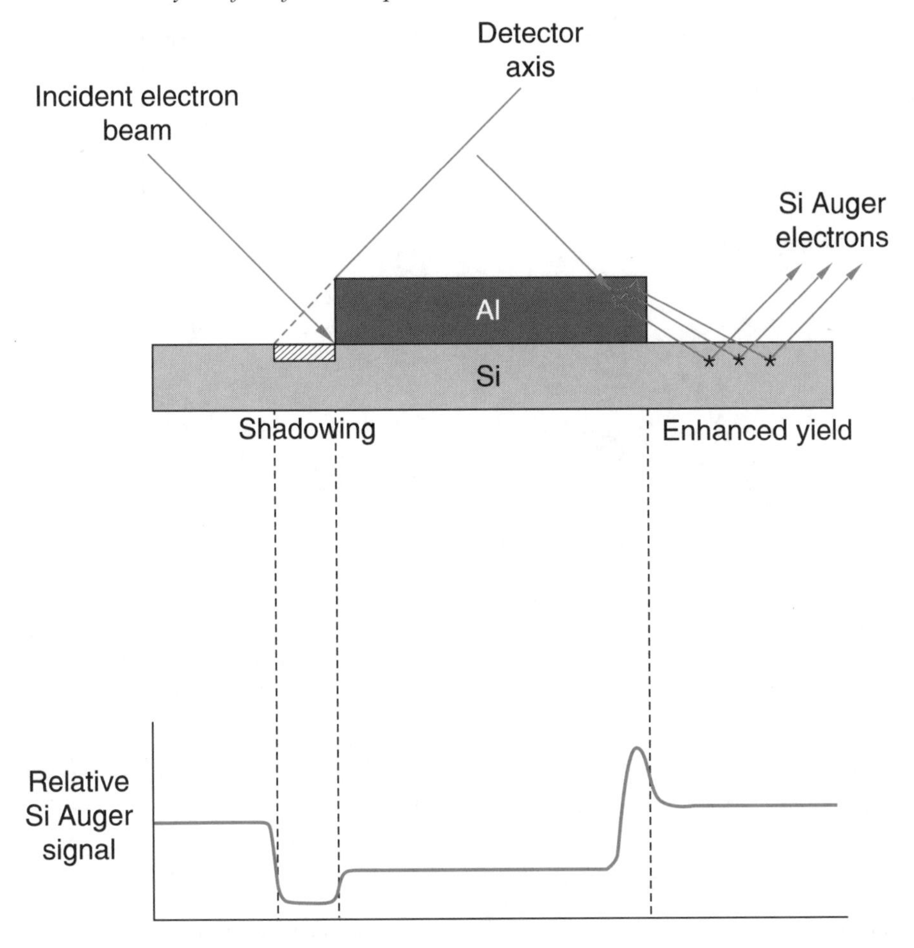

Figure 6.13 Schematic representation of the influence of shadowing, beam penetration and back-scattering on an Auger line-scan intensity from silicon partially covered by aluminium

bathed in a flux of low-energy electrons travelling in a cylindrically symmetric magnetic field. The *sputtering yield* in SIMS depends on a wide range of factors, which include the *mass* of the impinging ion, the *chemical activity* of the target species, the *incident angle* of the beam on the surface, the *take-off angle* of the sputtered ions, and the *composition* of the target.

The *yield dependence* for a particular species on the *composition* of the target makes *quantitative analysis* impossible in the absence of *calibration standards*. For example, the yield of copper ions from an aluminium alloy containing 2 % of copper may exceed the yield from a pure copper target exposed to the same sputtering conditions. The wide variations in *yield* from pure targets as a function of their

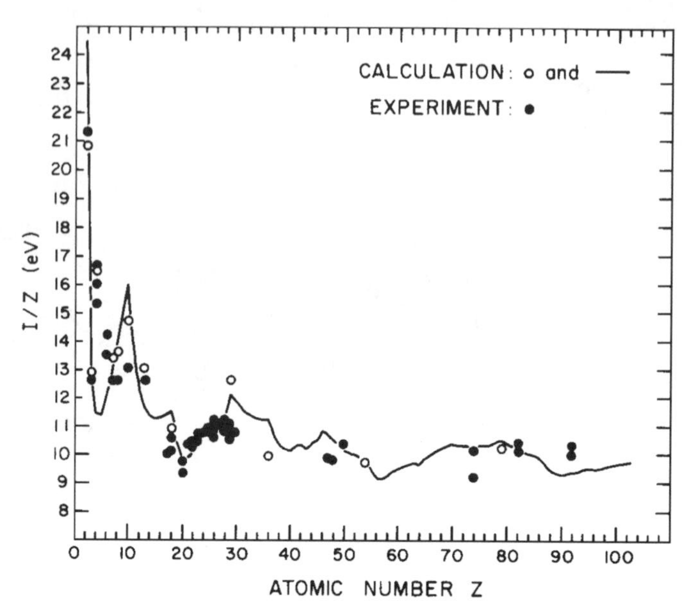

Figure 6.14 The wide variation in ion yield from pure targets as a function of their atomic number: (•) experimental; (O, —) calculated, (Reprinted from Feldman and Mayer, *Fundamentals of Surface and Thin Film Analysis*, Copyright 1986, with permission from Elsevier Science)

atomic number is clearly related to the periodic variations in *ionization energy* across the Periodic Table of the elements (Fig. 6.14).

6.3.1 Sensitivity and Resolution

The ions sputtered from the surface of the target specimen have low kinetic energies and must be accelerated into the spectrometer, where an electrostatic analyser is used to limit the energy spread. Three types of mass analyser have been used to achieve this.

A *quadrupole analyser* has good *mass resolution* but is limited to particles with less than 10^3 atomic mass units (amu). The *sensitivity* is correspondingly low. *Magnetic-sector analysers* have rather poorer mass resolution but can analyse *all* particle masses and are far more sensitive. Both of these analysers are *sequential* instruments, and the signal must be *scanned* across a detector slit, recording each *mass peak* separately.

The third type of analyser is the *time-of-flight* (TOF) spectrometer, which records the time taken for a sputtered ion to reach the detector from the target. The resolution can be as good as a quadrupole and there is no limitation on the detectable mass range. The sensitivity is excellent, and practically *every* particle entering the TOF spectrometer can be detected. Moreover, the system operates in *parallel*, with ions of all masses being detected simultaneously. The major disadvantage is the need to *pulse* the signal sent to the flight tube, either by pulsing the incident ion beam (for example, by switching on a *magnetic field* to deflect the beam off axis), or by deflecting the sputtered ions from the detector. Since the *flight times* in the spectrometer require *nanosecond* resolution, care must be taken to ensure that capacitance effects are minimized. Pulsing the sputtered ions may therefore be preferred, even though this means that a large fraction of the signal is lost.

A second disadvantage is fundamental to the actual technique. Since *all* of the sputtered ions are accelerated through a *constant* voltage, ions of *higher charge* are recorded at *shorter times*, and the spectrum is recorded as a function of the *mass-to-charge-ratio*, e/m, rather than the mass. Inevitably there are some coincidences and hence ambiguities of interpretation. However, most of these ambiguities concern elements with more than one isotope. The TOF spectrometer is quite capable of resolving to better than 0.1 amu, so that a knowledge of the *isotopic abundance* of the elements is usually sufficient to remove any ambiguity.

An example of a TOF spectrum is given in Fig. 6.15, in which a silicon wafer was examined *before* and *after* an organic cleaning treatment. The peak at $m/e = 18$ in the spectrum from the uncleaned wafer is probably associated with H_2O^+ ions. Under suitable conditions, the *sensitivity* of the SIMS technique will enable surface and sub-surface impurities to be detected at *below* the ppm level, while suitably calibrated data may be interpreted quantitatively to of the order of 10 ppm.

6.3.2 Calibration and Quantitative Analysis

As noted previously, the only satisfactory method for converting the relative intensities of the *mass peaks* in SIMS into *mass concentrations* is by recording data from calibration samples. This can be carried out accurately by injecting ions of the impurity of interest at high kinetic energies into a *pure* sample of the matrix. The sub-surface distribution of the impurity can be calculated from a knowledge of the ionic mass and energy and the matrix density, using *Monte Carlo simulation*, and compared directly with changes in the SIMS signal for the impurity as the surface of the *ion-implanted standard* is sputtered away.

While quantitative analysis for impurity or dopant concentrations of up to a few percent can be satisfactorily treated by the use of calibrated standards, there is no such possibility for *higher* concentrations, which are susceptible to major changes in ion yield as a function of concentration. However, it is precisely in this range that the techniques previously discussed, in particular *Auger spectroscopy*, are capable of achieving accurate and unambiguous results.

6.3.3 SIMS Imaging

The sputtered ions from a specimen target can be accelerated and *electromagnetically* focused on to a suitable detector to yield an *image* of the target in which only ions of a *specific mass* will be in focus. Suitable positioning of an aperture at the back focal plane of the collecting 'lens' can be used to select those ions having the

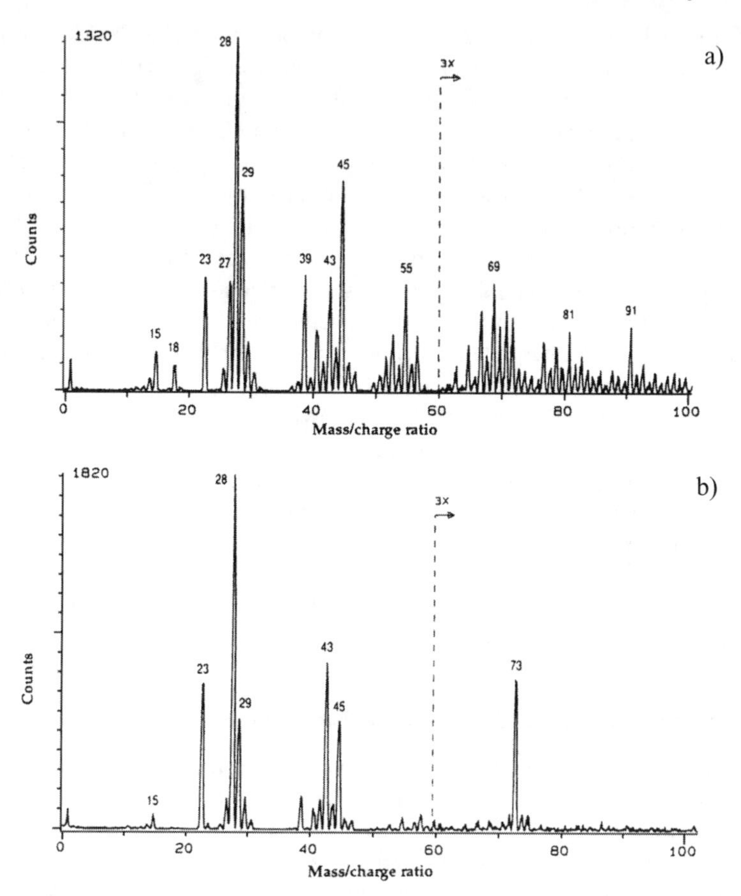

Figure 6.15 An example of a TOF spectrum obtained from a silicon wafer examined (a) before and (b) after an organic cleaning treatment. (Reproduced by permission of ASM International from Carmia, *Advanced Materials and Processes*, **12**, 18 (1992))

required mass from the signal, usually after the signal has first been electrostatically filtered. An electromagnetic projection lens then forms the final image.

There are many problems with this scheme, most notably *radiation damage* to the detection system by the energetic incident ions and the poor *electron optics* available to focus heavy particles over an extended area.

The alternative is to *scan* a focused probe of the sputtering ions in a television raster over the surface, accelerating and collecting the sputtered ions for analysis, and then selecting the ion signal having the required mass. As in scanning electron microscopy, the final image is obtained on a monitor with the same time-base as the scanning coils of the beam probe after the signal has been amplified and processed. The *lateral resolution* is poor (perhaps 10 μm), but the *sensitivity* is high, and successive scans can reveal changes in the sample with a depth resolution of better than one monolayer.

A possibly more rewarding technique is to compromise between the 'point' analysis of a *static* SIMS analysis and the poor definition of a *dynamic* SIMS image by using a *line-scan*. As in X-ray microanalysis (Section 4.3.2.1), *line analysis* gives far better counting statistics and very often provides a very satisfactory insight into the relationship between the *microchemistry* and the *microstructural morphology*, particularly when combined with some alternative method of imaging the sample. The range of concentrations which may be detected by SIMS covers over *four orders of magnitude* with a depth resolution of *better than 10 nm*. No competitive technique presently available could hope to achieve this same combination of resolution and sensitivity.

Finally, and very recently, *high-resolution SIMS* has been developed. In this technique, the ion beam is formed from Ga^+ ions, usually extracted from a liquid Ga source. The focusing system used to form the Ga^+ probe provides a fine, ion-probe beam which has an analytical image resolution capability of 50 nm!

Summary

The *chemical composition* is the third class of information, following the *crystal structure* and *microstructural morphology*, required to complete the microstructural characterization of a material, but the sensitivity of many material properties to the surface condition has led to the more specific requirement for an assessment of the *surface chemistry*, that is the composition of the first few atomic layers of the sample.

Examples range from the *segregation of impurities* at the surface of a brittle fracture, the composition of *catalytically active layers*, and the chemistry of thin-film *semiconductor components* and *optronic detector devices*. The chemical analysis of such a surface is generally beyond the reach of the microanalytical methods discussed in Chapter 5. A high-energy electron probe generates a *characteristic X-ray signal* over a thickness of the order of 1 μm, while the *energy-loss spectrum* detected in transmission requires a sample thickness of the order of the extinction distance; typically at least 10 nm.

Of the wide range of physical phenomena used to derive chemical information from a sample, only *three* result in a signal which is sufficiently *localized* at the surface of the sample to ensure that only the surface layers of the atoms contribute. These three methods are *X-ray photoelectron spectroscopy*, *Auger electron spectroscopy* and *secondary-ion mass spectrometry*.

In *X-ray photoelectron spectroscopy* (XPS), a secondary *photoelectron* signal is excited by an incident *X-ray beam*. Since low-energy secondary electrons can *only* escape from the sample if they originate from the first few layers of atoms, only these first layers are sampled, while the *energy* of these photoelectrons is sensitive to the local *work function* and the *binding energy* of the photoelectron. The technique is able to distinguish different *chemical binding* states, for example those associated with the different valency states of a polyvalent cation. Although *X-ray photoelectron spectroscopy* has no useful *lateral resolution*, the sensitivity to the chemistry of a single surface layer of atoms and the ability to detect changes associated with the early stages of *adsorption* (corresponding to partial coverage by an adsorbate) combine to make this a very useful tool for surface science.

Auger electron spectroscopy differs from XPS in that the decay of an excited state of an atom results in the emission of an electron whose kinetic energy is determined by the different energy states of the excited atom. The energy of the Auger electron is *independent* of that of the radiation used to excite the emission (either an energetic electron beam or X-rays), unlike the energy of a photoelectron. While the range of the Auger electrons is similar to that of photoelectrons (both originate from a depth of two to five atoms at most), the *lateral resolution* achievable in Auger spectroscopy is limited only by the statistics of Auger excitation. As in microanalysis using characteristic X-radiation, the cross-section for Auger excitation is small, and, with electron beam excitation, comparatively large beam currents are required to obtain adequate counting statistics. For a conventional electron source, the *minimum* electron beam diameter needed to acquire adequate Auger statistics is not much less 1 μm, although far better resolution, down to perhaps 10 nm, is achievable by using a LaB_6 or *field-emission source*.

Auger spectra are conventionally presented as a *differentiated signal*, in which the peak position is determined by the cross-over of the differentiated signal. The *shape* of the differentiated peak is sensitive to the *chemical state*, so that the Auger spectrum can be an important diagnostic tool for analysing the nature of the chemical bonding. *Quantitative* Auger analysis is based on a comparison of measured integrated intensities relative to known standards, and the correction procedures vary. In general, quantitative Auger analysis may not be very accurate, partly because of the difficulties of *calibration*, but also because the signal is usually from a comparatively large, thin area, so that the composition may well vary *within* the area analysed.

In practice, *Auger spectroscopy* is often combined with *ion sputtering* of the surface, so that the Auger spectrum is obtained as a function of sputtered depth. The application of '*depth profiling*' to research on thin-film devices is well established. Again, *quantitative* results are not easy to obtain, primarily because it is difficult to ensure a uniform rate of sputtering over the area being analysed. *Depth profiling* is a standard procedure in the development of *electronic* and *opto-electronic devices*, and has also proved extremely useful for the study of embrittlement associated with trace-impurity *segregation* in structural materials.

When using a high-intensity electron source, the area analysed may correspond to a spatial resolution of the order of 100 nm or better. It is then possible to obtain an *Auger image* by scanning the probe beam over the surface and collecting the Auger electrons characteristic of the element of interest. Unfortunately, a statistically adequate image requires large numbers of Auger electrons, and it is only rarely that the signal intensity is sufficient to justify Auger imaging.

Secondary-ion mass spectrometry (SIMS) is far more sensitive than *either* XPS *or* Auger spectroscopy, and it is usually possible to detect *trace impurities* or *dopants* present at concentrations of the order of 10 ppm. However, the *ions* detected are sputtered from the surface, so that the technique is *destructive*, with successive atomic layers being removed during analysis. The SIMS technique is not easy to calibrate, since the yield of a particular mass species depends quite sensitively on the concentrations of other elements present in the sample. In addition, since it is the *mass per unit charge* on the ion which is determined, while most elements have more than one *isotope* and often more than one *valency*, a knowledge of the *isotopic abundance* and *ionic charge* is sometimes necessary in order to eliminate ambiguities. Since very small numbers of ions are detected, it is not usually possible to *localize* the signal to better than *ca* 10 µm. Although it is possible to obtain *secondary-ion images*, in conventional SIMS this is only at the cost of losing the *depth resolution*, and with rather poor *lateral resolution*. Recently, *high-resolution SIMS* has been developed, with a spatial resolution of *ca* 50 nm.

It is important to recognize the primary difference between the three *surface analysis* techniques discussed in this present chapter and the methods of *microanalysis* discussed in Chapter 5. *Neither* X-ray microanalysis *nor* electron energy-loss spectroscopy are capable of providing analytical information localized to the immediate vicinity of the *solid surface*, namely the first few atomic layers. The methods of *microanalysis* discussed in Chapter 5 are certainly influenced by the presence of *thin surface films* and *contamination* (which are therefore to be avoided), but the objective of microanalysis is to determine *bulk concentration* on the microscale level. The methods of *surface analysis* introduced in this present chapter are tailored to determining compositional changes at or near the surface itself.

Bibliography

1. J.M. Chabala, K.K. Soni, J. Li, K.L. Gavrilov and R. Levi-Setti, *High-Resolution Chemical Imaging with Scanning Ion Probe SIMS*, *Int. J. Mass Spectrom. Ion Processes*, **143**, 191 (1995).
2. L.C. Feldman and J.W. Mayer, *Fundamentals of Surface and Thin Film Analysis*, Elsevier, London (1986).
3. J.B. Wachtman, *Characterization of Materials*, Butterworth-Heinemann, Oxford (1993).
4. J.M. Walls and R.S. Smith (eds), *Surface Science Techniques*, Elsevier, Oxford (1994).

Worked Examples

In order to demonstrate some of the techniques discussed in this chapter, we focus on characterization of samples prepared by chemical vapour deposition of aluminium on to a $TiN/Ti/SiO_2/Si$ stack. We begin with *Auger spectroscopy*, used to examine the surface of the aluminium, as well as the relative thickness of the different layers.

Fig. 6.16 show a non-calibrated *Auger sputter profile* through the aluminium, the TiN and finally terminating in the titanium. The spectrum was taken from a specimen after deposition of a very thin aluminium film, which is *discontinuous*

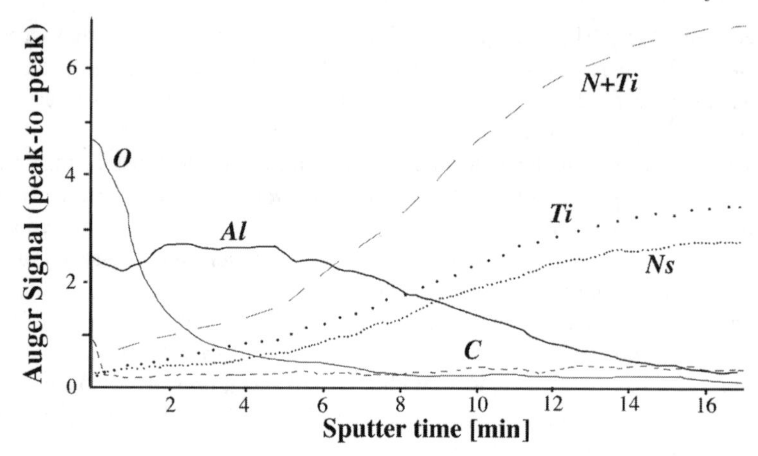

Figure 6.16 Auger sputter profile obtained from a multilayer specimen of aluminium deposited on a TiN layer on titanium. The spectrum was taken from a specimen after deposition of a very thin aluminium film which is discontinuous, that is islands of aluminium rather than a continuous film

(*islands* of aluminium on the TiN). A relatively strong *oxygen* signal is detected during the initial sputtering time (<2 min). This is due to the *native oxide* which formed on the aluminium when it was exposed to air. In addition, a surface layer of *carbon* is evident, again a *contaminant* associated with exposure to air. Both the oxygen and the carbon signal drop to background levels after the initial sputtering, although the oxygen signal persists to a deeper level, presumably because of oxygen adsorption on the TiN regions which were not covered by aluminium.

Signals from *titanium* and *nitrogen* are both evident, but it is impossible to differentiate between TiN and titanium, due to overlap between the Auger peaks, and the fact that the TiN and titanium films are so thin. Fig. 6.17 is a calibrated *Auger profile* of the same specimen in which an increase in the titanium and nitrogen signals is clearly seen as the aluminium signal decreases.

We can also use Auger to differentiate between *titanium metal* and *titanium cations* in TiN. Fig. 6.18 shows standard Auger signals from the various atomic species in our sample. A clear difference is evident in both the *peak* position and the *peak shape*.

Fig. 6.19 shows an *Auger sputter profile* from the same type of specimen, but now in the form of a much thicker (*continuous*) aluminium film. Here we see a definite drop in the oxygen signal after sputtering through the initial surface layers. (The increase in oxygen at *ca* 24 min of sputtering corresponds to the SiO_2 layer beneath the Al/TiN/Ti sandwich.) In addition to checking the *purity* and *concentration depth profiles* of our films, we can also measure the *thickness* of the individual layers, by *calibrating* the sputtering rate for the individual layers. This is quite easy to do, since we already have TEM results for these materials (see Chapter 4), and we need only to compare Fig. 6.19 with a cross-sectional transmission electron micrograph of the same specimen (Fig. 6.20).

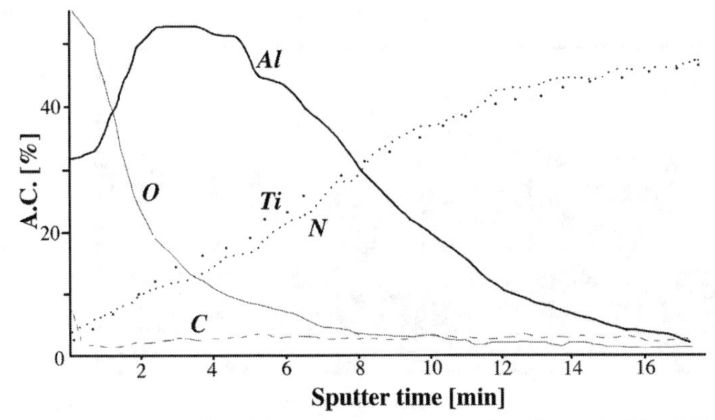

Figure 6.17 Auger profile of the same specimen as that shown in Fig. 6.16, but calibrated for atomic concentration

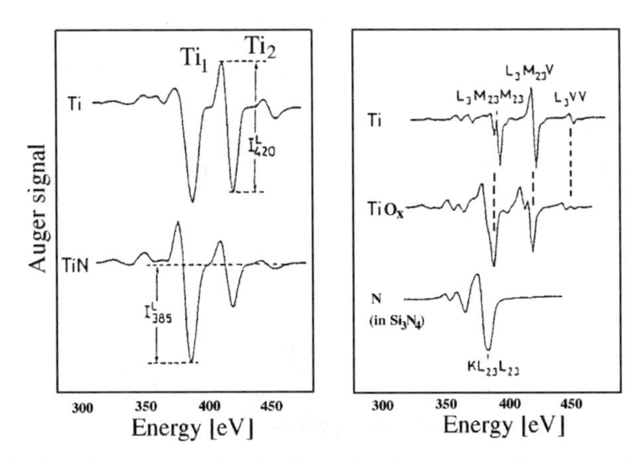

Figure 6.18 Standard Auger signals from the various atomic species in the multilayer sample

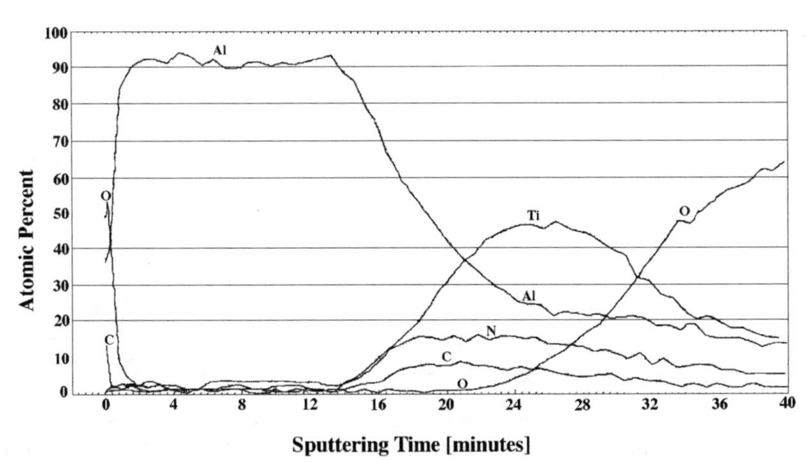

Figure 6.19 Auger sputter profile obtained from the same type of specimen as that shown in Fig. 6.16, but now in the form of a much thicker (continuous) aluminium film

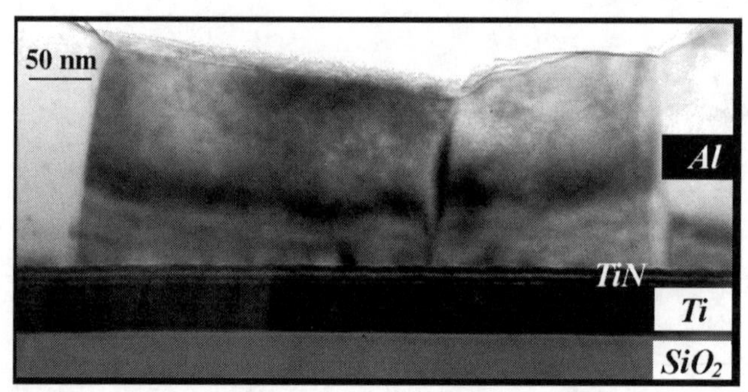

Figure 6.20 Cross-sectional transmission electron micrograph of the same specimen from which the Auger sputter profile shown in Fig. 6.19 was obtained. This micrograph can be used to calibrate the sputtering rates

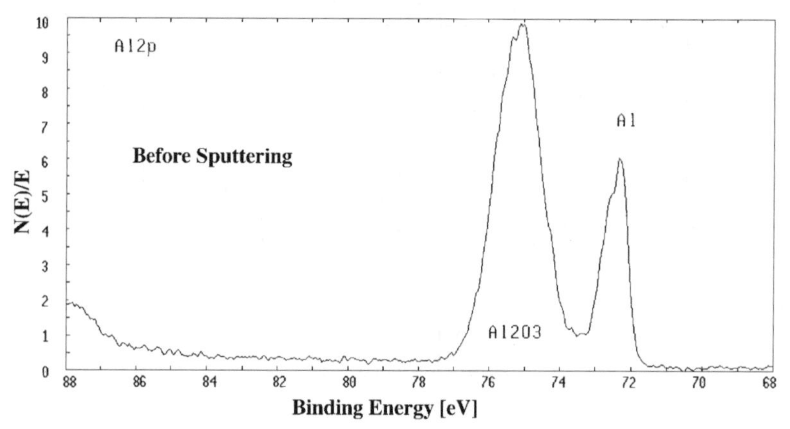

Figure 6.21 XPS spectrum obtained from the aluminium specimen before sputtering, showing the presence of Al_2O_3 in addition to aluminium

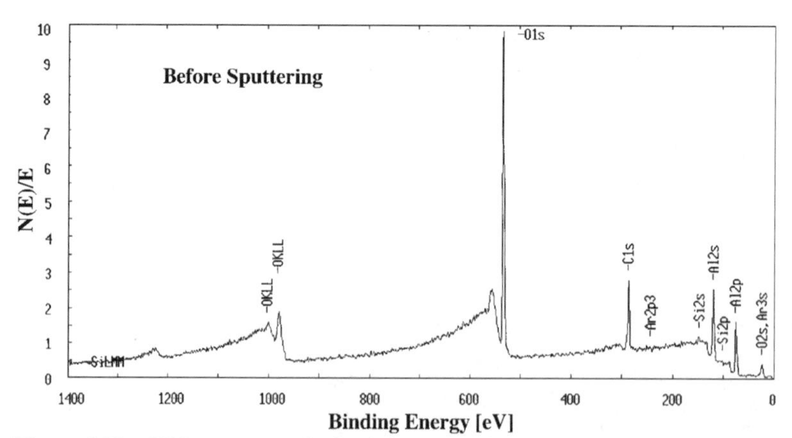

Figure 6.22 XPS spectrum obtained from the aluminium specimen over a larger energy range than that shown in Fig. 6.21, displaying the presence of carbon on the surface

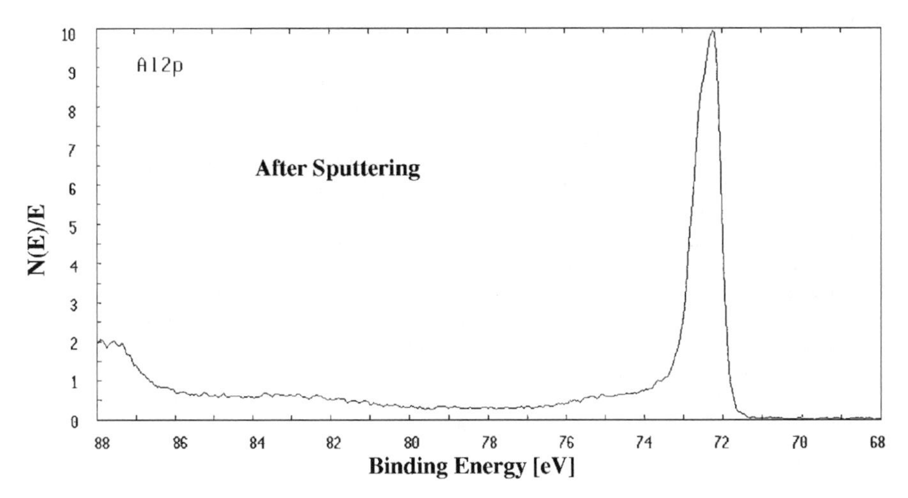

Figure 6.23 XPS spectrum obtained from the aluminium specimen after sputtering for 5 min; the Al_2O_3 signal is now no longer detected

Finally, we confirm the oxidation of the specimen surface by XPS. Fig. 6.21 is an *XPS spectrum* obtained before sputtering. A strong signal from Al_2O_3 (74–75 eV) is clearly evident, in addition to the metallic Al 2p peak (72 eV). There is also clear XPS evidence for surface contamination by *carbon* in the XPS spectrum taken over a *wider* energy range (but with a slightly *lower energy* resolution) (see Fig. 6.22). The carbon 1s XPS peak (285 eV) confirms the Auger findings. After sputtering the specimen for 5 min, the Al_2O_3 is now completely removed (Fig. 6.23).

Problems

6.1 How 'thick' is a free surface? Give reasons to justify your answer. Is there a distinction to be drawn between the *physical* thickness, corresponding to the electronic structure, and the *chemical* thickness, corresponding to the composition?

6.2 Give *three* examples in which you believe that surface analysis of a solid sample would be more significant than microanalysis of a polished section.

6.3 Analytical information can be obtained from a wide range of signals. What properties of *photoelectrons*, *Auger electrons* and *secondary ions* make them the signals of preference for surface analysis?

6.4 Compare the limits of detection for chemical analysis when using *photoelectrons*, *Auger electrons* and *secondary ions*. What different factors limit the confidence in a quantitative analysis in each case?

6.5 The XPS signal contains information on the chemical binding state of the elements detected. Suggest how this type of information might be useful in a study of *metallurgical* failure.

6.6 Auger analysis is often used for *depth profiling* in thin-film devices. Discuss the sources of error involved in plotting concentrations determined by Auger spectroscopy as a function of the sputtered depth. (Consider in particular the topography of the sputtered surface, 'knock-on' damage, and the calibration of the Auger signal.)

6.7 *Argon* (atomic weight 40) is usually used to sputter-clean surfaces for Auger studies and to remove successive atomic layers in depth profiling. Compare the expected sputtering efficiency of aluminium, iron and tungsten (atomic weights 27, 56 and 184, respectively) by argon and suggest some possible strategies that might be used to improve the reliability of depth profiling.

6.8 *Secondary-ion mass spectrometry* frequently gives a larger signal for copper in an age-hardened copper alloy (2.8 % Cu) than is obtained from a pure copper calibration standard. Suggest some possible physical reasons for this observation.

6.9 What are the advantages and disadvantages of a *time-of-flight* mass spectrometer for secondary ions, as opposed to a *quadruple* or *magnetic-sector analyser*?

6.10 The *spatial resolution* in conventional secondary-ion mass spectroscopy is limited to *ca* 10 μm, but in principle this could be improved by taking a line-scan rather than a TV-raster image. Assuming that the only limitation is in the counting statistics, estimate the expected resolution of a line-scan.

6.11 Multilayer films are often sectioned at an inclined angle to the surface prior to Auger analysis. Explain the advantages of a *wedge-shaped* specimen geometry.

Chapter 7

Quantitative Analysis of Microstructure

In Chapters 5 and 6, we have outlined some of the methods associated with converting *spectroscopic data* into *quantitative estimates* of *chemical composition*, but we have not yet attempted to define the *microstructural parameters* associated with *image data*, nor the methods which are available to estimate *numerical values* for these parameters from measurements taken from images.

In part, this difference in treatment is a consequence of the way in which the two types of information associated with *microstructural chemistry* and *microstructural morphology* are generally regarded. An *image* of the microstructure is, for many purposes, an end in itself (and 'worth a thousand words', at least to Confucius). The image is often described as though it *were* the object—*pearlite*, a *eutectic* or *dendrites*, to give but a few examples of terms used to describe both image features and microstructural constituents. This is *not* the case with *spectroscopic data*, and only in the event that the spectrum is used to identify the presence of an element (provide a *fingerprint* of the element), irrespective of its concentration in the matrix, are we likely to be satisfied with the simple identification of a characteristic intensity peak.

Crystallographic data are also frequently used to identify the presence of a phase, irrespective of its *volume fraction* in the sample (the *phase concentration*) or its distribution in space (the *phase morphology*). Any relationship between the crystal orientation and the geometrical axes of the component, corresponding to the existence of *preferred orientation*, is then seen as a topic requiring special investigation. A determination of the *particle size* of crystals and the presence of *micro-stresses*, determined from measurements of X-ray *peak width* and *peak shift* for the different reflections, are also techniques of *quantification* which, while available when required, is generally seen as subsidiary to the main business of diffraction studies, which are generally considered to be *phase identification* and *structure determination*.

In this last chapter, we will summarize the *quantification of microstructural morphology*. The study of the spatial relationships between objects is the science of *stereology*, and the basic concepts have been known (and appreciated) for over a century, both in the fields of medicine (anatomy and its microstructural counterpart, histology) and in geology (again, on the grand scale of the structure of the earth's crust, and on the fine scale associated with mineralogy). In recent years, the availability of ever increasing computational power has lead to a wide range of software programs devoted to the two key aspects of the quantitative interpretation of image data. The first of these is *image processing*, while the second is *image analysis*. Image processing is concerned with transforming raw, digitized data into a more 'useful' form, by removing random *background*, enhancing or reducing *contrast*, and possibly *smoothing* the data over neighbouring pixels (averaging data). Image analysis is concerned with extracting *quantitative* measurements from the image, and is the principle topic of this last chapter. In effect, we assume that the *recorded* image has already been 'processed', and we then focus on *image analysis*.

7.1 BASIC STEREOLOGICAL CONCEPTS

The quantitative analysis of *all* image data is effectively based on the 'one step back, two steps forward' principle. In other words, a *three-dimensional object* is reduced to a *two-dimensional image* ('one step back'). The image data, recorded in two dimensions (possibly as a function of time), are then analysed. The expectation is that *image analysis* will then result in a *quantitative* estimate of microstructural parameters which are relevant to the *three-dimensional* object, in effect, the 'two steps forward'. In some cases, image analysis can be achieved unambiguously, and we refer to the parameter as being '*accessible*' from the two-dimensional image, although with varying degrees of accuracy. In other cases, we can only estimate the three-dimensional parameters on the basis of an *assumed model* for the microstructural features; the parameter is then said to be '*inaccessible*'.

However, before we discuss these complications we need to examine some of the factors that affect a *stereological analysis*.

7.1.1 Isotropy and Anisotropy

Microstructural anisotropy can take two forms, since the term *anisotropy* is applied to both the microstructural *morphology* and the crystallographic *orientation* of the phases present. In both cases, our coordinates of reference are those of the *component geometry*. If the microstructure is *isotropic*, with respect to both microstructural morphology and crystallographic orientation, then the *diffraction spectra* and the *microstructural image* will be independent of the plane of the sample section or of any direction within that plane.

Crystallographic anisotropy is best termed *preferred orientation*, in order to avoid confusion with *morphological anisotropy* on the one hand, or *crystalline anisotropy* on the other. (Crystalline anisotropy is associated with the orientation dependence of the physical properties within the crystal lattice, for example the variation of the *elastic modulus* in tension measured along different directions in a single crystal.) Preferred orientation is usually determined from an analysis of the diffraction peak intensities for diffraction data measured in different spatial directions (see Section 2.4).

Morphological anisotropy implies that one or more microstructural parameters vary with the orientation of the direction or plane in which the parameter is measured in the sample. An obvious example is the *elongation* of the grains in a ductile metal which results from plastic extension of the specimen (Fig. 7.1). In this case, the change in the ratio of the grain length to grain width (the *aspect ratio*) will depend on the total plastic extension of the specimen and the plane of the section studied. If the sample plane is defined by the direction of extension and its normal, then the aspect ratio observed for the grains seen in the section will be maximized, while the *distribution* of the aspect ratio (maximum and minimum values in the plane) will be related to variations in ductility and the *stereological constraints* between the grains in the material.

The grain aspect ratio in a metal sheet will depend on the processing history. If it was rolled as a continuous strip, in a series of individual passes, then the grains will be *elongated* along the direction of rolling, but if the reduction in thickness is the result of rolling by equal amounts in two directions at right angles (*cross-rolling*), then the grains will be *flattened* in the plane of the sheet, rather than elongated. In general, we require *at least* two sections to characterize such morphological anisotropy, for example *perpendicular* and *parallel* to the plane of the sheet, so that the sample sections contain all three principle directions in the product.

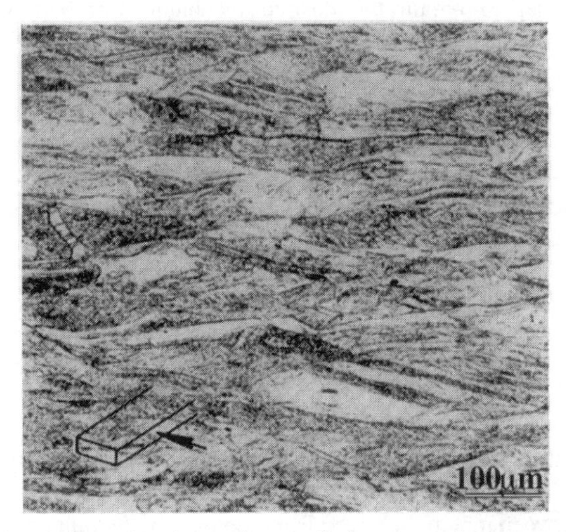

Figure 7.1 Anisotropic grain shapes in a rolled copper sheet, revealed by etching a specimen sectioned parallel to the rolling direction. (From *Metals Handbook*, Eighth Edition, Vol. 7, *Atlas of Microstructures of Industrial Alloys*, (1972), ASM International, Materials Park, OH 44073-0002, (formerly The American Society for Metals, Metals Park, OH 44073))

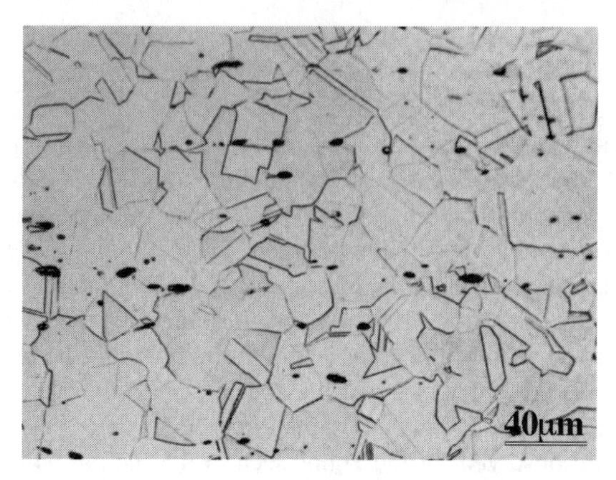

Figure 7.2 The same copper sheet as shown in Fig. 7.1 after annealing. Oxide inclusions are visible aligned along the rolling direction, again demonstrating anisotropy. (From *Metals Handbook*, Eighth Edition, Vol. 7, *Atlas of Microstructures of Industrial Alloys*, (1972), ASM International, Materials Park, OH 44073-0002, (formerly The American Society for Metals, Metals Park, OH 44073))

A similar example might be the distribution of non-metallic inclusions in a metal sheet, but in this case it is also the *distribution* of the inclusions (a second phase in the system) which is anisotropic, and not only their *shape* (Fig. 7.2).

Most composite materials exploit *mesostructural anisotropy* in order to optimize their engineering properties and minimize the weight or the dimensions of the structural components in the engineering system. Composite *lay-ups* of resin bonded sheets of fibre can be oriented with the fibres at specific angles in a variety of geometries in order to obtain the desired mechanical properties (strength and stiffness, usually). In short-fibre-reinforced composites, the lengths of fibre are randomly distributed in the plane of the product, but all lie within that plane.

Finally, we should note that an *individual* particle may exhibit anisotropy associated with *either* its shape *or* its crystal structure (or more often, both), but still be distributed *randomly* in the sample. If the *morphology* is unaffected by the plane of the section, then the sample is *isotropic*, irrespective of the shape of the individual grains or particles.

7.1.2 Homogeneity and Inhomogeneity

Processing technology is usually geared to ensuring the *homogeneity* of a product, so that any sample selected from a component will have the *same* structure and properties. However, this is not always the case. Cast ingots of steel and other alloys may be hot-rolled in a series of stages to produce a *homogenous* finished product, such as sheet, rod or profiled bar, but at the earliest stages of processing a distinction must be made between material derived from the *top*, the *bottom* or the *middle* of the ingot. The heavier non-metallic inclusions in the metal tend to be concentrated towards the *bottom* of the ingot, the higher impurity levels and alloy contents are usually found in the last fraction of the liquid metal to solidify, while the lighter inclusions end up in the *top* portion of the ingot. The end product is therefore *inhomogeneous*, with respect to both inclusion content and alloy concentration.

We can distinguish between several types of *inhomogeneity*, and in the example of the ingot we note that the *inclusion count* and the *alloy composition* may both vary independently. The first is a *morphological* inhomogeneity, while the second is a *chemical* inhomogeneity. *Crystallographic* inhomogeneity is also possible, most commonly in the distribution of preferred orientation, as in a ductile metal bar which will undergo extensive *shear* during extrusion (Fig. 7.3). The amount of shear is a function of the distance from the axis of the bar, so that the resultant *preferred orientation* varies across the section, showing a maximum shear component in the outer layers. This crystallographic inhomogeneity is retained after *annealing* the metal and is seen in the annealing texture (the preferred orientation which is characteristic of annealing a deformation texture).

Other mechanical forming operations have similar effects. Forged components retain a *macrostructural* inhomogeneity on the scale of the dimensions of the component, which may be very beneficial to the resistance to failure of the component, particularly in mechanical fatigue. In Fig. 7.4, a forged steel camshaft shows lines of macrostructural flow that strongly inhibit fatigue failure.

It follows that there are many instances for which a single sample or a single section cut from a component cannot adequately characterize the *microstructure*. An extreme situation is that of a *semiconductor device* consisting of a large number of constituent components, densely arrayed on a single chip. In such a case, *several* cross-sections may be required for *transmission electron microscopy* (Section 4.2.1.6), each of which may include a sequence of interfaces between the different active and passive components. The same considerations clearly apply to many situations where a sample has been *coated* or otherwise *surface-treated*, and the

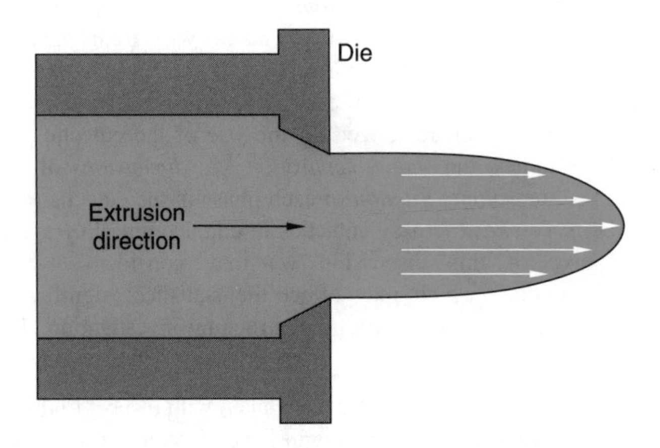

Figure 7.3 Schematic representation of the distribution of shear across an extruded rod

Figure 7.4 The macroscopic development of 'grain' as a result of forging (From Samuels and Lamborn in *Metallography and Failure Analysis*, McCall and French (eds), Plenum Press)

different surface layers vary in *microstructure* or *composition* from one region to another of the component.

7.1.3 Sampling and Sectioning

The question most generally posed concerns the extent to which an *observed* microstructure is significantly the same or substantially different from some other microstructure. Given that microstructures are examined on a diminishing *length scale* of increasing *magnification*, it is convenient to distinguish between microstructural variability *within any given sample* taken for analysis, and the variability observable *between different samples*. Moreover, within any observed region of a sample, the *microstructural features* will each have their own characteristic length scale, which may correspond to the *particle size*, the *particle separation* or the dimensions of other specific features. For example, a *eutectic structure* (Fig. 7.5) may be characterized by the *size* of the eutectic colonies, the *separation* of the colonies, the *shape*, *separation* and *dimensions* of the lamellae within a colony, and the *volume fraction* of each phase in the eutectic (bear in mind that there may be *three-phase* ternary eutectics in engineering alloy systems).

For the purposes of this discussion we may consider *three* scales of microstructural 'sampling' which may affect the statistical significance of any attempt to quantify the conclusions of a microstructural investigation. These are as follows:

(a) The *macroscale* on which the samples were taken with respect to the component being studied. On this scale, someone will have to decide if a *single section* is sufficient, or whether sections should be taken in *different orientations* (testing for *anisotropy*) or from *different regions* (testing for *inhomogeneity*). If several

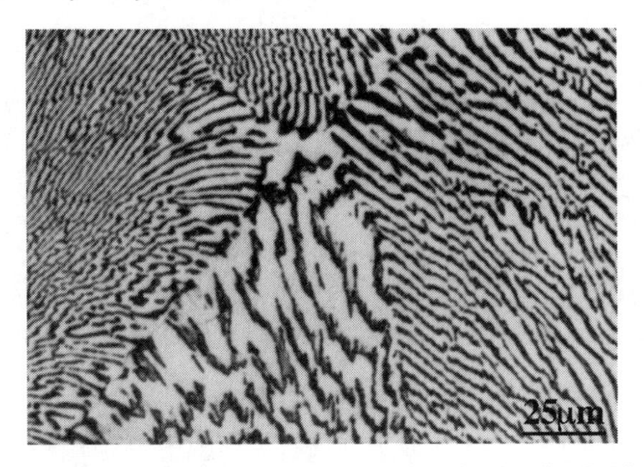

Figure 7.5 Eutectic colonies in a tin alloy (From *Metals Handbook*, Eighth Edition, Vol. 7, *Atlas of Microstructures of Industrial Alloys*, (1972), ASM International, Materials Park, OH 44073-0002, (formerly The American Society for Metals, Metals Park, OH 44073)

samples are to be taken (from a set of components, for example), then is it sufficient to take a *single* sample for each condition, or should several, nominally '*identical*' samples be taken to ensure the *uniformity* of the product? Perhaps it would be wise to take *several* screws, for example, from each of a number of production batches and test for a uniform grain size?

It is well to remember that the *statistics* of sampling require a minimum of *three* samples if an estimate is to be made of both a *mean* and a *variance* for any parameter (corresponding to a Student's-t distribution, with the total number of degrees of freedom equal to n–2).

(b) For any given microstructural sample it is important to check the variability of the microstructure across the *sample section* (the *mesostructural* variability). At this stage of the investigation, the *magnification* will have to be selected, which will determine the diameter of the *field of view*. It is obviously important that the features of interest should be visible in the image, or *resolved*, but at the same time a sufficient *number* of features must be available to ensure adequate sample statistics for each local area.
The statistics of this mesostructure will determine whether or not we can make a statement about the uniformity of the morphology within a given sample with any confidence. A good policy is to select a number of fields of view, equally spaced across the area of the sample. In cases where *inhomogeneity* is to be determined, it is usually straightforward to select a set of the possible fields of view, in order to derive a *statistically significant* data set. For example, a *surface coating* can be sampled by using a set of images recorded along a line on a section normal to the coating, but at a shallow angle to the plane of the coating. A set of three or more such lines will improve statistical estimates of the depth dependence for the relevant parameters (Fig. 7.6).

(c) Finally, the distribution of the microstructural features *within* any given field of view will determine the *statistical significance* of any estimates based on a single

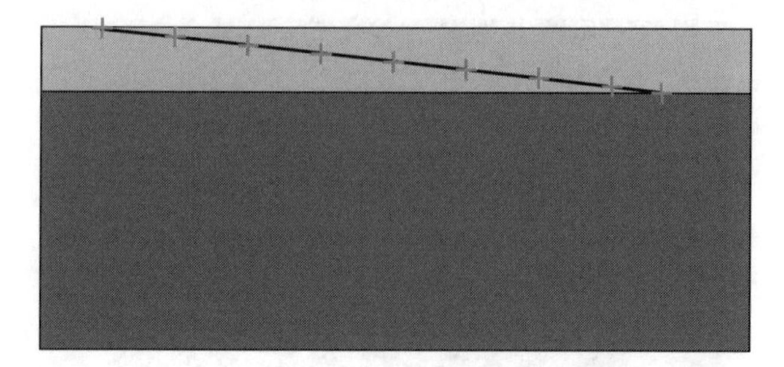

Figure 7.6 Schematic representation of a possible array of sites for high-magnification sampling of a coated sample

field of view. In general, the '*field of view*' may be considerably *larger* than any region imaged individually. This may be achieved by recording a 'panorama' of *overlapping images* from neighbouring regions, or by scanning an *image 'strip'* along a particular direction in the plane of the section.

Such procedures may be essential if the features to be recorded are *small* or *widely separated*, since it may not be possible to resolve a sufficient number of features in a single image and still accumulate sufficient counting statistics.

A balance has to be struck between these three scales of sampling, which we can designate as the *macro-*, *meso-* and *microstructural scales*, respectively. In practice, visual inspection (or indeed common sense) may allow us to dispense with sampling on the *macroscale*, and this is perfectly legitimate providing that we are aware of what we have done. It is much *less* legitimate to dispense with *mesosampling*. At the very least, the different regions of a sample section should be scanned for visually significant differences in the microstructure, while, ideally, steps should be taken to ensure that the *variance* associated with *mesostructural* sampling is no more than that associated with any single *microstructural* sample. In other words, the *statistical error* involved in comparing *different* regions should be no more than that observed in any *one* region.

One further point should be made. It is usual to select microstructural regions for quantitative analysis from within an imaged region, leaving a well-resolved *microstructural 'border'* outside of the area to be analysed (Fig. 7.7). This practice avoids selecting a region for analysis which is adjacent to an obvious defect (due either to processing or sampling). It also enables the observer to assess the importance of possible *edge effects* (such as grain boundaries and grain boundary junctions). Typically, the border should be selected to have an area equal to the area which will be quantitatively sampled, corresponding to a *linear dimension* which is $\sqrt{2}$ greater than the actual analysed area and a border width which is *ca* 20 % of the image dimensions. However, this criterion is useful rather than critical, providing that the microstructure in the border region is clearly visible.

Figure 7.7 The region to be analysed quantitatively should lie within a well-resolved microstructural 'border'. (See text for details.)

7.1.4 Statistics and Probability

A number of *statistical functions* have proved useful in the quantitative analysis of microstructural data, but the description of these functions is beyond the scope of this present text. Nevertheless, a *brief* account of some *statistical tools* is in order. In general, the measured values of any parameter will be distributed about a *mean* or *average* value, and the *width* of this distribution can be measured to determine the spread of measured values in terms of a '*statistical error*'. If the distribution is *skewed* (asymmetric), then this also may be described by an appropriate statistical function. Other statistical tests may also be applied, providing that the appropriate mathematical conditions are met, to decide, for example, whether two *different* samples come from the *same* population, or whether the values of two parameters are *correlated* with one another.

The concept of *probability* has always presented problems, particularly since there is considerable overlap between '*probability theory*' (the study of parameters that are related by a *stochastic*, that is a *non-deterministic* function) and *statistical analysis* (the estimation of a parameter based on the sampling of a population). In practice, *all* measurements of microstructural parameters are subject to *errors* associated with three separate factors, as follows:

(a) the inherent *variability* of the parameter in the bulk material (for example the *grain size* of a polycrystal);

(b) the *statistical errors* associated with selecting a sample, in particular the *number*, *position*, *orientation* and *size* of the features selected for analysis in the specimen;
(c) the errors associated with the *methods of observation* and *specimen preparation*, as well as the method of measurement itself, namely the polishing and etching procedures used to prepare the *section*, the performance and characteristics of the *microscope*, the properties of the *recording media*, and any intrinsic error associated with the *definition* of the microstructural parameters.

It is not always easy to separate the inherent variability of the parameter of interest (that is that associated with the bulk material) from errors due to specimen preparation or the methods of observation and recording. The essential requirement is to reduce these secondary *artifacts* to the point where they no longer affect the significance of the measurements. If this is *not* possible, then clearly it is impossible to determine the *inherent* variability associated with the material.

7.2 ACCESSIBLE AND INACCESSIBLE PARAMETERS

The distinction between *accessible* and *inaccessible* parameters is often inherent in the quantitative interpretation of morphological data, but is often ignored. There are a very limited number of parameters which are characteristic of the bulk structure that can be derived *unambiguously*. The *volume fraction* of a second phase is one of these. However, the *particle size*, the *volume per particle* and the *number of particles per unit volume cannot* be derived without making considerable assumptions about the particle shape, for example that all the particles are spherical.

Fig. 7.8 shows clearly why this is the case. A section through a doughnut or any other feature with a *concave interface* can appear as *two disconnected areas* of the second phase on the plane of the section. *Dendritic* grains are an excellent example of this class of feature, and dendritic dimensions *cannot* be derived from a single metallographic section.

Serial sectioning, on the other hand, can provide additional information perpendicular to the plane of the section. In *serial sections*, thin layers of *known thickness* are removed sequentially, recording the images from the *same* area at each stage. Providing that the features of interest in the microstructure have linear dimensions which are *large* compared to the thickness of the layers removed at each stage, then each feature observed on the nth section can be related to the *cross-section* of the same feature on the $(n-1)$th and $(n+1)$th sections.

The '*resolution*' *normal to the surface* now depends on the *thickness* of the serial sections. *Mechanical polishing* can be accomplished to an accuracy of better than $\pm 10\,\mu m$ with no special precautions, while, in the limit, *precision ion milling* can remove controlled thicknesses of less than ± 100 nm. Clearly, the latter technique is only justified for an electron microscope investigation, since the *best* lateral resolution in the optical microscope is only of the order of 300 nm. Very little use has

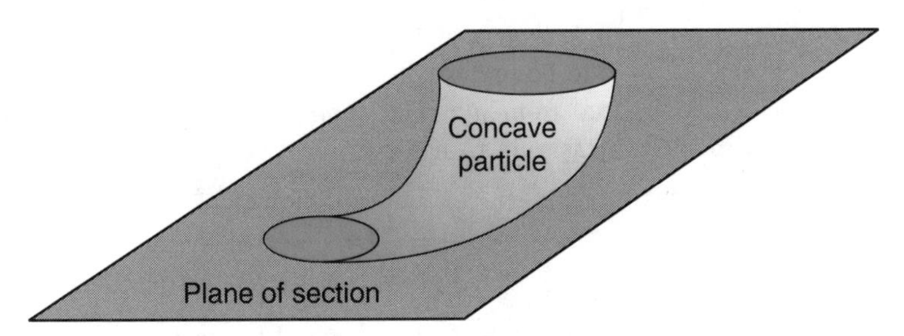

Figure 7.8 A section taken through a partially concave particle may appear as two separate intersected areas on the section

been made of precision ion milling for quantitative thin-film transmission electron microscopy, but a real need exists for this method to scan '*through*' the thin-film devices which are now under development for applications in microelectronic systems.

7.2.1 Accessible Parameters

While only a few *bulk* microstructural parameters can be accessed from a *two-dimensional section* without prior assumptions, an option does exist to *define* some parameters so as to make them accessible. Two examples of these are *grain size* and *particle size*, as we shall see below. Furthermore, only *one* assumption is required to make the transition from '*inaccessible*' to '*accessible*' for a wide range of parameters. This simple assumption is that all interfaces are *convex*, since *no* completely convex particle can intersect a planar section discontinuously. Therefore, for *convex particles*, every area of the second phase (or of the grain) observed on a section belongs to a *single* particle (or grain), and no *other* area on the section can be associated with any other area.

Of course, several parameters characteristic of a specific planar section can be determined *quantitatively*, such as the number of particles of a second phase per unit area of section. Such 'accessible' parameters are of limited value, but they do have their uses. For example, inclusion counts made *parallel* and *perpendicular* to a principle axis of the component may be written into the criteria for acceptance of a product by a consumer, or into a *quality control standard*.

7.2.1.1 Phase Volume Fraction

The *volume fraction* of a second phase, f_v, is a basic parameter of the microstructure which, in the case of a fully crystalline material, can often be determined by *quantitative X-ray diffraction analysis* (Section 2.4). However, many materials contain *non-crystalline* phases which cannot be detected with any degree of reliability by using X-rays, while there may be other reasons why the determination of f_v from a *planar section* is to be preferred.

A remarkable result obtained from *stereology* considerations helps to determine the choice of a method for measuring f_v, namely:

$$f_v = \frac{V}{V_0} = \frac{A}{A_0} = \frac{L}{L_0} = \frac{P}{P_0} \tag{7.1}$$

where V, A, L, and P are respectively the total *volume* of the second phase, the *area* of the phase observed on a *random planar section*, the *length of the line* traversing the phase for a *randomly oriented line*, and the *number of points* falling within the phase for a *random array of points*. The use of the subscript '0' then refers correspondingly to the *total volume* of the sample, the *total area* of the section, the *total length of the line* examined and the *total number of points* in the test array.

It follows that the *same* result is to be expected, regardless of whether f_v is determined by sampling a *volume fraction*, an *areal fraction*, a *line fraction* or a *point fraction*.

We can construct an 'experiment' to confirm this result (Fig. 7.9): First, a micrograph of a '*random*' *section* of the sample is recorded. Then, the *total area* associated with sectioning of the particles per unit area of the section is assessed (by scanning the image in a raster and selecting an appropriate contrast function to determine the total number of pixels associated with sections through the particles). Now, a random array of lines (with respect to both orientation and position) is superimposed on the sample, and the *length of line falling within the particle sections* is determined per unit length of the test line. Next, a random array of points is scattered over the random test section and the *proportion of the points falling within the areas corresponding to sections of second-phase particles* is determined (by simple counting). Finally, a unit volume of the sample is immersed in a suitable medium to dissolve the matrix and the particles are collected for a determination of their *volume* (by weighing, if the density is known).

Which of these methods is the 'best'? Dissolving away the matrix is only likely to be preferred in *very* exceptional circumstances, and is likely to be both time-consuming and prone to experimental errors which will be difficult to gauge.

Areal analysis is a natural choice for any image which is scanned in a raster, since all that is necessary is to set *intensity thresholds* and determine the proportion of the scanning time falling within the set intensity window. The plane of the section should be chosen so as to coincide with the *principal axes of the component*, and attention must be paid to statistically significant variations in the results derived from *different* sections, associated with microstructural *anisotropy* or microstructural *inhomogeneity*.

Linear analysis could also be based on a scanning raster. In this case, a predetermined *fraction* of the total number of line-scans would be used and the number of times a *selected* threshold intensity is reached would be recorded. Normally, the scan lines will be separated by a *set interval* and they would all be in the direction determined by the raster, so that anisotropy in the plane of the section could be assessed by scanning in *different* directions (Fig. 7.10).

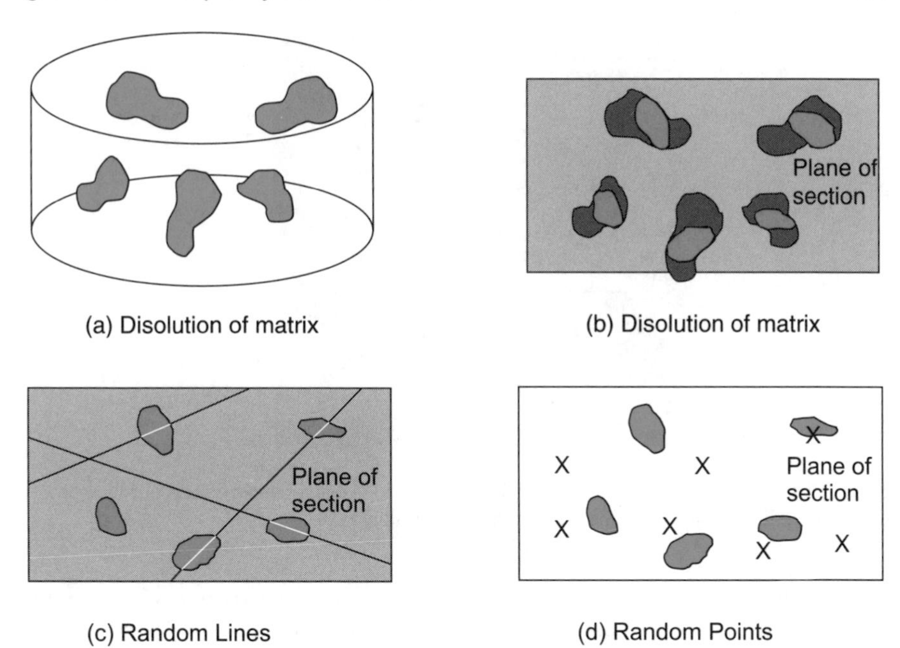

Figure 7.9 Possible methodologies for phase-volume-fraction analysis, based on (a) volumetric, (b) areal, (c) linear and (d) point analyses

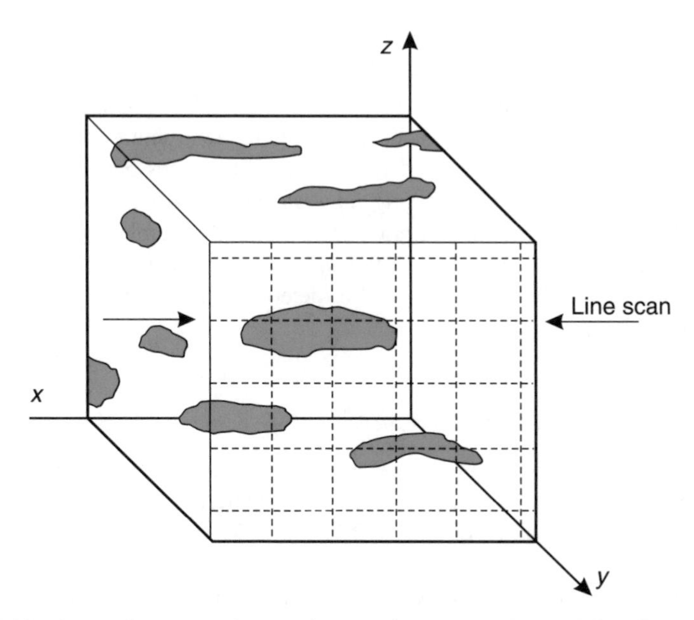

Figure 7.10 An areal-scan can detect anisotropy between sections, while a line-scan detects anisotropy in the plane of a section

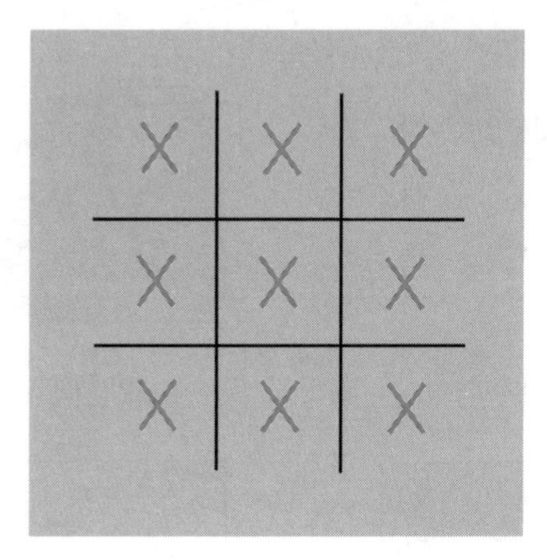

Figure 7.11 A 3 × 3 grid of test points can be used to visually assess and record the number of points falling on the sectioned second-phase particles

Before the introduction of digitized imaging systems, *point analysis* was by far the most effective method of assessing *volume fraction*, since a grid of points placed over an image could be used to accumulate numerical data rather rapidly. Most observers only require minimal experience in order to assess and record (without actually counting) the number of points in a nine-point grid (noughts-and-crosses or '*tic–tac–toe*', see Fig. 7.11) which lie *within* the particles of a second phase. A regular array of test points (rather than a random array) actually reduces the statistical error.

These three methods for assessing the volume fraction of a second phase should all yield the *same* result if the sampling probe is *randomly positioned* (which is only true in exceptional cases), and providing that the microstructure is itself 'random', which is often approximately true. The observed differences between the methods may themselves be useful in assessing *microstructural order*, and, with care, this assessment can also be made *quantitative*. The guiding principle can be very simply stated: we require *minimum error for minimum effort*, and we will see below (Section 7.3) how this principle can be used to determine a strategy for *quantitative microstructural analysis*.

7.2.1.2 Particle Size and Grain Size

The measurement of both *grain size* and *particle size* must be based on the assessment of the *area of interface* present in the material, just as the measurement of *volume fraction* was based on the assessment of the *volume* of the second phase. In the case of particle size, we wish to know what *area of interphase boundary* is present *per unit volume of the second phase*, while in the case of grain size it is the *area of grain boundary per unit volume of the sample* which is of interest. Both of these parameters have the dimensions of *inverse length*, and are measures of the *surface-to-volume ratio*, in effect, they are an estimate of the *total curvature of the interface*, defined as dS/dV. The simplest example is that of a spherical particle of radius r, whose volume is $4/3\pi r^3$ and whose surface area is $4\pi r^2$. It follows that $dS/dV = (dS/dr)/(dV/dr) = 8\pi r/4\pi r^2 = 2/r$.

That this is indeed the parameter most commonly of interest becomes clear when the *driving force* for reducing the *internal energy* of the system is considered. The *total surface energy* of the system is γS, so that the driving force for a *reduction* in internal energy is $d(\gamma S)/dV$. If the surface energy per unit area is a constant, then this reduces to $\gamma dS/dV$. For the case of a stable soap bubble, the *surface tension force* is balanced by the difference in pressure between the *inside* and the *outside* of the bubble, leading to the well-known Laplace equation:

$$\Delta P = \frac{2\gamma}{r} \tag{7.2}$$

For the general case, the driving force for *any* reduction in total surface energy is given by the relationship $d(\gamma S)/dV$. A reduction in γ may result either from dopant or impurity *segregation* or, in the case of *anisotropic* systems, by *facetting* on low-energy crystallographic planes, which will *reduce* the driving force by reducing the average interface energy. For the case where γ is a constant, any *increase* in curvature of a *single* particle will *increase* the driving force, since small particles are *less* stable than large ones. In a polycrystalline sample, *grain growth* occurs by *reducing* the curvature, that is by a reduction in total *grain boundary area*, so that dS/dV is then *negative*.

Unfortunately, the grain size, rather than the curvature, is the parameter which is easiest to visualize, and several alternative definitions of '*grain size*' have been given. In a material with morphological anisotropy, the measured data are often interpreted as an *orientational dependence* of either the *mean intercept length*, or the *mean caliper diameter*, with maximum and minimum vales quoted for the plane of a selected section. We prefer to select an *anisotropy-independent* parameter and concentrate on a definition of grain size based on the *inverse* of the total *curvature*, or dS/dV.

We start with a *line element* of length Δl on a section containing a *test grid* of parallel lines separated by a distance d (Fig. 7.12). Averaging over all angles between the line segment and the *normal* to the grid, the *probability* of *intersection*, p, is represented by the equation:

$$p = \frac{2\Delta l}{\pi d} \tag{7.3}$$

For a *total length* of line, l, made up of randomly oriented segments, the *average number of intersections*, $\bar{N}$, will be given by:

$$\bar{N} = \frac{2l}{\pi d} \tag{7.4}$$

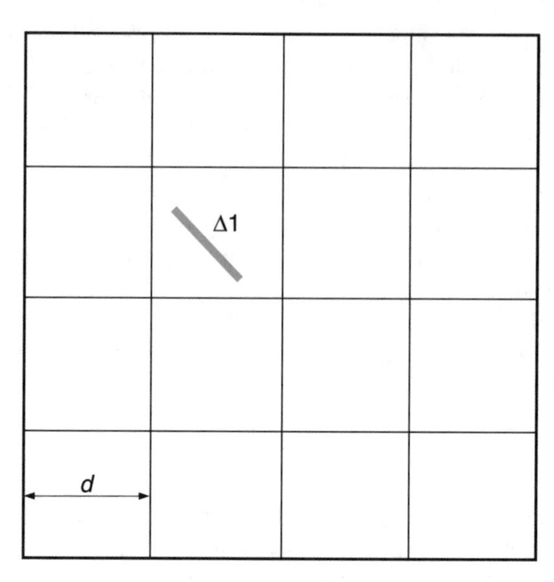

Figure 7.12 A line element projected on to a test grid in the same plane has a fixed probability of intersection

and if the *test area* is A, then the *total length of test line* will be $A/d = L$, so that the average number of intersections per unit length of test line is linearly related to the total length of intercepted interface per unit area of the section, thus:

$$\frac{N}{L} = \frac{l}{A}\left(\frac{2}{\pi}\right) \tag{7.5}$$

Now consider the intersection of an *irregularly shaped particle* of volume V and surface area S with a set of *parallel test planes* of separation d (Fig. 7.13). The average area in a section is given by the relation:

$$A = \frac{V}{d} \tag{7.6}$$

while the *average length of the intercept boundary* on the section is equal to:

$$\bar{l} = \frac{\pi S}{4d} \tag{7.7}$$

It follows that the surface-to-volume ratio is given by the expression:

$$\frac{S}{V} = \frac{4}{\pi}\left(\frac{\bar{l}}{A}\right) \tag{7.8}$$

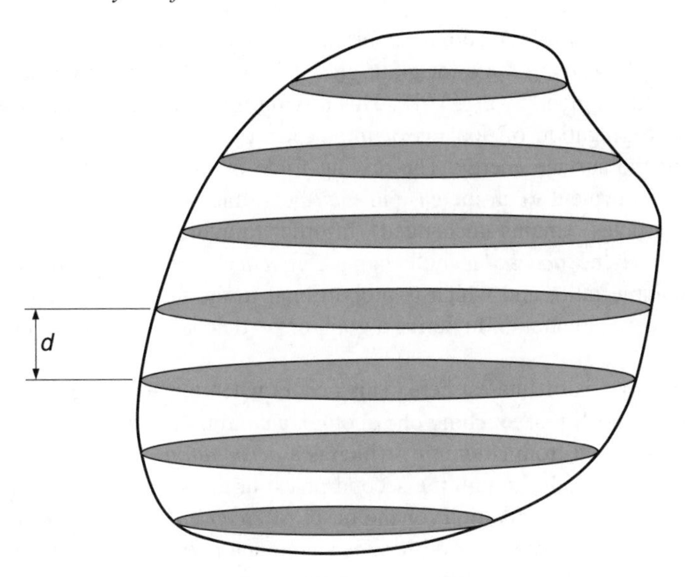

Figure 7.13 Schematic representation of the sectioning of an irregularly shaped particle with planes having a spacing d

By adapting the relationship for the length of intercept boundary per unit area of section, as derived above, for the case where we are *only* interested in that portion of the test grid which falls *within* the area of the second-phase particles intercepted by the section, $\bar{L}$, we obtain the following expression for a given phase α:

$$\left(\frac{S}{V}\right)_{\alpha} = 2\left(\frac{\bar{N}}{L}\right)_{\alpha} \tag{7.9}$$

that is the *surface-to-volume* ratio (the curvature of that phase) is just *twice* the number of intercepts that a unit length of a test grid makes with the trace of the interface on the *surface section*, where only the length of the grid falling *within* the particles is counted.

In the measurement of *grain size*, the test grid covers the *total area* of the section, while each boundary trace is shared by *two* grains (i.e. each *intercept* is shared by two grains), so that the relationship for the *curvature* or inverse grain size, S/V, becomes:

$$\frac{S}{V} = \frac{\bar{N}}{L} \tag{7.10}$$

In the literature, it is often difficult to know exactly what has been measured, since very often 'correction' factors are introduced in order to convert a *mean intercept length* into a 'grain size'. A common correction factor which is used is $4/\pi$ (see above). If it is the *driving force* for grain growth or particle coarsening which is of interest, then this is given by $d(\gamma S)/dV$. This driving force is *reduced* if γ is reduced, either by the segregation of dopants or impurities, or by facetting associated with anisotropy of the surface energy. The driving force is *increased* by a reduction in particle size (equivalent to an increase in *total* curvature, assuming that the *volume fraction* of particles remains unchanged). In order to avoid confusion, it is usually advisable to use the *accessible* bulk parameter, *curvature*, corresponding to the surface-to-volume ratio, and which is proportional to the *inverse* of the particle or grain size, rather than attempt to derive a grain or particle size defined in some other way.

One further case is of interest here. This occurs when the second-phase particles have a finite probability of touching one another, a condition termed *contiguity*. This must be distinguished from *continuity*, which is a *percolation condition* corresponding to a continuous path through the second phase in the sample. A good example would be the continuity condition for the onset of *electrical conductivity* when the matrix is insulating and the second phase is a conductor. *Two* types of interface are now of interest, namely those between two touching particles of the *same* phase, which are *shared* between the contacting particles, and the interphase boundaries between the particles and the matrix, which are *not shared* with neighbouring particles. The relationship for the *surface-to-volume ratio* (*curvature*) is now modified to become:

$$\left(\frac{S}{V}\right)_{\alpha} = \frac{(\bar{N}_{\alpha\alpha} + 2\bar{N}_{\alpha\beta})}{\bar{L}_{\alpha}} \tag{7.11}$$

For spherical particles, the *percolation threshold* for continuity occurs at a volume fraction close to 0.3, thus leading to *three* distinct morphologies for a two-phase mixture. When the volume fraction is below 0.3, the *minor* phase is dispersed in the *major* phase, predominantly as *isolated particles*. For volume fractions between 0.3 and 0.7, the two phases form two *interconnected* networks. Above 0.7 (a *second* percolation point), the phase roles are *reversed*, with the 'second' phase becoming the matrix while the original phase is now predominantly present as *isolated particles*.

Both *stress corrosion cracking* and *comminution processes* are good examples of twin percolation thresholds. In stress corrosion cracking, isolated *microcracks* join up to create a continuous '*leakage*' path through the component (the *first* percolation threshold). Eventually, *all* grain contacts are broken across a failure surface, corresponding to the *second* percolation point. Similarly, in comminution an applied pressure generates *microcracks* which can join together at the *first* percolation threshold to form a continuous *crack network*. A *second* percolation point is reached when the cracks begin to isolate the *fragments* of the solid (crushing). On a *two-dimensional section*, only a *single* percolation threshold exists, corresponding to the transition point at which the 'minor' phase becomes the continuous matrix. The wide region of *two interpenetrating continuous phases* can only exist in three dimensions.

7.2.2 Inaccessible Parameters

We have already touched on problems associated with *morphological anisotropy*, and have attempted to bypass them by using definitions of grain and particle sizes which are *independent* of any anisotropy in the material, and determined only by the *surface-to-volume ratio*. Nevertheless, *some* measure of morphological anisotropy is often required, and we will therefore discuss some of the limitations in the determination of *inaccessible* parameters.

As a preliminary example, we will use *dislocation density*, which has been quantitatively evaluated from thin-film transmission electron microscopy samples by two complementary methods. In the first, the *number of intersections* of the observed dislocations with the top and bottom surfaces of the foil sample is determined, while in the second the number of intersections of the dislocation images with a *superimposed test grid* is used. In the first instance, the *test area* is $2A$, since intersections with both the *top* and *bottom* of the sample foil are counted, while in the second method the *test area* is Ld, where L is the total length of line in the test grid and d is the *sample thickness*.

Leaving aside serious problems associated with changes in the *dislocation contrast* as a function of the parameter $\mathbf{g} \cdot \mathbf{b}$, there are many problems associated with determining the *sample thickness* and the effect of image forces on *dislocation rearrangement* at the surface. The two possible test areas defined above are *orthogonal* to one another, and should not be expected to give the same values for the dislocation density. Dislocations *normal* to the plane of the surface will *not* be counted on a superimposed *test grid* (if they appear as points), while dislocations *parallel* to the surface will *not* intersect the foil surfaces, but *will* intersect the test grid (unless they are also *parallel* to this grid!). It follows that the dislocation density, correctly defined as *dislocation line length per unit volume*, can be *estimated* from the number of intercepts per *unit area* (using either of the above methods), but this estimate is liable to be strongly *biased* by the counting method. The origin of errors in *quantitative microstructural analysis* and their classification is treated below (Section 7.3).

7.2.2.1 Aspect Ratios

As already noted, the estimation of *morphological anisotropy* in three dimensions is not readily accessible. In particular, both *particle* and *grain shape* are difficult to reduce to a *single* parameter. On the other hand, we have seen that an *unambiguous* determination of grain and particle size *is* accessible in terms of the surface-to-volume ratio or *curvature*, a parameter which is *not* sensitive to morphological anisotropy. A common solution is to seek a *single* measure of shape, even though it

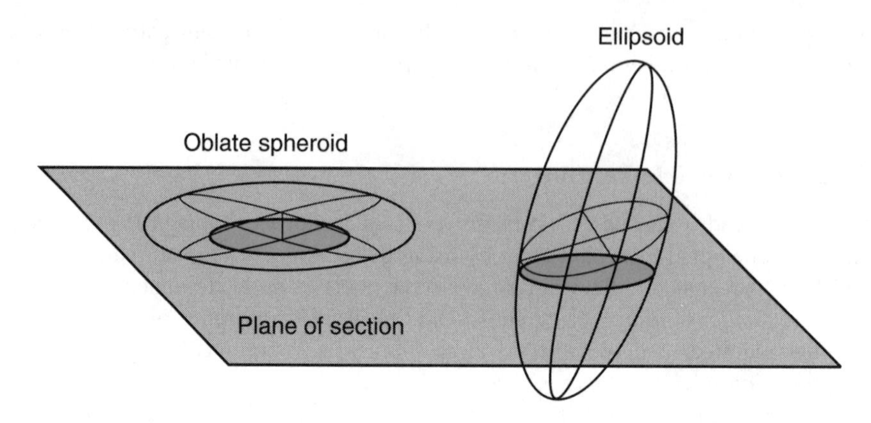

Figure 7.14 An oblate spheroid and an elongated ellipsoid may give identical sections in the plane of the image

is recognized that any such *shape factor* involves serious assumptions about the shape.

The shape factor usually chosen is the grain or particle *aspect ratio*, and the common assumption made is that the particles can be treated as *either* oblate spheroids (*lenticular* particles with an aspect ratio *less* than 1) *or* elongated ellipsoids (*accicular* particles with an aspect ratio *greater* than 1). In both cases, the particles are assumed to preserve *cylindrical symmetry*, but the distribution of shapes observed on a planar section will be quite different (see Fig. 7.13). Nevertheless, in a *single* section there is no way of distinguishing between an *elongated ellipsoid* and an *oblate spheroid* (Fig. 7.14).

Crystallographic anisotropy is the most common cause of *shape anisotropy*, and this being the case it would be unusual for an ellipsoidal approximation to be valid. In practice, second-phase particles are usually either *plate-like* or *needle-like*, but again the same problem exists, namely on a *surface section* any oblong features could represent sections of *either* shape, even though the section probability distributions (see Fig. 7.13) are quite different.

The ratios of the maximum to minimum *caliper* dimensions for particles observed on a section (Fig. 7.15), have been used to estimate the bulk aspect ratio, by making the (justified) assumption that the *maximum* value for the experimental parameter determined on a given section coincides with the maximum value of the *bulk* parameter.

7.2.2.2 Size and Orientation Distributions

While *average* values of microstructural parameters can usually be determined to quite high statistical accuracy (of the order of a few percent, using readily available

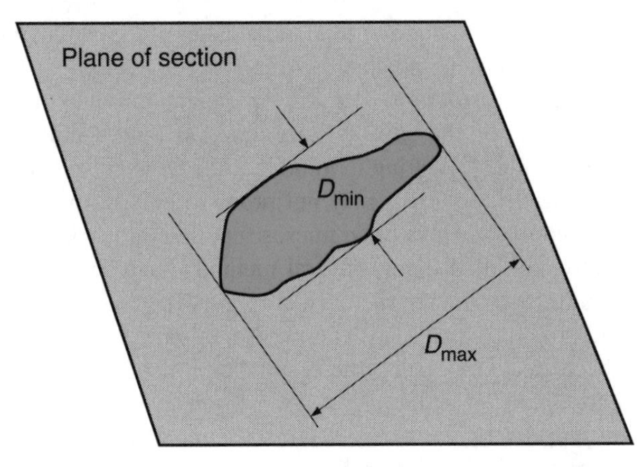

Figure 7.15 The maximum and minimum caliper dimensions of a sectioned particle can be determined, and the maximum value of the ratio of the two then used to estimate the bulk shape factor (aspect ratio)

computerized methods), the determination of a *size* or *orientation distribution* is much more difficult.

There are *two* basic problems. In the first place, the volume fraction of the *smallest* particles may be low, but the *detection errors* associated with their small size will almost certainly lead to large errors. (In the limit, when the smallest particle size falls below the resolution limit, the errors become indeterminate!) At the other end of the scale, the contribution to the volume fraction made by a few of the largest particles may be very significant, but the poor *counting statistics* associated with the limited number of particles will severely limit the accuracy. It follows that, in the limit of small particles, *detection* errors will limit quantitative microstructural analysis, while, in the limit of the largest particles, the *counting statistics* will be highly unfavourable to an accurate assessment.

A third problem is associated with the *methodology* of the analysis of particle size data from a surface section, which results in the *cumulative* propagation of errors in the measurement of the *largest* size fractions down to the *smallest*. We can define a *sectioning function* for particles of a given size D, which describes the probability P_d that the size observed on a surface section will be *less* than d. By assuming that we have *spherical* particles, the simplest *sectioning function* is given by the following:

$$(1 - P_d)^2 = 1 - (d/D)^2 \qquad (7.12)$$

We now examine the raw data for the number of particles observed on the surface section for *each size group*. Knowing that the *largest* areas of particle seen on the section can *only* be due to the *largest* particles, we can calculate, using the sectioning function, the contribution of the *largest particle fraction* to the *smaller* intercept sections. In order to optimize the accuracy, we adjust the *size intervals* so that

$d_1/d_2 = d_2/d_3 = \ldots, d_n/d_{n+1}$, in order to avoid a disproportionate number of small particles. It is also common practice to choose $d_1/d_2 = 1/\sqrt{2}$, for visual convenience. Clearly, the *errors of analysis* propagate through the distribution, and will be a *maximum* for the *smallest* size fraction (and quite often result in unacceptable *negative* values). Bearing in mind that the *volume fraction* varies as the *cube* of the particle size, these errors may not prove to be serious.

Orientation distributions are also often microstructurally important, as in a forged bar or a directionally solidified ingot. Several methods have been described in the literature for analysing such microstructures. It is important to distinguish between some very different possible cases. The general case, of *partial directionality in three dimensions*, has been treated by using the orientational dependence of an *intercept analysis* made with a *parallel array* of test lines. Some *symmetrical* cases (such as partial orientation *parallel* to one axis but with a random distribution in the plane *normal* to the symmetry axis) can be analysed more easily.

The presence of *two* populations of grains or particles, one *randomly oriented* and the other *partially oriented*, has also been of interest, although no generally accepted methodology exists. An example would be a *partially recrystallized sample*, containing both *elongated*, cold-worked grains and *equiaxial*, recrystallized grains.

Composite materials, in which *anisotropically distributed* reinforcement is used to achieve the desired mechanical properties, present their own problems, and include a variety of *woven* reinforcements in which the distribution of the fibres within the *weave* is an important variable.

7.3 OPTIMIZING ACCURACY

The *quantitative analysis* of microstructural data can be a tedious and unrewarding exercise, but this does not necessarily need to be the case. In the first place, *digitized image processing* and *analysis* has made data collection much faster and has largely eliminated the need for photography (although not the need for *visual judgment*!). In the second place, a wide range of *software packages* are now available (many in the public domain) for improving both the *quality* of the image and the *accuracy* of analysis, and in particular for improving the *counting statistics*.

The accuracy of *quantitative microstructural analysis* depends on a wide range of factors, but it is convenient to start by asking three key questions:

(a) *Is the sample representative of the object*? We have already discussed the problem of *sample selection* (Section 3.3.1). In the following, we will assume that this question has been satisfactorily answered, and that we are concerned with the accuracy of the analysis of the *sample* alone, and *not* the object from which it was taken.

(b) *Is the image representative of the sample*? This is the question which we will address before we discuss the different contributions to *errors of analysis*. The problem is to identify the *artifacts* which may appear in the image, determine their *cause* and *minimize* their presence.

(c) *How can we optimize the quantitative analysis of the image*? There is a two-part answer to this question, and we will address this next.

First, we must know what accuracy is *feasible*. In *Poisson statistics*, the error in determining the mean of a distribution is equal to $1/\sqrt{N}$, where N is the number of measurements. Other statistical functions also show a similar *inverse dependence* of the errors on the number of measurements made. Assuming that each measurement takes the *same* finite time, a significant *reduction* in the errors of analysis can *only* be achieved by devoting more time to *data collection*.

To be able to answer the second part of our question, we must recognize that the microstructure of the material is usually sampled on several *scales*. On the *macroscale,* the microstructure of a series of cast components, for example, may vary depending on the flow rate of the molten metal and its preheating temperature, and we will require a *number of sections* taken from *different* components if we wish to detect these differences. On the other hand, the microstructure may also vary across the plane of an *individual* section prepared for examination, and these *mesoscale* variations will have to be detected by collecting a *number of images* from across the section. However, when the microstructure of an *individual* image is analysed it is only the *microscale* which is being sampled, and the individual image will not provide any information on *meso*- or *macroscale* variations.

In order to optimize the *efficiency* of analysis, we must decide whether it is necessary to sample on all three scales, macro-, meso- and micro-, or whether it is sufficient to *assume* that the microstructure is *homogeneous* on the macroscale, and restrict the analysis to a *single* sample. In this case, it is then important to ensure that the errors associated with analysis of a *single* image (*microscale*) are similar to or less than those involved in comparing a *set* of images from different regions (*mesoscale*).

We should also emphasize again the importance of sample preparation. For example, the *artifacts* in optical microscopy are commonly associated with poor *polishing* and *etching* and may include the following features:

(i) *scratches* and the *traces of scratches* subsequently revealed by etching, which automated systems usually count as 'boundaries';
(ii) *occluded particles* of polishing media in soft, ductile materials, seen as 'inclusions' or 'second-phase particles' in automated systems;
(iii) '*pull-out*' of grains during polishing of brittle, polycrystalline materials, often interpreted as 'porosity';
(iv) *rounding of edges*, associated with elastic mismatch or occurring at the edges of pores, increasing the apparent areal fraction of a microstructural feature;
(v) poor contrast and failure to reveal some grain boundaries, leading to an *underestimate* of the *surface-to-volume ratio* and an *overestimate* of the grain or particle size;

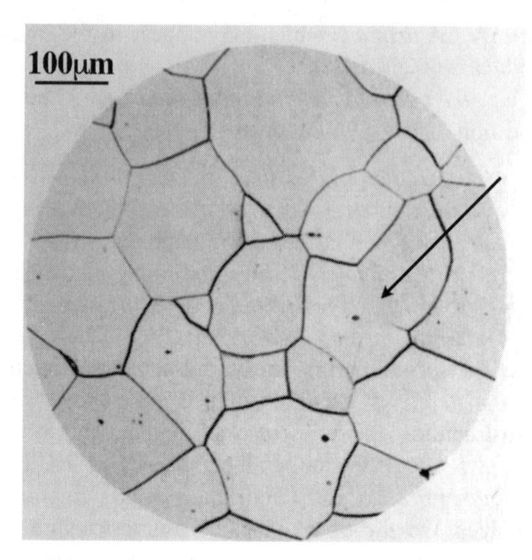

Figure 7.16 A micrograph of low-carbon steel, containing a grain boundary (arrowed) which will not be detected by a computer program used for automatic grain size measurements. (From *The Metals Handbook*, Vol. 7, 1976. Published by ASM)

(vi) *over-etching* of features, leading to increased *resolution errors* and large *sectioning errors* (see below).

Fig. 7.16 shows just one example of the above. If the specimen is polished and etched by using standard procedures, some grain boundaries may not be revealed, while the individual grains may vary in contrast due to orientational dependent chemical staining. Heat-treating the sample in order to decorate the grain boundaries with small precipitates has no effect on the grain size but can improve the uniformity of the etching response of the boundaries. Lightly polishing the sample *after* etching can then remove the chemical staining of individual grains, thus giving *maximum contrast* for the grain boundaries.

A great deal depends on the operator's awareness of the limitations of quantitative analysis, and of the conflicting requirements of *accuracy* and *time*. We emphasize here the need to optimize the *accuracy* of any analysis procedure, while minimizing the *effort required* to collect and analyse data, but on the understanding that sample preparation is *no less important*.

7.3.1 Sample Size and Counting Time

In any statistical analysis, the *larger* the size of a sample, the *smaller* the statistical error, but it is important to recognize that all methods of analysis involve *more than one* source of statistical error. For example, in determining the *volume fraction* of a second phase, both the *number* of second phase particles in the field of view and the number of *measurements* made are equally important. At *higher* magnifications, the *detection* errors may be *smaller,* if only because of the higher numerical aperture of the objective lens which has been used, but a *single field-of-view* may contain a very limited *number* of particles (or none at all!), thus leading to a very large *sampling error* dominated by the small number of particles which have been imaged.

Each measurement made will require *time,* so that, as the data is accumulated, the statistical errors *decrease,* while the time required for the analysis *increases.* With an *automated system*, the time required to analyse any single image *must* include the time required to *select* the area, adjust the *contrast* and the *focus,* and correct the *background,* and is not just the time required to collect data from the field-of-view. Data accumulation is an *interactive* process in which the operator *cannot* rely solely on the computer to make the decisions.

It is important to avoid *statistical bias*. Data collection should *never* be started at a *non-random* point, such as a grain corner. Most microstructural parameters involve a *ratio*, and it is important to predetermine the *denominator* and not the *numerator.* For example, in an *areal analysis of volume fraction*, the *total area sampled* should be kept constant, and *not* the area of the second phase observed within the section. In the determination of *grain or particle size*, it is the *total length* of *test grid*, and not the *total number of intersection points*, that should be fixed.

Statistical errors are most conveniently summarized in terms of the *coefficient of variance* (*CV*), which is defined by the following:

$$CV = \frac{\sigma^2}{x_0^2} \cong \frac{s^2}{\bar{x}^2} \tag{7.13}$$

where σ is the *standard deviation* (the spread of values in the *population* being sampled), s is the *standard error* (the measured spread of values observed in the sample), x_0 is the *average value* of the property in the *population*, and $\bar{x}$ is the *average value* of the property in the *sample*. In other words, s and $\bar{x}$ are *estimates* of σ and x_0.

In terms of *Gaussian statistics* (the *normal distribution*), we have the following:

$$\bar{x} = \frac{1}{N}\sum_{1}^{N} x_i \tag{7.14}$$

where N is the number of measurements made, and:

$$s^2 = \frac{1}{N-1}\sum_{1}^{N} (\bar{x} - x_i)^2 \tag{7.15}$$

While the *normal distribution* is a good approximation in many cases, some care is required and standard texts on *statistics* should be consulted for *alternative* statistical functions. This is particularly the case where only a *small number of measurements* are being made. For the present, it is sufficient to recognize the *limitations* of the normal distribution. It is limited to *large* samples where the variables may take *any* rational value (*non-integer* or *integer, positive* or *negative*). Note that *none* of the variables of interest to us may take *negative* values, while, in many cases, we are concerned with *integral values* only. A good example is a count of the *random distribution* of surface features (integers from zero to infinity) on a *surface section.* The appropriate statistics are actually a *Poisson distribution*, for which the *coefficient of variance* of the section is $CV = 1/N$. If the section is subjected to an *areal analysis* to determine the *volume fraction* of the second phase in the bulk, then the coefficient of variance associated with the sectioning of the *particles*, CV_{A}, must also be included in the areal analysis, as follows:

$$CV_{\mathrm{AA}} = \frac{(1 + CV_{\mathrm{A}})}{N} \tag{7.16}$$

For randomly distributed uniform spheres, $CV_{\mathrm{A}} = 0.2$.

For a *random point count*, the proportion of points falling within a *given* area follows a *binomial distribution*. If a second phase occupies an areal fraction A_{α}, then *some* of these points, P_{α}, fall within this area, and the *coefficient of variance* for the random point count is now expressed by:

$$CV_{\mathrm{PR}} = \frac{1}{P_{\alpha}(1 - A_{\alpha})} \tag{7.17a}$$

This value of CV only applies to the *single area* being analysed, and a term needs to be added to take account of the *total number of features*. The final result is then given by the following equation:

$$CV_{\mathrm{PR}} = \frac{1}{P_{\alpha}(1 - A_{\alpha})} + \frac{(P - 1)}{PCV_{\mathrm{AA}}} \tag{7.17b}$$

For a random distribution of the areas of the *second phase*, we can substitute for CV_{AA} and neglect the factor $1/P$, thus leading to an *estimated* coefficient of variance for a *point count* which is given by the relation:

$$CV_{\mathrm{PR}} = \frac{1}{P_{\alpha}}(1 - A_{\alpha}) + \frac{1}{N}(1 + CV_{\mathrm{A}}) \tag{7.17c}$$

where N is now the *total number* of second-phase features in the area being analysed. Clearly, *optimum* efficiency requires that $P_{\alpha} \cong N$, so that for a minimum error attached to a given effort (a given number of counts), the *number of counts* should be of the order of the *number of features* that are visible.

This result is, strictly speaking, limited to a *random* point count and a random distribution of features, and is therefore a 'worst-case' scenario. Since the second-phase features cannot overlap (so that a second feature has to be 'placed' in a smaller available area) and point counts are performed in an *ordered* array (so that *clusters* of points are excluded), the actual *CV* will be *smaller* than estimated here. Nevertheless, the conclusion remains valid and *statistical accuracy* will always be limited by the *number of features sectioned*, so that there is *no* advantage in increasing the number of points counted much beyond this value.

Linear analysis of boundary and interface traces is similarly limited. The grid of lines used to probe a microstructure should be regularly spaced, and the *line separation* should be of the order of the *grain size* or the *particle separation*. The statistical accuracy is limited by the *number of interfaces sampled on the section*, and the number of *intercept counts* should be of the same approximate magnitude. This will be the case when the *grid spacing* approximates to the particle spacing or grain size.

In the absence of an automated system, *counting* is by far the easiest method of assessing *accessible* microstructural parameters, such as the following:

(i) *point features* (etch pits, particle density, dislocation intercepts) which are estimated as the *number of points per unit area*;
(ii) *linear features* (dislocation density, grain size, particle size) which are estimated from the *number of intercepts with a superimposed regular test grid* having a line separation similar to the features in the image;
(iii) *areal features* (volume fraction of precipitates or inclusions) which are estimated by using a *systematic point count*, with a point spacing comparable to the spacing of the features.

7.3.2 Resolution and Detection Errors

While the *resolution limit* of the microscope constitutes the ultimate limit on the accuracy with which the coordinates of a feature can be located, it is seldom the controlling factor. In optical microscopy, it is *sample preparation* which usually determines the *detection error*, and common *chemical etchants* usually develop steps and grooves which scatter light out of the objective aperture over a width of the order of 1 μm. *Thermally etched ceramic* samples show similar grooving (Fig. 7.17).

Another source of *resolution error* is the *contrast mechanism*, and in *diffraction contrast* images of thin crystalline films taken in the transmission electron microscope the width of a dislocation image is typically of the order of 10 nm, even though the *resolution limit* for the microscope is better than 0.2 nm. The parameter $\mathbf{g} \cdot \mathbf{b}$ plays a dominant role in determining both the width and the position of a *diffraction contrast image* (Section 4.2.3.5), but it is possible to *improve* the resolution by using *weak-beam*, dark-field imaging, in which elastic scattering is from the dislocation core region. In this case, the width of the image is only of the order of 2 nm, while the position corresponds to that of the dislocation core itself.

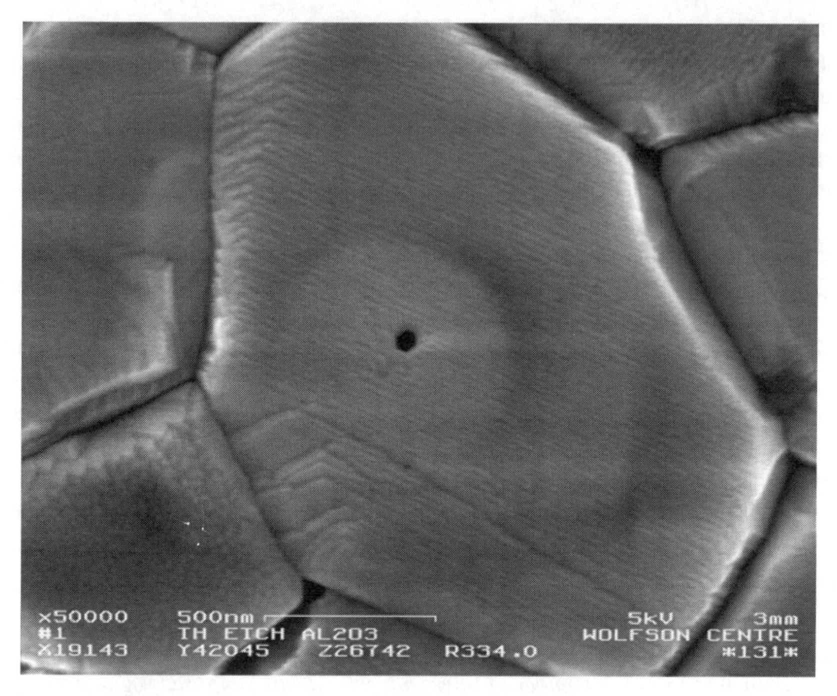

Figure 7.17 Thermal grooving of a ceramic sample typically limits the resolution in an optical micrograph to over 1 μm. The use of secondary-electron scanning electron micrographs significantly improve this limit

Finally, the *recording media* may also be a limiting factor. In digitized images, the resolution limit is 3 pixels (the minimum number needed to record a contrast variation between two features). When processing digitized image data, it is important to use sufficient pixels in order to record the required *resolution* (the separation of features in the image) and the intended *field-of-view* of the image (that is the *total* number of pixels necessary to record *all* of the resolved information in the image). This consideration may be paramount in the *transmission* of image data, for example by the *Internet* or by *electronic mail*, since there are severe time limitations on the rates of data transmission which electronic communications systems can handle. *Compression algorithms* are available to optimize the efficiency of image data transmission and processing, and we will touch on this later (Section 7.4.1).

The *apparent width* of a linear feature, δ, can be used to estimate the error in an areal fraction which is associated with poor image definition. The *areal fraction* of the boundary traces can then be represented by the following:

$$A_L = \frac{\pi}{2}\delta\frac{N}{L} \tag{7.18a}$$

where N is the number of intercepts and L is the total length of the test line. However, the estimated *particle size*, D, determined from a linear analysis, is given by the relation:

$$D = \frac{2L_\alpha}{N} \tag{7.18b}$$

where L_α is the length of the test line lying in the particles. By inserting the *volume fraction* of the particles, $f_\alpha = L_\alpha/L$, and assuming that the error in f_α is given by $\Delta f_\alpha = A_L$, we obtain the equation:

$$\frac{\Delta f_\alpha}{f_\alpha} = \frac{\pi\delta}{D} \tag{7.18c}$$

In other words, the error in determining a *volume fraction* becomes large as the *particle size* approaches the effective resolution limit.

It follows that attempting to improve the *counting statistics* for small particles or other features may be unrewarding, since the accuracy will in any event be limited by *image resolution* and *not* by the size of the sample taken.

7.3.3 Sample Thickness Corrections

Errors associated with the *sample thickness* in a thin film or, the equivalent parameter for a planar section, the *sectioning errors*, are a major factor limiting the accuracy of *quantitative image analysis*. Just as the *resolution* and *detection errors* limit the accuracy in determining the coordinates of a feature in the x–y plane, so the *sectioning errors* limit the accuracy associated with an uncertainty in the position of the section along the *z-axis*. Fig. 7.18 shows schematically the effect of this *slice thickness* for both an *etched* surface viewed in *reflection* and a *thin film* viewed in *transmission*.

There are *two* corrections required, namely:

(a) the *increase* in the measured area of a second phase from *particle projections* lying within the slice rather than just in the plane of the section;
(b) the *overlap* of particles within the slice, *obscuring* part of the projected area.

Since the sample is viewed from *one* side, the contribution from internal surfaces amounts to just *half* of the surface area which lies within the slice, so that the correction is $1/4S/(Vt)$, where S/V is the surface-to-volume ratio and t is the slice thickness. Using *linear intercept analysis* to estimate the *surface-to-volume ratio*, $S/V = 2N_\alpha/L_\alpha$, gives a first-order correction for this *additional* area, so that the *corrected volume fraction* of the second phase takes the following form:

$$f_\mathrm{n} = \frac{A_\alpha}{A} - \left(\frac{N_\alpha}{L_\alpha}\frac{t}{2}\right) \tag{7.19}$$

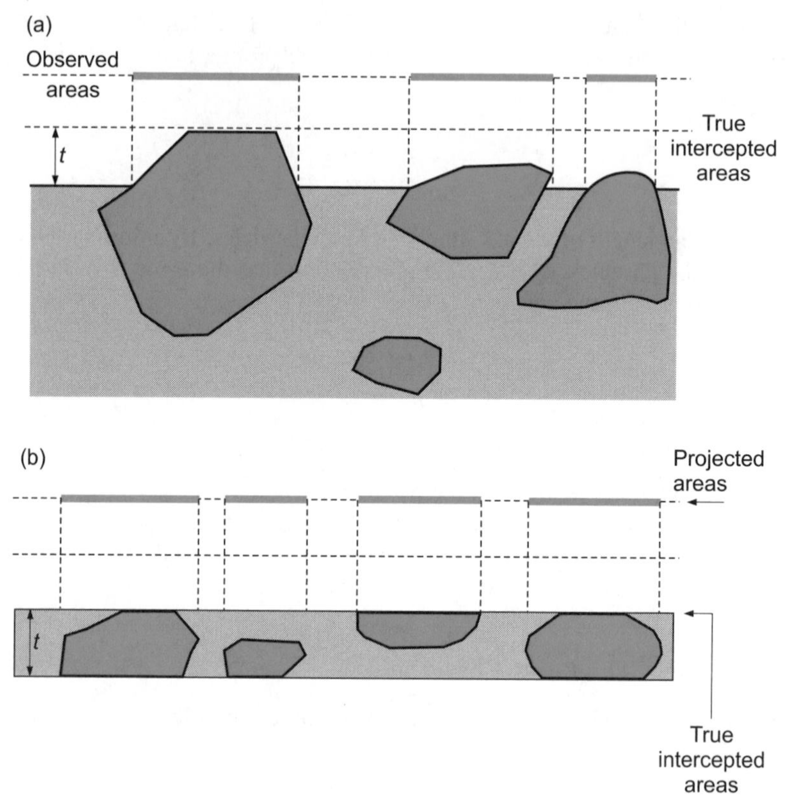

Figure 7.18 True and observed intercepted areas are not the same, but are determined by the thickness of the slice of material which contributes to the projected image. The corrections required are similar for both a reflected image of an etched surface (a) and a thin film seen in transmission (b), but the overlap of features is only possible in the thin-film transmission image

The correction is clearly *particle-size dependent*, and will be negligible for *large* particles and *small* slice thicknesses.

As the *particle size* approaches the *slice thickness*, particles will either be etched away (for observations made in *reflection*) or tend to overlap in the projected image (in thin-film transmission electron microscopy). This *overlap correction* will become important for *large* volume fractions. An approximate relationship which takes into account *both* the contribution from surfaces within the slice *and* that due to overlap has been suggested by Hilliard, as follows:

$$1 - f_{\mathrm{a}} = \left(1 - \frac{A_{\alpha}}{A}\right) \exp\left(\frac{S}{4Vt}\right) \tag{7.20}$$

This correction is actually an *overestimate*, since it does not allow for the '*exclusion*' *volume* which surrounds each particle, and which significantly *reduces* the extent of overlap estimated on the basis of random positioning.

7.3.4 Observer Bias

It is remarkable how much difference the experience and training of the observer may play in determining the accuracy of any quantitative procedure, independent of the equipment being used. While the introduction of *computerized image analysis* has improved rates of data collection by several orders of magnitude, it has had little effect on *observer bias*, since it is the observer who prepares the samples and determines the computer settings. It is important to recognize the possible forms this bias may take. The following example may help here.

Let us assume that the *grain size* of a ceramic is to be determined from a *thermally etched sample* (see Fig. 7.17). *Observer A* is worried that some boundaries may not be clearly visible, so he *increases* the annealing time in order to improve the visibility of the boundary grooves. *Observer B*, however, considers the resolution error to be excessive, due to the width of the grooves, and so *reduces* the annealing time accordingly. *Observer C* wishes to improve the contrast in the image, and to do so evaporates a film of a reflecting metal on to the thermally etched surface. We may confidently predict that each of these three observers, using exactly the *same* microscope and recording system, as well as the *same* data analysis procedure, will arrive at a *different* average grain size. Providing that sufficient data are collected, these three observers will be able to *prove* that the grain sizes they have determined are, to a high degree of probability, *not* the same!

One more example is appropriate as a further illustration of observer bias. The *same* sample has been supplied to the *same* three observers, but now the sample has *already* been prepared. *Observer A* is careful to keep the magnification in the microscope to a *minimum*, in order to ensure that a large number of grains are recorded in the field-of-view. *Observer B*, on the contrary, wishes to ensure that the full detection limit of the instrument is utilized, and records a series of images at a *higher* magnification. *Observer C* is contrast-conscious, and decides to use a dark-

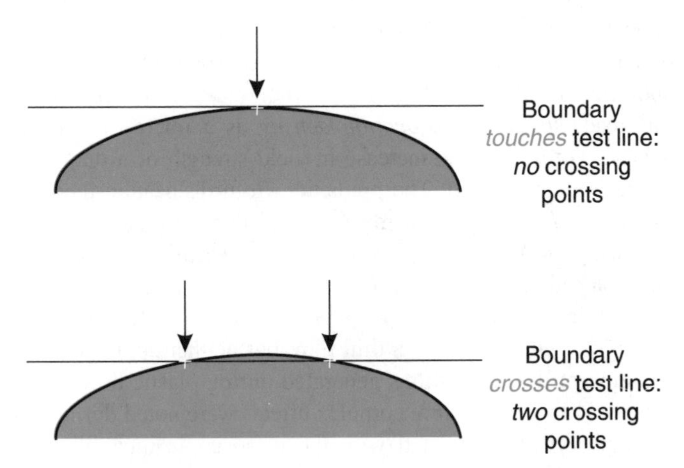

Figure 7.19 Schematic representation of how an interface may be judged to cross a test line by one observer, but not by another

field objective in order to reduce background to a *minimum* and thus highlight the grain boundaries.

All of the above choices can be justified and they reflect differences in the *professional judgement* of the individual observers. In practice, it is not usually possible to identify the precise reasons for the choices made by an observer. Does the trace of an interface (see Fig. 7.19) *cross* and then *recross* a test line (add *two* points to the tally of intercepts!), or does it merely *touch* the test line (do *not* increase the tally of intercept points!)? It is clear from this example that there is always some observer bias in such experimental procedures.

7.3.5 The Case of Dislocation Density

We have already noted some of the complications associated with attempting to determine *dislocation densities* from diffraction contrast images in transmission electron microscopy (Sections 4.3.2.5 and 7.3.2). In this present section, we will summarize the problems associated with the determination of dislocation density and the methods which have been used to overcome them.

One of the first attempts at quantitative analysis of dislocation sub-structures was made well over forty years ago by using *X-ray line broadening*. Line broadening is associated with both *lattice strain* (changes in *d*-spacings) and decreased *particle size* (sub-grain formation). The basic difficulty is that an array of dislocations in a slip plane (a *pile-up*) introduces *strain* into the lattice, while an array with the same average spacing in a sub-grain boundary results in *misorientation*, and it is not straightforward to separate the two. X-ray measurements of dislocation density were subsequently followed by a series of *etch-pit analyses*, pioneered by Gilman, using ionic single crystals. In this work, dislocation arrays generated by grit particles impacting the surface were analysed and their distribution was shown to confirm the predictions of *dislocation theory.*

Although etch-pitting was used to reveal dislocation sub-structures in a wide range of semiconductors and ionic materials, as well as in metals (including iron), it was always limited by the *resolution limit* of the method (typically a few microns), and it was not until the advent of thin-film transmission electron microscopy that a serious attempt was made to define *dislocation density* as a microstructural parameter in theories of *work-hardening* (the increase in yield strength of a ductile material with increasing plastic deformation). The prediction from dislocation theory was that the increase in the *yield strength* of a metal should depend on the square root of the dislocation density, as $\Delta\sigma \propto Gb\sqrt{\rho}$, where G is the *shear modulus*, b is the *Burger's vector* of the dislocations and ρ is the *dislocation density* (line length per unit volume).

As noted previously, two strategies were employed to 'count' the dislocations revealed by *diffraction contrast* in a thin film, but neither could account adequately for the relaxation of the dislocations generated during plastic flow into dense arrays (a dislocation *cell* structure). More complex effects were noted during *stress reversal* (the *Bauschinger effect*, associated with the mesoscopic internal stress field) and *mechanical fatigue* (thousands of stress reversals at a stress level well below the yield stress).

It required some 20 years of research effort before workers finally realized that the *artifacts* involved in estimating dislocation density from transmission electron micrographs taken in diffraction contrast would *always* limit the usefulness of any general concept of 'dislocation density'.

Today, the importance of determining the *morphology* of the dislocation sub-structure and identifying the dominant dislocation interactions is fully recognized, but the concept of *dislocation density*, is used sparingly. Consider the following question: why is the *residual* density of *isolated* dislocations important in a semiconductor single crystal?

As an alternative to dislocation density, workers in the field of plastic deformation have evolved a complete library of concepts to describe the complex *dislocation morphologies* which result from dislocation interactions during plastic flow of ductile polycrystalline materials, including *kink bands*, *shear bands*, *cell structure*, *sub-grains*, *pile-ups* and *dipoles*. It is sufficient here that the reader should be aware of the limitations of the *concept* of dislocation density, in addition to understanding the problems associated with determining this parameter.

7.4 AUTOMATED IMAGE ANALYSIS

The computer revolution, and in particular the twin benefits of increased data handling capacity and reduced prices, has placed options for automating *quantitative image analysis* within the reach of *any* research, teaching or industrial laboratory.

The first automated systems were constructed *in-house* some forty years ago. These were superseded by commercial systems for image analysis of ever increasing sophistication. Today, good *image scanners* are available for less than the cost of some 'holiday packages' (less than $1200), while the *software* necessary for *image analysis* can be downloaded from the public domain.

In practice, there are *three* options for converting either three-dimensional data (such as a fracture surface) or a two-dimensional projection of a three-dimensional body (such as a TEM specimen) into a data set, depending on whether the 'scanning' is of the *source*, the *object*, or the *image*. The following are presented here by way of an explanation (see Fig. 7.20):

(a) In the *scanning electron microscope*, the electron beam is focused on to the surface of the sample to form an image of the electron source (the *probe*), and this probe is scanned across the surface of the sample in a raster. It is the focused image of the *source* which is scanned across the surface, and this is known as a *source-scanning system*. The *image data set* is collected as a function of the probe *position*, that is as a function of *time*. Some *light-optical* systems have been built in which an inverted optical microscope geometry has been used to focus a light beam on to the specimen with a probe diameter determined by the numerical aperture of the probe-forming lens. If the *source* is a cathode ray tube (CRT), then the point of light on the CRT is scanned and the screen is imaged as a scanning probe of light on the sample surface.

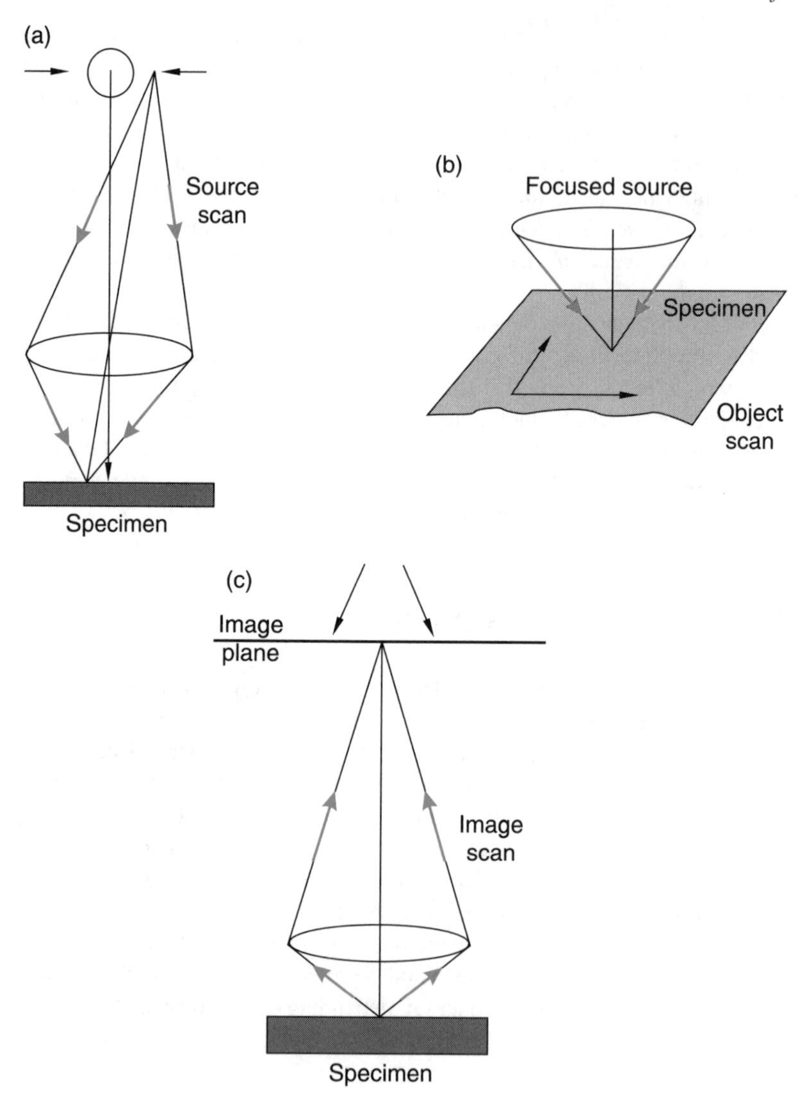

Figure 7.20 Various options available for image data collection, based on (a) source, (b) object or (c) image scanning

(b) In some systems, it is the *object* which is scanned beneath the probe, for example in the (semi-continuous) analysis of a powder sample. Many commercial *particle analysers* work on this principle, although instead of the data from a *focused image*, they detect the scattering of light as individual particles, carried in a gaseous or fluid stream, as they traverse an aperture illuminated by a *laser beam*, so that the analysis is performed in *reciprocal space* (Section 2.2). In such systems, *thousands* of particles are rapidly counted and sized, according to the scattering angle, and the data are presented as the *occupancy probability* for the different size fractions, typically in a range from a few tenths of a micron up to tens of microns.

(c) Finally, optical images, typical of the *optical microscope* and the *transmission electron microscope*, can be scanned in the *image* plane, by using a suitable detection system. A *charge-coupled device* (CCD) camera is the most effective of the systems available, although it would come as little surprise if the CCD camera were to be replaced eventually with an alternative solid-state device offering more robust performance, a better spatial resolution and a higher signal gain. The CCD camera has replaced earlier devices based on vidicon and other television cameras, and routinely allows for the capture of over 10^6 pixel points at frame speeds suitable for *real-time*, video recording (16 frames s^{-1}). A wide range of both *monochrome* and *colour CCD* cameras are commercially available. The colour versions are based on a *three-colour code* (red, green and blue—the RGB system of primary colours) which permits access to most regions of the *chromaticity triangle* (Fig. 7.21 (see Plate 2 facing page 19)). The chromaticity triangle is a convenient way of analysing colour. By varying the three intensity levels, *any* colour within the triangle defined by the three *primary colours* can be simulated, including white, over a wide range of intensities.

In addition to the CCD camera, which can be used on-line with an optical microscope assembly, *image scanners* are available which combine an electronic *x-scan* with a mechanical *y-scan* (as in a photocopier) and can convert image information from photographic negatives or positive prints over dimensions from that of an 8 mm still from an old 'home movie' to an A3 glossy plate.

In the following, we will assume that scanning is of either the *source* (as in the scanning electron microscope) or the *image* (as with a CCD camera or image scanner), and that we are less interested in collecting data scattered in reciprocal space (as in a *particle size analyser*).

7.4.1 Digital Image Recording

Several problems have to be solved if digitized data is to be quantitatively analysed to provide an *unbiased* measure of microstructural parameters. The first condition to be fulfilled is that the *integer value* assigned to a particular pixel should be *linearly* related to the physical processes occurring in the plane of the specimen. In the *optical microscope* it is the distribution of the intensity of light in the image plane which determines the *contrast*, in *transmission electron microscopy* it is the distribution of the electron current across the image plane, while in *scanning electron microscopy* the signal in the 'image' plane (the signal recorded by the *collector* as the probe beam rasters across the specimen surface) may correspond to either secondary-electron emission, back-scattered electrons or characteristic X-rays.

Photographic recording is essentially a *non-linear* process, and will only give a *linear* correspondence between the blackening of the emulsion and the incident intensity over a very limited range (Section 3.2.4.2). Light excitation of the emulsion is a two-stage process, and the blackening curve is usually expressed as a function of the *logarithm* of the dose (incident intensity multiplied by the exposure time). The linear region of this curve depends on the '*gamma*' (γ) for the emulsion and is best for low-contrast emulsions. High-energy radiation (X-rays and electrons) *does* give a linear photographic response (the number of silver grains in the developed emulsion is linearly dependent on the incident dose), but only as long as there is no *overlap* of the silver grains. Digitized images obtained by scanning a photographic recording should be treated with caution. The linear range of response available from a *transparency* ('negative') viewed in *transmission* is about an order of magnitude greater than that which can be obtained from a photographic *print* viewed in reflection.

The resolution of an image scanning system is usually quoted in *dots per inch* (dpi), which indicates the *maximum* density of pixels that the system can record, irrespective of either the density of pixels in the image to be scanned, or the density of pixels needed to retain the best resolution of the microscope. For *quantitative image analysis*, the integer value assigned to each pixel should be regarded as an estimate of the information present in the image. The choice of *pixel density* depends on the following three factors:

(i) The *minimum density* of pixels required to resolve the features of interest.
(ii) The *accuracy* desired for the value to be assigned to *each* pixel. (This usually depends on the image acquisition device, rather than the pixel density.)
(iii) The *speed and storage capacity* of the computer and data transmission facilities. Given the low price of storage media and computer random-access memory (RAM), most microscopists tend to be conservative and use the *maximum pixel density* that the system can provide!

It is important to recognize that the digitized image, in which each pixel is a *data point*, is only the *raw data*. Prior to undertaking any quantitative analysis, the raw data should be 'pre-treated', and a number of *algorithms* are available for achieving this. The *contrast* may be adjusted, *background* can be subtracted, the data can be '*smoothed*', and edge effects can be *enhanced*, with various mathematical functions being available for more complex manipulation of the digitized data. It is also possible for the operator to 'manipulate' the image data by deleting *artifacts*, such as *scratches*, or inserting 'missing' features, for example *unetched boundaries*. In all of these operations there is wide scope for *operator bias*—up to and including wishful thinking!

We can summarize as follows:

- There is *no* substitute for good sample preparation!
- The digitized image should be recorded so that the digitized values are *linearly related* to the physical phenomena responsible for image contrast.
- The digitized image data must be pre-processed in order to optimize the extractable information content *before* undertaking *quantitative image analysis*.

7.4.2 Statistical Significance and Microstructural Relevance

At several points in this chapter we have been at pains to point out the difficulties attached to *quantifying* information contained in a micrograph. The transition from a *value judgment* ('fine-grained') to a *quantitative statement* ('a grain size of 0.4 μm') represents *additional* information, and any attempt to improve accuracy is to be applauded. Nevertheless, the imaging of a specimen is a complex process, and the relationships between the mechanism of *contrast formation* in the image, the *microstructural features* responsible for the contrast, and the *physical*, *mechanical* and *chemical properties* of the original component are in no way self-evident.

The results of *quantitative analysis* may by highly significant, in terms of the *statistical distribution* of the data, and yet still prove to be largely *irrelevant*. An excellent example is to be found in those *mechanical properties* which are associated with the presence of '*defects*', namely fracture strength, ductility, notch impact energy and fracture toughness. These properties are primarily sensitive to the *size* of the defect and the angle between the plane of the defect and the axis of the applied load. Since the *largest* defects are those that have the *greatest* effect, no measurement of the *average* defect size or the *size distribution* can possibly be correlated with these mechanical properties. Furthermore, each *class* of defect (microcracks, soft inclusions, hard inclusions or porosity) will affect the properties differently, posing an even more stringent condition for the relevance of certain microstructural data.

Of course, all is not lost. Quantifying *grain* and *particle size*, measuring the *volume fraction* of a second phase, and determining the extent of morphological *anisotropy* are all important activities, not only for the fulfilment of basic *research objectives*, but also in *material* and *process development*, the improvement of criteria for *materials acceptance* and product *quality control*. However, quantitative analysis should not be undertaken lightly. *Microstructural characterization* has an essential role in technological development, and *quantitative characterization* should be an ultimate objective, but only if the results are *genuinely relevant* to the objectives of the investigation.

Summary

The quantitative analysis of image data requires an understanding of the *stereological relationship* between the two-dimensional information recorded on an image and the three-dimensional microstructure of the sample before sectioning. To this knowledge of stereological relationships must be added an appreciation of the *statistical errors* associated with sample selection and sectioning, and data collection and recording. The objective is to optimize the quantitative analysis of image data with respect to *both* the effort involved in collecting the data *and* the statistical accuracy of the final analysis.

The *isotropy* or *homogeneity* of a material are important microstructural characteristics. The properties of most engineering materials are to some extent *anisotropic* and *inhomogeneous*, and this is reflected in their microstructure. *Crystallographic anisotropy* is associated with the preferred alignment of specific crystallographic planes and directions with respect to the coordinates of the component, while *morphological anisotropy* refers to the spatial alignment of elongated grains or particles, or ordered arrays of particles. Inhomogeneity may be associated either with variations in *chemical composition* or with regional variations in grain or particle size, that is *morphological inhomogeneity.*

A single section is frequently insufficient to characterize a microstructure. The *position* and the *orientation* of any chosen section must be known and should be selected with respect to the *principal axes* of the component. Microstructure can be expected to vary *across* the section of a component (inhomogeneity) as well as with the *angle* the section makes with the principal axes (anisotropy). The results of statistical analysis of a microstructural parameter may reflect *macroscopic* variability throughout the *bulk* of the component, *mesoscopic* variability over any given *cross-section,* and *microscopic* variability within the field of view covered for any given magnification.

The *statistical variability* of an analysis arises from several sources. That of primary interest is generally the *inherent variability* of the parameter, for example the spread of grain size or the lack of uniformity in the distribution of a second phase. *Sampling errors*, over which some control is possible, are associated with the number of recorded measurements, the position of the samples selected and the chosen plane of the section. There will also be *experimental errors*, such as the quality of the specimen preparation (polishing and etching), the resolution of the microscope, and the recording of the image.

It is important to distinguish between *accessible parameters*, which can be determined unambiguously from a two-dimensional section without making any stereological assumptions about the microstructural morphology, and *inaccessible parameters*, which require some geometrical model for the structure (for example spherical particles) for their interpretation. The *volume fraction* of a second phase is a straightforward example of an accessible parameter, and can be determined from a surface section to any degree of accuracy, limited only by the resolution of the imaging system. Providing that *grain size* and *particle size* are defined in terms of *surface-to-volume ratio*, these parameters are also *accessible*.

The *surface-to-volume ratio* is an accessible parameter, but it is independent of whether particles are elongated or have re-entrant angles. *Needles*, *platelets* and *dendritic* shapes are not distinguished, and any measure of *particle shape* taken from a two-dimensional image requires a *stereological model* of the bulk geometry if it is to be interpreted in three dimensions, such as an *aspect ratio*. The same applies to *distributions* of particle or grain sizes, as well as to dislocation arrays. In the case of particles, it is sometimes useful to assume *convexity* (no regions of negative curvature), since for such a case no particle can intercept a section more than once. Composite materials, which include highly textured, woven reinforcement, present a particular problem in this respect.

Errors of analysis associated with *specimen preparation* and *recording* can be minimized by careful attention to detail and an awareness of the origin of the possible artifacts. *Counting errors* are usually minimized by ensuring that the sample selected is large enough to ensure that any microstructural variability of the sample is detectable above the random statistical background. Since both the *microscopic* variability, within a given area, and the *mesoscopic* variability, from one area to another, contribute to the statistical errors, it is important to ensure that sufficient areas are sampled. For example, the average number of *particles* (or grains) observed in a specific area should be comparable to the number of *sample areas* selected, in order to ensure the minimum *statistical variance* for any given counting effort.

The *resolution errors* will limit the available accuracy when the size of the features of interest approaches the resolution limit of the method being used. Specimen preparation always introduces some *thickness* or *sectioning error*, associated with the depth over which the sample contributes contrast to the projected image. In optical microscopy, this *slab thickness* is of the order of the resolution, since the depth-of-field is of the order of the resolution. In electron microscopy (both scanning and transmission), the depth-of-field is orders of magnitude greater than the resolution, and *thickness corrections* can be very important indeed.

Observer bias is a major source of variability, and may reflect rather more than varying levels of experimental competence. Two observers can arrive at significantly different estimates of the same microstructural parameter as a result of entirely justifiable differences in *professional judgement*.

Automated image analysis has not eliminated *observer bias*, since the settings for any computer program have to be based on the *professional judgement* of the operator, but it *has* increased the rate of data collection by orders of magnitude, thus making quantitative stereological analysis of microstructural morphology readily available and reducing the effort required to attain statistical significance. However, the proliferation of computer software programs has led to some confusion. In several cases, these programs do *not* indicate the stereological assumptions on which they are based, and the ease with which the data are processed by the computer may obscure the intermediate steps involved in passing from *image data collection* (the statistical significance) to *data processing* (digitizing of the image) and the presentation of the final results of the *data analysis*. Major errors are often associated with specimen selection and preparation, and no amount of automated image analysis system can 'correct' for poor specimen preparation.

Bibliography

1. J.C. Russ, *The Image Processing Handbook*, 2nd Edn, CRC Press, Boca Raton, FL (1995).

Worked Examples

Three examples of size measurements from different microstructures, on different length scales, will demonstrate the principles of quantitative measurement at the *meso-*, *micro-* and *nano*-morphological levels. The first of these is the size of the *alumina grains* in a sintered body, expected to be in the micron range; for these measurements, we employ SEM. The second example is the size of *aluminium grains* developed by nucleation and growth during chemical vapour deposition (CVD). The final example explores the limit of detection of TEM in the measurement of *ordered domains* in a disordered matrix of $Pb(Mg_{1/3}Nb_{2/3})O_3$.

Fig. 7.22 shows two micrographs of *sintered alumina*; the first after sintering at 1400°C for 2 h, and the second after sintering at 1600°C for 10 h. The sintering of ceramics represents a compromise situation. On the one hand the *density* should be maximized, which generally requires high sintering temperatures and long periods of time. However, on the other hand, the final *grain size* should be minimized, which is usually achieved by limiting the sintering temperature and time. Thus optimizing a sintering process usually involves measuring the *residual porosity* and the *grain size* as a function of sintering time and temperature.

As far as the grain morphology is concerned, two questions need to be answered—how does the *average grain size* depend on the sintering parameters, and what is the *aspect ratio* of the alumina grains? In order to quantify the *grain size*, we apply the *linear intercept* method, discussed in Section 7.2.1.2. Fig. 7.23 shows the processed images and the *average values* of the grain intercepts. As expected, a significant *increase* in grain size has occurred after sintering at 1600°C for 10 h, as compared to sintering at 1400°C for 2 h. By quantifying the grain size as a function of the process parameters, we can characterize the sintering mechanisms empirically, and so optimize the sintering process.

We now turn to our CVD aluminium samples. We need to determine the *size* of the aluminium grains as a function of the *deposition time*, in order to assess the influence of certain process parameters. Scanning electron micrographs of the aluminium grains on two *different* TiN substrates, as a function of the deposition time, are shown in Fig. 7.24, while Fig. 7.25 shows an example of the *processed* image, in which the aluminium grains are differentiated from the TiN background by their *contrast*. The aluminium grain size (now measured as the projected *area* of each grain) as a function of the deposition time for the two TiN substrates is given in Fig. 7.26. As in the case of the alumina sample, the use of a computer program greatly increases the sample size, and thus improves the *statistical significance* of the results.

A final example from a completely different material system demonstrates the combined problem of *quantitative microstructural analysis* and *elemental detection limits*. The material is $Pb(Mg_{1/3}Nb_{2/3})O_3$ (PMN), which has been doped with varying amounts of lanthanum. PMN has the *perovskite* (cubic) structure, shown in Fig. 7.27. In the ideal structure, cations of type A occupy the corner positions of the lattice $(0, 0, 0)$, while cations of type B occupy the body-centred positions $(1/2, 1/2, 1/2)$, and oxygen anions are located in the face-centred positions of the cubic unit cell $(1/2, 0, 0)$. In disordered PMN, there are *two* types of cations located at the type-A positions (Mg and Nb), while Pb occupies the type-B positions. However, under certain conditions *chemical ordering* can take place to form a *superlattice*, in which distinctive (111) planes, containing either Mg or Nb cations,

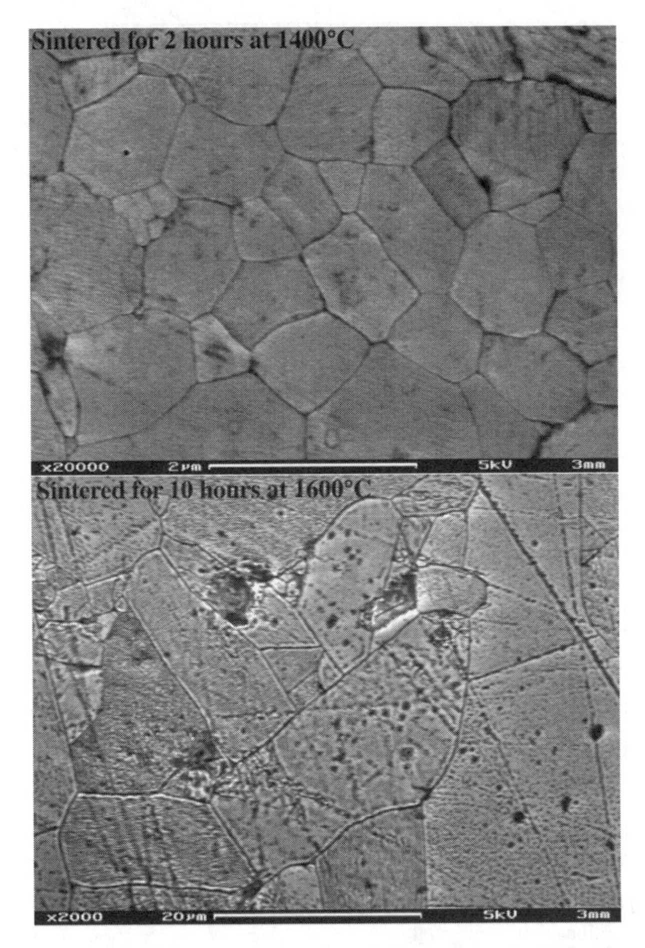

Figure 7.22 Scanning electron micrographs of thermally etched alumina after different periods of sintering: (a) 2 h at 1400°C; (b) 10 h at 1600°C

form a *new* face-centred cubic unit cell with a lattice parameter of $2a_0$, where a_0 is the lattice parameter of *disordered* PMN, as shown in Fig. 7.28.

The difference between the *disordered* and *ordered* crystal structures is easily detected by *selected-area diffraction* after orienting the specimen into a [110] zone axis, as shown in Fig. 7.29. If the *ordered* regions are sufficiently large, then *dark-field diffraction contrast* images can be recorded by using the {111} *ordered* reflections. Fig. 7.30 demonstrates such a *dark-field image* in which the ordered regions, separated by *antiphase boundaries*, are clearly visible.

PMN may also form small *ordered* regions, with length scales of the order of nanometres, in which increased doping with lanthanum appears to increase the size

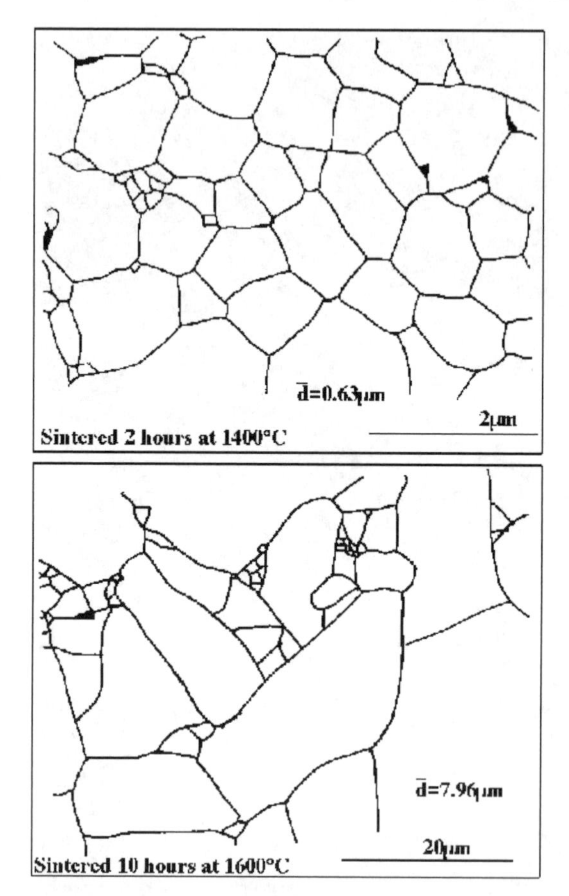

Figure 7.23 The micrographs from Fig.7.22 after processing for image analysis; the results obtained from the linear intercept method for grain size measurements are indicated

of the nano-ordered regions. In this example, we will use transmission electron microscopy and image processing to analyse this effect quantitatively.

The first step is to select a *method* to observe the nano-ordered regions. Since the minimum size of the ordered regions is a few nanometres, *any* good transmission electron microscope should have the required resolution. However, we have to detect and differentiate the *ordered regions*, from the *disordered* PMN grain (the background). We will compare two methods quantitatively, namely *dark-field diffraction contrast imaging* and *high-resolution transmission electron microscopy* using phase contrast.

We have already noted that large ordered regions can be detected in a *dark-field image* taken with the ordered {111} reflections, while *nano-ordered* regions can be detected in a similar way. Micrographs showing the influence of varying lanthanum

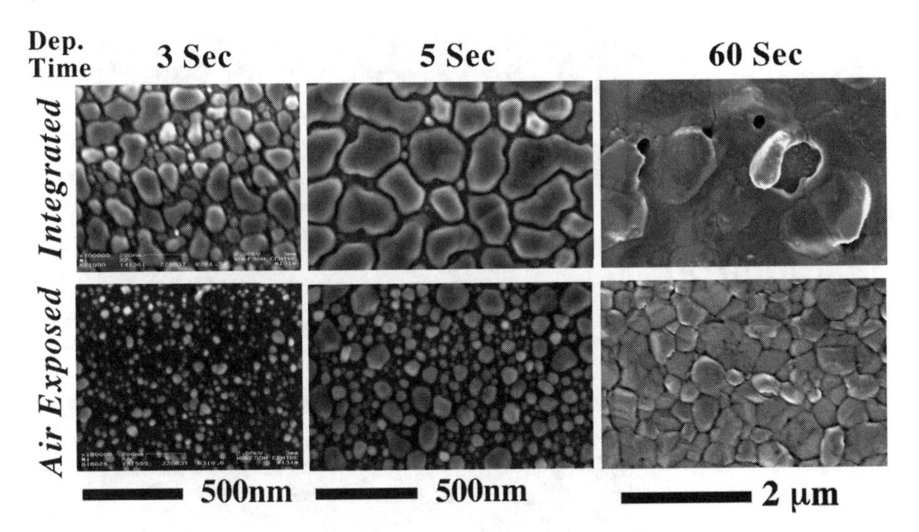

Figure 7.24 Scanning electron micrographs of the aluminium grains on two different TiN substrates, as a function of the deposition time

concentrations on the *size* of the ordered regions are shown in Fig. 7.31, in which the *bright* regions are the ordered domains. In order to quantify the size of these ordered regions, we must first process the image (Fig. 7.32(a)) to remove the background (the disordered matrix), thus obtaining the image shown in Fig. 7.32(b). We then measure the *projected area* of each ordered region. The data are now summarized as the *average projected area* of the ordered regions (and corresponding *standard deviation*) as a function of *lanthanum* concentration, as shown in Fig. 7.33.

From Fig. 7.34 we can immediately identify a problem with the data. The *average domain area* shows a tendency to *increase* with *increasing* lanthanum content, but the *standard deviation* values of the data are too large for the effect to be judged as being *statistically significant*. A major reason for the large standard deviations is that the ordered regions are *smaller* than the thickness of the TEM specimen, with frequent *overlap* of the ordered regions, thus invalidating any quantitative conclusions. We *could* collect data only from the *thinnest* regions of the specimen, providing that we could determine the thickness accurately, but *instead* we prefer to use *high-resolution transmission electron microscopy* (HRTEM), and optimize the microscope parameters to *detect* the ordered regions preferentially.

What do we mean by *preferential detection* of the ordered regions? The *periodicity* and *contrast* in HRTEM lattice images depends not only on the *crystal structure*, but also on the microscope *contrast transfer function* (CTF). In order to detect and differentiate the ordered regions, the microscope operating conditions should maximize the intensities of the *superlattice reflections*. Fig. 7.34 shows the CTF for the microscope at the *Scherzer defocus*, as well as the relative values of the *structure factor* for the various crystallographic planes in the PMN crystal

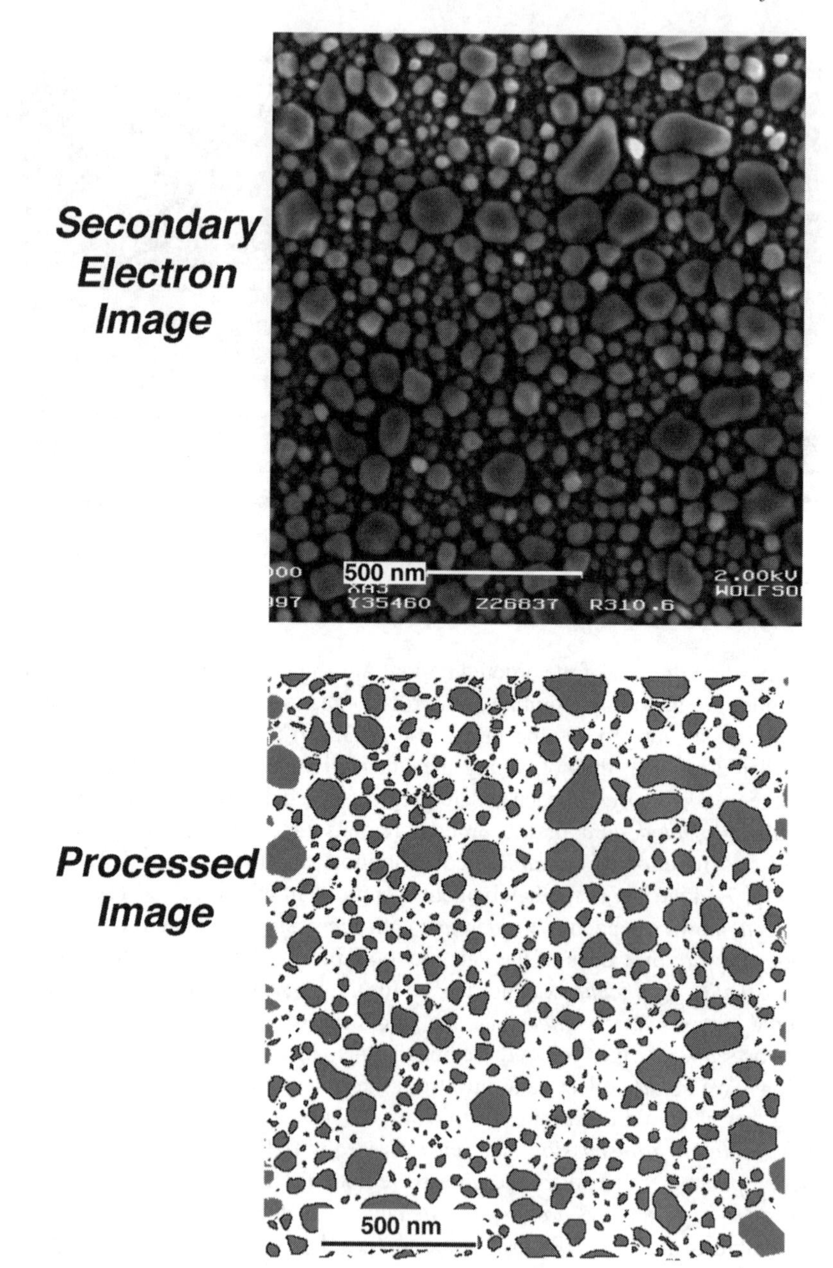

Figure 7.25 The original micrograph (a) and the processed image (b), in which the aluminium grains are differentiated from the TiN background by their contrast

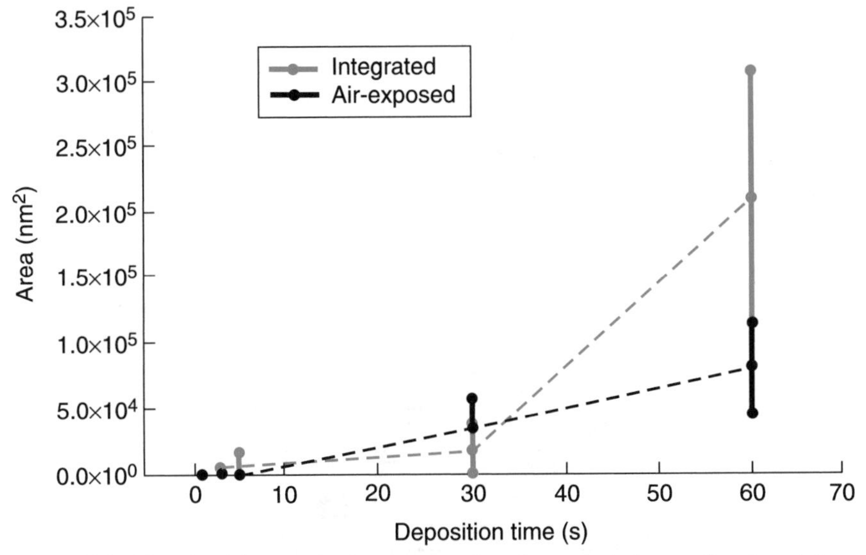

Figure 7.26 The aluminium grain size (measured as the projected area of each grain) as a function of the deposition time for the two TiN substrates

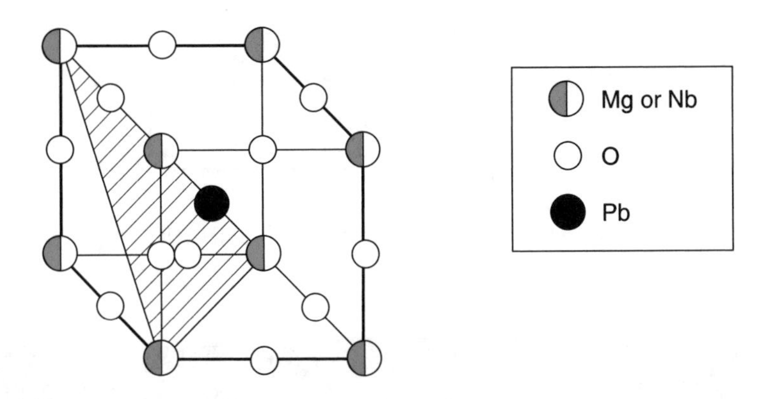

Figure 7.27 The perovskite (cubic) structure of PMN

superlattice. Most of the diffracted beams which contribute to the [110] lattice image are from the {202}, {222}, and {004} matrix lattice planes, while the relative *structure factors* of the ordered (superlattice) reflections, {111} and {113}, are of much lower intensity than those of these matrix reflections. Therefore, while the *point resolution* is optimized by setting the objective lens current to the Scherzer defocus value, the detection of *ordered* planes from the superlattice will be extremely difficult using this setting. However, we can modify the CTF by *changing*

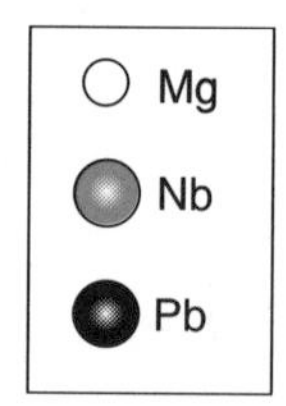

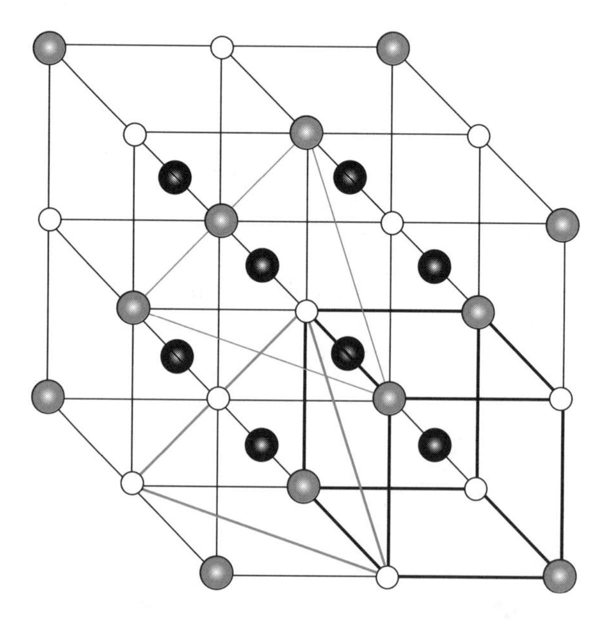

Figure 7.28 The face-centred cubic unit cell of ordered PMN. The oxygen ions are not shown for clarity

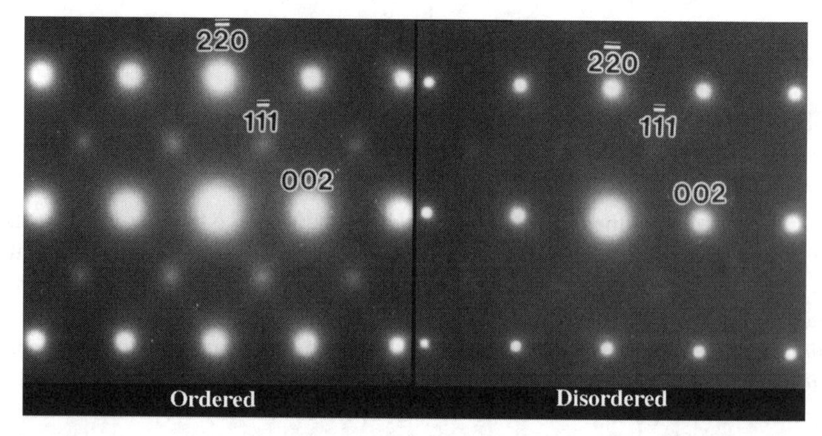

Figure 7.29 Selected-area diffraction patterns, in a [110] zone axis, of the (a) ordered and (b) disordered PMN structures

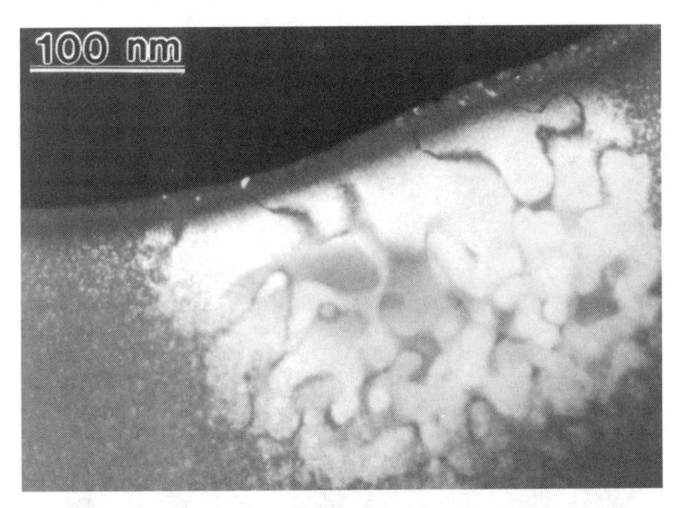

Figure 7.30 Dark-field transmission electron micrograph recorded by using the {111} ordered PMN reflection; large ordered regions, separated by antiphase boundaries, are clearly visible

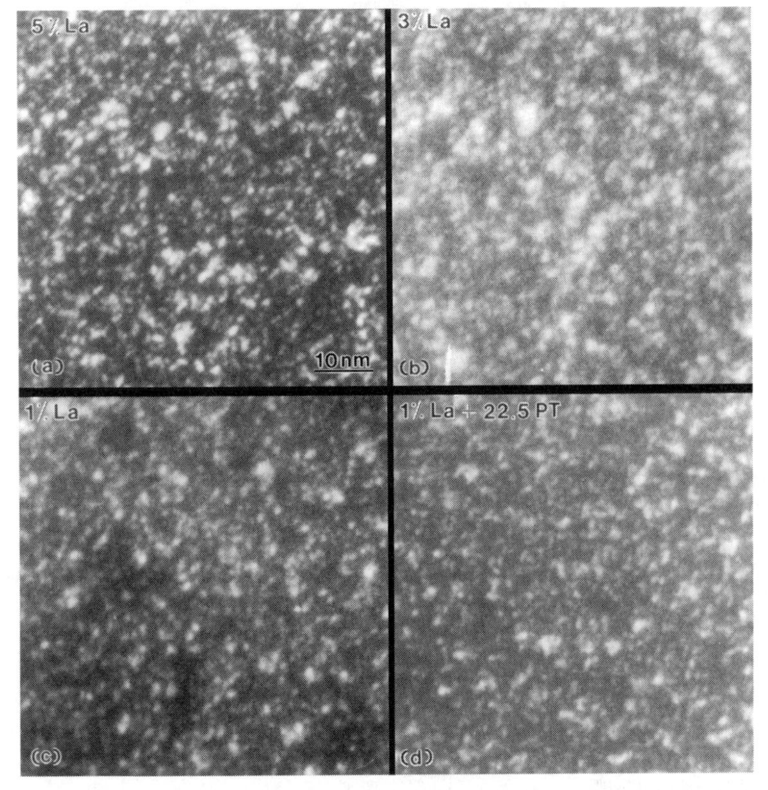

Figure 7.31 A series of dark-field transmission electron micrographs showing the influence of lanthanum concentration on the size of the nano-ordered regions; these ordered regions display a bright contrast

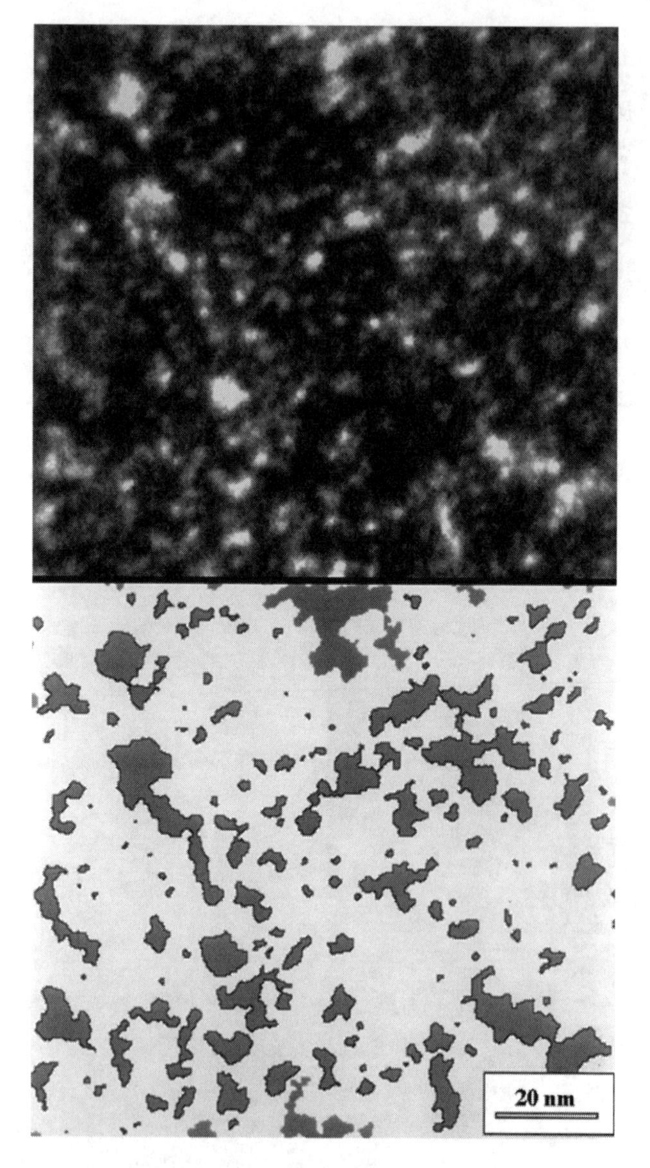

Figure 7.32 Processing of the dark-field TEM image (a) in order to remove the background (the disordered matrix), gives the image (b) in a form suitable for image analysis

the object lens defocus and *amplify* the contribution of at least one of the superlattice reflections to the image, simultaneously *suppressing* the dominant disordered (matrix) reflections, as shown in Fig. 7.35, in which the CTF has been optimized for the {111} ordered reflections.

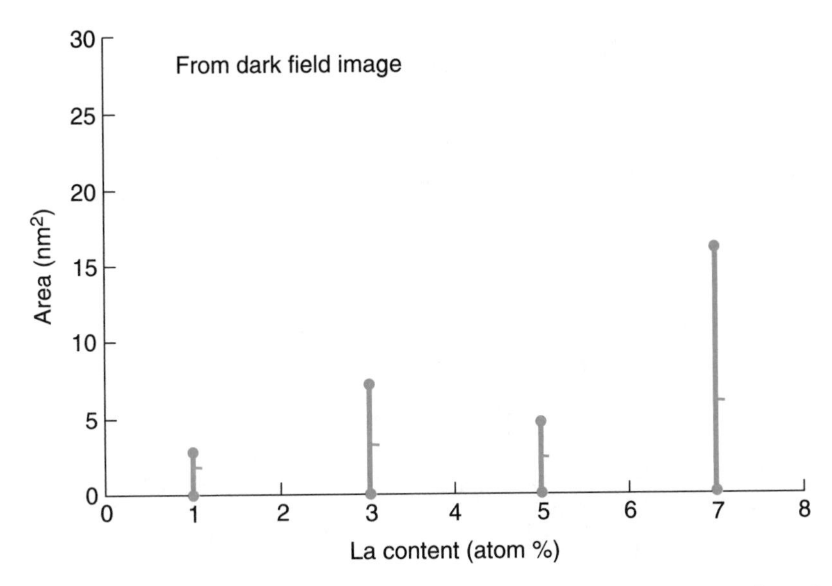

Figure 7.33 The average projected area of the ordered regions as a function of the lanthanum content; corresponding standard deviations are also indicated

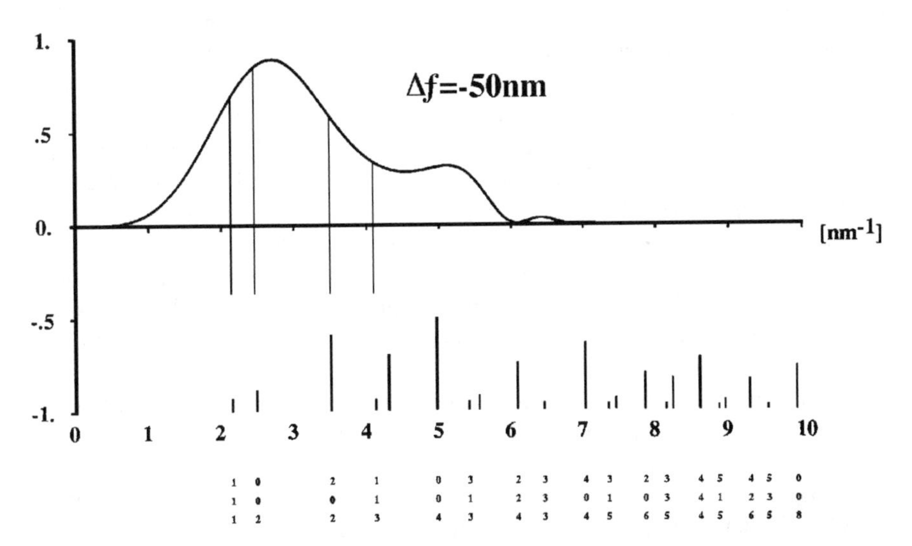

Figure 7.34 The CTF for the microscope used to study the nano-ordered regions at the Scherzer defocus, as well as the relative values of the structure factor for the various crystallographic planes in the PMN superlattice

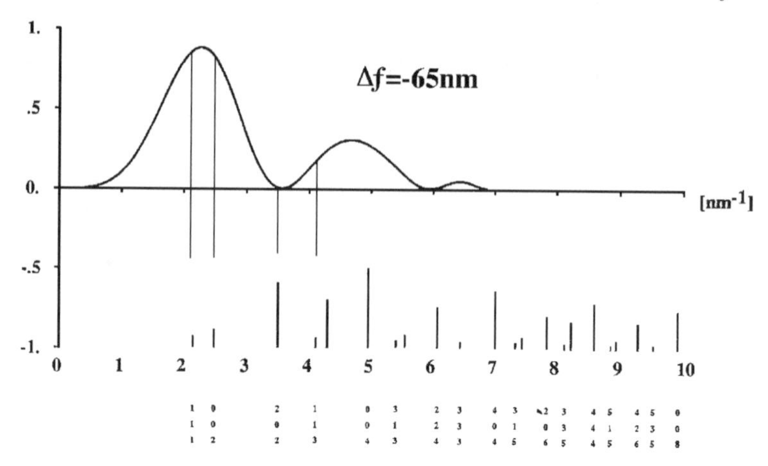

Figure 7.35 The CTF for the microscope used to study the nano-ordered regions at the optimum defocus for detection of these regions, as well as the relative values of the structure factor for the various crystallographic planes in the PMN superlattice

We have used this optimized defocus to record the lattice image shown in Fig. 7.36(a). Although this figure shows ordered PMN domains under optimized conditions, it is *still* difficult to differentiate the *ordered* from the *disordered* regions, but this can be improved by *Fourier filtering*. Fig. 7.36(b) is a Fourier transform (FT) of the image shown in Fig. 7.36(a). By applying a mask to all regions (frequencies) in the FT, *except* for the periodic reflections from PMN (both ordered and disordered), we can remove all of the background noise in the image (Fig. 7.36(c)). Now the ordered regions are easily visible. If we further mask all of the regions in the FT other than the {111} ordered reflections, the contrast from both the background noise *and* the disordered matrix is suppressed (Fig. 7.36(d)). From images such as the latter, we can measure the nano-ordered projected area and compare the results with those obtained from the nano-ordered regions seen in dark-field images (see Fig. 7.33). The results obtained (see Fig. 7.37) again show the *average ordered area* and *standard deviation* values as a function of lanthanum content, but optimizing the HRTEM lattice imaging has significantly reduced the error in the measurements, and there is no doubt now that lanthanum *has increased* the size of the ordered regions.

The calculations required to determine the optimum *defocus*, the Fourier filtering and the image processing, may appear complex, but many computer programs now exist for such calculations, both commercial and 'free-ware'. Use of these computer programs is not particularly difficult, and can also be a lot of fun! The important point to take into account is that computerized image processing *still* requires an *understanding* of the mechanism of image formation.

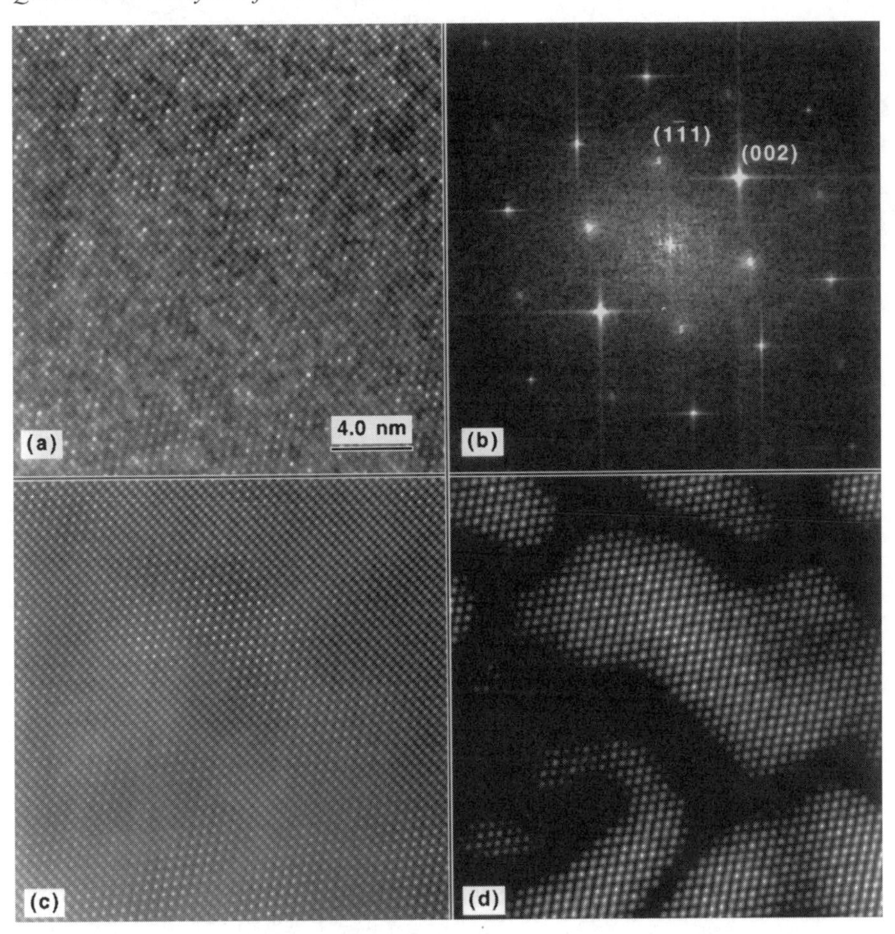

Figure 7.36 (a) Lattice image of the ordered PMN domains under optimized conditions. (b) A Fourier transform (FT) of the image shown in (a). (c) By applying a mask to all regions (frequencies) in the FT (b) except for the periodic reflections from PMN (both ordered and disordered), we can remove all background noise in the image. (d) If we further mask all regions in the FT other than the {111} ordered reflections, the contrast from both the background noise and the disordered matrix can be suppressed

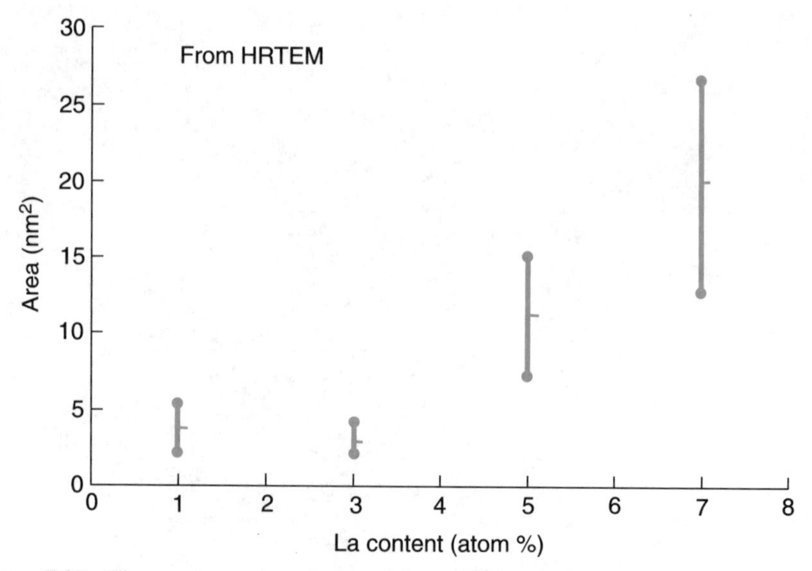

Figure 7.37 The average projected area of the ordered regions as a function of lanthanum content, as obtained from optimized HRTEM lattice images; corresponding standard deviations are also indicated

Problems

7.1 A pixel in a digitized image is assigned specific (x,y) coordinates, while the resolution of a printer is commonly given in dots per inch (dpi). What dpi value would you choose in order to ensure that the intensity assigned to a pixel is accurate to within 10 % if the separation of the pixel points in the digitized image is 0.2 mm?

7.2 Given that the resolution of the human eye is *ca* 0.2 mm and that images are most comfortably viewed from a distance of *ca* 200 mm, estimate the number of pixels needed to ensure that a digitized image does not appear discontinuous.

7.3 Distinguish between morphological and crystallographic anisotropy and describe some experimental tests that are required to determine *both* forms of anisotropy.

7.4 Define the term, *sampling error*. The number of microstructural features in a given field-of-view and the number of fields-of-view selected for quantitative microscopy both contribute to the statistical errors of measurement. Explain how you would ensure maximum statistical accuracy for minimum effort.

7.5 The number of particles of a second phase per unit volume of the sample *cannot* be estimated from a planar section without making an assumption about the particle shape. What is this assumption and why is it necessary?

7.6 The volume fraction of a second phase and the surface-to-volume ratio are both described as *accessible* microstructural parameters. What is meant by this term?

7.7 How is the surface-to-volume ratio related to *grain size*, and what factors contribute to the ambiguity of the concept of 'grain size'?

7.8 The *aspect ratio* of a feature (particle, grain, inclusion, cluster, etc.) on a surface section bears no obvious or simple relationship to the shape of the feature in three dimensions. Give *three* examples of microstructural features illustrating this 'unfortunate' fact.

7.9 *Sectioning errors* occur in the quantitative analysis of both polished sections and thin films. Assuming that an etched boundary appears twice as wide as the depth of etch, model the sectioning error as a function of grain size by assuming a boundary curvature equal to the grain size. (*There is no one 'correct' answer to this question, and various assumptions are possible about what constitutes an intersection of the boundary with a test line.*)

7.10 List the errors associated with estimating the density of dislocations observed by diffraction contrast in a thin film. Compare the errors associated with a count of the intersections of the dislocation images with a test grid as opposed to a count of the number of dislocation intersections with the surfaces of the thin film.

Index